INDUCED RADIOACTIVITY

INDUCED

‘Πᾶν μέτρον ἄριστον.’

‘In all things moderation is best.’

RADIOACTIVITY

MARCEL BARBIER

CERN, GENEVA

1969

NORTH-HOLLAND PUBLISHING COMPANY

AMSTERDAM · LONDON

This book is published and distributed
in the United States by
AMERICAN ELSEVIER PUBLISHING COMPANY, INC.
52 Vanderbilt Avenue, New York, N.Y. 10017

Library of Congress Catalog Card Number: 76–83086

SBN 7204 0145 3

Publishers:

NORTH-HOLLAND PUBLISHING COMPANY - AMSTERDAM
NORTH-HOLLAND PUBLISHING COMPANY, LTD - LONDON

Sole Distributors for the Western Hemisphere:

WILEY INTERSCIENCE DIVISION
JOHN WILEY & SONS, INC. - NEW YORK

PRINTED IN THE NETHERLANDS

PREFACE

The number of accelerators and reactors capable of producing radioactive nuclei in large quantities is rapidly growing. Sometimes these radioactive nuclei are desired entities produced under optimal conditions for various applications in science, medicine and industry. Often, however, their formation is a serious disadvantage leading to health hazards and complicated and expensive handling problems. In both cases it is of extreme importance to be able to estimate the amounts of different nuclides formed. The author of this book, Dr M. Barbier, has undertaken the task to collect and systematize all information available on the formation of radioactive nuclei, filling the gaps in our knowledge with his own experiments wherever needed. The results of his survey are presented in a large number of curves and tables covering virtually all conceivable conditions. Consequently, all those who are in one way or another concerned with induced radioactivity will find valuable information in the book which enables them to make reasonably accurate estimates of the amounts of radioactive species formed and of the radiation danger involved.

Gösta Rudstam

The Swedish Research Council's Laboratory,
Studsvik, Nyköping

FOREWORD

The author hopes that the present work will be useful in various branches of research and technology, such as radiochemistry, high-energy physics, accelerator and reactor engineering, space research and radiobiology. The radiochemists and nuclear physicists, in examining the production cross-sections and properties of radioactive elements, have laid the basis for all quantitative work in induced activity. The field of high-energy physics and accelerators is the one for which this work has been initiated. Space research is concerned with radioactivity induced by cosmic ray and solar flare particles in man-made objects or celestial bodies. Radiobiology depends on induced activity as a means of measuring spallations in cell atoms and damage to tissue, and also for a-posteriori dose evaluation.

With such a variety of subjects, it will not be surprising if the different chapters are uneven in length and scientific rigour. Indeed, as knowledge and theories are rather scarce in a number of the cases considered and as experimentation is long and difficult in this field, some time will elapse before all the gaps in this study can be filled. The writer asks the readers for comprehension and indulgence, as he has endeavoured to give a first review of a topic which is still in the course of development.

A large number of persons have helped with guidance and criticism, in particular Professors B. Forkman, V. di Napoli, A. C. Wahl, Drs G. Bathow, G. Bührer, J. Dutrannois, C. B. Fulmer, H. H. Hubbell Jr., G. Rudstam. I would also like to thank Prof. V. K. Weisskopf and Drs P. Lapostolle, G. Brianti, E. Michaelis, of CERN, who have encouraged this work. It is a pleasure to acknowledge the help of Mr F. Hoffmann in irradiation experiments, computer programming, drawing of the graphs and of Miss C. Kostorz for careful typing of the manuscript.

Geneva, March 1969 M. B.

CONTENTS

LIST OF SYMBOLS

A	Atomic weight of material in grams, $A=N+Z$, also called mass number
A_i	Atomic weight of isotope i (g)
A_T	Atomic weight of target material (g)
a	Specific activity, either in disintegrations per second per gram or in curie per gram
D	Danger parameter: dose rate existing in a cavity embedded in an infinite volume of active material per unit flux (rad/h per particle/sec cm^2)
d	Absorbed dose (rad or rem)
E	Energy of incident nucleon (MeV)
E_0	Energy of the electron initiating an electromagnetic shower
E_π	Pion energy (MeV)
f_ν	Flux of photons of energy E_ν per square centimetre giving a dose of one rad in soft tissue (cm^{-2}rad^{-1})
$g_0(k)\,\mathrm{d}k$	Total track length travelled by all photons with energies between k and $k+\mathrm{d}k$ in electromagnetic shower (g cm^{-2})
i	Total current of particles (sec^{-1}), also index for isotopic species
K	Constant depending on target and irradiation energy in Rudstam's formula (millibarn)
k	'k-factor' of radioactive isotope, dose rate in rad/h at 1 m from a point source of strength 1 curie
k	Photon energy (MeV)
k_0	Energy of giant resonance reaction (MeV)
$k_{\max}$	Maximum energy of photon generated by an electron of energy E_0
L	1/e attenuation length of gamma radiation through matter (cm or m)
N	Number of neutrons in the nucleus, $N=A-Z$
N_0	Avogadro's number, 6×10^{23} atoms per mole
N_T	Number of target atoms per cm^3 of target material

N_ν	Number of radioactive atoms of isotope ν per cubic centimetre of target material
$n=\sum_\nu n_\nu$	Total number of atoms of radioactive elements present in one gram of matter
n_T	Number of target atoms per gram
n_β	Number of beta particles (negative electrons) emitted per second per gram of radioactive substance, improperly called beta disintegration rate ($\mathrm{sec}^{-1}\,\mathrm{g}^{-1}$)
n_γ	Number of photons (including positrons, counted as 2 photons of 0.51 MeV each) emitted per second per gram of radioactive substance, improperly called gamma disintegration rate ($\mathrm{sec}^{-1}\,\mathrm{g}^{-1}$)
n_ν	Number of atoms of the radioactive element ν present in one gram of matter
$-\mathrm{d}n/\mathrm{d}t$	Time derivative of n with changed sign, total number of atoms of radioactive elements per gram of matter decaying per second
$-\mathrm{d}n_\nu/\mathrm{d}t$	Time derivative of n_ν with changed sign, number of atoms of the radioactive element ν per gram of matter decaying per second
P	Parameter related to energy of incident nucleon
p_i	Percentage of isotope i in target material
R	Parameter related to width of charge distribution
R	Dose rate existing at 1 cm from a (supposed) point source of radioactive material of given weight or activity (rad/h or millirad/h per gram or curie)
R_0, r_0	Radius of a given sphere
R_el	Range of electrons ($\mathrm{g\,cm}^{-2}$)
R_n	Mean free path of neutrons ($\mathrm{g\,cm}^{-2}$)
R_p	Range of protons ($\mathrm{g\,cm}^{-2}$)
r	Counting rate
S	Parameter defining peak value of charge distribution in Rudstam's formula
T	Parameter defining peak value of charge distribution in Rudstam's formula
t	Depth in electromagnetic cascade (cm)
$t_\mathrm{c}, t_\mathrm{cool}$	Cooling time (sec)
$t_\mathrm{i}, t_\mathrm{irr}$	Irradiation time (sec)
t_ν	Mean life (decay time constant) of isotopic species ν ($t_\nu=1.44t_{\nu,\frac{1}{2}}$) (sec)

V	Total volume of radioactive substance (cm^3)
X_0	Radiation length in material containing the cascade ($g\,cm^{-2}$)
x	Coordinate (cm)
Y	Yield, i.e. total number of photonuclear reactions in a cascade initiated by one electron
Y_{dp}	Number of interactions which proceed by the deuteron photodisintegration mechanism for one incident electron of energy E_0
Y_{gr}	Total number of giant resonance photonuclear reactions in a cascade initiated by one electron
Z	Number of protons in the nucleus, also called atomic number
Z_i	Number of protons in the nucleus of isotope i
Z_T	Number of protons in the target nucleus
Z_{peak}	Peak value of charge distribution
z	Coordinate along an axis, or distance (cm)
$\gamma(k,t)\,dk$	Average number of photons with energies between k and $k+dk$ at depth t in electromagnetic cascade
$\varepsilon_{i,\nu}$	Number of photons of energy E_ν emitted by isotope i per disintegration (positrons are counted as 2 gammas of 0.51 MeV)
$\varepsilon'_{i,\mu}$	Number of beta particles (negative electrons) of energy E_μ emitted in each disintegration of isotope i
$\varkappa$	Index for species of radiation
$\mu_{\nu,T}$	Narrow beam linear absorption coefficient of gamma rays with energy E_ν in target material T (cm^{-1})
ν	Index for isotopic species
ϱ	Density ($g\,cm^{-3}$)
Σ	Macroscopic cross-section (cm^{-1})
σ_A	Cross-section for a reaction involving photodisintegration of a deuteron within the nucleus
σ_D	Cross-section for photodisintegration of a free deuteron
σ_{inel}	Total inelastic cross-section
$\sigma_{T,i}$	Cross-section for production of isotope i from target material T (cm^2 or millibarns when so specified)
$\sigma(Z,A)$	Independent cross-section for the spallation product of atomic number Z and mass number A (cm^2 or millibarns, when so specified)
τ	Integration variable
Φ	Flux of activating particles ($cm^{-2}sec^{-1}$)

Φ_γ	Flux of photons ($sec^{-1}cm^{-2}$)
$\varphi(E)$	Flux density per unit particle energy interval ($sec^{-1}cm^{-2}$ MeV^{-1})
Ω	Solid angle (sterad)

SHORT REVIEW OF THE PHYSICAL PHENOMENA INVOLVED IN ACTIVATION

The activation of substances, that is the action of making them radioactive by irradiation (bombardment by particles or gamma rays) can take place by a large number of different physical processes. We shall try to mention them all briefly in order to give a general view of the physical phenomena involved. More detailed attention will be later given to each of them in turn wherever possible, in the subsequent chapters.

Basically, activation is a transformation of the nucleus of an atom by the incoming particle or radiation. Such a phenomenon is called nuclear reaction. This change can lead to the formation of another isotope of the same chemical species, or, more generally, of a nucleus which is completely different chemically.

Once the nuclear reaction has taken place, one is left with one or more new atoms, which can be unstable and undergo radioactive decay.

Thus, we will have two kinds of phenomena to examine: the nuclear reaction and the radioactive decay. We shall make here a brief survey of the various kinds of nuclear reactions which can occur, and then explain what are the different modes in which a radioactive atom can decay. With the description of the different nuclear reactions we will include a review of the principal sources of activating radiation available now or occurring in nature, as each of them has a particular type of flux of particles which we will relate to the nuclear reactions they produce.

1 A survey of the observed types of nuclear reactions

The types of nuclear reactions are so numerous that it is appropriate to say according to what features and in which order we will consider them in this survey.

We shall start with elastic scattering for completeness' sake as it occurs for all particles at all energies, although it does not alter the nuclei and contribute to induced activity. The other types of nuclear reactions are listed below by order of increasing mass and energy of the particle considered as the incident one (nuclear reactions induced by gamma rays are taken last):

Thermal and slow neutron reactions (radiative capture, fission, other neutron-induced reactions);

Charged-particle induced reactions in the medium energy range (above the Coulomb barrier from a few to some hundreds of MeV);

High energy nuclear reactions by neutrons or protons (intranuclear cascades giving spallation products, high energy fission, fragmentation reactions, π-meson production, etc.);

Deuteron and heavy ion reactions;

Photonuclear reactions induced by gamma rays.

1.1 *Elastic scattering*

A collision event is called an elastic scattering if the particles do not change their identity during the process, and if the sum of their kinetic energy is the same before and after the event. It is therefore similar to the collision of billiard balls. The forces that are responsible for the change of the directions of flight of the two particles are of two sorts. First, if both particles are charged, we have the electrostatic Coulomb force. Second, we have the so-called nuclear forces, which are the peculiar short range forces which hold the nucleons together in a nucleus, and do not depend appreciably on electric charge. At energies large in comparison with the Coulomb potential of the atoms involved, and also in the case where at least one of the interacting particles is not charged (a neutron for instance), the nuclear forces will determine the process.

Elastic scattering is important because it is always present, independent of the energy and kind of the particles undergoing the collision. A percentage of all possible nuclear interactions will always be elastic scattering.

Elastic scattering events do not contribute to activation. As the atoms are left in their original state, they are not made unstable and do not decay. Hence there is no activity to be expected.

In general, the word scattering is used to describe any interaction of one particle with another. All processes different from elastic (or billiard-ball) scattering are then grouped under the name of 'inelastic' scattering. Inelastic scattering can thus mean equally well scattering involving a change in the energies of the interacting particles without a change in their nature, or their complete transformation to new particles. Of course, only processes leading to the production of new particles can contribute to induced activity, if the new particles produced turn out to be unstable and radioactive.

1.2 *Thermal and slow neutron reactions*

Neutrons are particles that are not charged and so will not be repelled by the electrostatic charge of the atom nucleus they are aimed at (the so-called Coulomb barrier). Thus they will be able to come within reach of the nuclear forces practically independently of their energy, unlike charged particles which need a certain energy to overcome the Coulomb barrier of the nucleus, which makes itself felt at a much larger distance than the nuclear forces. To give an example, an energy of about 5 MeV is necessary for a proton to arrive into an iron nucleus so that a nuclear reaction of the type (p,n) can begin to be expected (the first particle mentioned in the parenthesis is the incoming one, the second the outgoing one). It is customary to write the reaction in a more complete form by putting down the target nucleus before and the product nucleus after the parenthesis, giving in this case:

$$^{56}\mathrm{Fe(p,n)}^{56}\mathrm{Co}.$$

However, a neutron approaching the same nucleus will interact down to zero relative velocity, by the reaction:

$$^{56}\mathrm{Fe(n,}\gamma)^{57}\mathrm{Fe}.$$

The product is an atom of the same kind, but heavier by one neutron, and the reaction takes place with emission of a gamma ray. Reactions of this kind are called 'radiative capture'.

The neutron velocities in this case are called thermal, because they are obtained by letting the neutrons be slowed down by scattering on nuclei that have only their thermal motion. A typical order of magnitude is 2200 m/sec which corresponds to a kinetic energy of 0.025 eV.

With a small number of very heavy elements that have so many nucleons that they do not hold together very well, thermal neutrons are found to have also quite another type of interaction. The nucleus hit by the neutron can become unstable and break into two or more fragments, a phenomenon called fission. The fragments are themselves rather large nuclei.

Apart from radiative capture and fission, reactions of the type (n, p) or (n, α), i.e. reactions by which a neutron enters the nucleus, and knocks a proton or an alpha particle out of it, begin to be observed in the case of light target elements in this velocity range. With heavier elements they occur at a higher kinetic energy of the incident neutron. These phenomena will be studied in the next section, together with nuclear reactions induced by charged particles.

1.3 *Medium energy reactions*

The Coulomb (or electrostatic potential) barrier makes it impossible for charged particles to induce nuclear reactions at low kinetic energies. The incident particle, as stressed in the last section, must have enough energy to overcome this barrier. In this section we will thus be concerned with an energy region which extends roughly from a few MeV up to about 50 MeV kinetic energy of the incident particle which we will take to be a proton. We can mention that the kinds of reactions possible with protons in this energy range can also be produced by neutrons of similar or possibly lower kinetic energy.

Due to its relatively high energy, the particle entering the nucleus is now capable of knocking out one or more nucleons, or even a part or fragment of the target nucleus. We find with protons reactions of the types (p, n), (p, pn), (p, 2n), (p, p2n), (p, α) and so forth, in which a neutron, a proton and a neutron, two neutrons, a proton and two neutrons, an alpha particle (helium atom) are expelled. Each of these reactions begins at a so-called threshold energy of the incident proton which is higher than that of the preceding one. We have quite a similar situation for the neutron-induced reactions (n, p), (n, pn), (n, 2p), (n, n2p), (n, α) and so forth, which are also observed.

To describe the nature of all these phenomena with some unity, one assumes that the incident particle is absorbed by the target nucleus, and distributes its energy in a random manner to the various nucleons, forming a so-called 'compound nucleus' in an excited state. When the energy of the incoming particle is not too great, none of the nucleons in the compound nucleus has enough energy to escape immediately. Only after some time, statistical fluctuations in the energy distribution will concentrate enough energy on a nucleon, and possibly at a later time on others, to allow it or them to escape. So there can be a sequential emission of several particles, each with a relatively low kinetic energy. The similarity of this behaviour to that of molecules escaping from hot liquid has caused the emission of particles from excited nuclei to be called 'evaporation'.

1.4 *Nuclear reactions at high energies*

At still higher energies, the picture gets more complicated. The nucleons struck by the incident particle in the target atom obtain enough energy themselves to travel through the nucleus and to hit other nucleons of this same atom in the same way, thus giving birth to an 'intranuclear' cascade of fast nucleons. These can either escape at some moment from the nucleus or be captured and give up their energy to excite the whole nucleus. One sees that a much wider variety of new atoms can be formed in this way than by the channels previously considered as a result of the large number of cascade nucleon events which can take place in the nucleus.

After the directly ejected nucleons have gone the excited residual nucleus can de-excite in several ways. One is for it to evaporate nucleons or even whole groups of nucleons in a process similar to the evaporation of particles from the compound nuclei mentioned in the previous section. The products formed in this manner are called 'spallation products', from 'to spall' – Middle English (old German 'spellen') – meaning to break up or chip with a hammer. Another way the excited nucleus can de-excite is for it to divide into two or more pieces in a manner similar to the fission process considered with thermal neutrons. This is then called 'high-energy fission', and the products have other probabilities of formation and distribution than in the previous case.

Other processes than intranuclear cascades causing spallation or fission are also observed at high energies. Direct ejection of high energy ions or light nuclei, called fragmentation reactions, do not fit in the two-step process outlined and are also observed.

Also the production of new particles by high energy nucleons striking other nucleons or target nuclei, as a consequence of Einstein's law of equivalence of mass and energy, is a possible role for an incident or an intranuclear cascade particle. As an example, π-mesons are produced by nucleons from 300 MeV upwards, K-mesons from 1 GeV, nucleons and antinucleons from 4.5 GeV. Among these newly generated particles, the π-mesons, nucleons and antinucleons are able to develop intranuclear cascades in their turn, and eventually π-mesons and antinucleons can transfer their total energy to nucleons in the target nucleus, thus making the number of all possible products from the nuclear reaction still larger.

1.5 *Deuteron and heavy ion reactions*

Deuteron reactions are peculiar in that they are found well below the Coulomb barrier of the target nucleus. This is due to the fact that in a number of cases it is the neutron end of the deuteron that is turned towards the nucleus and enters first, the proton end being repelled by the Coulomb force. The threshold energy is thus lowered for deuterons. Typical reactions are the stripping reaction, in which either the proton or the neutron is stripped off by the nucleus while the other particle continues essentially in the original direction, and compound nucleus reactions in the medium energy range as mentioned before.

Heavy ions must have a considerable kinetic energy before they can surmount the Coulomb barrier of the nucleus. At energies lower than this barrier, reactions are observed where the heavy ion makes a grazing collision with the target nucleus, whereby one or more nucleons can be transferred from one to the other, in either direction, the heavy ion continuing in the forward direction. Such reactions are known as stripping (as above) or pick-up reactions. As the energy of the heavy ion increases, a compound nucleus is thought to be formed, with the subsequent evaporation of several particles because of the high excitation energies involved.

1.6 *Photonuclear reactions*

These are the nuclear reactions induced by high energy gamma rays. One can distinguish between giant resonance reactions, explained as vibrations excited by the electromagnetic wave of all the protons in a nucleus against all the neutrons, and photospallations, where nucleons or fragments are knocked out directly by the incoming photon.

2 A review of the existing kinds of activating radiation

One can distinguish between natural and artificial sources of activating radiation.

Natural activating radiations have, fortunately, very low flux densities at the surface of the earth. One is only faced with the natural radioelements of the earth's crust, the fraction of the primary cosmic radiation which has not been attenuated by the atmosphere, and some high energy secondaries produced therein. The natural radioelements are mainly alpha sometimes neutron emitters, the cosmic radiation includes protons with the highest energies ever discovered (10^{10} GeV) and also heavy ions, the secondary radiation produced in the atmosphere are mainly π-mesons, electrons, and gamma quanta.

As one goes above the protective layer of the earth's atmosphere, one encounters first the Van Allen radiation belts (protons and electrons trapped by the earth's magnetic field). Then one is subjected to the radiation field of the sun (mainly corpuscular, with proton energies up to a few GeV). This field is liable to sudden rises by orders of magnitudes produced by solar flares. The primary cosmic radiation is, of course, present as well.

Man-made sources of activating fluxes are becoming very numerous.

The first sources in chronological order were accelerators. The first type to appear, electrostatic accelerators, are today still constructed up to energies of the order of 15 MeV and accelerate electrons, protons, or ions. Cyclotrons were then produced up to 1000 MeV for protons, followed by synchrotrons which are reaching 70 GeV for protons and 10 GeV for electrons at this moment. Parallel development of linear accelerators has made it possible to push the final energy for electrons much higher (40 GeV machine now in construction). Electron high energy accelerators constitute large sources of both gamma rays and neutrons. Neutron generators became available quite soon, using the deuteron reaction on lithium at 400 keV.

The next big step in the production of vast fluxes of neutrons was the appearance of fission reactors, which give a continuous spectrum of neutrons up to about 15 MeV, depending on the type of reactor and the fuel. With these it became possible to produce large quantities of artificial radioelements.

Another form of fission fluxes appeared with nuclear explosions, which are accompanied by the release in the open of vast quantities of both neutrons and heavy ions (fission products), the latter being themselves often particle emitters.

It will be seen that one can be faced with induced activity in quite a number of ways.

3 The various modes of radioactive decay

After having looked at the various types of possible nuclear reactions, it is appropriate to review all the known processes by which radioactive, i.e. unstable, nuclei made in these reactions can decay. We will examine briefly in turn alpha and beta decay, gamma transitions and spontaneous fission.

Binding energy considerations for the nucleons in the nucleus show that the heavy nuclei (with $A > 140$) are usually unstable with respect to alpha decay. This is so because the emission of an alpha particle lowers the Coulomb energy which is the principal negative contribution to the binding energy of heavy nuclei, but changes the nuclear binding energy very little, since the alpha particle itself is almost as tightly bound as a heavy nucleus. The alpha particles are either all emitted with the same energy, or distributed among a few monoenergetic groups, related to the energy levels of the different states of excitation in which the product nucleus can be left.

Any radioactive decay process where the mass number A of the nucleus remains unchanged but the atomic number (charge of the nucleus) changes, is classed as a beta decay. A beta decay can occur with the emission of a negative electron or of a positive electron (positron). The naturally occurring beta emitters all emit negative electrons (beta-minus decay). In such a decay the atomic number Z increases by one unit. It is considered at present that this comes about by the transition of a nucleon from its neutron to its proton state. This explains the fact that negative electron decay occurs on the neutron excess side (right-hand side) of the line of stable nuclei in the nuclide chart.

Conversely positron (beta-plus) decays arise from the transition of a proton into a neutron in the nucleus. The charge of the nucleus Z decreases by one unit. This mode of decay takes place in nuclei that are proton-rich in relation to their most stable isobars (nuclei with the same mass). Whenever a positive electron encounters a negative one they interact in such a way that they both disappear and give rise to two quanta of electromagnetic radiation, each of energy equal to

$mc^2=0.51$ MeV (the electron rest mass times the square of the velocity of light). This process is referred to as 'annihilation'. It occurs principally with positrons that have nearly come to rest by ionization processes in matter.

There is another process which is equivalent to the positron emission process, by which the charge of the nucleus decreases by one unit without change of the mass of the nucleus. It is the capture of a negative electron from the external electron shells, usually from the nearest or K-shell. This so-called electron capture is a very common mode of decay among neutron-deficient (or proton-rich) nuclides on the left-hand side of the line of stable nuclei on the nuclide chart. One can show that this is the only decay mode possible for such a nuclide when the decay energy (mass difference between decaying and product atom) is less than $2mc^2$.

Beta particles, either positive or negative, are emitted with a continuous energy distribution, and not with discrete kinetic energies as alpha particles.

An alpha or beta decay process leaves the product nucleus frequently in an excited or unstable state. This means that the nucleus has a supplementary energy, called excitation energy, which it must give up before returning to its normal or ground state. It may do this in a variety of ways, the most common of which is the emission of electromagnetic radiation (gamma rays). The energies of the gamma rays emitted by any particular atom have one single or several discrete values which are characteristic of the excitation levels of this atom and make it possible to recognize it. The emission of a single gamma quantum is a so-called 'transition' from one energy level to another. The observed energy range is between about 10 keV and 7 MeV.

In some cases the emitted gamma rays can interact with the outer electrons with the result that the gamma quantum is absorbed in the electron shell and that an electron is expelled from the shell. It has then an energy equal to the difference between the energy of the gamma quantum and the binding energy of the electron in the atom. This process is called internal conversion, because it happens still inside the atom.

In most cases the gamma transitions occur so quickly that their half-lives have not been measured. However, in some cases the excited states can have quite a long life-time (from seconds to days). They are then called 'metastable' (meaning half-stable) or 'isomeric' (meaning with equal shares) states. One has simply the same atom, but it has not yet discharged its energy.

The de-excitation processes described above are all characterized by a change in energy of the nucleus, without change in mass (A) or charge (Z).

Spontaneous fission is another de-excitation mode for excited nuclei. In this process the nucleus breaks up into two fragments of approximately equal masses. Since the binding energy per nucleon decreases with increasing mass number A, all nuclides above $A \approx 60$–100 are in fact liable to spontaneous fission provided the excitation energy is high enough to overcome the high nuclear potential barrier for the emission of the fission fragments. It is this nuclear barrier that keeps the heavy natural elements together. Some measurable spontaneous fission of the natural elements has been observed, however, among the heaviest ones ($A > 230$).

CHAPTER I

GENERAL THEOREMS AND PROPERTIES COMMON TO ALL REACTIONS

1 The cross-section of nuclear reactions

It is usual to characterize nuclear reactions quantitatively by their cross-section. The definition of the cross-section is based on the simple picture that the probability for the reaction between a nucleus and a flux of impinging particles evenly distributed in space is proportional to a cross-sectional target area presented by the nucleus. More precisely the cross-section σ is defined by the equation

$$N = iN_{\mathrm{T}}\,\sigma x \qquad (1.1)$$

where N is the number of processes of the type considered occurring per unit time in the whole target (sec^{-1}); i is the total particle current per unit time (sec^{-1}); N_{T} is the number of atoms per cubic centimetre of target (cm^{-3}); σ is the cross-section (cm^2); x is the target thickness (cm).

The target thickness is considered to be small enough, so that the beam intensity does not decrease significantly when passing through it.

The cross-section σ varies with the kinetic energy E and also with the type of the impinging particles. This function $\sigma(E)$ is called the excitation function of the nucleus for the particular process and bombarding particle. A lot of work in activation studies generally goes into the determination of this cross-section and its dependence on energy. Basically, one has to measure experimentally the number of radionuclides of the type considered produced by a known number of impinging particles and calculate the cross-section out of the experimental data. For this, separation of the relevant radionuclides from all the others produced at the same time is necessary, be it by chemical or other means. The work must be repeated at other energies, often in other laboratories which have the type of accelerator giving this energy. As an example proton cross-sections can be measured up to 12 MeV with Van de Graaff electrostatic generators, up to 50 MeV with linacs, up to several hundred MeV with cyclotrons, and above this level with synchrotrons. Calibrated neutron fluxes of known and discrete energies are difficult to obtain. Up to this date, one has obtained very accurate neutron excitation functions only up to about 15 MeV, and this has been done with Van de Graaff generators. Knowledge on cross-sections of photonuclear reactions is also rather scarce, especially if the reaction is complicated (photospallation), although giant resonances in the 15–20 MeV photo energy region have been fairly well investigated. More information about cross-sections will be given for the most general types of reactions in the following chapters along with empirical methods of predicting their order of magnitude and energy dependence whenever such methods exist.

Once some knowledge is available on a cross-section for a given process, it is possible to predict the amount of radionuclides produced and consequently the radioactivity induced. All that can be meant by 'activity' must now be explained.

2 The definition of activity

We have seen in the previous section that one can calculate the number of atoms having undergone a particular nuclear reaction in a target provided one knows the cross-section for that reaction and the number of incident particles. The number of atoms which have undergone the nuclear reaction is also the number of atoms of the new species given by the reaction considered that have been produced. This is not yet what one calls induced activity. If the atoms of the new species happen to be stable, there is no activity. If they are unstable, they are said to be radioactive, and they disintegrate or decay. This disintegration is a random process and can happen at any time after the radioactive atom has been produced. When a large number of such atoms are present, experiment shows there is a certain number of disintegrations occurring per unit time, and that this number varies slowly with time, decreasing according to an exponential law. It is this number of disintegrations per unit time, also termed decay rate, that we call the activity of the sample considered. This is in fact a number of radioactive decay processes occurring per unit time.

Supposing the bombarding particle and energy are such that a large quantity of nuclear reactions can take place, leading to the production of various elements with different decay schemes, the activity is still the total number of the disintegrations of all sorts occurring per unit time in the sample.

One often distinguishes however between alpha, beta, gamma, neutron and fission activity and considers then only the number of those particular processes per unit time which correspond to an alpha, or beta, or gamma, or neutron, or fission decay.

In practice, activity is often referred to as the number of detectable, or easily detectable, decays per unit time in which a particular particle or radiation is actually emitted. As an example, the so-called gamma activity is the total number of photons emitted by unit time. It often includes the annihilation radiation from a positive electron emitted in a beta-plus decay. The beta activity is usually taken as the total number of positive and negative beta particles emitted per unit time. Here the electron capture phenomenon, in which an electron from the electron shells of the atom falls into the nucleus, is considered not to contribute to the activity. Thus ^{7}Be, which actually emits a gamma ray in 12% of cases, decaying by internal conversion in all other cases, has an apparant gamma activity which is only 12% of the activity according to the first definition.

As for practical purposes (measurements or protection) it is the apparent activity of a particular type that is relevant, we shall adopt this point of view in the following discussions and assume that the activity is the rate of emission of the corpuscular or electromagnetic radiation we are interested in.

So, for example, we will include under gamma activity all the photons emitted per second including those from the positive beta particles, actually emitted per second, unless we distinguish between beta-plus and beta-minus activity, and so on.

The emitted radiation can of course be absorbed in the active sample itself, if this is large enough. But this absorption will be examined in detail in one of the following sections and all activity figures will be given in principle as basic data without considering this absorption.

3 The activation formula

We will now proceed to calculate the activity of a sample of material bombarded by a given flux of particles, assuming we know the cross-sections of the possible nuclear reactions and the time constants characterizing the exponential decay curves of all the isotopes produced.

We shall start with the formula defining the cross-section and modify it to find the number of atoms of the new species produced in a volume element of the target, or better, in the unit volume. For this purpose, we must introduce the flux Φ of incident particles, defined as the number of particles incident on the sample per unit surface (cm^2) and unit time (sec).

Let N_T be further the number of target atoms per unit volume (cubic centimetre), and $\sigma_{T,\nu}$ the particular cross-section leading from target nucleus with subscript T to the desired isotope with subscript ν. Then the number of radioactive nuclei of this isotope produced per unit volume in unit time will be

$$N_\nu = \Phi N_T \, \sigma_{T,\nu}. \tag{3.1}$$

This equation still holds if we divide both sides by the density ϱ of the target material. We have then instead of N_ν and N_T the number of radioactive nuclei n_ν and the number of target atoms n_T per gram of target material:

$$N_\nu/\varrho = n_\nu, \qquad N_T/\varrho = n_T. \tag{3.2}$$

The number of atoms per gram of target material can be expressed as Avogadro's number $N_0 = 6 \times 10^{23}$ atoms per mole divided by the atomic weight (or mass number) A_T of the target material

$$n_T = N_0/A_T. \tag{3.3}$$

The number of radioactive nuclei of isotope ν per gram of target material produced per unit time is thus

$$n_\nu = \Phi \frac{N_0}{A_T} \sigma_{T,\nu}. \tag{3.4}$$

The point of taking the numbers per gram instead of cubic centimetre of target material is that the density disappears from the formulae altogether because of the property of n_T to be represented by N_0/A_T.

Once these radioactive nuclei have been produced, their number decays exponentially with time. Usually it is the half-life $t_{\frac{1}{2},\nu}$ that is quoted for each isotope ν. For mathematical reasons we shall use the mean life t_ν (time to decay to the fraction 1/e of the initial numbers)

$$t_\nu = \frac{1}{\ln 2}\, t_{\frac{1}{2},\nu} = \frac{t_{\frac{1}{2},\nu}}{0.693} = 1.442 t_{\frac{1}{2},\nu} \tag{3.5}$$

and write the law of decay as a function of time t

$$n_\nu(t) = n_\nu(0)\, \mathrm{e}^{-t/t_\nu}. \tag{3.6}$$

Let us now irradiate a substance during a length of time t_i (the so-called irradiation time) and see how many radioactive nuclei we shall have at the instant the irradiation stops. Obviously, at the instant t_i, taking the moment the irradiation starts as the origin of a time coordinate τ, we shall have a number $n_\nu(t_\mathrm{i})$ of radioactive nuclei of isotope ν equal to

$$\begin{aligned} n_\nu(t_\mathrm{i}) &= \Phi\, \frac{N_0}{A_\mathrm{T}}\, \sigma_{\mathrm{T},\nu} \int_0^{t_\mathrm{i}} \exp[-(t_\mathrm{i} - \tau)/t_\nu]\, \mathrm{d}\tau \\ &= \Phi\, \frac{N_0}{A_\mathrm{T}}\, \sigma_{\mathrm{T},\nu} t_\nu (1 - \exp[-t_\mathrm{i}/t_\nu]). \end{aligned} \tag{3.7}$$

If we now let a defined 'cooling' time t_c elapse without irradiating further, this number will have decreased to the value

$$n_\nu(t_\mathrm{i}, t_\mathrm{c}) = \Phi\, \frac{N_0}{A_\mathrm{T}}\, \sigma_{\mathrm{T},\nu} t_\nu (1 - \exp[-t_\mathrm{i}/t_\nu])\, \exp[-t_\mathrm{c}/t_\nu]. \tag{3.8}$$

We are interested in the activity, that is the instantaneous disintegration rate of this isotope. We obtain this by differentiating the quantity n_ν with respect to t_c and we change its sign, because the decay rate is defined as a positive quantity.

Thus we arrive at the expression

$$-\frac{\mathrm{d}n_\nu}{\mathrm{d}t_\mathrm{c}} = \Phi\, \frac{N_0}{A_\mathrm{T}}\, \sigma_{\mathrm{T},\nu} (1 - \exp[-t_\mathrm{i}/t_\nu])\, \exp[-t_\mathrm{c}/t_\nu] \tag{3.9}$$

which is the so-called 'activation formula' for one particular isotope. It gives the value of the activity or disintegration rate of a particular isotope ν in one gram of a target material T which has been irradiated during a time t_i and left alone to decay during a time t_c.

The total specific activity a of the target material will be the sum of the specific activities of all the particular isotopes ν producible

$$a = -\sum_\nu \frac{\mathrm{d}n_\nu}{\mathrm{d}t_\mathrm{c}} = \Phi \frac{N_0}{A_\mathrm{T}} \sum_\nu \sigma_{\mathrm{T},\nu}(1 - \exp[-t_\mathrm{i}/t_\nu]) \exp[-t_\mathrm{c}/t_\nu]. \quad (3.10)$$

This specific activity is expressed in disintegrations per second per gram.

The activity of a sample of arbitrary mass is equal to the product of its specific activity and its mass. It is expressed in disintegrations per second.

An activity can also be expressed in curies. The unit of 1 curie (1 Ci) is the activity of a source having a disintegration rate of 3.7×10^{10} disintegrations per second. (This is the activity of 1 gram of radium.)

There exists a simple theorem in radioactive decay which permits one to find easily from the curves presented in figs. I.1 and I.2 the instantaneous activity in the case of any finite irradiation and cooling times. It has its origin in the form of the radioactive build-up or decay formula (3.10) which can be written

$$-\sum_\nu \frac{\mathrm{d}n_\nu}{\mathrm{d}t_\mathrm{c}} = \Phi \frac{N_0}{A_\mathrm{T}} \sum_\nu \sigma_{\mathrm{T},\nu}(\exp[-t_\mathrm{c}/t_\nu] - \exp[-(t_\mathrm{i} + t_\mathrm{c})/t_\nu]) \quad (3.11)$$

where

$$\Phi \frac{N_0}{A_\mathrm{T}} \sum_\nu \sigma_{\mathrm{T},\nu} = -\sum_\nu \left(\frac{\mathrm{d}n_\nu}{\mathrm{d}t_\mathrm{c}}\right)_{t_\mathrm{i}=\infty, t_\mathrm{c}=0} \quad (3.12)$$

corresponds to the extreme value. One sees that the activity for any irradiation and cooling time can be interpreted as the difference of the activity values along the decay curve for the case of infinite irradiation taken at the times t_c and $t_\mathrm{i}+t_\mathrm{c}$. This will spare us the effort of drawing graphs, as the graph for $t_\mathrm{i}=\infty$ can now be interpreted for finite irradiation times also. We arrive thus at the following theorem.

Theorem. *The intensity of radioactive product radiations following a finite period of irradiation is simply the difference between two values on the decay curve for the case of infinite irradiation. These are:*

a) *the activity value at the time elapsed from the end of the exposure period to the moment in question (the time normally considered to be the 'cooling' time following a run), and*

b) *the activity value at a hypothetical cooling time equal to the time elapsed from the beginning of the exposure period to the moment in question (the sum of the irradiation time and the cooling time).*

Isotope weight per curie

It is easy to calculate the number of grams of any isotope which will give 1 curie of activity. In fact, this depends only on the life-time of the particular isotope. From the exponential decay law (3.6) we find the activity per gram at a given moment, which we can take to be $t=0$,

$$-\frac{dn_\nu}{dt} = \frac{n_\nu}{t_\nu} = \frac{N_0}{A_\nu t_\nu}, \tag{3.13}$$

A_ν being here the mass number of the isotope, t_ν the mean life and N_0 Avogadro's number. The decay rate should be equal to 3.7×10^{10} per second to give one curie. Hence the quantity G of grams necessary is found from

$$G = 3.7 \times 10^{10} A_\nu t_\nu / N_0. \tag{3.14}$$

If we make use of the half-life $t_{\frac{1}{2},\nu}$ and introduce the numerical value of N_0 we arrive at

$$G = 8.9 \times 10^{-14} A_\nu t_{\frac{1}{2},\nu}\ \mathrm{g/Ci}, \tag{3.15}$$

where $t_{\frac{1}{2},\nu}$ has to be expressed in seconds.

4 Dependence of activity on irradiation and cooling times

Let us now examine a few features of these formulae in particular as regards their time dependence.

4.1 *Saturation activity of a particular isotope*

One sees from (3.9) that if the irradiation time t_i is large compared with the decay constant t_ν of a particular isotope, the decay rate or activity at a zero cooling time for this isotope has a limit, the so-called 'saturation value':

$$\left(-\frac{dn_\nu}{dt_\mathrm{c}}\right)_{\mathrm{sat}} = \Phi \frac{N_0}{A_\mathrm{T}} \sigma_{\mathrm{T},\nu}. \tag{4.1}$$

This is due to the fact that isotope nuclei that have been produced during a part of the irradiation time have had time to decay before the irradiation stops. This saturation value is the natural limit of the specific activity one can obtain with a given particle flux.

Formula (4.1) shows also that this saturation activity does not depend on the decay constant (or the half-life) of the particular isotope, but only on the reaction cross-section and the irradiating flux. Of course, the number of radioactive nuclei present in the sample at saturation depends linearly on this decay constant (see formula (3.7) of the preceding section). However, it is normally not the number of atoms present, but the decay rate, which will be of interest to experimenters.

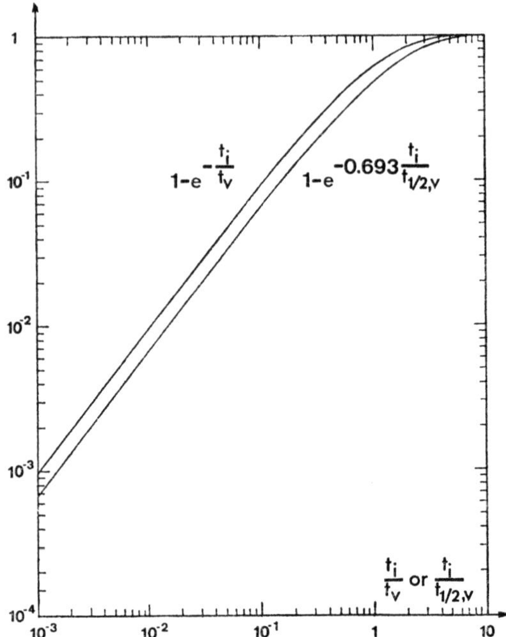

Fig. I.1 Growth of radioactivity for a single isotope. The functions $1 - \exp[-t_i/t_\nu]$ and $1 - \exp[-0.693 t_i/t_{\frac{1}{2},\nu}]$ are plotted versus t_i/t_ν and $t_i/t_{\frac{1}{2},\nu}$.

Fig. I.1 shows the way the activity builds up to saturation as a function of irradiation time. We have plotted the irradiation time t_i in relative units with respect to both the half-life $t_{\frac{1}{2},\nu}$ and the mean life t_ν. This is convenient since it is the half-lives that are usually quoted in practice.

Fig. I.2 also shows the exponential decay curves with the cooling time t_c plotted in units of both half-life and mean life. With this graph it is possible to determine the fraction of the saturation activity of a particular isotope for a chosen irradiation time at any cooling time as a percentage of saturation activity. One has only to multiply the ex-

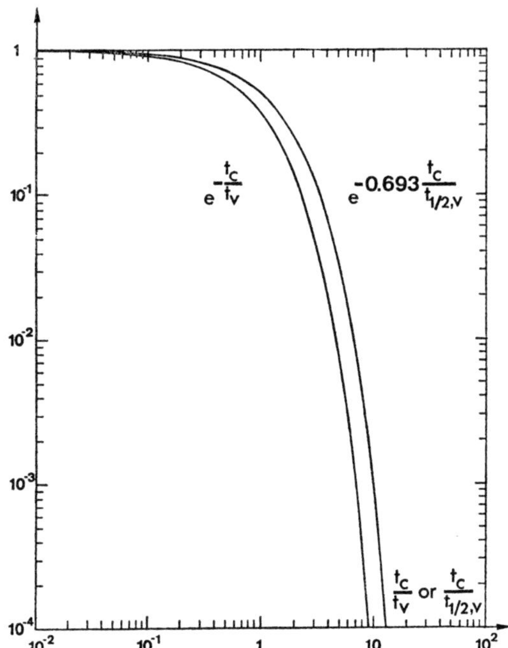

Fig. 1.2 Decay of radioactivity for a single isotope. The functions $\exp[-t_c/t_v]$ and $\exp[-0.693t_c/t_{\frac{1}{2},v}]$ are plotted versus t_c/t_v and $t_i/t_{\frac{1}{2},v}$.

pression $1-\exp[-0.693t_i/t_{\frac{1}{2},v}]$ giving the fraction of the saturation activity obtained at the end of the irradiation time t_i, by the decay factor $\exp[-0.693t_c/t_{\frac{1}{2},v}]$ giving what remains after the cooling time t_c. This is of particular interest in irradiation experiments aimed at the production of special isotopes.

4.2 *Growth and decay of the activity of a large number of isotopes*

In general one has to deal with a mixture of a large number of isotopes produced at the same time in the target, especially if this target has a high atomic weight. If something is known of the distribution of the numbers of isotopes produced with respect to their half-lives, and if this distribution is sufficiently smooth, one can venture to derive mathematical formulae for the approximate production of build-up and decay of the total activity, which can be of general interest, at least in a limited time range, by assuming they are all formed with equal cross-sections. However, if accurate information is required at all times, it will be more proper to evaluate exactly the activation formula (3.9) on a digital computer (computer calculations of this kind will be given in ch. II).

It is a matter of general interest, at any rate, to examine the distribution of the number of all known isotopes with respect to their half-lives. As the range of these half-lives covers a very large span on the time scale, it is preferable to use a logarithmic scale and plot the number of isotopes found in an arbitrary interval chosen on this logarithmic scale. This counting was done using radionuclides charts, with a total of 603 isotopes of half-lives above 5 min.

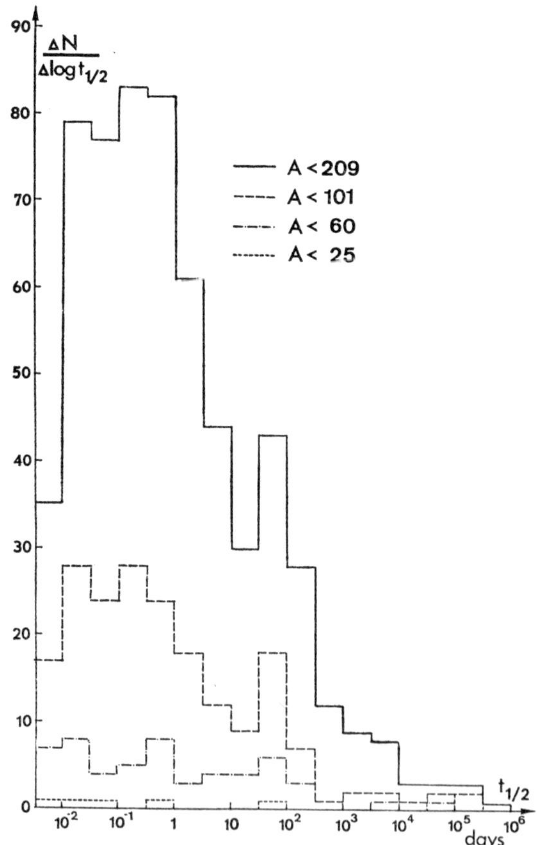

Fig. I.3 Distribution of radioactive isotopes below a given atomic number with respect to their half-lives.

As time unit we have taken one day, as interval in which the isotopes are counted the span in which the common logarithm of the half-life increases by $\frac{1}{2}$, i.e. the half-life is multiplied by $\sqrt{10}=3.16$.

The upper curve in fig. I.3 shows the result, i.e. the distribution over the half-lives of all isotopes up to an atomic number of 209 (bismuth region) with half-lives comprised between 3.16×10^{-2} (5 minutes) to 10^6 days. As one sees the distribution has a broad maximum in the region 10^{-1} to 10^2 days half-life, and falls on either side rather rapidly. Fig. I.3 also shows the distributions obtained when taking only isotopes up to a lower atomic number, i.e. with $A < 101$ (molybdenum region) and $A < 60$ (iron region). The bulk of the latter distributions falls at the same place as the first one, without being so broad.

Another presentation of the same data can be found in fig. I.4, which

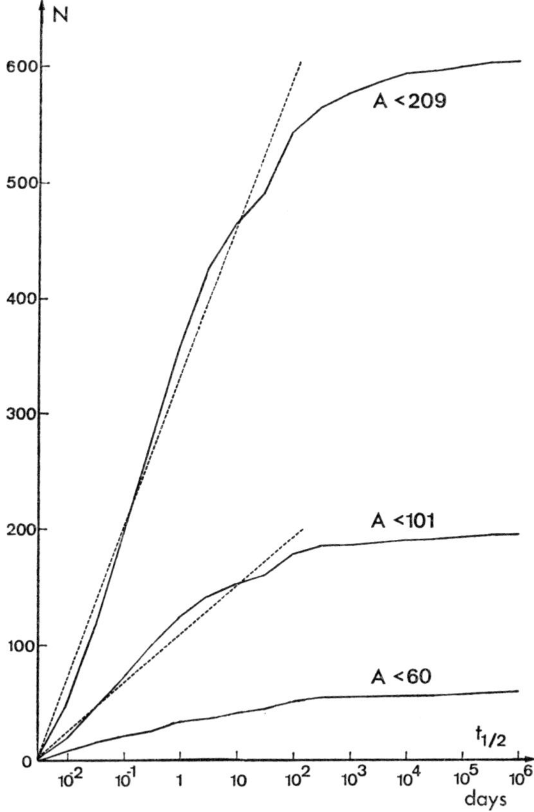

Fig. I.4 Total numbers of isotopes up to a given half-life as a function of this half-life (cases $A < 209$, $A < 101$, $A < 60$).

shows the integrals of the distributions, in other words the total number of isotopes found up to a given half-life as a function of the logarithm of this half-life. It is seen by inspection that these curves have a fairly linear rise in the interval corresponding to the bulk of the distributions given in fig. I.2. The straight broken lines drawn in the figure are an attempt at linearization over this range. On fig. I.5 are plotted an en-

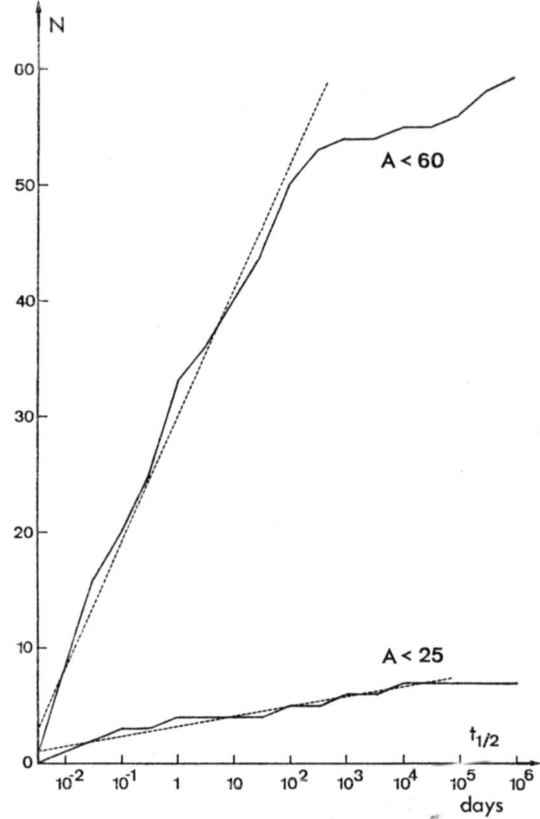

Fig. I.5 Total number of isotopes up to a given half-life as a function of this half-life (cases $A<60$ and $A<25$).

largement of the curve for $A<60$ and the curve for $A<25$, showing what a good fit the straight line is between the limits 3.16×10^{-3} and 10^3 days for the $A<60$ curve (isotopes in the iron region). It is this property which has led various authors to derive a general formula for the obtained activity as a function of irradiation and cooling times in the case of a mixture of a large number of isotopes, on the assumption that they are all produced with the same cross-section. This is done in the manner outlined below.

Let $N=N(t_{\frac{1}{2},\nu})$ be the number of isotopes found to have a half-life smaller than $t_{\frac{1}{2},\nu}$. The curves found indicate that a good approximation for this function is

$$N(t_{\frac{1}{2},\nu}) = a + b \log t_{\frac{1}{2},\nu} \tag{4.2}$$

where $\log t_{\frac{1}{2},\nu}$ is the natural logarithm of $t_{\frac{1}{2},\nu}$ and where a and b are appropriate constants. Therefore we find for its derivative in the interval where the approximation holds

$$\frac{\mathrm{d}N}{\mathrm{d}t_{\frac{1}{2},\nu}} = \frac{b}{t_{\frac{1}{2},\nu}} \tag{4.3}$$

where $\mathrm{d}N$ is the number of isotopes with half-lives lying in the interval $(t_{\frac{1}{2},\nu}, t_{\frac{1}{2},\nu}+\mathrm{d}t_{\frac{1}{2},\nu})$. Referring to eq. (3.10) of this chapter, we see that the activity of a whole mixture of isotopes assuming equal production cross-sections can be written

$$a = \text{const} \int_{t_{\nu_1}}^{t_{\nu_2}} (1 - \exp[-t_\mathrm{i}/t_\nu]) \exp[-t_\mathrm{c}/t_\nu] \,\mathrm{d}N(t_\nu) \tag{4.4}$$

where we have now, as t_ν is always equal to $1.44t_{\frac{1}{2},\nu}$,

$$\mathrm{d}N(t_\nu) = \frac{b}{t_\nu}\,\mathrm{d}t_\nu \tag{4.5}$$

and where t_{ν_1}, t_{ν_2} are the lower and upper limits considered for the mean lives of the elements of the mixture. It is convenient for ease of calculation to introduce the decay constant

$$\lambda = 1/t_\nu \tag{4.6}$$

with its differential

$$\mathrm{d}\lambda = -\frac{1}{t_\nu^2}\,\mathrm{d}t_\nu. \tag{4.7}$$

Eq. (4.4) becomes thus

$$a = \text{const}\; b \int_{\lambda_2}^{\lambda_1} [\exp -\lambda t_\mathrm{c} - \exp -\lambda(t_\mathrm{i} + t_\mathrm{c})] \frac{\mathrm{d}\lambda}{\lambda} \tag{4.8}$$

where

$$\lambda_1 = \frac{1}{t_{\nu_1}} > \lambda_2 = \frac{1}{t_{\nu_2}} \tag{4.9}$$

are the new integration limits. We see that we arrive at integrals of the type

$$\int_x^\infty \frac{e^{-u}}{u}\, du = -\mathrm{Ei}(-x), \qquad 0<x<\infty \tag{4.10}$$

which are known in mathematics as exponential integrals. They can be expanded as follows

$$\mathrm{Ei}(-x) = C + \ln x + \sum_{n=1}^{\infty} (-1)^n \frac{x^n}{n \cdot n!}, \qquad 0<x<\infty, \tag{4.11}$$

where C=0.5772 is the so-called Euler constant. Fig. 1.6 shows the

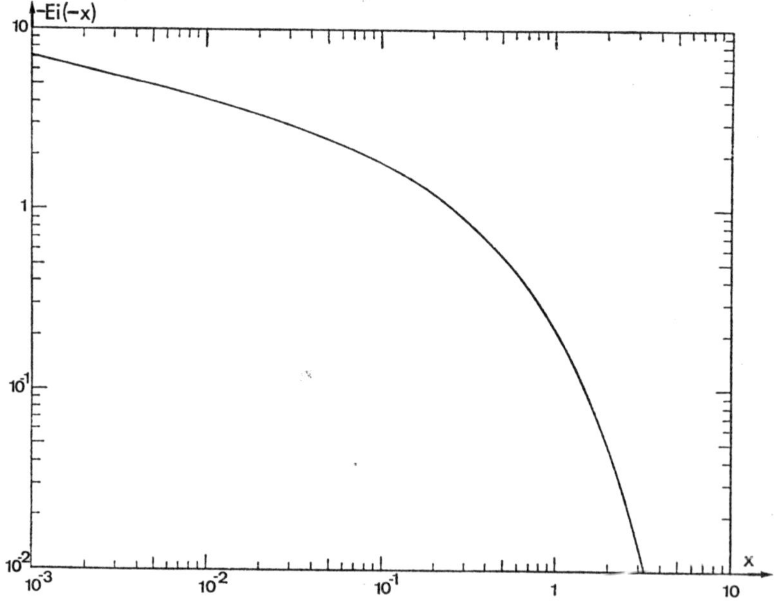

Fig. 1.6 Plot of the exponential integral $-\mathrm{Ei}(-x) = \int_x^\infty (e^{-u}/u)du$, for real and positive x.

function $-\mathrm{Ei}(-x)$ plotted as a function of x. As one sees, the function tends towards infinity when x tends towards zero, and towards zero when x becomes infinite. Using the exponential integral, the expression (4.8) for the activity becomes

$$a = \text{const}\, b\{\mathrm{Ei}(-\lambda_1 t_c) - \mathrm{Ei}(-\lambda_2 t_c) - \mathrm{Ei}[-\lambda_1(t_c + t_i)] + \mathrm{Ei}[-\lambda_2(t_c + t_i)]\}. \tag{4.12}$$

It is appropriate now to introduce the numerical values for λ_1 and λ_2 as found from the curves in figs. I.3 and I.4. It appears that half-lives of the order of 3×10^{-3} (5 min) and of 10^3 days (2.7 years) are reasonable limits for the nearly linear section of the curves, at least in the case $A<60$, which presents the greatest interest for induced radioactivity in machines. One finds then the decay constant values

$$\begin{aligned} \lambda_1 &= \frac{1}{t_{v_1}} = \frac{1}{4.31\times 10^{-3}} = 231\ (\text{day})^{-1}, \\ \lambda_2 &= \frac{1}{t_{v_2}} = \frac{1}{1.44\times 10^{3}} = 6.93\times 10^{-4}\ (\text{day})^{-1}. \end{aligned} \tag{4.13}$$

Clearly, phenomena involving to a large extent isotopes with half-lives shorter than 5 minutes or longer than three years cannot be treated correctly with this formula. Such phenomena include, for instance, studies of very short-lived activities or predictions on the increase of radioactivity beyond periods of the order of three years, as radiation age studies.

It is immediately apparent from the form of the function in fig. I.6 that the terms including λ_1, which is large, will practically vanish, provided t_c and t_c+t_i are not too small. Admitting the value $x=2$ as the limit beyond which the function $\text{Ei}(-x)$ is practically equal to zero, we obtain a lower limit for t_c and t_c+t_i above which we can neglect the terms including λ_1. These are given by the inequalities

$$\lambda_1 t_c > 2, \qquad \lambda_1(t_c + t_i) > 2 \tag{4.14}$$

which together bring us to the common condition

$$t_c > \frac{2}{\lambda_1} = 8.6\times 10^{-3}\ \text{day} = 12\ \text{min}. \tag{4.15}$$

For cooling times greater than this, which will be generally the case in practice, we can neglect the terms containing λ_1.

Let us now examine the terms with λ_2. For values of x which are small with respect to 1, the exponential integral can be approximated by the simple function

$$-\text{Ei}(-x) = -C - \ln x, \qquad 0<x\ll 1. \tag{4.16}$$

This approximation is already reasonable for $x\leq\frac{1}{3}$. As a consequence we find the limits

$$\lambda_2 t_c < \tfrac{1}{3}, \qquad \lambda_2(t_c + t_i) < \tfrac{1}{3} \tag{4.17}$$

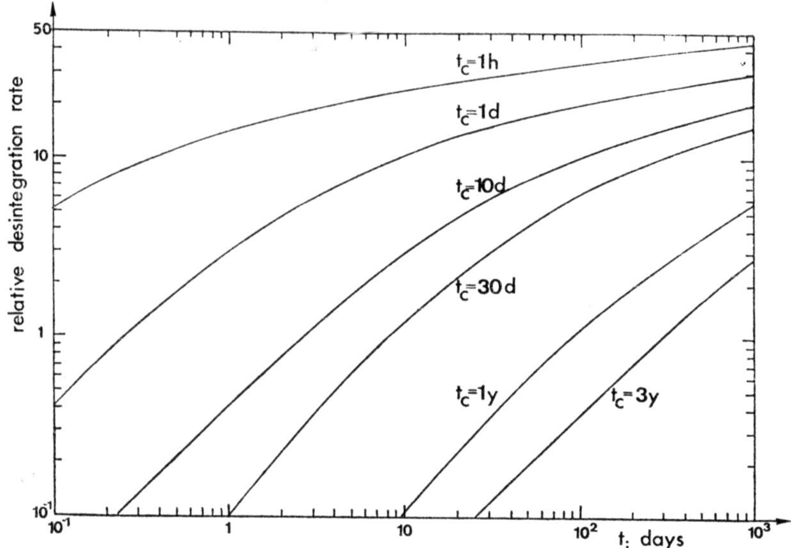

Fig. 1.7 Predicted build-up of induced radioactivity assuming constant intensity of irradiation.

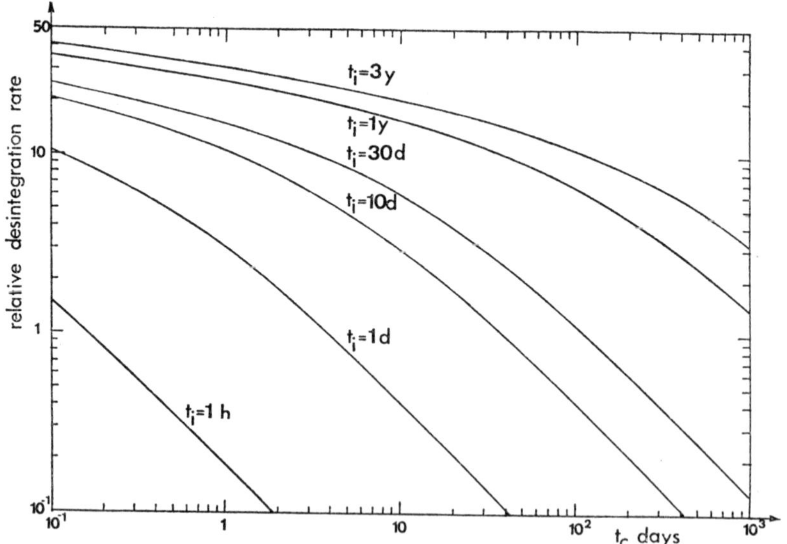

Fig. 1.8 Predicted decay of induced radioactivity after constant irradiation.

leading to the condition

$$t_c + t_i < \frac{1}{3\lambda_2} = 5 \times 10^2 \text{ days.} \tag{4.18}$$

The limits found in both cases

$$t_c > 12 \text{ min}, \qquad t_c + t_i < 5 \times 10^2 \text{ days} \tag{4.19}$$

define a region where formula (4.12) can be simplified and written as follows

$$a = \text{const } b \ln \frac{t_i + t_c}{t_c} \tag{4.20}$$

which is the expression used by Sullivan and Overton. These authors present experimental evidence which fits well with the theoretical predictions. The reader is also invited to check with the data computed in ch. II, where growth and decay curves are presented in the case of high energy nuclear interactions for which a rather good estimate of the production cross-sections exist.

Figs. I.7 and I.8 predict the build-up and decay of activation after irradiation with constant flux in relative disintegration rates using simply the relationship $10 \log_{10}(t_i + t_c)/t_c$. The scales and parameters chosen have been roughly matched to the region of validity of the simplified formula which fits in rather well with the range of half-lives in which the linear approximation in figs. I.4 and I.5 holds.

A useful application of formula (4.20) is found in the case of activation studies of rather heavy elements at high energies, where the number of isotopes formed is expected to be large in that it permits to reduce substantially the necessary irradiation time. As a time of 1 day appears to be right in the middle of the span of existing half-lives, it is sufficient to make an irradiation of that length and to measure the decay curve in order to fix the ordinate scale (disintegration rates) of figs. I.7 and I.8. The graphs permit then to determine the disintegration rates one would have for much longer irradiation times with a fair precision.

4.3 *Very short irradiation or sudden release of a large number of isotopes in fairly equal quantities*

Another interesting situation in which one can derive a simple decay formula is when a large number of isotopes are produced in equal quantities. In practice such a situation can be encountered in two cases: a very short irradiation time, that is, small in comparison to the shortest relevant half-lives, and the sudden release of a large number of different isotopes in fairly equal quantities. The first of these is found when irradiating with a short burst of particles, the second happens in a nuclear explosion. Both can be treated mathematically in exactly the same way.

Let us start from the expression (3.4) in the preceding section which gives the number of isotopes of kind ν produced by a given irradiating flux Φ per unit time and let us write the number $\mathrm{d}n_\nu$ of isotopes of kind ν produced in the time differential $\mathrm{d}t$ as

$$\mathrm{d}n_\nu = \Phi \frac{N_0}{A_\mathrm{T}} \sigma_{\mathrm{T},\nu}\, \mathrm{d}t. \tag{4.21}$$

If the total irradiation time t_i is short compared to the half-life, they will not have time to decay, and we will have at the end of the irradiation the following number of isotopes of the kind ν

$$n_\nu(t_\mathrm{i}) = \int_0^{t_\mathrm{i}} \mathrm{d}n_\nu = \frac{N_0}{A_\mathrm{T}} \sigma_{\mathrm{T},\nu} \int_0^{t_\mathrm{i}} \Phi\, \mathrm{d}t. \tag{4.22}$$

We assume here that the flux Φ can vary, but that we know its time integral. As we see, this number does not depend on the mean life t_ν; it is sufficient to assume that the cross-sections are fairly equal for all isotopes ν in order to get equal numbers produced for all kinds. Thus we get in the case of a large number of isotopes a sum total of

$$\sum_\nu n_\nu(t_\mathrm{i}) = \frac{N_0}{A_\mathrm{T}} \sum_\nu \sigma_{\mathrm{T},\nu} \int_0^{t_\mathrm{i}} \Phi\, \mathrm{d}t. \tag{4.23}$$

The decay of these products follows an exponential law with the appropriate mean life t_ν in each case. We thus get for the activity

$$a = -\frac{\mathrm{d}}{\mathrm{d}t_\mathrm{c}} \sum_\nu n_\nu(t_\mathrm{i}) = \frac{N_0}{A_\mathrm{T}} \int_0^{t_\mathrm{i}} \Phi\, \mathrm{d}t \cdot \bar{\sigma} \sum_\nu \frac{\exp[-t_\mathrm{c}/t_\nu]}{t_\nu} \tag{4.24}$$

where t_c is the cooling time and $\bar{\sigma}$ a mean cross-section.

We will consider now only the time-dependent term and rewrite it as an integral, taking advantage of the knowledge of the distribution function of the isotopes over the mean life-times gained in the preceding subsection. This brings us to the expression

$$\sum_{\nu} \frac{\exp[-t_c/t_\nu]}{t_\nu} = b \int_{t_{\nu_1}}^{t_{\nu_2}} \exp[-t_c/t_\nu] \frac{dt_\nu}{t_\nu^2}. \qquad (4.25)$$

If we introduce the decay constant $\lambda = 1/t_\nu$ as before we obtain

$$\sum_{\nu} \frac{\exp[-t_c/t_\nu]}{t_\nu} = b \int_{\lambda_2}^{\lambda_1} \exp[-\lambda t_c]\, d\lambda = b\, \frac{\exp[-\lambda_2 t_c] - \exp[-\lambda_1 t_c]}{t_c}. \qquad (4.26)$$

This is the function of cooling time t_c which describes the decay of a large number of isotopes produced in about equal quantities. It is not valid however for times less than the shortest or greater than the longest half-lives considered in the linear approximation.

As before, the formula can be simplified by considering the numerical values of λ_1 and λ_2. For isotopes of atomic number $A < 60$, we had $\lambda_1 = 231$ (day)$^{-1}$, $\lambda_2 = 6.93 \times 10^{-4}$ (day)$^{-1}$. Evidently the first term of the numerator can be equated to 1 if $\lambda_2 t_c \ll 1$, i.e. if $t_c < 0.1/\lambda_2 = 1.44 \times 10^2$ days (5 months). The second term can be equated to zero if at the same time $\lambda_1 t_c \gg 1$, i.e. if $t_c > 3/\lambda_1 = 1.3 \times 10^{-2}$ days $=$ $= 19$ min. In those limits of cooling time, the overall decay follows to a good approximation the law

$$\sum_{\nu} \frac{\exp[-t_c/t_\nu]}{t_c} = \frac{b}{t_c}. \qquad (4.27)$$

Experimental evidence is presented in fig. 2 of Sullivan and Overton, where the measured decay curve is given of a mixture of various metals after a one minute irradiation. The decay follows very clearly the $1/t_c$ law after the first few minutes in the cooling time range in which experimental points are present, i.e. up to $t_c = 100$ min.

In the case of nuclear explosions, it is appropriate to consider the distribution with respect to half-lives of the slow neutron fission products from ^{235}U. This has been done in fig. I.9 which shows both the distribution and the integral curve over half-lives of over 108 fission products which are gamma emitters.

Experimental decay curves of activity from fission products have been found to follow the law

$$a = \text{const}\ \frac{1}{t_c^{1.2}} \tag{4.28}$$

during a long interval of time. The fact that we find an exponent equal to 1.2 instead of 1 is undoubtedly due to the slight but marked curvature of the integral curve in fig. I.9.

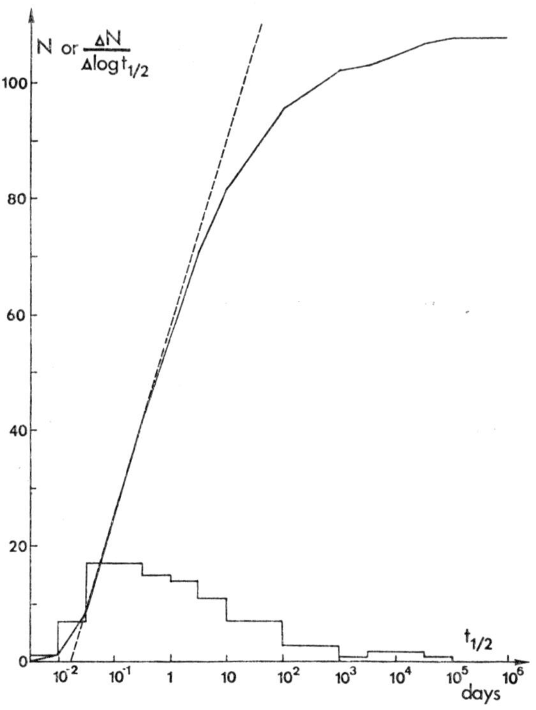

Fig. I.9 Distribution of isotopes according to half-lives and integral curve for the slow neutron fission products of ^{235}U.

5 Dose rates from radioactive isotopes

The object of the present study being obviously to define the hazards to which personnel are exposed because of induced radioactivity, it is appropriate, after having explained briefly how the activity is produced, to examine the effect of the various radiations emitted in terms of doses or dose rates.

5.1 *Definition of absorbed dose*

There are various ways of considering doses.

The physical definition of the dose is in terms of the energy given up by the radiation considered in the unit mass of a given absorber. The dose unit is then the rad, a dose corresponding to an energy deposit of 100 erg per gram of this absorber, irrespective of how the energy is deposited, what kind of radiation is incident and what the absorber is. As it is especially the hazard to personnel we are concerned with in this study, we shall take here as reference absorber soft tissue (or water, which gives practically the same results).

5.2 *The concept of radiobiological effectiveness*

This physical dose concept is however insufficient to establish the radiation hazard to man. Some kinds of radiation produce, for equal energy deposits per gram, greater damage to living tissue than others. An attempt has been made to take this into account by introducing a factor for multiplying the energy deposition in tissue to arrive at the real hazard or damage experienced by a human being. This is called the radiobiological effectiveness (RBE). Thus the physical dose in rads, multiplied by the RBE, should give the biological dose. This will be expressed in units called rem (from 'röntgen equivalent man', the röntgen being a previous unit associated to the energy deposition). It is in rems that all permissible radiation doses are indicated. Measured doses are however usually expressed in rads, as it it what the experimental apparatus available can measure.

What makes the evaluation of a dose in rems difficult in general is the fact that the RBE depends on the kind and energy of the incident radiation, on the organ in the body, and is only known with some accuracy in limited energy ranges. Also the concept of biological damage varies as time goes on as does in fact the opinion one has of the damage inflicted to living tissue by the particular radiation involved. If one considers further that such damage can be reparable or irreparable one will understand how difficult it is to estimate such a factor, or to define a 'tolerance level' considered as permissible.

In this study, we will always use the physical dose in rads, as this is what we will try to compute or measure from the physics point of view. However, we shall also give the RBE factors as they are known at present in order to enable the reader to carry out a calculation of the biological hazard produced by radioactive bodies whenever possible. Also for a comparison of the various types of radiation from radioactive substances, a knowledge of the biological hazard in rems that they produce is necessary.

5.3 *The definition of dose rate*

Besides the concept of dose, that of dose rate is also used. Whereas the dose refers to the energy absorbed from the radiation, the dose rate indicates in what length of time this occurs. It is usually expressed in rads per hour. A dose rate of 1 rad/h is the dose rate of a radiation field of such an intensity that the dose of 1 rad is deposited in a period of 1 hour.

5.4 *The absorption of energy in tissue for radiation emitted by radioactive substances*

When considering the chart of all radionuclides, one sees that the radiation they emit can be alpha and beta particles, gamma rays and in some cases neutrons (there is practically no proton radioactivity).

For these radiations we have first to give the energy deposited as a function of energy. This is done conveniently by indicating the number of incident particles or photons per unit surface (1 cm²) which will give an absorption of one rad per unit mass (1 g) of tissue. This number, which has the physical dimension $\text{cm}^{-2}\text{rad}^{-1}$ is plotted in fig. I.10 for gamma quanta (photons), alpha and beta particles and neutrons, according to the values found in the literature. For neutrons we have used the first collision dose curve in the NBS Handbook 63.

For the gamma ray curve, good approximation formulae, which are practical when using computers, can be found to be

$$f = 1.45 \times 10^{13} E^2 \text{ cm}^{-2} \text{ rad}^{-1} \qquad \text{for } E < 0.053 \text{ MeV} \tag{5.1}$$

$$f = 2.1 \times 10^{9} E^{-1} \text{ cm}^{-2} \text{ rad}^{-1} \qquad \text{for } 0.053 < E < 5 \text{ MeV} \tag{5.2}$$

where f is the number of photons per cm² to produce 1 rad in soft tissue and E the photon energy expressed in MeV.

For the electron curve one can write approximately

$$f = 3 \times 10^{7} \frac{E}{0.17} \text{ cm}^{-2} \text{ rad}^{-1} \qquad \text{for } 0.01 < E < 0.17 \text{ MeV} \tag{5.3}$$

$$f = 3 \times 10^{7} \text{ cm}^{-2} \text{ rad}^{-1} \qquad \text{for } 0.17 < E < 10 \text{ MeV} \tag{5.4}$$

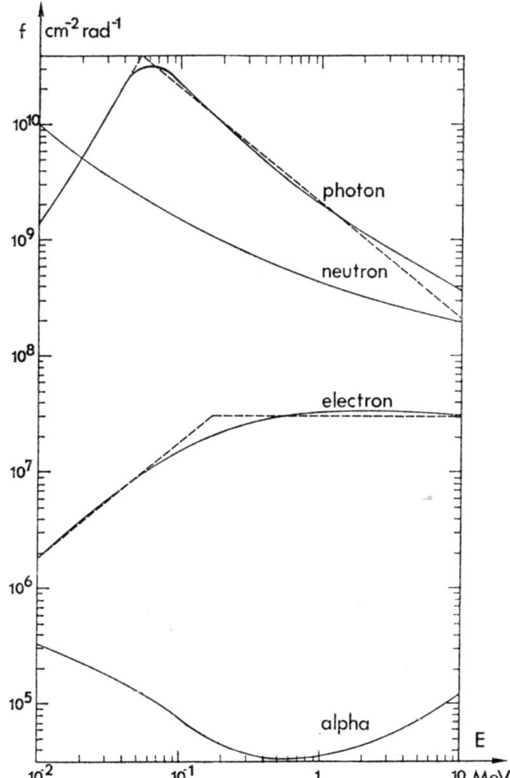

Fig. I.10 **Number of incident particles and photons per cm² to produce 1 rad in soft tissue.**

f being now the number of electrons incident per cm² to produce 1 rad and E the electron energy in MeV.

Dotted lines corresponding to these approximations have been drawn in fig. I.10.

From fig. I.10 we see that in the major portion of the energy range plotted (0.1 to 10 MeV) gamma quanta deposit less energy than neutrons, neutrons less than the electrons, and electrons less than alpha particles, per unit mass of tissue. So there are more gamma quanta needed to produce 1 rad than neutrons, more neutrons than electrons, more electrons than alpha particles. This is not however the order in which one would place the various kinds of radiation if one wished to consider the damage they inflict. This is because of the radiobiological effectiveness which assumes different values for each sort of radiation.

5.5 *Radiobiological effectiveness of radiation from radioactive substances*

In the range of energy to be considered for radiation emitted in radioactive decays, which does not extend above 10 MeV, the following can be said about the RBE.

X-rays or gamma quanta have an RBE of 1 independent of energy. Electrons, in this low energy region, interact much as X-rays and are assumed to have also an RBE of 1.

Neutrons have an RBE which is taken to depend on energy according to the curve in fig. I.11. Extreme values of the RBE are 3.5 and 10.5 in this range.

Alpha particles, in the whole range of energy from radioactive decays, have an RBE taken to be 10.

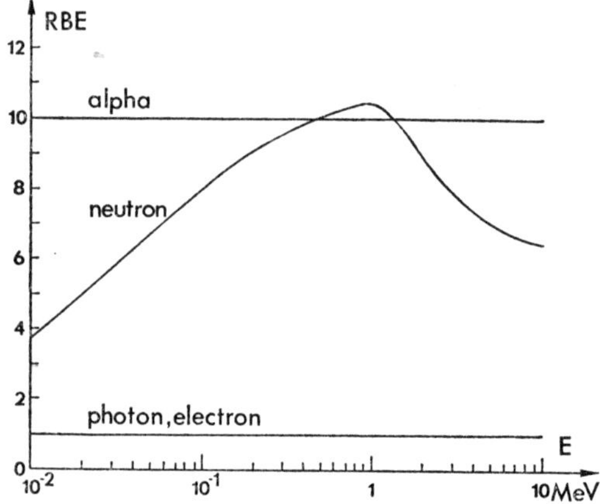

Fig. I.11 Radiobiological effectiveness of alpha particles, neutrons, photons and electrons in the range 0.01 to 10 MeV kinetic energy.

With the knowledge of the RBE for the various radiations from radioactive decay, we can start constructing curves which will give us the number of incident particles or photons per cm^2 to produce a 1 rem dose, as shown in fig. I.12. We will have thus related our physical findings to the units in which the maximum biologically permissible doses are calculated. We shall thus be in a position to evaluate for each particular radioactive substance the hazard to personnel involved, provided we can foresee or measure the value of the flux of the emitted radiation to which the experimenters are subjected, that is, provided we can compute the disintegration rate per gram of emitter and know the distance to the source, if it is a point source, or the configuration of the radiation field, if the radioactive objects are distributed in space.

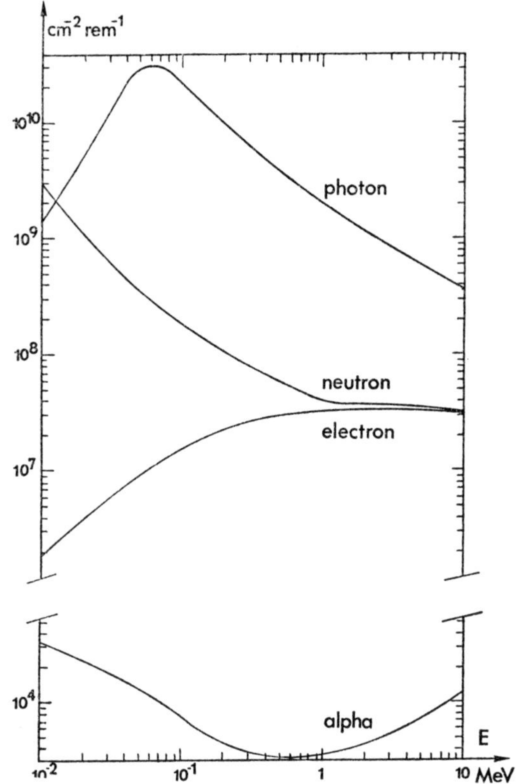

Fig. I.12 Number of incident particles and photons per cm² to produce a biological dose of 1 rem in soft tissue.

5.6 *Definition of the k-factor*

In work with radioactive substances, a quantity which is obviously needed is the radiation level corresponding to a given activity at a given distance. This depends on the intensity of the various kinds of radiation emitted, their type and energy, and the geometrical arrangement of the radioactive objects. Let us leave geometrical considerations aside for the moment and assume one point source only. Let us further assume that this source has unit activity, i.e. one curie (3.7×10^{10} disintegrations per second). We then arrive at a figure characteristic of the type of source considered, in fact a material constant. This quantity is colloquially known as the '*k*-factor' and gives the radiation level from a point source of unit activity at unit distance. We shall express it here in rad/hour in soft tissue at 1 metre distance from a point source of 1 curie. Absorption of the emitted radiation in the source material is ignored here.

5.6.1 *The k-factor for gamma rays.* We have seen that radioactive substances emit gamma quanta, beta and alpha particles, and neutrons. What is generally used is the k-factor of the radiation level caused by gamma rays, it being assumed in the case of positron emitters that all these are annihilated in the source with the emission of two gamma quanta of energy 0.51 MeV.

However, it is of interest to introduce also k-factors for the radiation levels caused by electrons, alpha particles and neutrons. We will thus speak of the factors k_γ, k_β, k_α, k_n, respectively. In practice, the k-factors are useful for two purposes. First, they make it possible to establish the hazard to which the experimenter is subjected when working with a source of a given strength. Second, in the case of a source of known composition but unknown strength, they make it possible to determine the source strength by measurement of the radiation field at a known distance.

The k_γ-factors vary over a wide range, owing to the large variety of gamma energies and probabilities of emission possible. Apart from 0, when there is no photon or positron emitted in the decay, typical k_γ-values range from 0.002 (^{127}Te) to 2.67 (^{86}Zr). A list of these factors computed using the decay schemes found in the literature is given in appendix A for the most-used isotopes up to $A=209$. Isotopes with half-lives of less than 5 minutes have been omitted, as well as a number of very long-lived ones. For the user's convenience all isotopes present in the naturally occurring radioactive families including the neptunium series have been added, as well as a few transuranic elements, as they become available slowly.

Dependence of the k_γ-factor on the mass number of the isotope. The results given in the table in appendix A can be represented graphically. It is interesting to plot the k_γ-factor against atomic mass number, to see the distribution over the mass of the elements. This is done in fig. I.13, where we have summed for each mass number the k_γ-values corresponding to all the isobars having this mass number. To limit the work of plotting we have restricted ourselves to gamma-emitting isotopes with a half-life longer than a day and with an atomic mass inferior to 210. From the preceding subsection 4, where the distribution of radioisotopes over their life-times was studied, we know that we retain about one half of the existing radioisotopes by this procedure. There is also another, more practical reason, to commend this choice. One usually waits some time after shut-down before starting work on installations that are radioactive. A waiting period of one or a few days is quite an acceptable average in practice, so that the hazard from isotopes with a life-time shorter than a day can be considered as faded away when work actually begins.

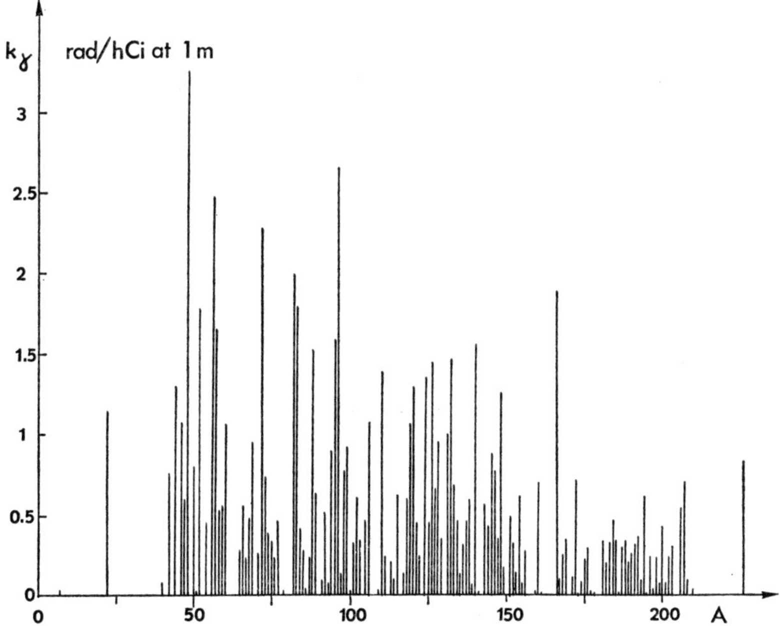

Fig. I.13 Projection of the k_γ-values of artificial radioisotopes with half-lives longer than 1 day on the mass number scale. The value for ^{226}Ra in equilibrium with its decay products is shown for comparison at the extreme right.

An interesting fact which strikes one immediately is that up to atomic mass number 42, there is a gap with practically no gamma emitters of half-life larger than a day, excepting ^{7}Be and ^{22}Na. Just above begins a series of isotopes that are the most dangerous of all. The k_γ-values then fall slowly along the mass number scale up to bismuth, so that there is no significant gap in which the k-values would be greatly reduced or absent.

This gives an interesting indication, when one is looking for absorbers or shield materials that are both dense and not too radioactive. It is clear that the heaviest elements which one can take as shield material against high energy particles without having to fear too high radioactivity are limited to about $A=42$ as the radioelements produced by spallation are all below the target mass number. The only practical element found in that zone is calcium. In nature calcium exists usually in the form of carbonate in limestone or crystallized in calcite (marble). Whereas limestone always has inclusions of various oxides (Si, Mg, Al), marble is extremely pure and has also been found to be the mineral having the smallest own residual radioactivity (Pensko et al.), so that it is ideally suited for low background laboratories. More on this subject will be found in the special subsection of ch. II.

A more elegant presentation of the data included in appendix A would be a three-dimensional one, plotting k_γ as a function of A and Z, instead of summing up for all isobars and plotting against A. We come so to a model where the k-values appear as spikes projecting from the usual Z, N chart of nuclides (Z, charge number; N, neutron number). Such a model has been constructed to get a general impression of the distribution of radiation hazard over all the nuclides and is shown in the photograph presented in fig. I.14 facing p. 50. Here also only gamma emitters with a half-life longer than a day and atomic mass inferior to 210 have been considered.

Number of photons emitted per disintegration. The table in appendix A also includes the number ε of photons emitted per disintegration. It is in fact the sum of the emission probabilities for each gamma quantum in the decay scheme. With this data and the counting rate measured by gamma counters of known efficiency, one can determine the activity in curies of a given sample.

Mean value of k_γ-factor. It is of interest to calculate the mean value of the k_γ-factors given in appendix A. In the cases where a large number of radioisotopes are produced at the same time (spallation by very high energy particles, fission, etc.) this mean value is likely to give the right order of magnitude for the k_γ-factor of the mixture obtained.

In table I.1 are listed the mean k_γ-factor values of all radioelements up to an atomic number A of 60, 100 and 209. Also the mean k_γ-factor values for all radioelements with half-lives smaller and larger than a day are given separately. The table also includes the number ε of photons emitted per disintegration.

TABLE I.1

Mean values of k_γ and ε for various groups of radioelements.

Group	$A<60$	$A<101$	$A<209$	$A<209$ $t_{\frac{1}{2}}<1$ d	$A<209$ $t_{\frac{1}{2}}>1$ d
k_γ (rad/h Ci) at 1 m	0.66	0.5	0.32	0.34	0.29
ε photons per disintegration	1.6	1.42	1.10	1.15	1.03

One notices a variation of the mean value of the k_γ-factor with the group chosen: isotopes of lower mass numbers tend to be more dangerous and to emit more photons. This reflects the trend noticed in figs. I.13 and I.14 for k_γ-factors to decrease slowly from $A=60$ onwards as one moves up in the periodic system towards higher mass number. In contrast there is no apparent dependence of the mean values on lifetime.

Milking systems. In some cases a radioactive isotope has a daughter product which is itself radioactive with a different half-life. This second product remains mixed with its parents and what one measures is the sum of the radiation from both. Such systems tend to be more and more used nowadays, as the parent isotope can be easily prepared in reactors and have a long life, whereas the daughter has a short one. Thus, relatively short-lived isotopes can be stored at the place of use. These systems are often called 'milking systems' and the parent isotopes 'cows' (Greene). The point here is to know the ratio of the two activities in the same sample. The following calculation shows how to arrive at the result.

Let us assume we have an activity A_{10} at time zero for the parent (subscript 1) and that we have no activity for the daughter (subscript 2). The activity of the parent decreases exponentially with an index λ_1, that of the daughter with λ_2. We have thus

$$A_1(t) = A_{10}\,\mathrm{e}^{-\lambda_1 t}. \tag{5.5}$$

One calculates the activity of the daughter as a function of time by noting that the number of daughter atoms produced per unit time is equal to $A_1(t)$, whereas the number of daughter atoms decaying per unit time is λ_2 times the number $N_2(t)$ of daughter atoms present at any moment so we have the differential equation for N_2:

$$\frac{\mathrm{d}N_2}{\mathrm{d}t} = A_1 - \lambda_2 N_2 \tag{5.6}$$

or if we introduce the expression for A_1

$$\frac{dN_2}{dt} + \lambda_2 N_2 = A_{10}\,e^{-\lambda_1 t}. \tag{5.7}$$

The exact solution is

$$N_2(t) = \frac{A_{10}}{\lambda_2 - \lambda_1}\,(e^{-\lambda_1 t} - e^{-\lambda_2 t}). \tag{5.8}$$

This is valid for $\lambda_2 \neq \lambda_1$, i.e. the decay constants of parent and daughter are different. The exact solution for the case $\lambda_1 = \lambda_2$ is also known, but need not be considered here.

To obtain the activity of the daughter we must multiply this with λ_2:

$$A_2(t) = \lambda_2 N_2(t) = \frac{A_{10}\lambda_2}{\lambda_2 - \lambda_1}\,(e^{-\lambda_1 t} - e^{-\lambda_2 t}). \tag{5.9}$$

For milking systems the daughter has a short life compared to the parent, so $\lambda_2 > \lambda_1$, and we get

$$A_2(t) = A_{10}\,e^{-\lambda_1 t}\,\frac{\lambda_2}{\lambda_2 - \lambda_1}\,[1 - e^{-(\lambda_2 - \lambda_1)t}]. \tag{5.10}$$

The factor $A_{10}e^{-\lambda_1 t}$ is the parent activity $A_1(t)$. The second term in the square brackets decreases to zero after some time. We are left with the following simple expression for the ratio of the daughter to the parent activity at any moment:

$$\frac{A_2(t)}{A_1(t)} = \frac{\lambda_2}{\lambda_2 - \lambda_1} = \frac{t_1}{t_1 - t_2}, \tag{5.11}$$

where t_1 and t_2 are the mean life-times of parent and daughter. It can happen that the daughter product has a longer half-life than the parent ($\lambda_1 > \lambda_2$). From eq. (5.10) one can easily show that after a time large compared with the half-life of the parent, there is only the daughter left, with an activity equal to

$$A_2(t) = A_{10}\,\frac{\lambda_2}{\lambda_1 - \lambda_2}\,e^{-\lambda_2 t}. \tag{5.12}$$

A comprehensive and detailed treatment of problems connected with the growth and decay of radioactive series is given, for instance, by Evans.

5.6.2 *The k-factor for beta rays (negative electrons).* Besides the k_γ-factor which is in wide use, it is appropriate to introduce the k_β-factor which is the corresponding constant for the emission of negative electrons (positrons having already been taken care of in the k_γ-factor). In view of the fact that, in similar numbers, negative electrons are far more dangerous than gamma rays (see fig. I.12), it is relevant to be able to estimate the hazard relative to the beta activity, neglecting absorption in the source and in the surrounding air. This estimate is then an upper limit, but in many cases it can be approached, if, for instance, one has foils and if the distance is not too large (see beta absorption in air as a function of energy in the next section). In contrast with gamma emission, where the emission probability of a gamma quantum varies considerably according to the decay scheme, and where positron emission competes with electron capture, the total number of negative electrons emitted per beta decay is always one (Evans, ch. 17, 1). Sometimes there are two values of maximum energies. It is also apparent from fig. I.10 that the absorbed dose is independent of energy with electrons over a range of energies (roughly from 0.17 to 10 MeV) in which nearly all the maximum beta energies from radioactive decay lie. Thus, we can dispense with a numerical table and compute the k_β-factor in the energy range mentioned above to be

$$k_\beta = 35.3 \text{ rad/h Ci} \quad \text{at} \quad 1 \text{ m}. \tag{5.13}$$

This gives at least the upper limit for the hazard from beta sources. In fact, as electrons are emitted with a continuous energy distribution extending from zero to a maximum energy there will be many electrons with smaller energy than the maximum. The average energy is about one third of the maximum energy. For lower electron energies ($E<0.17$ MeV) we would have according to fig. I.10

$$k_\beta = 35.3 \frac{E}{0.17} \text{ rad/h Ci}, \qquad E<0.17 \text{ MeV}, \tag{5.14}$$

where E has to be expressed in MeV.

One sees that the highest k_β-value is one order of magnitude above the highest k_γ-value found. This draws attention to the hazard from beta radioactivity near large radioactive bodies for uncovered parts of the body (face, eyes, hands etc.).

5.6.3 *The k-factor for alpha rays.* We continue in a logical manner by introducing the k_α-factor for alpha emitters. Here also there is one alpha particle per disintegration. Its energy deposition in tissue, however, varies markedly with its kinetic energy. One has thus to compute the k_α-values as a function of energy. The result is given in fig. I.15. It is seen that k_α varies between the limits 3.4×10^3 to 2.87×10^4 rad/h Ci at 1 m. This is at least two orders of magnitude higher than the k_β-values. Because of their limited range, however, they are absorbed in the dead layer of the skin and not considered as a hazard for external irradiation, except for the eyes. Moreover, they are absorbed in air within distances of the order of 10 cm, so that the definition of k_α as a radiation field value at 1 m from the source is academic and can only serve to scale down to much smaller distances. Also self-absorption in the sample should be taken into account here.

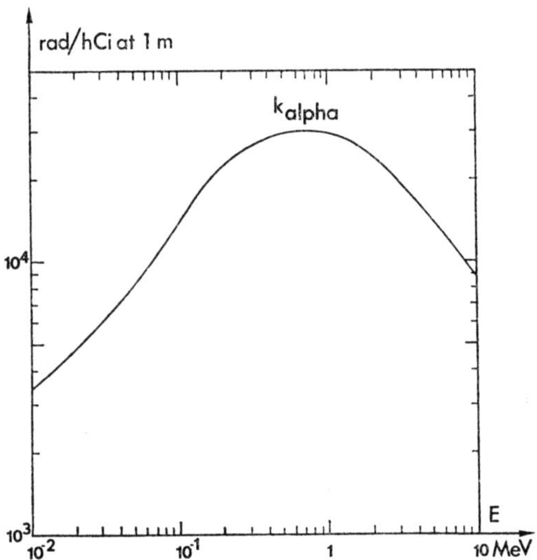

Fig. I.15 Variation of factor k_α as a function of the kinetic energy of the alpha particle.

5.6.4 *The k-factor for neutrons.* Less speculative is the k_n-factor for neutrons, although neutron emitters are very few. For instance ^{17}N is found in the cooling circuits of reactors, because of the presence of the stable isotope ^{17}O in water. Table I.2 gives the list and some properties of known neutron emitters excluding fissile material. It is seen that the rad doses from neutron emitters are comparable in magnitude to those from gamma emitters. However, the biological effect is much greater because of the RBE of the neutrons. Also neutrons are much more difficult to shield off.

TABLE I.2

Properties of neutron emitters.

Element	Half-life	Decay products	E_n mean (MeV)	k_n at 1 m (rad/hCi)
^{17}N	4.14 s	n, β^-	1	2.52
^{87}Br	55.6 s	n, β^-, γ	0.25	1.41
^{89}Br	4.5 s	n	0.5	1.77
^{137}I	22 s	n, β^-	0.56	1.9

6 Absorption of the radiation from radioactive decay in matter

In practical cases, one does not have to deal with point sources, but with massive radiating bodies. A source can be considered as massive as soon as there is some absorption of the emitted radiation in it. This absorption is called the self-absorption in the emitting body. Self-absorption is an important factor to consider for two reasons. First, for specific activity measurements of sources or active samples, it introduces an important correction. Second, for the appreciation of the hazard to an experimenter standing beside an extended active body, the knowledge of the fraction of radiation absorbed inside the emitting body itself is essential. So we have to review the basic properties of absorption in matter of the radiation likely to be emitted in radioactive decays, if we want to be able to make a step further in estimating radiation fields. We shall examine in turn how to treat gamma, beta, alpha rays and neutrons from this point of view in the energy range met in this application.

6.1 *Attenuation of gamma rays in matter*

The absorption of electromagnetic radiation in matter has the feature of being exponential in narrow beam geometry (well-collimated beam). This is because photons are absorbed or scattered out of the beam in a single event. Those photons which pass through the absorber considered have had no interaction. As soon as the beam is broad two things happen. First, photons scattered out by Compton effect from other places than the line of sight from the source to the supposedly small detector, can fall on the detector, and the radiation level existing is always higher than the one found in the thin well-collimated beam. Second, the spectrum of gamma ray energies transmitted is quite different from the incident spectrum, as the photons have a smaller energy after the scattering event.

6.1.1 *Build-up factors.* The first effect, the increase of the radiation level by Compton-scattered photons from other parts of the beam, is taken care of by the introduction of the so-called build-up factor. A build-up factor always refers to some measurable property of the photon beam (e.g. number of photons, or energy flux, or dose absorbed in soft tissue) and is defined as the ratio of this quantity when the effect of all quanta are included to that one would have when only the uncollided flux (photons which have not had any interactions) is considered.

Because of the degradation in energy of the transmitted spectrum due to the addition of these scattered photons, there will be a different build-up factor for each property of the radiation considered: overall number of photons, or energy content, or absorbed dose, for instance.

Extensive theoretical calculations and experimental investigations have been made on this subject, and information is usually given as energy or dose build-up factors for isotropic point sources or plane monodirectional sources as a function of absorber material, thickness and energy of the gamma ray emitted in the radioactive decay process (Price, ch. 2, § 7).

As a consequence, the procedure for using build-up factors is as follows. With point sources, for instance, one uses the formula

$$\frac{N}{N_0} = B_N \frac{\mathrm{e}^{-\mu x}}{4\pi r^2}, \tag{6.1}$$

x being the absorber thickness, r the distance from the source, and B_N the build-up factor for the number of quanta. If it is the energy build-up factor B_E that is given, one finds the initial energy E_0 by multiplying N_0 by the original gamma ray energy E_γ and has then

$$\frac{E}{E_0} = B_E \frac{\mathrm{e}^{-\mu x}}{4\pi r^2}. \tag{6.2}$$

If the dose build-up factor B_d is given, one finds the dose d_0 related to the N_0 original quanta of energy E_γ from fig. I.10 and has

$$\frac{d}{d_0} = B_d \frac{\mathrm{e}^{-\mu x}}{4\pi r^2}. \tag{6.3}$$

More about build-up factors will be said in one of the later subsections of this chapter.

6.1.2 *Narrow beam mass attenuation coefficient.* However, the basic physical material constant that interests us from the point of view of gamma ray absorption is the linear attenuation coefficient μ which gives the total attenuation in narrow beam geometry as a function of material and gamma ray energy. It includes the scattering effect, scattered-out particles being counted as lost. It makes it possible to calculate the fraction N/N_0 of original photons present after passage through a thickness x of material

$$N/N_0 = \mathrm{e}^{-\mu x}. \tag{6.4}$$

To get rid of the density the thickness of material x is usually replaced by the number ϱx of grams per cm² traversed, ϱ being the density. One has then the law

$$N/N_0 = \mathrm{e}^{-(\mu/\varrho)\varrho x} \tag{6.5}$$

and μ/ϱ is found not to vary very much from one element to another over a wide energy range. It is however strongly energy dependent. The quantity μ/ϱ in $g^{-1}\,cm^2$ is plotted for various elements in fig. I.16.

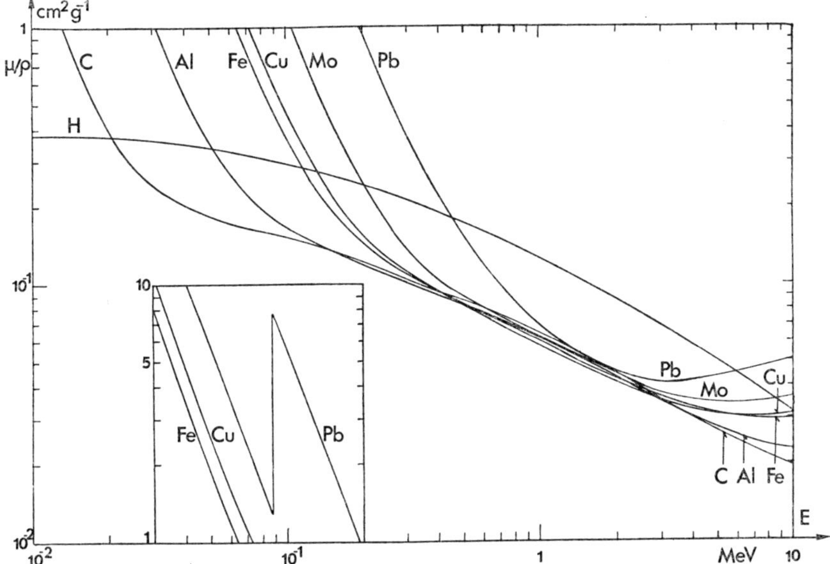

Fig. I.16 Narrow beam mass attenuation coefficient for photons in various materials plotted as a function of photon energy.

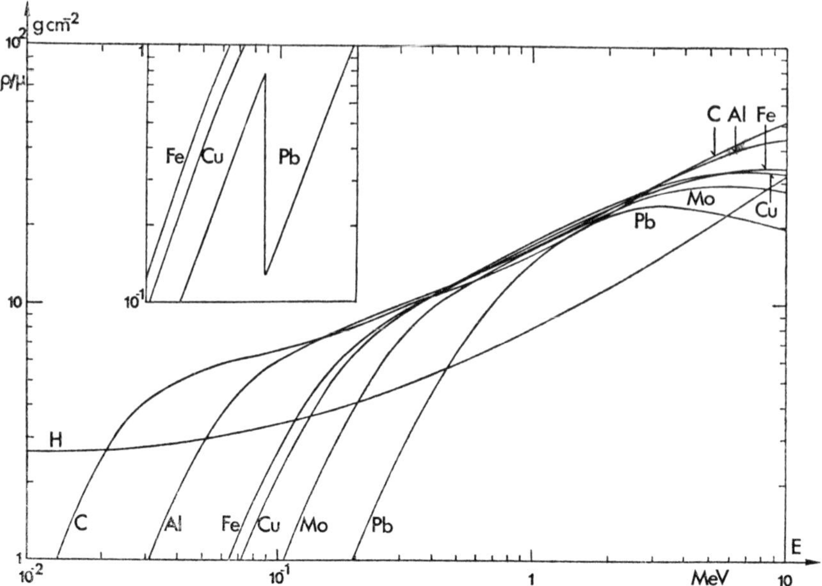

Fig. I.17 Mean attenuation path for photons in various materials plotted as a function of photon energy.

One sees that μ/ϱ decreases with energy: harder gamma rays are less well absorbed. Another quantity of practical interest is the mean attenuation length X, for which the fraction N/N_0 has decreased to $1/e$. It is also usual to express it in grams per cm^2, taking instead of X the quantity

$$\varrho X = \varrho/\mu \text{ g cm}^{-2}. \tag{6.6}$$

This quantity is plotted in fig. I.17.

6.1.3 *Approximation formula for narrow beam mass attenuation coefficient.* For computer use it was felt necessary to derive a simple approximation formula for this mass absorption coefficient, as it is usually found in tables or graphs which do not lend themselves easily to storage for a large variety of photon energies and types of absorbers.

As is well-known, gamma rays can be absorbed or scattered out by three effects: the photoelectric effect in the lower energy range (emission of an electron from the atomic shell), Compton scattering and absorption (by setting an electron in motion) in the medium energy range, and electron–positron pair production in the high energy range (above 1.2 MeV).

From the graphs (see for instance Evans, ch. 23, § 2) it appears that one can approximate the photoelectric curve and the total attenuation curve due to the Compton effect by straight lines on log-log paper in the regions where each of them gives the main contribution to the absorption. So for instance, the sum of the Compton absorption and scattering gives nearly a straight line in the energy range where photoelectric effect and pair production are weaker.

The third contribution to the absorption, which comes from pair production, can be taken care of by a simple formula which is already well-known (Evans, ch. 24, § 2).

To take into account the variation of the coefficient with atomic weight A_T and electric charge Z_T of the target element considered, we have used the scaling formulae also indicated by Evans, and deduced the numerical coefficients from the values for lead.

So we come to the following approximation for the total attenuation coefficient of electromagnetic radiation in matter:

$$\frac{\mu}{\varrho}(E, A_T, Z_T) = 1.8\times 10^{-5}\,\frac{Z_T^{4.05}}{A_T}(10E)^{-3} + 0.1267\,\frac{Z_T}{A_T}(E)^{-0.45} + 3.52\times 10^{-4}\,\frac{Z_T^2}{A_T}[3.11\ln(3.92E) - 8.07]\ \mathrm{g^{-1}\,cm^2}, \qquad (6.7)$$

where μ/ϱ is in $\mathrm{cm^2 g^{-1}}$ and E in MeV.

The first term comes from the photoelectric effect, the second one from Compton absorption and scattering and the third from pair production. The third term should only be used above the energy which makes the quantity in square brackets equal to zero (about $E_\gamma = 2$ MeV), and below energies where screening corrections become large (about $E_\gamma = 1000$ MeV), as these have not been included in the third term.

The figure 3.52×10^4 in the last term is in fact the product of the quantities $\frac{1}{137}(e^2/m_0c^2)^2 = 5.8\times 10^{-28}$ cm²/nucleus and $N_0 = 6.025\times 10^{23}$, the Loschmidt–Avogadro number.

The whole approximation formula should be applicable for all elements including hydrogen in the range from the K-edge (see table I.3) to 1000 MeV and have an overall precision of better than 30%.

As an example, the cases of lead ($A_T = 208$, $Z_T = 82$), iron (56, 26) and hydrogen (1, 1) have been computed and are shown as dotted lines in fig. I.18 together with the exact values (full lines).

TABLE 1.3

Energy of the K-edge in various elements.

Element	Z	E_K (keV)
Al	13	1.56
Fe	26	7.09
Ni	28	8.3
Cu	29	8.96
Zn	30	9.64
Sr	38	16.1
Pd	46	24.3
Ag	47	25.45
I	53	33.15
Ba	56	37.2
Ta	73	67.2
W	74	69.2
Pt	78	78.2
Au	79	80.6
Pb	82	87.8
U	92	115.4

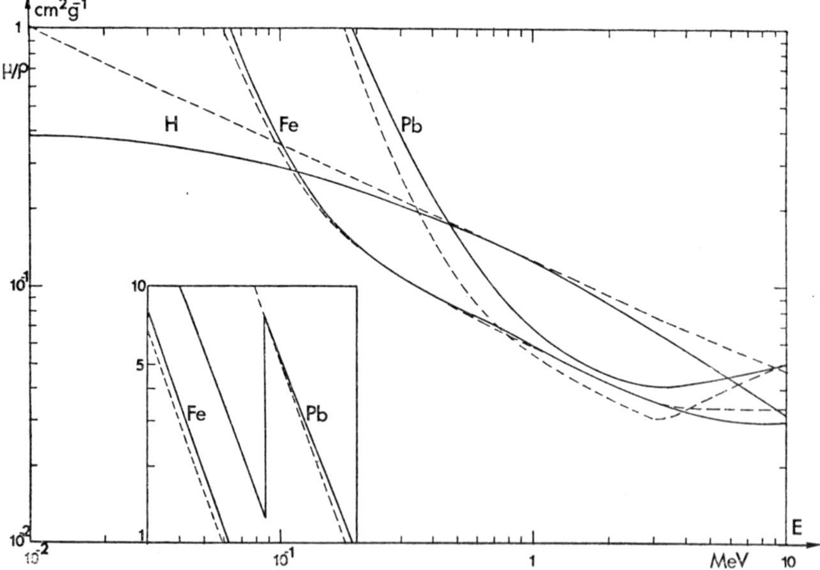

Fig. 1.18 Comparison of the experimental values of the narrow beam mass absorption coefficient for H, Fe, Pb (full lines) with the values obtained using the approximation formula.

6.2 *Absorption of electrons in matter*

In contrast to gamma rays, electrons are bodies that move through matter, losing their energy by bits in successive collisions with atom shells until they come to rest somehow. So they have a so-called range, meaning that there is a certain distance beyond which none will be found. This range is a function of the initial energy of the electron when entering the absorber, but hardly varies with the substance itself (Evans, ch. 21, § 3 and 22). The range of electrons in matter as a function of initial energy is plotted in fig. I.19 which is the range curve in aluminium given by Katz and Penfold.

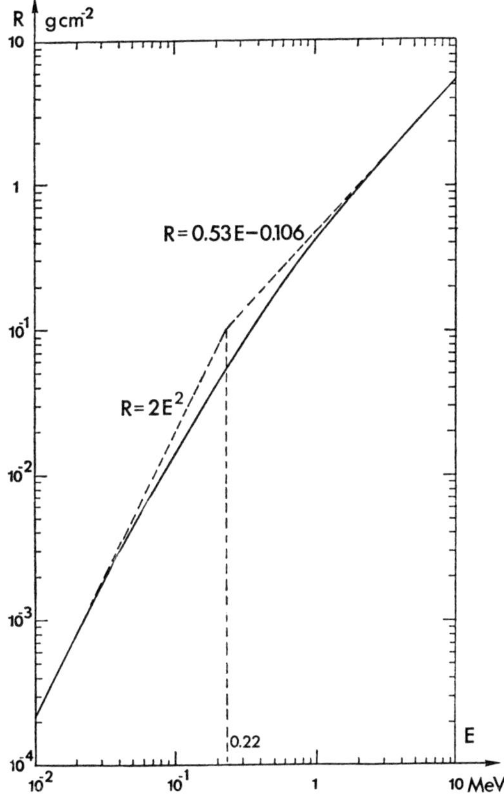

Fig. I.19 Range R of electrons in matter as a function of initial energy.

Now radioactive beta sources emit electrons with a continuous energy spectrum of a shape varying from element to element. What is generally indicated is the maximum beta energy. Absorption experiments with electrons from radioactive sources show however that the number of electrons of all energies present after a given absorber thickness decreases exponentially with this thickness. This they do only up to the range corresponding to the maximum electron energy present from the decay, where the electron number curve drops sharply to zero. It is of interest to know the mean attenuation length of this exponential decay up to the maximum range of the electrons present in order to be able to make the necessary corrections in the case of rather thin absorbers (thickness smaller than the maximum range). Particularly when measurement is made by an ionization chamber, the transmission curve is nearly exponential over the majority of its length and can be represented by

$$I/I_0 = \mathrm{e}^{-(\mu/\varrho)\varrho x}, \tag{6.8}$$

where I/I_0 is the fraction of ionization remaining after passage through an absorber thickness x. It is now found experimentally that the mass absorption coefficient μ/ϱ for electrons is nearly independent of the atomic weight of the absorber, rising only slightly with increasing Z. One can use the empirical relation

$$(\mu/\varrho)_{\mathrm{electrons}} = 17/E^{1.14}, \tag{6.9}$$

where E is the maximum electron energy in MeV. Conversely the thickness in $\mathrm{g\,cm^{-2}}$ for reduction by a factor $1/\mathrm{e}$ is

$$(\varrho/\mu)_{\mathrm{electrons}} = E^{1.14}/17. \tag{6.10}$$

This mean attenuation thickness is plotted as a function of E_{max} in fig. I.20. The remarkable fact about the attenuation of beta ray spectra is that the mass absorption coefficient seems to be determined only by E_{max}. If after a certain thickness half of the radiation is absorbed, the remaining radiation retains the absorption coefficient characteristic of the parent beta ray spectrum.

Bleuler and Zünti have given curves for finding the absorption of whole beta ray spectra from radioactive isotopes (Fermi spectra) in absorbers, which are often useful for finding the corrections mentioned in case of self-absorption in the source. These curves are valid within a few percent for absorbers independently of their atomic number Z. From fig. I.21 one finds for a given maximum energy of the spectrum and a given absorber thickness ϱx in $\mathrm{g\,cm^{-2}}$ the exponent n from which

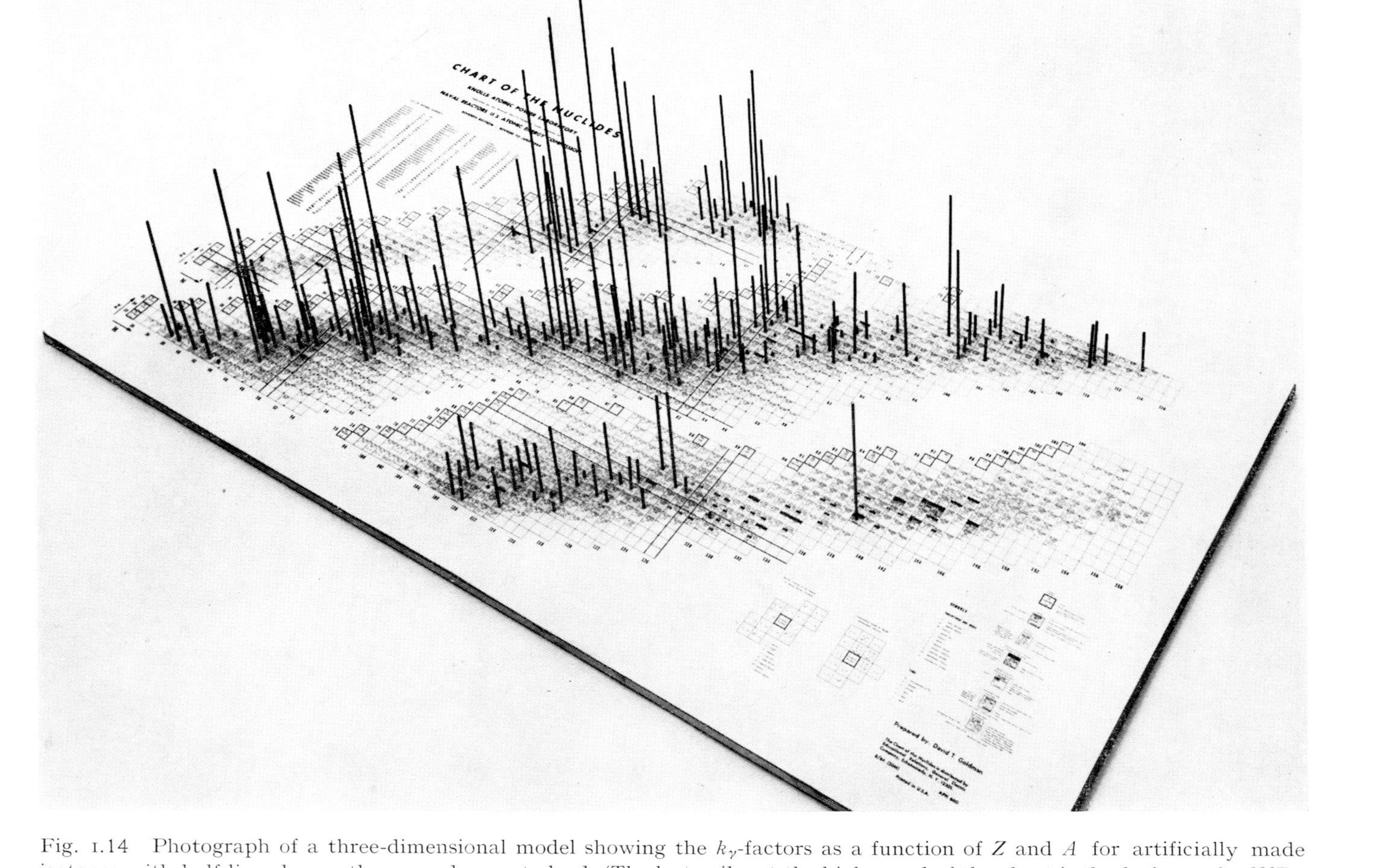

Fig. 1.14 Photograph of a three-dimensional model showing the k_γ-factors as a function of Z and A for artificially made isotopes with half-lives larger than one day up to lead. (The last spike at the higher end of the chart is the k_γ-factor for ^{226}Ra for comparison.) For text, see p. 38.

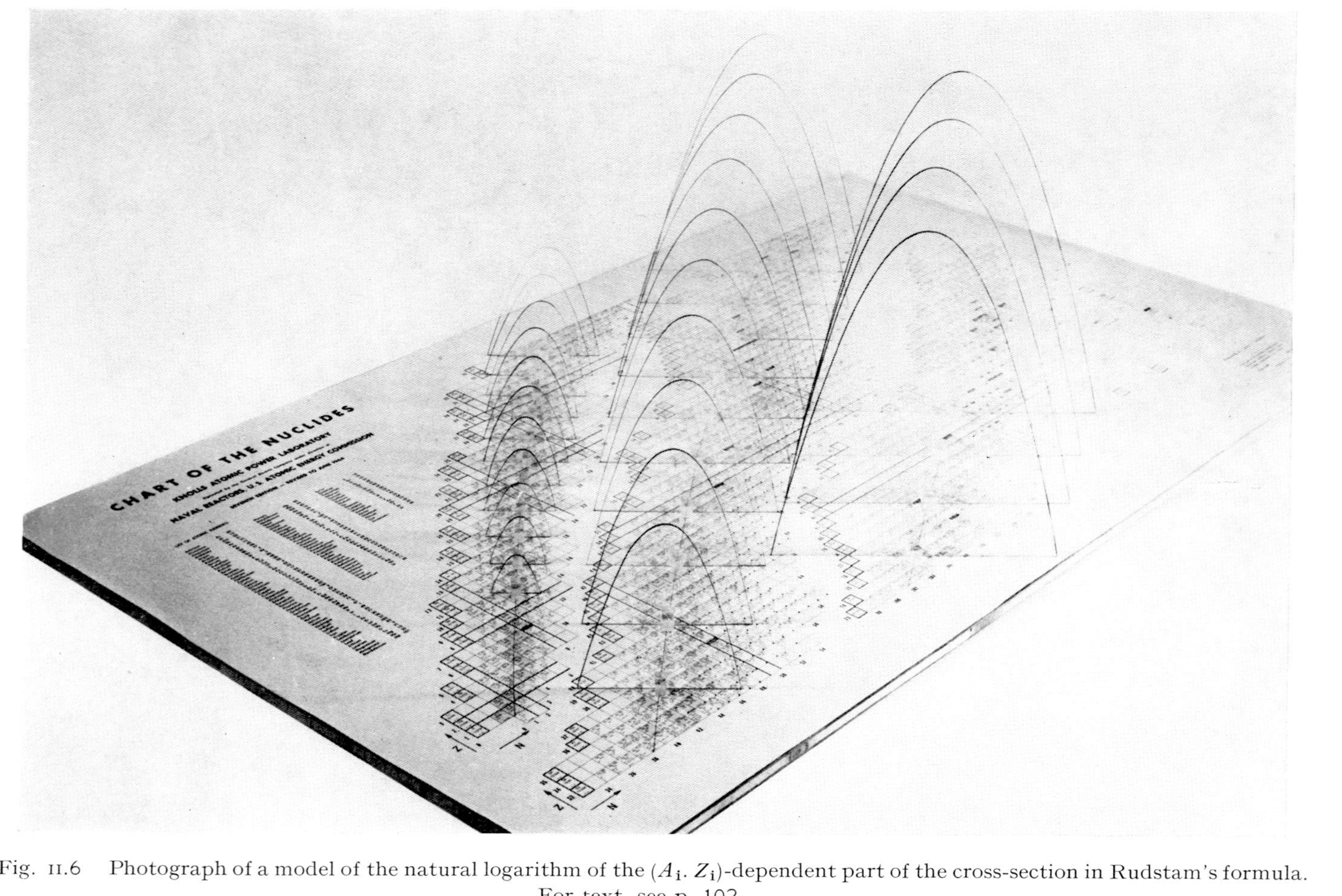

Fig. II.6 Photograph of a model of the natural logarithm of the $(A_i.\ Z_i)$-dependent part of the cross-section in Rudstam's formula. For text, see p. 102.

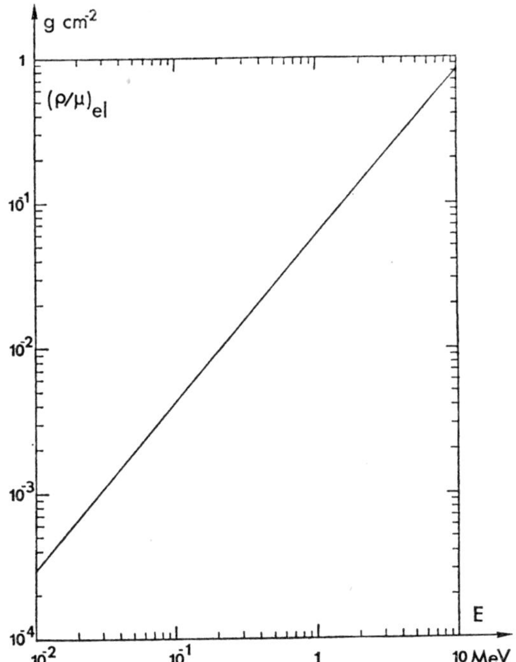

Fig. I.20 Mean attenuation thickness in matter for an electron spectrum with maximum energy E_{max}.

one can calculate the attenuation A by the formula

$$A = 2^n. \tag{6.11}$$

6.3 *Absorption of alpha particles in matter*

Like electrons, alpha particles also lose their energy in bits and have a definite range. What is usually given is the stopping power of the material, which is a function of the energy of the alpha particle and of the atomic mass number of the absorber. The stopping power $-(\mathrm{d}E/\mathrm{d}x)$ is defined as the energy loss of the particle for one gram per cm² of absorber, so that x is the path length multiplied by the density. Northcliffe has made a comprehensive review of the absorption of ions in various materials. Fig. I.22 shows the stopping power for alpha particles in C, Al, Ni, Ag, Au taken from his work. With this data it is possible by numerical integration of the inverse of $-(\mathrm{d}E/\mathrm{d}x)$ over the

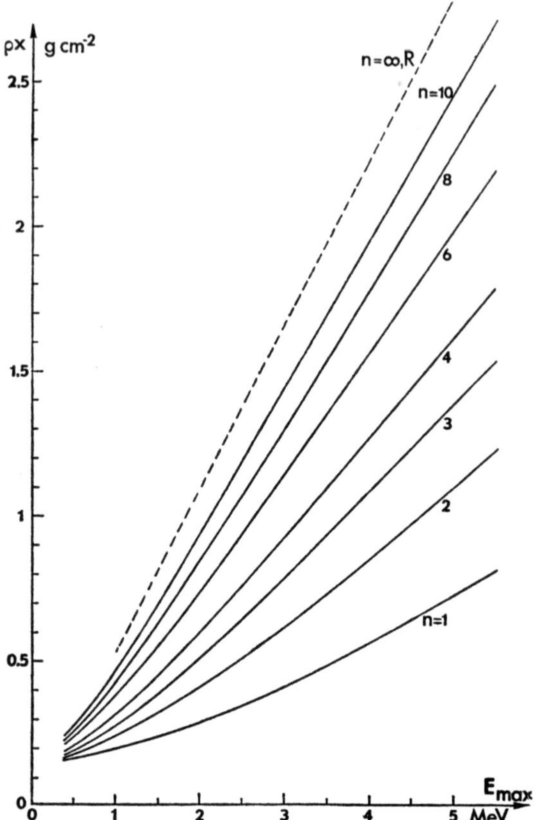

Fig. I.21 Absorber thickness to have a given attenuation of beta rays as a function of the maximum energy of the beta ray spectrum. The attenuation is 2^n, the value n can be read from the curves or interpolated. The curve $n=\infty$ corresponds to the range R.

energy E to compute the range R of the particle

$$R(E) = \int_0^E [-(\mathrm{d}E/\mathrm{d}x)]^{-1}\,\mathrm{d}x. \tag{6.12}$$

At low energies ($E<0.2$ MeV) it is seen from fig. I.22 that $-(\mathrm{d}E/\mathrm{d}x)$ can be equated to const $\cdot E^{\frac{1}{2}}$ and one has thus $R(E)=\text{const}\cdot E^{\frac{1}{2}}$. For higher energies the integration has been made graphically.

The range of alpha particles as a function of energy in the absorbers mentioned above is represented graphically in fig. I.23.

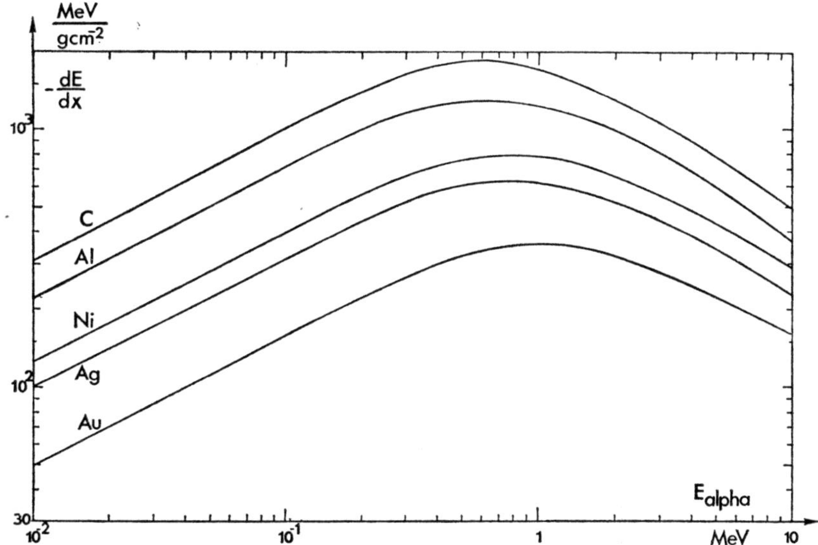

Fig. 1.22 Stopping power for alpha particles in C, Al, Ni, Ag, Au.

6.4 *Absorption of neutrons from radioactive decays in matter*

As we have seen in a previous subsection, there are only four radioactive isotopes emitting neutrons: ^{17}N, ^{87}Br, ^{89}Br and ^{137}I with neutron energies equal to 1, 0.25 and 0.5 MeV. Much has been written on neutron absorption in matter. In conformity with the preceding cases of gamma, alpha and beta particles where we have only considered narrow beam absorption we will indicate here the mean free paths for exponential attenuation of neutrons with the energies quoted above in various materials, as they can be calculated from the known total (elastic plus inelastic) neutron cross-sections given in the literature. Let σ be the total neutron cross-section in a material with atomic mass A and density ϱ. A neutron flux Φ is attenuated in a thickness dx by the amount

$$\mathrm{d}\Phi = \Phi\sigma \frac{N_0}{A} \varrho \, \mathrm{d}x, \tag{6.13}$$

where N_0 is Avogadro's number, 6×10^{23} atoms/mole. The quantity $(N_0/A)\varrho$ is the number of atoms per cubic centimetre. The exponential attenuation can be derived as follows

$$\Phi(x) = \Phi(0) \exp[-\sigma(N_0/A)\varrho x], \tag{6.14}$$

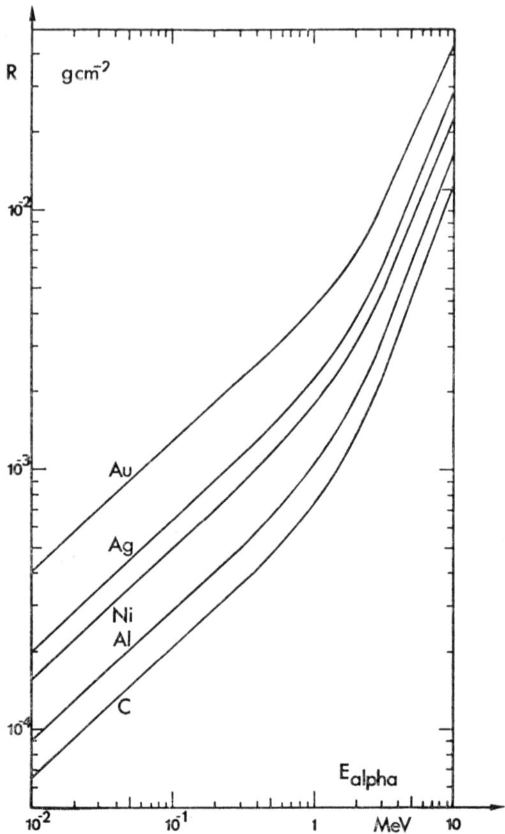

Fig. I.23 Range–energy relations for alpha particles in various absorbers computed by numerical integration from the stopping power curves of fig. I.22.

x being the length of absorber traversed and $\Phi(0)$ the incident flux. One sees that the neutron mean free path L_n will be, if we express it as usual in g cm^{-2}

$$L_n = \frac{A}{\sigma N_0} \text{ g cm}^{-2}. \tag{6.15}$$

The following table I.4 gives the neutron mean free paths L_n for the neutron energies 0.25, 0.5, and 1 MeV in various materials.

7 Definition of the danger parameter

For comparing materials from the point of view of induced activity, one has to find a figure of merit or at least a characteristic figure for

TABLE I.4

Neutron mean free paths
L_n in $g\,cm^{-2}$ in various materials.

Neutron energy (MeV)	0.25	0.5	1
H	0.181	0.27	0.397
Be	3.37	4.17	4.41
C	4.76	5.72	7.41
Al	16.92	11.55	15.85
Si	8.49	14.6	10.15
Ca	10.75	39.2	23.2
Fe	31.1	31.7	37.3
Cu	21.1	23.4	31
Nb	17.04	18.46	23.9
Mo	18.58	20.7	26.1
Ag	23.1	25	30.5
Ba	45.1	40.3	37.1
W	39.8	46.5	48.7
Au	38.2	45	59.7
Pb	46.9	99	55

describing the actual hazard to personnel having to work in the neighbourhood of radioactive installations built of these materials. Such a figure should include:

the magnitudes of all cross-sections for formation of the radioactive isotopes under the particular conditions of activation (particle energy),

the specific kinds of radiation emitted by the isotopes produced,

the self-absorption of the radiation emitted in the bulk of the radioactive material,

conversion to the dose from such a radiation absorbed in soft tissue.

All these items are specific material properties. It is felt that one has to take them all into account together in order to get at what one is really after, the actual hazard to the human being. The amount of hazard is usually expressed by indicating the level of a radiation field, given in rad/h (or rem/h if alpha particles or neutrons are involved). So we must aim at finding a material constant which can be given in rad/h (or rem/h) as well. This material constant can be different according to the irradiation conditions, i.e. particle type or energy, but can be made not to depend on flux, which can be taken as unity in all cases (one particle or photon per second per square centimetre).

The radiation fields from radioactive substances are complicated functions of irradiation and cooling times, according to the isotopic composition and the half-lives of all the radioactive isotopes produced. So these time functions are different from one irradiated material to another. One could get rid of these time dependences in a simple way by taking the characteristic figure for saturation (infinite irradiation time) and at zero cooling time. However, this time dependence is also a specific property of each irradiated material. It is often of great interest to know the residual radioactivity at particular cooling times, say 1 hour or 1 day, when work begins for instance, or to know it as a function of irradiation time, in order to be able to predict radiation levels or limit them by limiting the irradiation time. So it is a good thing to include these time functions in the characteristic figure, which should then be given always for specified irradiation and cooling times.

As each species of gamma radiation quanta or of beta particles emitted by a particular isotope has a different flux-to-dose conversion factor and a different absorption coefficient in the target material, according to their energy, we shall have to build up the sum of all the individual effects of all these species of gamma quanta or electrons.

We are then led to a figure which we will call the danger parameter D, as it indicates the hazard produced, and characterizes fully the properties of the target material under particular irradiating conditions from the point of view of induced radioactivity. We find it must read as follows:

$$D = 3600 \frac{N_0}{A_\mathrm{T}} \sum_{\nu} \sum_{\varkappa} \frac{\varepsilon_{\nu,\varkappa} \varrho_\mathrm{T}}{f_\varkappa \mu_{\varkappa,\mathrm{T}}} \sigma_{\nu,\mathrm{T}} [1 - \exp(-t_\mathrm{i}/t_\nu)] \exp(-t_\mathrm{c}/t_\nu)$$

$$\text{rad/h per particle/sec cm}^2, \qquad (7.1)$$

where the subscripts T, ν and $\varkappa$ stand for target, isotope and species of radiation emitted by isotope ν during its decay. The meaning of the symbols used is as follows:

N_0 is Avogadro's number, 6×10^{23} atoms/mole.

A_T is the target atomic mass number (number of nucleons in target atom).

$\varepsilon_{\nu,\varkappa}$ is the number of gamma quanta or electrons of type $\varkappa$ emitted by radioactive decay (as seen in subsection 5.6).

$f_\varkappa$ is the factor for converting a flux of the particular radiation quanta or particles emitted in the decay process to absorbed dose in rads (as seen in subsection 5.4). It is expressed in $cm^{-2} rad^{-1}$.

$\mu_{\varkappa,T}$ is the narrow beam linear absorption coefficient for gamma rays or the inverse mean free path (for electrons or neutrons). It is expressed in cm^{-1}.

ϱ_T is the density of the target material ($g cm^{-3}$).

$\sigma_{\nu,T}$ is the cross-section for producing isotope ν from the target element under the particular irradiation conditions considered (given type of incident particle and energy).

t_i is the irradiation time.

t_ν is the mean life of isotope ν (1.44 times its half-life).

t_c is the cooling time.

As it stands this figure is best calculated with a digital computer in view of the large number of factors involved.

It will be seen in the next subsection, devoted to the calculation of radiation fields for simple geometrical forms of radioactive bodies, that this danger parameter has also a very concrete meaning. It is equal to the radiation field actually existing inside a cavity of arbitrary form embedded in an infinite volume of radioactive substance with uniform distribution of activity which has been irradiated by a unit flux (one particle per second per square centimetre).

If the same substance has been irradiated by a flux Φ particles/sec cm^2, the radiation field level in such a cavity would be $D\Phi$.

It can also be shown that the radiation field experienced in front of an infinite and very thick wall which has been irradiated uniformly by the flux Φ is $\frac{1}{2}D\Phi$, independently of the distance from the wall. The danger parameter D thus gives directly the possibility of calculating the radiation field in front of radioactive massive bodies, provided the flux Φ that has been activating these bodies is known. It will be the scope of the following chapters to present computing methods or experimental data for the danger parameters of a large number of technically important materials under various irradiation conditions.

It is however of interest and already possible here to make a comparison in general terms of the hazards resulting from the emission of gamma radiation and from the emission of negative electrons, in order to see which kind of radiation can be suspected to be the most dangerous, at least for surface dose.

Comparison between the hazards due to gamma rays and electrons from radioactive decays

This comparison is important, as it will tell us the ratio of the radiation levels due to gamma and to beta emitters to be expected in front of an extended massive radiating body. Let us consider two kinds of isotopes only, one being a gamma and the other a negative electron emitter, and assume that both kinds of isotopes emit their gamma rays or electrons at a given moment in equal quantities. We have then to calculate the danger parameter in both cases. The result will be a function of A_T (the target element) and of the energy of the emitted radiation (gamma quantum or electron). We can relate our results either to 1 barn cross-section for the production of each isotope, by setting in formula (7.1)

$$\sigma_{\nu,T} = 1 \text{ barn} \equiv 10^{-24} \text{ cm}^2 \tag{7.2}$$

or to the total inelastic cross-section of the target element considered, by putting in the same formula

$$\sigma = \sigma_{\text{inel}} = 15.9\pi A^{\frac{2}{3}} \text{ millibarn} = 15.9 \times 10^{-27} \pi A^{\frac{2}{3}} \text{ cm}^2. \tag{7.3}$$

It is also convenient to get rid of the time functions by assuming that one is at saturation (infinite irradiation time) and at zero cooling time. The time functions then become identical to one

$$[1 - \exp(-t_i/t_\nu)] \exp(-t_c/t_\nu) \equiv 1. \tag{7.4}$$

What remains of the danger parameter is then

$$D_{\nu,T} = 3600 \frac{N_0}{A_T} \frac{1}{f_\nu} \left(\frac{\varrho}{\mu}\right)_{\nu,T} \sigma_{\nu,T}, \tag{7.5}$$

subscript ν standing for the kind of isotope and T for the target element. With the numerical values $N_0 = 6 \times 10^{23}$ and $\sigma_{\nu,T} = 10^{-24}$ cm^2 (1 barn cross-section) put in, e.q (7.5) becomes

$$D_{\nu,T} = \frac{2160}{A_T f_\nu} \left(\frac{\varrho}{\mu}\right)_{\nu,T}. \tag{7.6}$$

$D_{\nu,T}$ is clearly a function of the energy of the gamma ray or negative electron emitted, because both f_ν, the factor for converting flux per cm^2 to rads, and $(\mu/\varrho)_{\nu,T}$ the mass absorption coefficient, are. Introducing the values of f_ν and $(\varrho/\mu)_{\nu,T}$ given in subsections 5 and 6 of this chapter and plotted in figs. I.10, I.16 and also given by eq. (6.9), we can represent graphically $D_{\nu,\mathrm{T}}$ in both cases of gamma and electron emission as a function of the energy of the emitted radiation for a number of target elements per barn cross-section. This was done in figs. I.24 and I.25 for the target elements C, Al, Ni, Mo, Pb and an assumed cross-section of 1 barn. As will be seen, the danger parameter for both gamma rays and electrons increases as a function of energy.

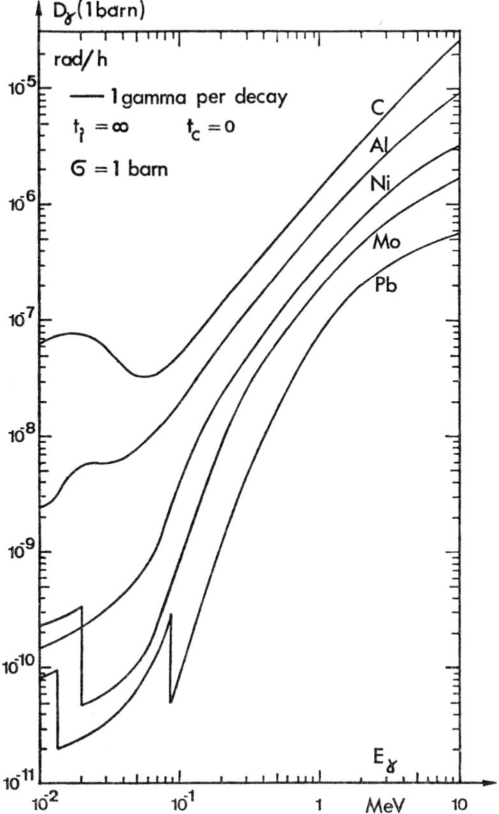

Fig. I.24 Danger parameter per barn cross-section for the emission of one gamma quantum per decay at saturation and zero cooling time as a function of gamma energy for various target elements.

They range from 10^{-11} to 10^{-5} rad/h for gamma rays and 10^{-9} to 10^{-6} rad/h for electrons depending also on the target material. The values are lower for target elements with higher atomic weights than for light elements. In order to put this property better into evidence, we have

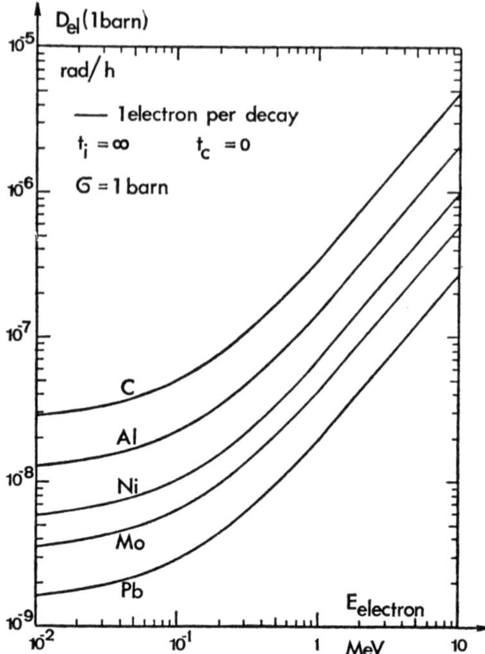

Fig. I.25 Danger parameter per barn cross-section for the emission of one negative electron per decay at saturation and zero cooling time as a function of electron energy for various target elements.

calculated the danger parameters anew, this time putting in the total inelastic cross-section of the target atom instead of the arbitrary cross-section of 1 barn. The result is shown in fig. I.26 for both gamma rays (full lines) and electrons (dotted lines). Although we do not obtain one single curve for all atomic weights in each case, the distance between curves corresponding to various atomic weights is greatly reduced. The values range now from 10^{-10} to 10^{-5} rad/h for gamma rays and 10^{-9} to 10^{-6} for electrons.

One notices in addition that at lower energies, electrons represent a higher danger than gamma rays. This property reverses for higher energies, where gamma rays give the greater hazard. The energy range where the two kinds of radiation appear to be equally dangerous is seen from fig. I.26 to be between 0.1 and 0.3 MeV. However, electron spectra emitted in radioactive decays have often much higher maximum energy. For this reason it is of interest to have an idea of how the gamma rays and electrons emitted by radioactive elements are

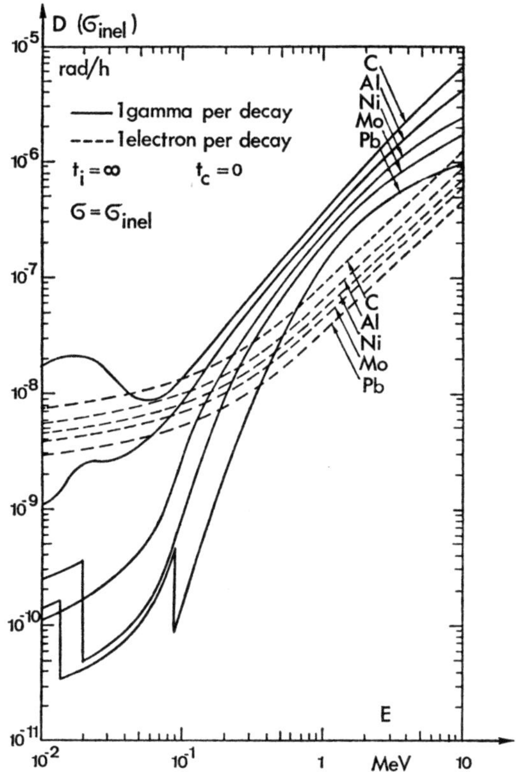

Fig. 1.26 Danger parameters for the emission of one gamma quantum or one negative electron per decay at saturation and zero cooling time, the production cross-section assumed to be the total inelastic cross-section of the target atom. (Full lines: gamma rays, dotted lines: electrons.)

statistically distributed in energy. Such an investigation was actually made and figs. 1.27 and 28 show the distribution in numbers per energy interval of all isotopes up to $A=60$, 101 and 209 with respect to the energy of the gamma rays or the maximum energy of the negative electrons emitted in their radioactive decays.

From figs. 1.27, 28 it appears that the majority of gamma emitters lie in the energy region 0 to 1.5 MeV whereas the maximum negative electron energies range up to 5 MeV. One can conclude that in the case of large bodies containing a large number of different isotopes, as accelerators or reactor parts, for instance, the surface dose due to negative electrons is not at all small in comparison to the gamma ray dose. To protect their eyes from the impact of electrons, radiation workers

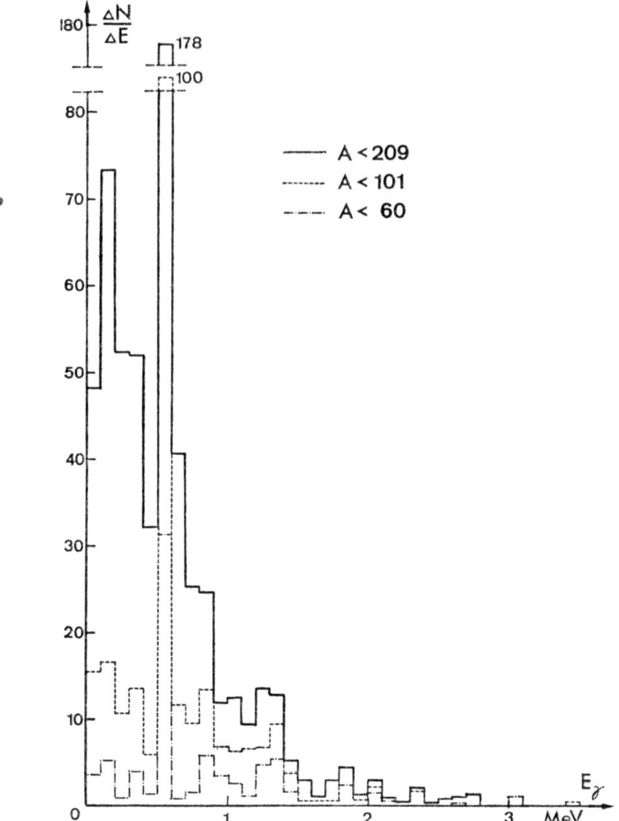

Fig. I.27 Distribution of radioactive isotopes over the energies of the gamma rays emitted in the decay process.

occupied around such installations should by all means wear glasses. The same is true when alpha emitters are present as is often the case in radiochemical work or in the neighbourhood of radioactive bodies of high atomic number.

8 Radiation field calculations for simple geometries

8.1 *General properties of radiation fields from uniformly activated bodies*

One of the purposes of our work is obviously to make it possible to calculate in advance the field of radiation expected in a given instal-

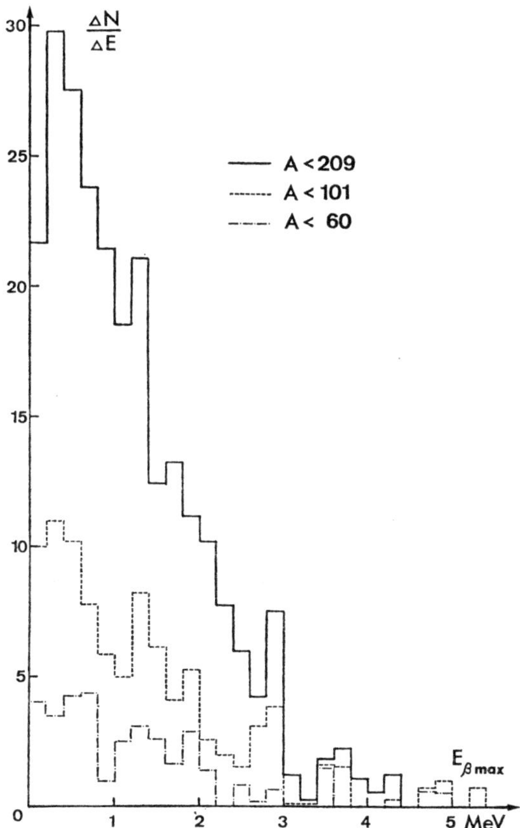

Fig. 1.28 Distribution of radioactive isotopes over the maximum energies of the negative electrons emitted in the decay process.

lation which has been exposed to the action of a given flux of activating particles.

In the preceding subsections we have tried to define the material constants of interest in the field of induced radioactivity such as specific activity, dose rate, self-absorption and danger parameter.

Assuming we know these material constants for a particular case, we shall now try to calculate the radiation levels present in the vicinity of activated bodies, taking into account their geometrical configuration and of course the attenuation of the radiation in the bulk of the active material. As the main contribution to the radiation field at some distance from thick active masses comes from gamma rays, we shall consider only these in the present subsection. Large contributions to the dose from electrons are found, however, in the case of finely dispersed radioactive dust or powder, which must then be treated as thin emitting layers and to which the following does not apply.

Let us recall the physical quantities we have so far defined and which we shall have to use here. We will always consider the contribution of one specific gamma ray from one isotope at a time and assume that all contributions are summed up later. Also we will drop the time functions in this development. So we have:

$$a = \Phi \frac{N_0}{A_\mathrm{T}} \sigma \tag{8.1}$$

where a is the specific activity in decays/g sec. If now

ε is the number of quanta of the particular energy emitted per decay,

f is the factor for converting flux of gamma rays per cm^2 to dose rate in rads ($\mathrm{cm}^{-2}\,\mathrm{rad}^{-1}$),

μ is the linear attenuation coefficient of gamma rays with good geometry (cm^{-1}),

ϱ is the density of activated material ($\mathrm{g\,cm}^{-3}$), then

$$D = 3600 \frac{N_0}{A_\mathrm{T}} \sigma \frac{\varepsilon\varrho}{f\mu} = 3600 \frac{a\varepsilon\varrho}{\Phi f\mu} \tag{8.2}$$

is the danger parameter in rad/h per unit flux (1 particle/sec cm^2).

When one has a block of active material of a certain volume it is evident that part of the radiation coming from the interior will be absorbed before reaching the surface. This absorption will be taken into account now. It is customary, as a first approximation to consider the linear absorption coefficient μ for narrow beams only and assume an exponential decay of the intensity of gamma radiation with the distance x it has to pass through in the active mass. In addition the photon flux issuing from the element of volume evidently decays in inverse proportion to the square of the distance. So the differential of the dose rate $\mathrm{d}R$ for a point P at a distance r from a volume element $\mathrm{d}V$ of a material of activity a and density ϱ can be written

$$\mathrm{d}R = \frac{3600 a\varepsilon\varrho}{f} \frac{\mathrm{e}^{-\mu x}}{4\pi r^2} \mathrm{d}V \text{ rad/h.} \tag{8.3}$$

This expression must then be integrated over the total active volume V

(see fig. 1.29) to give the dose rate

$$R = \frac{3600\varepsilon\varrho}{f} \int_{(V)} \frac{a\,\mathrm{e}^{-\mu x}}{4\pi r^2}\,\mathrm{d}V. \tag{8.4}$$

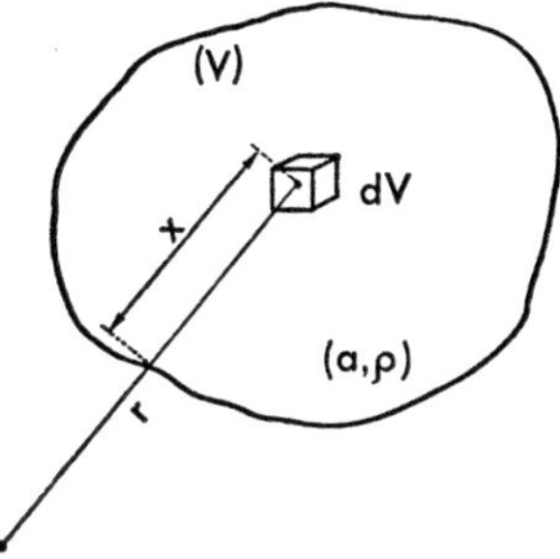

Fig. 1.29 Calculation of the radiation field from a radioactive body.

Here *a* can be a function of position. In the common case of uniformly distributed radioactivity the form of this integral will allow us to derive two general laws on the angular distribution of the radiation emitted from a plane surface, and on the constancy of the radiation field inside a cavity embedded in active material.

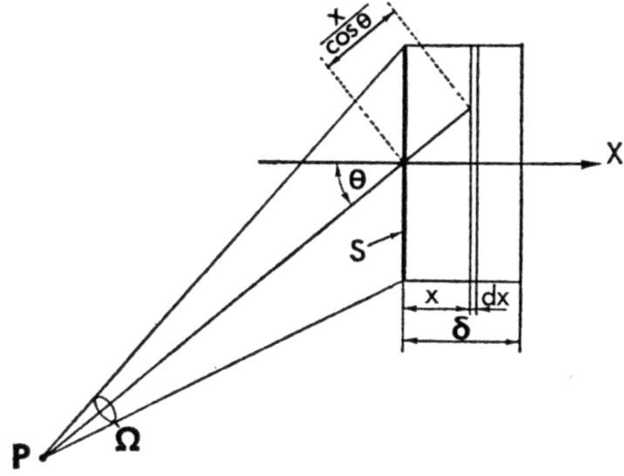

Fig. 1.30 Calculation of the distribution in space of gamma radiation from a plane bounded area.

Angular distribution of the gamma radiation emitted by a plane bounded area. In the case of evenly distributed activity, the specific activity *a* can be taken outside the integral of (8.4). Let us look at the radiation coming out of a plane bounded area as a function of direction. Fig. 1.30

shows the geometry considered. We find for the radiation field at point P at a large distance the dose rate

$$R_P = \frac{3600 a \varepsilon \varrho}{f} \int_{x=0}^{\delta} \frac{e^{-\mu x/\cos\theta}}{4\pi r^2} S \, dx$$

$$= \frac{3600 a \varepsilon \varrho S \cos\theta}{4\pi r^2 \mu f} (1 - e^{-\mu\delta/\cos\theta}), \tag{8.5}$$

where r is the distance of point P from the surface and S the area of emission considered; δ indicates the thickness of the active layer and θ the angle with respect to the normal.

For large values of δ, the angular distribution function is seen to tend towards

$$\lim_{\delta\to\infty} R_P = \frac{3600 a \varepsilon \varrho S \cos\theta}{4\pi r^2 \mu f} = \frac{3600 a \varepsilon \varrho}{f\mu} \frac{\Omega}{4\pi}, \tag{8.6}$$

where

$$\Omega = \frac{S \cos\theta}{r^2} \tag{8.7}$$

is the solid angle within which the area S is seen from P.

Eq. (8.6) is nothing else than Lambert's cosine law for the emission of light from a plane perfectly diffusing area (see particularly King, where the cosine law is derived by taking into account the absorption in much the same way as was done here). So we can state the following conclusion.

Theorem. *The dose rate from a uniformly radioactive body, whose dimensions are large compared with the mean attenuation length of the gamma rays it emits, is proportional to the solid angle under which this body is seen.*

In cases of thin active layers this property does not, of course, apply, and the emission should be taken as isotropic for each surface differential.

For a surface completely surrounding the point of observation P, as is the case inside a cavity embedded in an infinite body of uniform activity, we have

$$R_P = \frac{3600 a \varepsilon \varrho}{\mu f}. \tag{8.8}$$

This is constant and independent of the position inside the cavity, and also of the form of the cavity itself.

It can be shown that this statement is independent of the dimensions of the cavity and remains valid even if the cavity is small. The proof can be carried out if we go back to the integral expression in eq. (8.3) and start with a spherical cavity of radius R_0 for which we compute the radiation field at the centre C. It is obviously

$$R_C = \frac{3600 a \varepsilon \varrho}{f} \int_{r=R_0}^{\infty} e^{-\mu(r-R_0)}\, dr = \frac{3600 a \varepsilon \varrho}{\mu f}. \tag{8.9}$$

As one sees, this field is independent of the radius of the cavity.

Clearly this integral is also calculable if only a part of the surrounding material included in a cone of solid angle Ω as seen from the centre of the sphere is radioactive. The field value would then be

$$R_C = \frac{3600 a \varepsilon \varrho}{\mu f} \frac{\Omega}{4\pi} \tag{8.10}$$

and is also independent of the radius R_0, where the conical volume begins (fig. I.31).

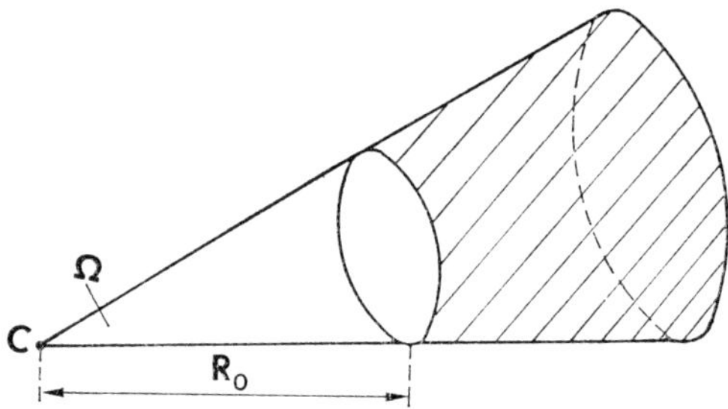

Fig. I.31 Calculation of radiation field at the projection centre of a conical active volume.

It is thus possible to modify at will the surface of the spherical cavity we had examined earlier by adding volume elements of conical form and arbitrary length projecting from the centre, without changing the value of the radiation field (see fig. I.32). In other words R_0 can be any function of direction and we arrive at the following conclusion:

Theorem. *The radiation field inside a cavity imbedded in an infinite volume of radioactive material of uniform activity is independent of the form of the cavity and constant from point to point therein.*

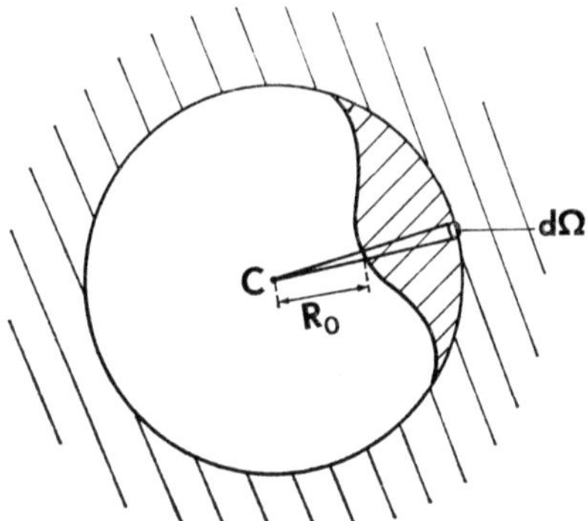

Fig. 1.32 Calculation of the radiation field in cavities of different shapes.

The value of the dose rate is according to eq. (8.9)

$$R_C = \frac{3600 a \varepsilon \varrho}{\mu f} = D\Phi \text{ rad/h}, \tag{8.11}$$

where D is our previously defined danger parameter (see eq. 8.2) and Φ the irradiating flux. This is obviously also the dose rate present in the material itself.

For any active body viewed within a solid angle Ω the radiation field will be, according to eq. (8.10)

$$R = \frac{\Omega}{4\pi} D\Phi. \tag{8.12}$$

In particular in front of infinitely thick and extended wall we will have

$$R = \tfrac{1}{2} D\Phi. \tag{8.13}$$

This quantity is independent of the distance from the wall.

Another question of general interest that should be raised here is whether the radiation field is a potential function or can be derived from a potential function by taking the gradient. It can be said immediately that a radiation level cannot be derived from a potential, as it is a scalar and not a vector. It remains to be seen if it can be assimilated to potential functions. The criterion is to fulfil Laplace's equation. Let us begin with a point source of gamma radiation. Laplace's equation is not valid in the case of a point source, be it in vacuum or in an absorbing body. One can show this easily in polar coordinates (r, θ, φ).

When

$$\frac{\partial}{\partial\theta} = \frac{\partial}{\partial\varphi} = 0, \tag{8.14}$$

there remains for the Laplacian of the dose rate R

$$\Delta R = \frac{1}{r^2}\frac{\partial}{\partial r}\left(r^2 \frac{\partial}{\partial r} R\right). \tag{8.15}$$

In the first case (point source in the open)

$$R = C/r^2, \qquad \Delta R = 2C/r^4 \tag{8.16}$$

and in the second case, taking the attenuation length as unit length

$$R = C\frac{e^{-r}}{r^2}, \qquad \Delta R = C\frac{e^{-r}}{r^2}\left(1 + \frac{2}{r} + \frac{2}{r^2}\right). \tag{8.17}$$

In both cases ΔR is not zero as would be required to have a potential function. It follows that radiation dose fields from extended radioactive bodies, which are in fact superpositions of such fields originating in each volume element, are also not strictly speaking functions obeying to Laplace's equation.

However, the solid angle property according to eq. (8.6) that we have derived for the case where the thickness of the radioactive objects was large compared to the gamma ray attenuation length makes it possible to assimilate dose rate levels outside radioactive objects to the scalar potential of so-called solenoidal (divergence-free) vector fields which arise around a turbulence line in a liquid or a wire carrying electrical current in space. Outside this line or wire, these solenoidal fields can be expressed as gradients of a scalar function, which in turn satisfies Laplace's equation. All this can be shown on the basis of the solid angle property expressed in eq. (8.12).

Let us consider the velocity field $\boldsymbol{v}$ from a single isolated turbulence line or the magnetic field from a current-carrying thin electrical wire (see fig. I.33). It can be shown (Lagally and Franz, ch. 3, § 7, 114, 115) that this vector field $\boldsymbol{v}$ can be expressed as gradient of a scalar potential u at all points in space which do not belong to the turbulence line (or current loop)

$$\boldsymbol{v} = -\text{grad}\, u, \tag{8.18}$$

where u can be calculated to be

$$u = \frac{\Gamma}{4\pi} \int_{(S)} \frac{\partial(r^{-1})}{\partial n} \, \mathrm{d}s, \tag{8.19}$$

where Γ is a constant, r is the distance from each surface element of any surface S spanned on the turbulence line to the point P where $\boldsymbol{v}$ is calculated, $\partial/\partial n$ expresses the derivative taken with respect to the normal to this surface, and ds is a surface element.

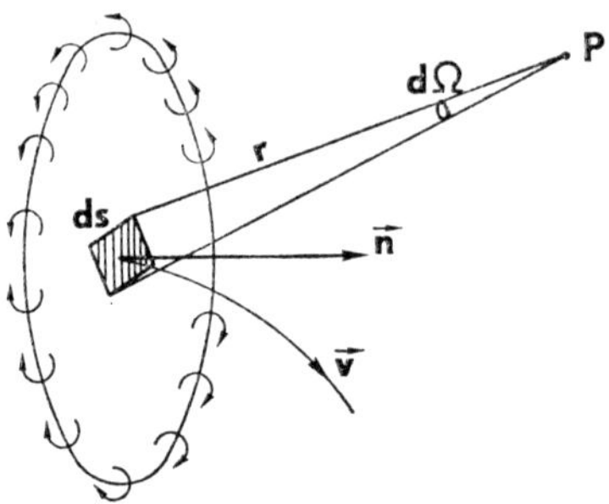

Fig. I.33 Vector field of an isolated turbulence line. Calculation of the radiation field using the solid angle property.

It can further be shown that

$$\frac{\partial(r^{-1})}{\partial n} \, \mathrm{d}s = \frac{\boldsymbol{r}}{|r|^3} \, \mathrm{d}\boldsymbol{s} = \mathrm{d}\Omega \tag{8.20}$$

is the differential of the solid angle inside which the turbulence line is seen from the point where the field is calculated. We arrive thus at the expression

$$u = \frac{\Omega}{4\pi} \Gamma \tag{8.21}$$

from which one sees by comparing with eq. (8.12) that the potential u can be assimilated to the dose rate

$$u \equiv R \tag{8.22}$$

if one cares to set

$$\Gamma \equiv D\Phi. \tag{8.23}$$

So we arrive at the following statement:

> Theorem. *Outside uniformly radioactive objects whose thickness is large compared with the absorption length of the emitted gamma radiation, the radiation levels are scalar potential functions satisfying Laplace's equation and are proportional to the solid angle inside which the radioactive object is seen from the point considered.*

This property makes it possible to calculate quite easily the distribution of radiation levels around complicated objects, by taking only their *contour* and working out the solid angle. Numerical tables of solid angles for various objects are available (Masket). If the radioactivity is distributed uniformly only in parts, these parts can be treated as separate bodies and the contributions to the dose added provided the dimensions of the various areas are again large with respect to the attenuation length of the gamma radiation in the material.

The theorem stated above permits an electromagnetical analogy and model work to determine experimentally the dose rate distribution outside massive active bodies of intricate forms. It is then correct to shape a wire according to the boundary of a radioactive surface region with uniform activity, to send an electrical current through this wire in such a way as to have a closed current loop, and to measure the magnetic field outside the wire. This magnetic field is then the gradient of the desired dose rate distribution function. One arrives at this dose rate distribution by integrating the magnetic field along a line of force. The integration constant must be chosen in such a way that the dose rate is $\frac{1}{2}D\Phi$ when arriving at the active surface. Because of the scalar additive property of potential and of the vector additive property of their gradients, one can use as many current loops at the same time as required to take account of the various radioactive surfaces present. However, one should be careful to choose the sign of the current in each loop so that all potentials add.

8.2 *Selected examples of radiation fields*

8.2.1 *Flat active layer of given thickness.* One case which occurs in practice is that of a radiation field in the vicinity of a flat active layer of indefinite extent and uniform activity, but with a given thickness T. Fig. 1.34 is a schematic diagram of the situation. Let A be the distance between the detector and the layer. As the layer covers an indefinite area the measurement will not depend on A, but A must be introduced for purposes of calculation. According to eq. (8.4) the dose rate in A will be

$$R = \frac{3600 a \varepsilon \varrho}{f} \int \frac{e^{-\mu x}}{4\pi r^2} \, dV \tag{8.24}$$

r being the distance from volume element dV and x the thickness of material through which the radiation passes. The element of volume becomes

$$dV = 2\pi r^2 \sin \alpha \, d\alpha \, dr. \tag{8.25}$$

One obtains:

$$R = \frac{3600 a \varepsilon \varrho}{f} \int \frac{1}{4\pi r^2} \, e^{-\mu(r - A/\cos\alpha)} \, dV \tag{8.26}$$

$$R = \frac{3600 a \varepsilon \varrho}{2\mu f} \int_{\alpha=0}^{\frac{1}{2}\pi} (1 - e^{-\mu T/\cos\alpha}) \sin \alpha \, d\alpha. \tag{8.27}$$

One recognizes in front of the integral a quantity equal to the previously defined danger parameter D multiplied by the flux Φ.

In the event of the thickness T being infinite (all the half space being active) the exponential function disappears and one is left with

$$R = \frac{3600 a \varepsilon \varrho}{2\mu f} = \tfrac{1}{2} D\Phi \tag{8.28}$$

as expected.

When the thickness T of the layer is finite, the integral can be integrated in parts and one finds

$$R = \tfrac{1}{2} D\Phi [1 - e^{-\mu T} - \mu T \, \mathrm{Ei}(-\mu T)], \tag{8.29}$$

where μ is the linear att enuation coefficient of the radiation to $1/e$ and where the function

$$-\mathrm{Ei}(-x) = \int_x^\infty \frac{e^{-t}}{t} \, dt, \qquad x > 0 \tag{8.30}$$

is well known in sine integral theory. Fig. I.35 represents the expression in square brackets which shows how the field of radiation grows with the thickness of the active layer.

8.2.2 *Cylindrical beam absorber.* Another case which occurs in practice is that of an absorber for a high energy particle beam. Let πR_0^2 be the

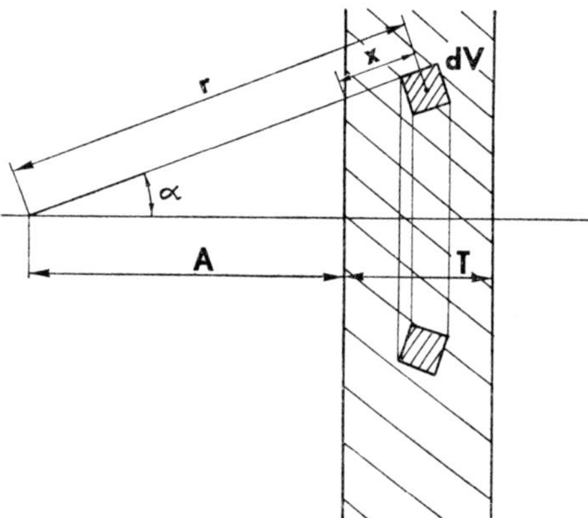

Fig. I.34 Calculation of the radiation field of a flat active layer.

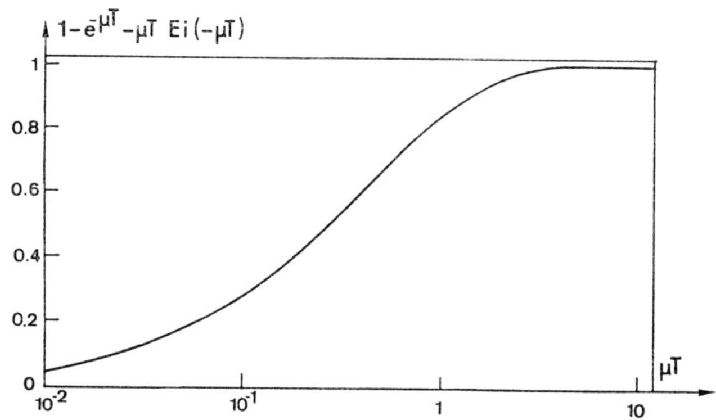

Fig. I.35 Relative radiation field versus thickness of active layer.

cross-section of the beam assumed to be circular and let it fall perpendicularly on to the absorber. A value which we would like to calculate is the intensity of the activation radiation on the surface of the absorber at the centre of the area on which the beam fell. Let us use a system of cylindrical coordinates (r, z) centred on the surface of the absorber with the axis of the beam as the z-axis (fig. I.36). Since the penetration of protons is generally much greater than the length of absorption of gamma rays, the active area can be taken to extend to infinity in the direction of the z-axis. One obtains

$$R = \frac{3600 a \varepsilon \varrho}{f} \int_0^{R_0} \int_0^{\infty} \frac{1}{4\pi(r^2 + z^2)} \exp[-\mu\sqrt{(r^2 + z^2)}]\, 2\pi r \,\mathrm{d}r \,\mathrm{d}z. \qquad (8.31)$$

After integration with respect to r one has

$$R = \tfrac{1}{2}D\Phi[1 - \int_0^\infty -\mathrm{Ei}[-\mu\sqrt{(R_0^2 + z^2)}]\,\mathrm{d}\mu z] \tag{8.32}$$

an expression which should be graphically integrated for different μR parameter values. Fig. I.37 gives the shape of the quantity in square brackets as a function of μR, that is to say the variation of the intensity of the radiation with the radius of the beam cross-section (the beam flux Φ but not its intensity is assumed to remain constant). It is seen that one reaches the same value $\tfrac{1}{2}D\Phi$ as in the previous case as soon as μR exceeds 1.

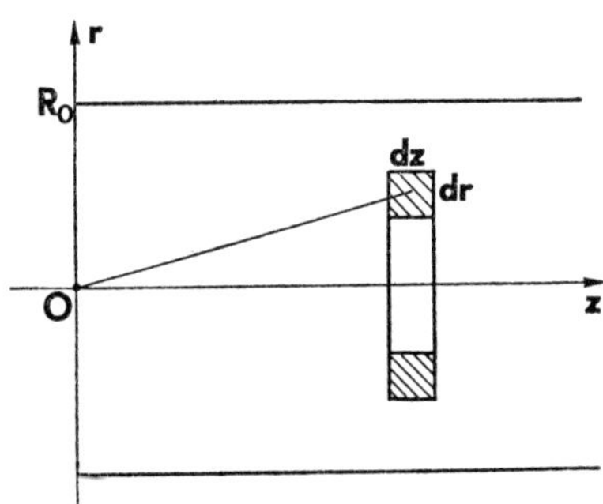

Fig. I.36 Absorber geometry for beam of circular cross-section.

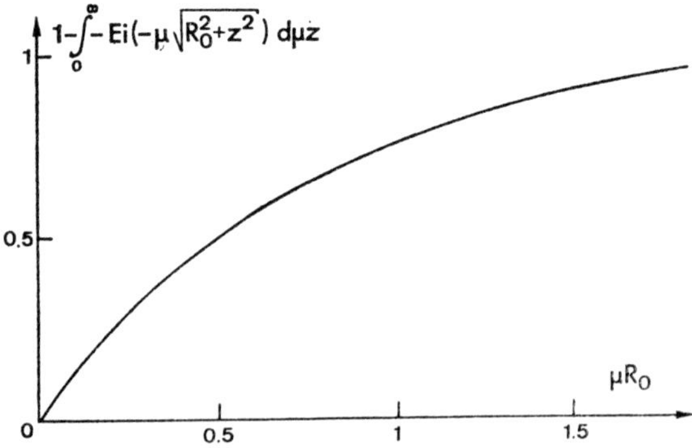

Fig. I.37 Relative intensity of radiation field at centre of entry face of absorber versus beam radius measured in gamma attenuation lengths.

The dose rate at the centre of the entry face of the absorber is the maximum one can be exposed to. It is interesting to know how the dose rate decreases when one walks away from the absorber. On the axis and at distances A large compared with R_0, one can assume that the cylindrical block constitutes a linear source for dose rate calculations (see fig. I.38), getting

$$R(A) = \frac{3600 a \varepsilon \varrho}{f} \int_{z=0}^{\infty} \frac{e^{-\mu z}}{4\pi(A+z)^2} \, dV, \tag{8.33}$$

where

$$dV = S \, dz; \tag{8.34}$$

$$S = \pi R_0^2 \tag{8.35}$$

being the section of the block. One can also consider the denominator of the quantity under the integral sign as a constant and equal to $4\pi A^2$, if A is large compared with the gamma ray attenuation length. One then gets

$$R(A) = \frac{3600 a \varepsilon \varrho}{\mu f} \frac{S}{4\pi A^2} = D\Phi \frac{S}{4\pi A^2} \tag{8.36}$$

in accordance with previous theorems, as $S/4\pi A^2$ is the solid angle within which the entry face of the absorber is seen from a distance A.

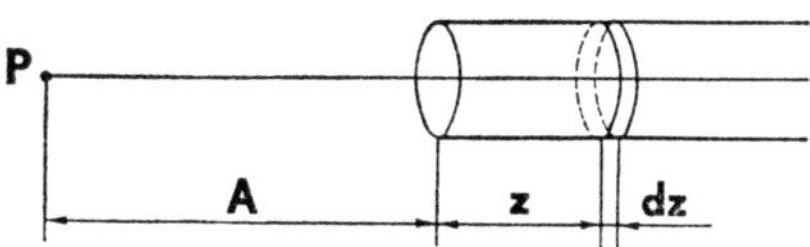

Fig. I.38 Calculation of the dose rate at a large distance A of a cylindrical beam absorber.

9 Some remarks on build-up factors of gamma rays and on Compton electrons

9.1 *Build-up factors for gamma rays in extended bodies*

In the following we shall have essentially to find in what way our danger parameter is affected by build-up. Let us start from the equation giving the dose rate in the centre of an active spherical shell embedded in non-active material (see fig. I.39). In its differential form we have for unit flux and one gamma ray energy

$$\mathrm{d}R = \frac{3600 a \varepsilon \varrho}{f} \frac{\mathrm{e}^{-\mu r}}{4\pi r^2} \mathrm{d}V, \tag{9.1}$$

where

$$\mathrm{d}V = 4\pi r^2 \, \mathrm{d}r. \tag{9.2}$$

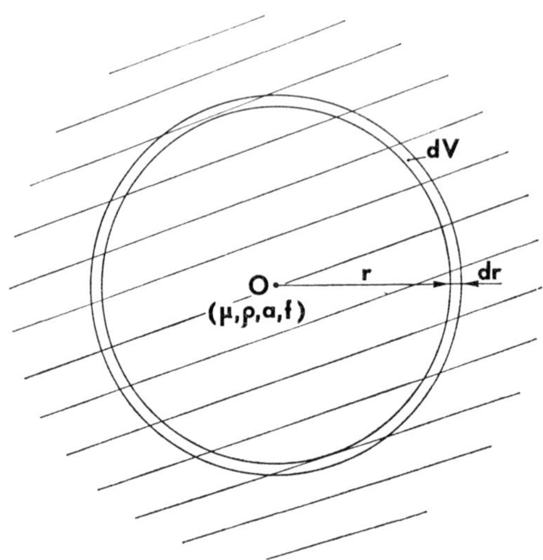

Fig. I.39 Calculation of the correcting factor for the danger parameter to account for the build-up of gamma rays due to Compton scattering.

What is usually given is the build-up factor $B^{pt}(\mu r)$ for a point source in an infinite absorber medium of attenuation μ as a function of the distance r. This factor is different for each gamma ray energy. Hence we have to integrate, in order to find the real dose, the expression

$$R = \frac{3600 a \varepsilon \varrho}{f} \int_{r=0}^{\infty} \mathrm{e}^{-\mu r} B^{pt}(\mu r) \, \mathrm{d}r$$

$$= \frac{3600 a \varepsilon \varrho}{\mu f} \int_{0}^{\infty} \mathrm{e}^{-x} B^{pt}(x) \, \mathrm{d}x. \tag{9.3}$$

We see that the build-up factor by which our danger parameter has to be multiplied is given by the value of the integral on the right. This

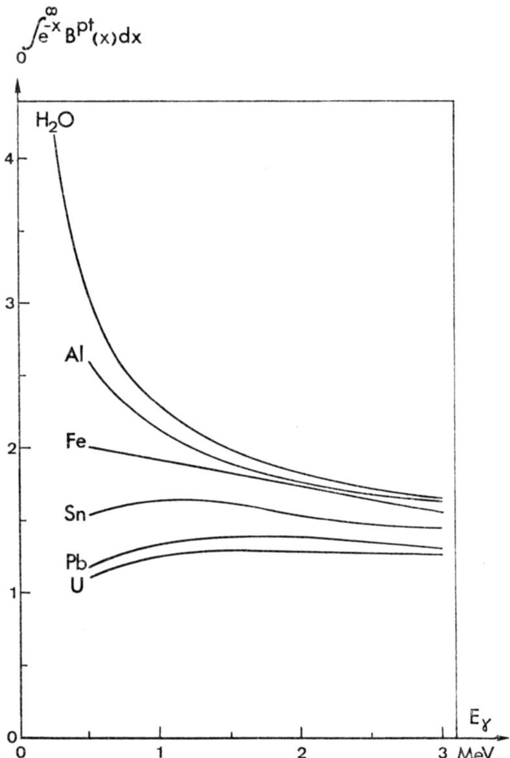

Fig. I.40 Correcting factors for the danger parameter for gamma rays to take account of Compton scattering, as a function of primary gamma ray energy.

value has been computed as a function of the gamma ray energy for a variety of materials from water to uranium, using the data for the dose rate build-up factors for point isotropic sources from Goldstein. The results obtained for energies between 0.025 and 3 MeV are shown in fig. I.40. The factor by which one has to multiply our danger parameter is seen to lie between 1 and 2 in the majority of cases, except for very light materials and low energies, where it can reach 4. These correcting factors have to be applied when the radiation field is computed in the neighbourhood of massive radioactive bodies subtending a substantial solid angle from the point where the measurement is made. They do not need to be applied on, for instance, the axis of slabs of small cross-section.

9.2 *Compton electrons from the scattering of gamma rays*

Another contribution to the dose with large radioactive bodies, which has not yet been considered, is that from the recoil electrons produced in Compton scattering of the gamma rays from radioactive decays.

Let us consider as usual a spherical cavity of radius R_0 embedded in a large volume of homogeneously active material and assume we have a dose d_0 from gamma rays in the cavity. Let us assume for the moment that all gamma rays have the same energy E_0, and let us try to calculate the dose d resulting from Compton electrons at the centre of the spherical cavity.

We begin by considering a small volume element $\mathrm{d}S\,\mathrm{d}A$ of area $\mathrm{d}S$ and thickness $\mathrm{d}A$ lying at a small depth A under the boundary of the cavity, the normal to the area $\mathrm{d}S$ being directed towards the centre O. We have now to calculate the integrated photon flux n_0 per unit area and unit solid angle, as can be inferred from the value of the dose d_0 and the value of the photon energy E_0 specified. One easily finds, as the photon flux is obviously equally distributed between all directions of space that

$$d_0 = 4\pi n_0 f_0^{-1} \text{ rad,} \tag{9.4}$$

where f_0 (in $\text{rad}^{-1}\,\text{cm}^{-2}$) is the number of photons per cm^2 to produce one rad at energy E_0. Hence

$$n_0 = f_0 d_0/4\pi \text{ cm}^{-2}\,\text{sterad}^{-1}. \tag{9.5}$$

This information is clearly needed to calculate the number of photons striking the surface $\mathrm{d}S$ inside an angle of incidence of ψ with respect to the normal and within a solid angle $2\pi \sin\psi\, \mathrm{d}\psi$ (see fig. I.41). This number of photons is then $(n_0\, \mathrm{d}S \cos\psi)(2\pi \sin\psi\, \mathrm{d}\psi)$.

Let us now consider the elementary Compton scattering process. The process which interests us is the one in which a photon of energy E_0, arriving from any direction, hits an electron in the volume element $\mathrm{d}S\,\mathrm{d}A$ in such a way that this electron goes towards the centre of the sphere. The direction of flight of this recoiling electron has thus to make an angle ψ with the direction of the primary photon. Let $\mathrm{d}\sigma_\mathrm{e}/\mathrm{d}\Omega$ be the differential cross-section for the recoil of a free electron at an angle ψ with the primary photon's direction of propagation into the solid angle $\mathrm{d}\Omega$. Its numerical value can be found as a function of E_0 and ψ, as well as other data on the Compton effect from Nelms. The solid angle $\mathrm{d}\Omega$ that interests us here is that under which one sees a given surface, say 1 cm^2, placed at the centre of the sphere from the

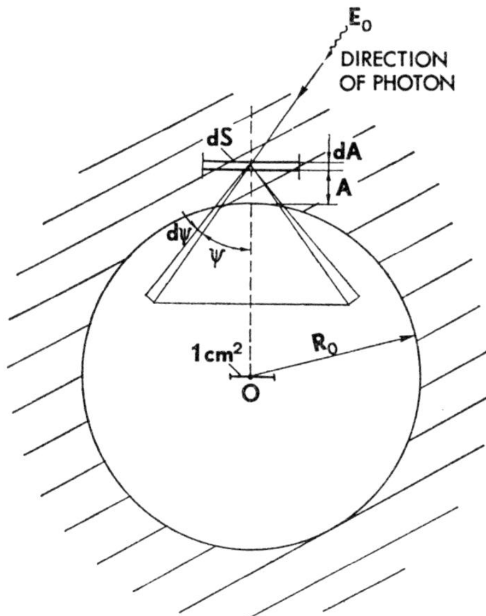

Fig. I.41 Calculation of the contribution to the danger parameter which is due to the Compton recoil electrons from the gamma rays emitted in radioactive decay processes.

volume element $dS\,dA$, i.e. we have approximately, R_0 being as large as one wishes,

$$d\Omega = R_0^{-2}. \tag{9.6}$$

With this differential cross-section and this solid angle we can calculate the number of recoil electrons from the volume element $dS\,dA$ reaching a surface of 1 cm^2 around O. We have only to remember that there are $N_0\varrho Z/A$ electrons per cm^3 in a material with atomic charge Z and mass A, N_0 being Avogadro's number and ϱ the density. In addition, a photon arriving inside the angle of incidence ψ passes through a thickness $dA/\cos\psi$. We have thus for the number of recoil electrons dn_e reaching 1 cm^2 around point O the expression

$$dn_e = (n_0\, dS \cos\psi)(2\pi \sin\psi\, d\psi)\, \frac{d\sigma_e}{d\Omega}\, \frac{1}{R_0^2}\, N\, \frac{Z}{A}\, \varrho\, \frac{dA}{\cos\psi}. \tag{9.7}$$

These recoil electrons scattered inside an angle ψ by primary photons of an energy E_0 have a definite kinetic energy E which is a function of E_0 and ψ only. The numerical values of E can be found in Nelms over a wide range of energies E_0 and for angles ψ between 0 and 90°.

There are no Compton electrons scattered backwards.

We now have to introduce the absorption of the Compton electrons by the layer of thickness A they must penetrate before reaching the surface of the material.

As we have to deal with initially monoenergetic electrons, a fair assumption is to take the transmission curve as linear with depth until one arrives at the electron range. One is thus left with a thickness of one half range for the equivalent layer emitting without absorption. The range in $\mathrm{g\,cm^{-2}}$ for initially monoenergetic electrons in matter was given in fig. I.19 of section 6 of this chapter as function of energy. We shall take one half of this value R_{el} (in $\mathrm{g\,cm^{-2}}$) and substitute it for the product $\varrho\,\mathrm{d}A$ which is also in $\mathrm{g\,cm^{-2}}$ in eq. (9.9). We thus arrive at

$$\mathrm{d}n_{\mathrm{e}} = n_0\,\mathrm{d}S(2\pi \sin\psi\,\mathrm{d}\psi)\,\frac{\mathrm{d}\sigma_{\mathrm{e}}}{\mathrm{d}\Omega}\,\frac{1}{R_0^2}\,N\,\frac{Z}{A}\cdot\tfrac{1}{2}R_{\mathrm{el}}. \tag{9.8}$$

In order to obtain the dose rate produced by these electrons we have now to multiply by a factor $1/f$, where f is the factor for converting flux to dose in $\mathrm{rad^{-1}\,cm^{-2}}$ for electrons given in fig. I.10 of section 5. It is a function of the electron energy E. When we have done this, we can integrate over all surface elements $\mathrm{d}S$, as doses add up independently of the direction of the incoming radiation. This leads us to replace $\mathrm{d}S$ by $4\pi R_0^2$ and we obtain the dose differential

$$\mathrm{d}(d) = 4\pi^2 n_0 f^{-1} \sin\psi\,\mathrm{d}\psi\,\frac{\mathrm{d}\sigma_{\mathrm{e}}}{\mathrm{d}\Omega}\,N\,\frac{Z}{A}\,R_{\mathrm{el}}. \tag{9.9}$$

Here f and $(\varrho/\mu)_{\mathrm{el}}$, being functions of the electron energy, are single-valued functions of ψ via the Compton relation. To obtain the total Compton electrons dose in the sphere from all possible directions of primary photons, it is now sufficient to integrate (9.9) with respect to $\mathrm{d}\psi$. Replacing n_0 by its value as a function of the primary photon dose one then gets the ratio of the dose from the Compton electrons to the primary photon dose

$$\frac{d}{d_0} = \pi N\,\frac{Z}{A}\,f_0 \int_{\psi=0}^{\frac{1}{2}\pi} \frac{1}{f}\,\sin\psi\,\frac{\mathrm{d}\sigma_{\mathrm{e}}}{\mathrm{d}\Omega}\,R_{\mathrm{el}}\,\mathrm{d}\psi. \tag{9.10}$$

This is based on the assumption that all primary photons had the same energy E_0.

This expression is practically independent of the material, as Z/A is a quantity lying between 0.4 and 0.5 for practically all elements.

Formula (9.10) has been computed and integrated numerically for various energies E_0 of the primary photons. The resulting d/d_0 values, i.e. the ratio of the Compton electrons doses to the original gamma ray doses, are plotted as a function of E_0 in fig. 1.42. The Compton electrons dose can thus be taken into account if desired in the calculation of the danger parameter.

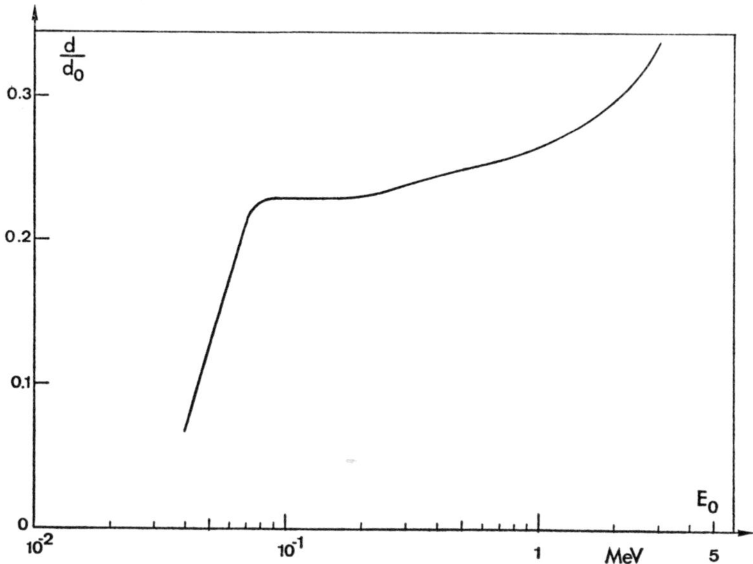

Fig. 1.42 Ratio of danger parameter due to Compton electrons to original gamma ray danger parameter (without build-up) as a function of primary gamma ray energy.

10 The aims of experiments on induced activity

At the end of this first chapter, in which we have presented a general discussion of the problems encountered in induced radioactivity, it is of interest to draw conclusions on what one can expect from experiments in this field and on the way to perform them.

10.1 *Cross-sections measurements*

As induced activity arises from nuclear reactions, the knowledge of their cross-sections is the basis of any forecast in this field. The determination of nuclear reaction cross-sections is the main object of nuclear chemistry which is basically concerned with the properties and reactions of atomic nuclei. Much has been written on experimental techniques in nuclear chemistry (see for example Friedlander, ch. 12). Usually, as well as exact calibration of the irradiating flux, a chemical separation is necessary to isolate the isotope of interest from all those produced at the same time. We shall however only outline here two cases, where cross-section measurements may well be made accurately without having recourse to chemistry.

The first is when one has a monitor reaction of known cross-section which in decaying emits a radiation of the same kind and energy as the element investigated. It is then possible to irradiate both simultaneously and measure them separately, comparing the counts for the radiation common to both. Thus one avoids any calibration of irradiating flux or counter efficiency.

The second case is that in which one spectral line from the isotope desired appears well isolated from the rest, either because one uses a gamma ray spectrograph of very high resolution (lithium-drifted germanium crystal for example), or because there are only very few lines in all. It is then possible, by counting the decay processes in each peak registered, to make fair estimates of the production cross-sections for each isotope present in the sample. Flux and detector calibration are then, of course, necessary.

Apart from cross-section measurements, the engineer or physicist engaged in designing, servicing or making safe radioactive installations or apparatus will often be faced with experiments involving the direct measurement of induced activity under given particular flux or irradiation conditions (time, intensity, materials). These experiments should serve the purposes of both determining, as far as one can, the activity induced under these conditions, and forecasting activities and radiation levels to be expected under other conditions (irradiation times, materials, fluxes) from similar kinds of particles. For this sort of work it is as well to make clear what are the quantities to be measured and the precautions to be taken, and to state a few facts which can help in drawing conclusions and making forecasts from the experimental data acquired.

10.2 *The physical quantities of interest in radioactive samples*

We have seen that the material constants which characterize a given type of induced activity in a material are three in number: the specific activity per gram, the dose rate in tissue from this activity at a given distance from a source without self-absorption, and the danger parameter, i.e. the radiation level in a cavity embedded in a large mass of uniformly active material. These three quantities fully determine the kind of activity induced or present in the type of material considered.

They are, of course, divided according to the nature of the types of radiation emitted: gamma rays, beta and alpha particles or neutrons. For each of these one can define in principle the three characteristic constants mentioned above. However, they are currently used in practice only for gamma rays, and possibly beta rays. Alpha particles are easily shielded, or absorbed in air so that the alpha emitters are considered dangerous especially as regards contamination or ingestion into the body. Neutron emitters are very few and the related problems are special. However, as neutrons are difficult to absorb, neutron fluxes experienced outside large volumes of neutron emitters are likely to come from everywhere inside the active mass.

Therefore we shall examine only the cases of gamma and beta radioactivity in some detail and try to work out the important points to be watched when experimenting.

For gamma and possibly beta activity, one should thus try to measure:

1) the apparent specific activity in curies per gram, given by the emission rate found by counting a small or thin sample of known weight and dividing by the efficiency of the counter (in fact one takes a spectrum and divides by the counter efficiency in each energy range),
2) the radiation field level in rad/hour per gram at 1 metre, found with a dose rate metre placed at a given distance from a sample thin enough to minimize self-absorption,
3) the danger parameter in rad/hour by measuring the dose rate at some distance from a thick sample in a good geometry arrangement, which could be as described below, and making the necessary calculation to arrive at the danger parameter figure.

To describe this procedure of determining the danger parameter from an experiment with a thick sample, let us assume we have irradiated a cylinder of material of a length that can be considered as infinite (2 or 3 times the mean attenuation length for gamma rays of the highest expected energy will be ample). According to the example treated in the preceding subsection on geometry the dose rate R in rad/h at a distance x_0 from a cylinder of section S and practically infinite length is (see eq. 8.36)

$$R = D\Phi \frac{S}{4\pi x_0^2}, \tag{10.1}$$

where D is the danger parameter, provided x_0 is large compared to the diameter of the cylinder and to the attenuation length in the material. From the measurement of the dose rate R one thus obtains

$$D = \frac{R}{\Phi} \frac{4\pi x_0^2}{S}. \tag{10.2}$$

If the transverse dimensions of the chamber used for measuring R are small in comparison with x_0, the geometry is good and one can well assume that the radiation measured escapes through the front face of the cylinder as required, the amount reaching the detector from the side of the cylinder being negligible.

It often happens that the irradiation in a uniform manner of a long massive cylinder of material is impossible in practice. In fact it is only possible in external beams. There is however a good method of determining the danger parameter when one only has a small sample at one's disposal, say a round thin pill (fig. I.43). One prepares several layers of absorbers of the same material, as the sample, but not radioactive, taking care to give them the same diameter as the pill, to avoid build-up. One takes the curve of the dose rate R at distance x_0 as a function of absorber thickness x. This curve is certainly not exponential, in view of the large mixture of gamma ray energies present. One can integrate this curve graphically and show that

$$\int_0^{x_0} R(x)\, \mathrm{d}x = D\Phi \frac{V}{4\pi x_0^2}, \tag{10.3}$$

where $R(x)$ is the dose rate measured with an absorber layer of thickness x and V is the sample volume. Thus one finds

$$D = \frac{4\pi x_0^2}{V\Phi} \int_0^{x_0} R(x)\, \mathrm{d}x. \tag{10.4}$$

All these measurements have, of course, to be repeated for a sufficiently long cooling period in order to find the decay of activity with time.

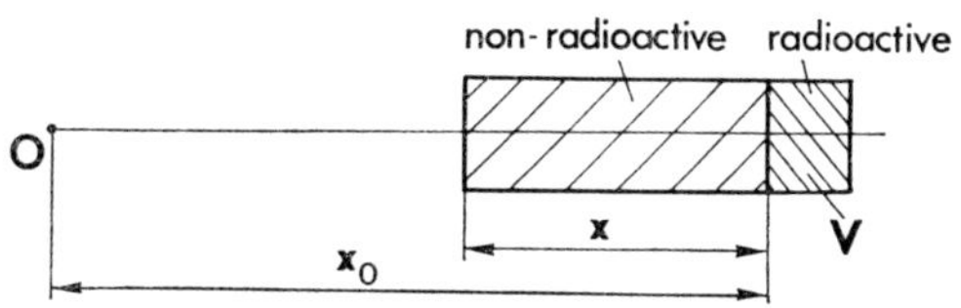

Fig. I.43 Experimental set-up for measuring the danger parameter with a radioactive sample of small volume and with non-radioactive absorbers of the same material.

10.3 *The derivation of the effective k-factor and the effective* 1/e *attenuation length of the sample from the physical quantities mentioned under section* 10.2

As soon as one has measured or computed the three physical quantities which describe an activity of some kind (beta or gamma, etc.) present in a particular material, i.e. the specific activity a (in dis/sec g or curie/g), the dose rate in tissue R (in rad/h per gram at 1 m) and the danger parameter D (in rad/h) all three for unit incident flux, it is possible to derive by calculation two other characteristic quantities which will be useful in all practical applications. These are the effective k-factor of the activity produced in the material under investigation and the effective 1/e attenuation length in the active material itself. In nearly all the cases the material considered contains a larger number of different radioactive isotopes and it is of interest to define effective k-factors and attenuation lengths valid for the particular mixture.

We refer to eqs. (8.1, 8.2, 8.3) defining the three quantities and, starting from eq. (8.1) which gives the specific activity per gram, we write, for unit flux,

$$\frac{a}{\Phi} = \frac{N_0}{A_T} \sigma \text{ dis/g sec} \qquad \text{for 1 particle/sec cm}^2. \tag{10.5}$$

The danger parameter was defined in eq. (8.2) as follows:

$$D = 3600 \frac{a\varepsilon\varrho}{\Phi f\mu} \text{ rad/h}. \tag{10.6}$$

We calculate now the dose rate existing at a distance r of a small active sample, by integrating eq. (8.3) over the sample volume. We take the section S of the sample to be constant and the thickness x_0 to be small with respect to r. We find then

$$\frac{R}{\Phi} = 3600 \frac{a\varepsilon\varrho S}{\Phi f} \int_0^{x_0} \frac{\mathrm{e}^{-\mu x}\,\mathrm{d}x}{4\pi r^2} \cong \frac{D}{4\pi r^2}\,\mu S x_0 \tag{10.7}$$

from which it is apparent that the effective 1/e attenuation length μ^{-1} is

$$L = \frac{1}{\mu} = \frac{R}{\Phi}\,\frac{4\pi r^2}{DV}\ \mathrm{cm} \tag{10.8}$$

where $V = Sx_0$ is the sample volume.

For the expression of the effective k-factor we have simply to divide the radiation field R taken at the distance of 1 m per gram of target element by the activity of the same quantity of the same substance expressed in curies. We arrive thus at the radiation field in rad/h per curie at 1 m, which is per definition the k-factor:

$$k = R/a \tag{10.9}$$

where R is now in rad/h at 1 m and a is expressed in curies.

10.4 *The flux of irradiating particles*

The ideal case would be to have only one sort of particle, of only one energy, and to be able to measure the incident flux or the number of particles which have fallen on the sample. Primary beams of accelerators are usually monoenergetic. They can be used on internal targets or extracted and aimed at external targets.

With internal targets one has usually the highest specific activities, due to the high internal beam intensity, the small beam size and multiplicity of target traversals by the same beam. However, the targets or samples must be thin, they are not uniformly activated, no powders or liquids can be used and it is not a flux that one measures, but rather a total number of incident particles. This measurement can only be done in the case of very high energy particles by activation of monitor foils using a reaction of known cross-section. At lower energies, Faraday cups and thermocouples are also used, if the beam can be absorbed in one traversal of the target.

In contrast to this case, with external targets the beam can be more uniformly distributed over a larger area. Also thick samples can be irradiated. Although the number of particles is reduced because of the extraction efficiency of the beam out of the machine and the lack of multiple traversals, it can be conveniently measured, as the beam is easily accessible, by such means as secondary emission chambers, current transformers, calorimeters etc. So primary extracted beams on external targets can be considered as the ideal for induced activity measurements. However, adequate shielding must be provided between the irradiated sample and the accelerator, to prevent secondary radiation (mainly neutrons), produced by the fraction of the beam which is not extracted and gets lost inside the machine, from hitting the sample.

In many cases, one has not a pure monoenergetic beam of particles. Secondary radiation from accelerator targets, for instance, is composed of a large variety of charged and neutral particles with a wide spectral energy distribution. As it is difficult to analyse the particles composing this radiation and measure the energy spectra, and as it would be difficult to compute all the activities due to the individual components, one is often obliged to measure the activation of typical materials under such circumstances. The results are then only valid for the particular type of flux dealt with. Sometimes one can sweep out the charged particles with an analysing magnet placed near the accelerator target.

Particles that are nearly always produced with a wide spectral energy distribution are, for example, neutrons, either in accelerators or in reactors. The knowledge of this distribution requires special measurements. All neutron activation problems require experiments with the particular kind of flux and energy distribution of interest.

It is also a good idea to use the maximum flux available, in order to be still able to measure the activity levels after long cooling times. The fact that the samples might get too radioactive and saturate the counters at the beginning can be easily dealt with if one places the sample at a large distance or divides it into smaller pieces; at any rate, this is easier than increasing the sensitivity of the measuring apparatus.

10.5 *The irradiation time*

Another item to be chosen with care for an induced activity experiment is the irradiation time. One case is where a particular isotope must be produced in such a way that it can be easily selected for measurement from among others which may be produced at the same time. The irradiation time is then chosen equal to about the half-life of the element of interest. If it is shorter, this element risks not being produced in large enough quantities and being masked by shorter-lived elements which have reached saturation. If the time is longer, long-lived elements might be produced in sufficient quantities to interfere. Least interference is also found at cooling times of the order of the half-life of the element selected.

Another case of interest is where one knows that a large number of different isotopes are produced at the same time, which are all of interest to the experimenter. An example is activation by very high energy particles which produce a vast amount of spallation or fission in materials which are reasonably above the very light elements in the periodic table. This happens when studying the activation of all construction materials above aluminium in atomic weight. It is then possible to use Sullivan and Overton's formula to make forecasts from one irradiation time to any other. The half-life distributions plotted in subsection 4 of this chapter reveal that about one half of all isotopes formed from the iron region upwards have a life time shorter than a day. This makes it possible to shorten the necessary exposure or machine running time to about a day and be confident that one has brought to a fair saturation level about one half of the activities involved. Forecasts can then be made for other irradiation times and any cooling times within the range of validity of the formula (up to about one year and down to about one hour of irradiation and cooling times). This simplifies the experimental work considerably by reducing the number of necessary exposures to one, if one is satisfied with estimates.

PROBLEMS

CHAPTER I

1. What are the nuclei formed by the following reactions on ${}^{56}_{26}Fe$:

(p,n), (p,d), (p,2n), (p,α);
(n,γ), (n,p), (n,d), (n,2p), (n,α);
(α,n), (α,p);
(γ,n), (γ,p), (γ,pn), (γ,2p 3n)?

2. What are the decay products of

${}^{11}C$, ${}^{22}Na$, ${}^{48}V$, ${}^{52}Mn$, ${}^{56}Co$, ${}^{65}Zn$	by beta-plus decay,
${}^{14}C$, ${}^{24}Na$, ${}^{32}P$, ${}^{60}Co$, ${}^{115}In$, ${}^{137}Cs$, ${}^{131}I$	by beta-minus decay,
${}^{7}Be$, ${}^{51}Cr$, ${}^{54}Mn$	by electron capture,
${}^{12}B$, ${}^{32}Cl$, ${}^{142}Ce$, ${}^{210}Po$, ${}^{226}Ra$	by alpha decay,
${}^{17}N$, ${}^{87}Br$, ${}^{89}Br$	by neutron emission?

3. What is the disintegration rate to be expected from ${}^{11}C$ formed in a carbon sample after high energy proton bombardment for several hours, per gram of sample, at end of irradiation? Assume the reaction cross-section is 25 mb and the irradiating flux is 10^6 particles per sec per cm^2. To how many curies does the activity correspond?

4. What is the total weight in grams of the following isotopes necessary to have 1 Ci of activity:

${}^{11}C$, ${}^{22}Na$, ${}^{24}Na$, ${}^{60}Co$, ${}^{226}Ra$?

5. What is the flux in particles per cm^2 per sec necessary to produce 1 mCi of ${}^{24}Na$ at saturation in an aluminium target of 1 cm^2 cross-section and 1 g weight? Assume the production cross-section is 20 mb. What would be the number of millicuries produced at the end of irradiation if the irradiation lasted only one half-life (i.e. 15 h), or one tenth of a half-life (1.5 h)? What time must one wait to see the activity decrease by a factor $1/e$?

6. A germanium transistor was found to fail after a high energy proton irradiation of 10^{13} protons/cm^2. Assuming the value of the total inelastic cross-section to be 850 mb calculate the percentage of germanium atoms which have been transformed into other elements by nuclear reactions during this irradiation.

7. In activation analysis one measures the activity of an isotope produced in a sample irradiated under standard conditions to determine the content of the element in the sample which will yield this isotope under the irradiation. Assuming the measured activity is 1000 disintegrations per second per gram of sample of this particular isotope of half-life 1 hour, 1 hour after an irradiation, which was long enough to reach saturation, calculate the number of parent nuclei present in the sample (irradiating flux 10^{15} particles/sec cm^2, nuclear reaction cross-section 200 mb, atomic number of the parent nucleus $A=100$).

8. What is the approximate mean half-life of the gamma-emitting radioisotopes with $A<60$, $A<101$, $A<209$ (the half-life below which fall the half-lives of one half of the radioisotopes present)? What would be a suitable irradiation time if one wanted one-half of the saturation activity of a large mixture of radioisotopes at one day cooling time, assuming the distribution of isotopes $\mathrm{d}N/\mathrm{d}\log t_{\frac{1}{2},\nu}$ to be constant between 10^{-2} and 10^{+2} days, and zero elsewhere?

9. Assuming one has the activity figures for samples irradiated during a) 5000 days, b) 500 days, at 1 h cooling time, find the factors for determining the activities of the same samples irradiated 1 h, 1 day, 1 week, 1 month, 1 year after the same cooling period, making use of the formula of Sullivan and Overton.

10. What is the dose rate produced by 10^4 Ci of ^{60}Co at 1 m? ^{60}Co emits, per disintegration, 2 gamma rays, one of 1.17 and the other of 1.32 MeV energy.

11. A block of uranium irradiated in a reactor core produces a gamma radiation field of 1 rad/h at 1 m. How many curies has it? How many photons would be counted per second by a detector with 20% efficiency presenting a surface of 10 cm^2 at 1 m? One may use the mean values for a large number of isotopes given in table 1.

A block of copper irradiated by a flux of very high energy protons produces the same radiation field as above. Calculate its number of curies, and the counts registered on a similar photon detector, assuming again one can use the mean values for a large number of isotopes.

12. A typical milking system is ^{132}Te–^{132}I with half-lives of 77 h and 23 h. Let us assume one provides at a certain time 100 mCi of pure ^{132}Te. After a period of 24 h, what will be the activities of ^{132}Te and ^{132}I, and how many atoms of each will one have?

13. How many curies of ^{17}N must one have to produce at 1 m 1 rem/h in neutrons? What is at the same time the dose rate due to negative electrons (neglecting absorption by air)? ^{17}N emits by disintegration 1 neutron of 1.1 MeV and 1 negative electron of 3.8 MeV kinetic energy. To how many atoms of ^{17}N does this number of curies correspond?

14. What is the alpha dose rate in rem/h at 1 cm from a milligram of ^{226}Ra in equilibrium with its daughter products? One can assume there are five alpha particles emitted per decay of one radium atom by the whole family, and that the mean kinetic energy is 6 MeV.

15. Find the 1/e attenuation length in cm of gamma radiation of energy (MeV) 0.5 (positron annihilation), 1.25 (^{60}Co) and 2.75 (^{24}Na) in the following materials: carbon (density $d=1.7$), aluminium ($d=2.7$), iron ($d=7.8$), tungsten ($d=19.3$) and lead ($d=11.3$).

16. Find the range in cm of electrons of maximum energy 1 and 3 MeV in air (NTP), aluminium, copper and lead.

17. Find the range in cm of alpha particles of 5 MeV energy in air (NTP), carbon, and gold.

18. A beta-minus emitter with a maximum beta ray energy of 3 MeV is uniformly embedded in a small copper plate of 0.2 mm thickness. What is the number of electrons which will reach a counter of 10 cm^2 area at 1 m distance in a direction normal to the plate, if the total source strength is 10 mCi.

19. What are the total inelastic cross-sections for high energy protons in germanium, selenium, iridium, arsenic?

20. The isotope ^{24}Na emits 1 negative electron of maximum energy 1.39 MeV and 2 gamma rays, one of 2.75 and the other of 1.37 MeV. What are the gamma and beta dose rates in front of a thick uniformly active aluminium wall in which the fraction 10^{-8} of all atoms have been turned to ^{24}Na, a) without considering build-up of gamma rays, b) including build-up of gamma rays, c) including the Compton electrons from the gamma rays?

21. What is the radiation field one would measure at 2 m from a thick plate of radioactive material of area 1 m^2, assuming the product of the danger parameter of the material and the irradiating flux to be $D\Phi = 1$ rad/h? What would have been the product of danger parameter and flux if one had measured 1 rad/h at the same place?

22. Assuming the moon is a radioactive body with uniformly distributed activity, how does its radiation field decrease in space as a function of the distance from its centre, normalizing to 1 the radiation field at the surface?

23. What is the radiation field at the centre of a spherical cloud of minutely dispersed radioactive dust (neglect absorption of gamma rays). The radius of the cloud is R_0, the density ϱ and the specific activity such that one has a gamma dose rate R rad/h at 1 m from a point source of 1 g.

24. What is the radiation field at the centre of a solid spherical shell of uniform activity? The radius is R_0, the thickness is T, the mass attenuation coefficient of the specific gamma rate is ϱ/μ, the danger parameter is D, the irradiating flux Φ.

CHAPTER II

ACTIVATION BY HIGH ENERGY PARTICLES PRODUCING SPALLATION

1 Introductory remarks

In the first chapter, we have attempted to present all that could be said in general on the subject of induced radioactivity. It is time now to devote oneself to the study of specific conditions, given the type of incident particle, the bombarding energy, the material to be activated, etc.

Installations where radioactivity is produced can, as we have seen, be roughly separated into two classes: accelerators and reactors. In reactors the activating flux is mainly contributed by neutrons with energies from thermal to 15 MeV. In accelerators, the primary particles can be protons, deuterons, alpha particles, heavier ions or electrons, but their energy can go well up into the GeV region for protons and electrons, and attain the 100 MeV region in the case of deuterons and alpha particles with present day techniques. Heavier ions can be accelerated to above 10 MeV per nucleon. All these particles generate secondary neutrons with energies possibly as high as the primary particle if it is a proton, and these secondary neutrons activate in turn the materials they pass through. Secondary neutrons are also produced as evaporation particles in spallation processes, their energy does not exceed then a few MeV.

Cosmic ray and solar flare particles are mainly protons and can be treated as the similar accelerator particles. Electrons activate in a peculiar fashion by generating gamma rays, which in turn give rise to nuclear cascades.

The difficulty in all activation problems is to have an idea of the cross-sections of the nuclear reactions taking place. From these cross-sections one can then deduce which isotopes are produced in larger quantities and find which of them emit dangerous radiations.

At present, the knowledge on cross-sections is far from covering the whole range of energies, and all the number of incident particles, target elements and radionuclides produced of interest. For neutrons, reaction cross-sections are fairly well known at thermal energies and even up to 15 MeV and experimental data on the activation of various materials in reactors is available. However, at higher energies cross-section data is scarce for other particles than protons, and even there does not include the variety of bombarding energies, target and product nuclides desirable. This is due to the fact that the nuclear reactions involved are of various types and very complicated. For instance, with protons we find with increasing energy first (p, n), (p, np), (p, 2n) reactions, then compound nucleus reactions, then π-meson production reactions, before spallation sets in with its intranuclear cascade, sometimes accompanied by some fission. Clearly there cannot be a theory or a formula to cover this whole set of phenomena. This is what makes predictions in the field of induced activity rather intricate and difficult.

For one of the above-mentioned reactions, however, the high energy spallation reaction produced by nucleons, an attempt was made to bring all the experimental data available from nuclear chemistry under one and the same hat. It gave birth to an empirical formula giving the order of magnitude of the production cross-section for a given radionuclide from a given target element, valid within broad limits either of energies (above 50 or 100 MeV) or of mass number (above 20 for the produced nuclides, and with some inaccuracy in the very last naturally occurring heavy elements). This is Rudstam's formula.

Thanks to the existence of this cross-section formula we are able to treat the activation by nucleons in the whole high energy region in a very neat and exhaustive manner. This is why we will begin by treating this category of phenomena, prior to the subsequent chapters, where we will examine in turn activation by lower energy nucleons and ions in the compound nucleus range, activation in reactors, and activation by electrons. In all these cases no general formulae are available and one is reduced to extrapolating from scarce experimental data eventually introducing some nuclear reaction models.

2 Estimation of spallation yields with Rudstam's formula

2.1 *The spallation reaction*

If an element is irradiated with high energy particles the most important kind of reaction (apart from fission for heavy elements) is spallation. This is a violent interaction in the nucleus resulting in the emission of various kinds and numbers of light particles and fragments and leaving behind a reaction product which is generally radioactive. The effect of all these different spallation products must be summed up if one wants to make an estimate of the total radiation emitted at the decay of an irradiated sample. An important factor entering such estimates is evidently the formation cross-section of the various spallation products.

A great deal of information about spallation yields has been accumulated during the past twenty years. The experimental conditions – kind and energy of bombarding particles and composition of the target – can be varied in so many ways, however, that it is rarely possible to find an experimental value for the cross-section of a particular nuclide under given conditions of irradiation. Nor will interpolations and extrapolations in the long lists of measured cross-sections give satisfactory results. One would rather prefer a closed formula to predict the cross-sections for specified experimental conditions. Attempts to describe the cross-section distribution by such a formula have been made a long time ago. The present treatment will be based on a recent systematic investigation (Rudstam).

2.2 *Theoretical model for the spallation reaction*

The theoretical model for spallation assumes that the reaction takes place in two steps: a fast nucleonic cascade followed by a slower evaporation step (Serber). The incoming high energy particle makes quasi-free collisions with the nucleons (or possibly nucleon aggregates) within the nucleus. The collision partners may make further collisions and in this way a cascade develops. Some of the particles taking part in this cascade reach the nuclear boundary and get lost. Others are caught and distribute their kinetic energy among the remaining nucleons in the nucleus. Both these processes contribute to the excitation of the residual nucleus. The second step of the reaction consists of a de-excitation of this intermediate nucleus by the evaporation of particles such as protons, neutrons, deuterons, and alpha particles, and the emission of gamma rays until the excitation energy has been fully used up and the final spallation product has been reached.

Formally we can write the spallation reaction as follows:

$$q + Z_{\mathrm{T}}^{A_{\mathrm{T}}} = Z_{\mathrm{i}}^{A_{\mathrm{i}}} + \nu_1 \mathrm{p} + \nu_2 \mathrm{n} + \nu_3 \mathrm{d} + \nu_4 \alpha + \ldots \qquad (2.1)$$

where the particle q hits a target nucleus (Z_{T}, A_{T}) producing a spallation product (Z_{i}, A_{i}) and besides ν_1 protons, ν_2 neutrons, ν_3 deuterons, ν_4 alpha particles, etc. Owing to the complexity of this reaction, a proper analytical treatment is hardly possible. One can, however, apply the Monte-Carlo method to the theoretical reaction model and thereby predict the outcome of the spallation reaction. It turns out that the theoretical model given above quite well reproduces the essential experimental features of spallation (Metropolis et al., Dostrovsky et al.). So far, however, theoretical calculations are not very useful for estimating spallation yields. It is better to fall back upon an empirical description of the yield distribution of the spallation products for such estimates. A suitable formula is constructed containing a number of parameters whose variation with the irradiation conditions can be studied. Then any particular spallation yield can be found after putting into the cross-section formula the parameter values that correspond to the given irradiation conditions.

2.3 *Form of the cross-section distribution*

The first thing to note is that many combinations of numbers ν_1, ν_2 will lead to the same spallation product. Furthermore, for every given combination, each particular order of emission of the different particles corresponds to a certain reaction path. The total number of reaction paths leading from a given target to a given product is therefore very large, and the formation cross-section will be built up by many small contributions. In such a case it is reasonable to expect a smooth variation of the cross-section from nuclide to nuclide. Specific nuclear properties might, of course, cause more or less important deviations from such a smooth behaviour.

The yield distribution among the spallation products can be visualized in terms of a yield–mass curve, i.e. the total isobaric yield versus the mass number, and charge distribution or cross-section variation versus the atomic number for a given mass number. Empirically it is found that the yield–mass curve is well represented by an exponential function e^{PA}, where the parameter P depends strongly on the irradiation energy (decreasing with increasing irradiation energy) up to about 1 GeV and then flattens out. On the other hand, P does not seem to depend strongly on the target, but lack of spallation data for heavy targets makes this point somewhat uncertain. Some target dependence is, in fact, to be expected.

The yield–mass curve is closely related to the excitation energy distribution of the residual nuclei remaining at the end of the nucleonic cascade. The charge distribution, however, is essentially determined by the evaporation step of the reaction and more specifically by the last stage thereof. This means that the memory of the initial phase of the reaction has been lost and, consequently, the charge distribution should be independent of the irradiation conditions.

Assuming the charge distribution to be symmetrical around its maximum value (obtained for $Z=Z_p$) we can write

$$\sigma(Z) = \text{const} \cdot \exp(-R\,|Z-Z_p|^a), \tag{2.2}$$

where the form of the distribution is determined by the constant a ($a=2$ corresponds to a Gaussian distribution), and the width depends on the parameter R.

We expect to find the locus of the peak at the place where the probability of evaporating a proton equals that of evaporating a neutron. Owing to the Coulomb barrier the loci should then be displaced to the neutron-deficient side of stability. One of the simplest assumptions which can be made is to put

$$Z_p = SA_i - TA_i^2, \tag{2.3}$$

where S and T are parameters which should be independent of the choice of target and bombarding particles.

The parameter R is found to decrease slowly, i.e. the width of the charge distribution increases, with increasing mass number of the spallation product. This simply reflects the fact that the width of the mass valley increases with increasing mass (the neutron and proton separation energies have a strong influence on the evaporation).

In order to find the best form to approximate the charge distribution one would need experimental cross-sections far out at the wings of the distribution. Such determinations are rare, however, and the constant a cannot yet be well determined. Thus it seems that the values $a=\frac{3}{2}$ and $a=2$ give about equally good agreement with experiments, possibly with a slight preference for the former.

The cross-section formula then takes the form

$$\sigma(Z_i, A_i) = K \exp[PA_i - R(Z_i - SA_i + TA_i^2)^{\frac{3}{2}}]. \tag{2.4}$$

The parameter of the formula which give the best fit with a given set of experimental cross-sections are determined. By applying the formula to various experimental spallation investigations, the variation of the parameters with the irradiation conditions can be studied. The parameter K is a normalization constant depending essentially on the target used. It is clearly related to the total inelastic cross-section which must result by summing up the cross-sections of all possible reaction products. Thus an approximate value can be obtained by integration. Empirically, there also seems to be some dependence on the irradiation energy, especially at low energies. In the formula given below the parameter K has been split up into one target-dependent part and one energy-dependent part, both of them determined empirically. The formula, valid for $PA_T \geq 1$, is

$$\sigma(Z_i, A_i) = f_1(A_T) f_2(E) \frac{P \exp[-P(A_T - A_i)]}{1 - 0.3/PA_T} \times \exp[-R(Z_i - SA_i + TA_i^2)^{\frac{3}{2}}], \tag{2.5}$$

with $S = 0.486$, $T = 0.00038$, and P, R, $f_1(A_T)$ and $f_2(E)$ taken from figs. II.1–4, respectively. Alternatively, P and R can be determined from the expressions:

$$\left.\begin{aligned} P &= 20E^{-0.77} && \text{for } E \text{ (in MeV)} \leq 2100 \text{ MeV} \\ P &= 0.056 && \text{for } E > 2100 \text{ MeV} \end{aligned}\right\} \tag{2.6}$$

and

$$R = 11.8A^{-0.45}. \tag{2.7}$$

The formula is valid for proton or neutron-induced spallation. It is fitted to the formula found by Ashmore et al. for the total inelastic reaction cross-section of protons on nuclei at high energies:

$$\sigma_{inel} = 15.9\pi A^{\frac{2}{3}} \text{ millibarn} \tag{2.8}$$

in such a way that the sum of the cross-sections of all elements produced from the target must amount to this total inelastic cross-section. The formula of Rudstam can be used also for other bombarding particles than protons if it is multiplied by the factor $\sigma_{inel}(q)/\sigma_{inel}(p)$ where $\sigma_{inel}(q)$ denotes the total inelastic cross-section for particle q and $\sigma_{inel}(p)$ the total inelastic cross-section for protons.

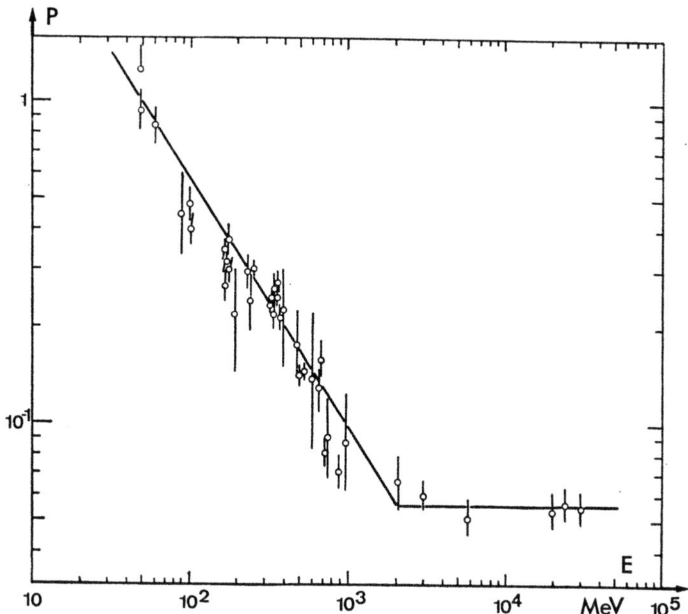

Fig. II.1 The parameter P versus the irradiation energy E. The solid line is P defined by eq. (2.6).

Experimental data on the total inelastic cross-section of neutrons, pions, deuterons, ^{3}He atoms and alpha particles at high energies are given by various authors (Bellettini, Longo, Wikner, Millburn, Horikawa etc.). Fig. II.5 shows the cross-sections in millibarns of the various particles plotted against the atomic number of the target element. One sees clearly that there are two groups of particles. First the protons, neutrons and pions of either sign which lie closely together on the line found by Ashmore et al. and which has been used in writing Rudstam's formula. Second, deuterons ^{2}H, ^{3}He atoms and alpha particles (^{4}He) lie closely together on a parallel line lying above the proton–neutron–pion line, the factor being 2.4. Hence, it is sufficient to multiply by 2.4 Rudstam's cross-sections for protons to get them for ^{2}H, ^{3}He and ^{4}He ions. It is felt, however, that the formula is not applicable to heavier ions than ^{4}He, as things then get more involved with the formation of a compound nucleus built by bombarding particle and target element instead of the intranuclear nucleonic cascade, which was the basis of the derivation of the formula.

If the charge distributions are assumed to be gaussian, i.e. a is chosen to be 2, the parameters of the formula will be slightly changed.

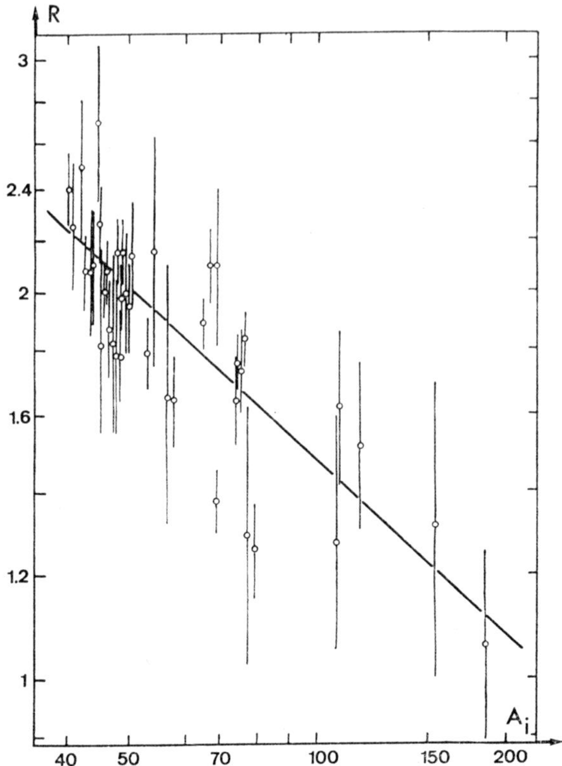

Fig. II.2 The parameter R versus the mass number A_i of the spallation product. The solid line is R defined by eq. (2.7).

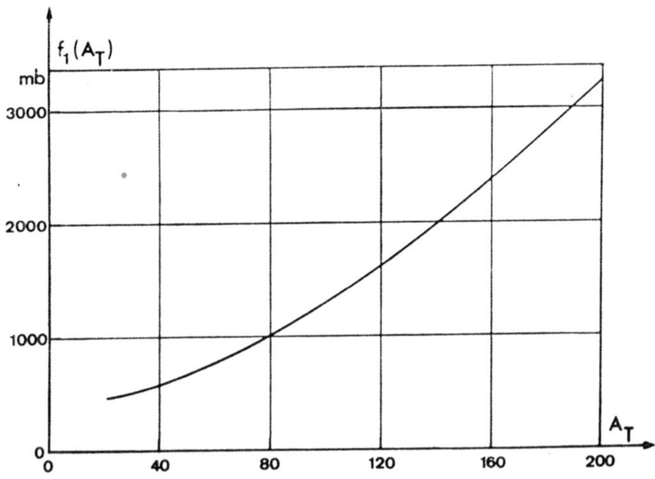

Fig. II.3 The function $f_1(A_T)$ versus the target mass number A_T.

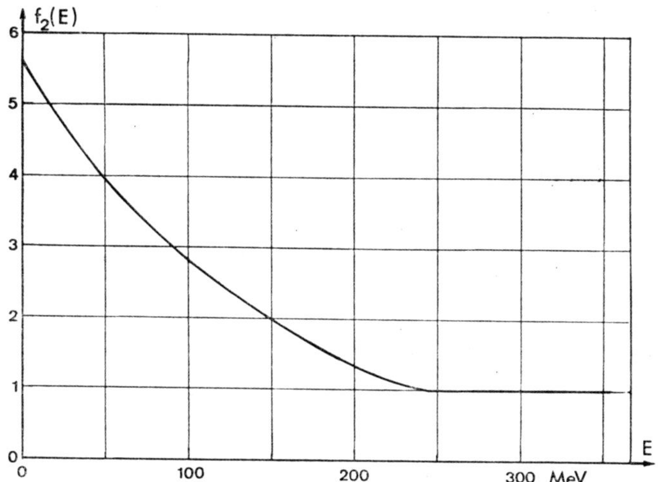

Fig. II.4 The function $f_2(E)$ versus the irradiation energy E.

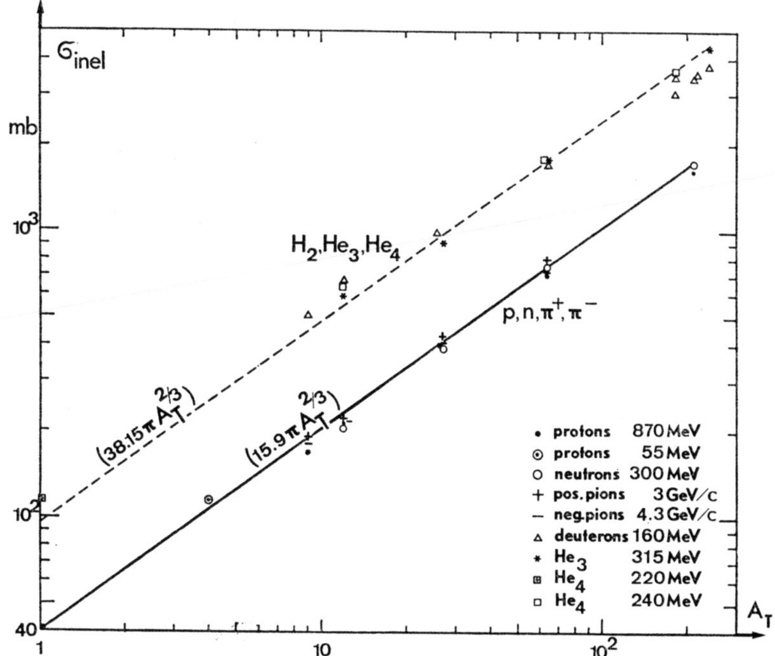

Fig. II.5 Total inelastic cross-sections of various particles on nuclei at high energies.

A set of curves similar to those of figs. II.1–4 can be constructed, however, and the experimental determinations are about as well reproduced by this kind of formula as by eq. (2.5). In this case the formula derived by Rudstam is

$$\sigma(Z_i, A_i) = K \exp[PA_i - R(Z_i - SA_i + TA_i^2)^2] \tag{2.9}$$

with

$$K = \frac{2.35PA_T^{-0.32}\sigma_{\mathrm{inel}}}{e^{PA_T}(1 - 0.32/PA_T) + (0.32/PA_T) - 0.68}, \tag{2.10}$$

where σ_{inel} is to be taken from eq. (2.8) and

$$P = 13E^{-0.69} \qquad \text{for } E<2900 \text{ MeV}$$

and (2.11)

$$P = 0.054 \qquad \text{for } E>2900 \text{ MeV}.$$

One has further

$$R = 17.4A^{-0.64} \tag{2.12}$$

and

$$S = 0.487, \qquad T = 0.00041. \tag{2.13}$$

This formula (2.9), which was chronologically the first derived, gives less good results far from the $Z_i = Z_{\mathrm{peak}}$ line, where the cross-sections are anyhow very small, so that the use of the formula (2.5) with the exponent $a = \frac{3}{2}$ is preferred. Graphs and tables of K, P, R and Z_{peak} for the case $a=2$ can be found in Barbier and Cooper (1965).

One can construct a three-dimensional model of this formula corresponding to a given irradiation energy (i.e. given value of P) and having the form

$$PA_i - R(Z_i - Z_{\mathrm{peak}})^a$$

for any cut along a constant value of A_i (remember that both R and Z_{peak} depend on A_i). Any point on this parabola of degree a will be proportional to $\log_e \sigma(Z_i, A_i)$ apart from the constant

$$f_1(A_T)f_2(E)\,\frac{P\,e^{-PA_T}}{1 - (0.3/PA_T)}$$

which can be obtained from other diagrams. Fig. II.6 facing p. 51 shows the photograph of a model of this sort which was constructed for an exponent value $a=2$ and an irradiation energy $E=600$ MeV.

2.4 *Limitations in using the cross-section formula*

Most spallation data forming the basis for the construction of the cross-section formula have been found with targets in the mass range 50–100. The usefulness of the formula for very light or very heavy targets has not been established because of lack of experimental data. For targets with mass above about 200 there is an additional complication, namely the competition with fission which will depress the spallation yields. It should also be noted that the cross-section formula cannot be expected to hold right up to the target, and no product nearer to the target than two mass units has been considered in the determination of the parameters. Furthermore, although the formula does not contain any explicit cut-off, it is clear that it should not be used below the threshold for the reaction under consideration.

A comparison with about 1200 experimental determinations shows that the cross-section formula predicts the cross-sections within a factor of 3, on the average (within a factor of 2 if only products with $A \leq 65$ are included). In judging this result it must be taken into account that the range of cross-sections covered is very large: a factor of more than one million. Furthermore, the experimental determinations are often quite uncertain, and a much better precision could have been obtained by basing the systematics on certain experimental investigations assumed to be the most accurate ones. Thus it is easy to pick out cases where the formula predicts the cross-sections within 10–20%, on the average. As such a method includes a certain amount of subjectiveness, it has not been adopted, however. Instead, practically all experimental determinations appearing in the literature have been included in the analysis. Returning now to the estimation of the total radiation emitted by an irradiated sample, this is expected to be rather more accurate than the estimation of individual cross-sections, the reason being that by summing up the contributions from many spallation products, the errors in the cross-sections are to a large extent averaged out.

Finally, it should be pointed out that the cross-section formula (2.5) is intended to give the primary cross-section of the spallation products. The effect of decaying parent nuclides must be added.

2.5 *Practical tables for the rapid computation of a cross-section with Rudstam's formula* (2.5)

As we have seen from the three-dimensional model representation,

Rudstam's cross-section formula (2.5) can be divided into two factors: one depending on the irradiation energy and target element, the other depending on the produced isotope. For ease of computation, when a particular cross-section is desired given the target element, the isotope of interest produced and the irradiation energy, it is convenient to be in possession of a table listing the $R(A_i)$ and $Z_{peak}(A_i)$ values as well as of graphs giving the functions

$$P(E), \quad \frac{P(E)}{1 - 0.3/P(E)A_T}, \quad f_1(A_T) \quad \text{and} \quad f_2(E).$$

The next table II.1 gives the R and Z_{peak} values as a function of A_i, whereas table II.2 gives some values of $P(E)$, which can also be read

TABLE II.1

The values of $Z_{peak} = SA_i - TA_i^2$ and R as a function of A_i.

A_i	Z_{peak}	R	A_i	Z_{peak}	R	A_i	Z_{peak}	R
20	9.568	3.065	45	21.101	2.128	70	32.158	1.744
21	10.038	2.998	46	21.552	2.107	71	32.590	1.733
22	10.508	2.936	47	22.003	2.087	72	33.022	1.722
23	10.977	2.878	48	22.452	2.067	73	33.453	1.712
24	11.445	2.823	49	22.902	2.048	74	33.883	1.701
25	11.912	2.772	50	23.350	2.029	75	34.312	1.691
26	12.379	2.724	51	23.798	2.011	76	34.741	1.681
27	12.845	2.678	52	24.244	1.994	77	35.169	1.671
28	13.310	2.634	53	24.691	1.977	78	35.596	1.661
29	13.774	2.593	54	25.136	1.960	79	36.022	1.652
30	14.238	2.554	55	25.581	1.944	80	36.448	1.642
31	14.701	2.516	56	26.024	1.928	81	36.873	1.633
32	15.163	2.481	57	26.467	1.913	82	37.297	1.624
33	15.624	2.447	58	26.910	1.898	83	37.720	1.615
34	16.085	2.414	59	27.351	1.884	84	38.143	1.607
35	16.544	2.383	60	27.792	1.869	85	38.564	1.598
36	17.004	2.353	61	28.232	1.856	86	38.986	1.590
37	17.462	2.324	62	28.671	1.842	87	39.406	1.582
38	17.919	2.296	63	29.110	1.829	88	39.825	1.573
39	18.376	2.269	64	29.548	1.816	89	40.244	1.566
40	18.832	2.244	65	29.985	1.803	90	40.662	1.558
41	19.287	2.219	66	30.421	1.791	91	41.079	1.550
42	19.742	2.195	67	30.856	1.779	92	41.496	1.542
43	20.195	2.172	68	31.291	1.767	93	41.911	1.535
44	20.648	2.149	69	31.725	1.755	94	42.326	1.527

A_i	Z_{peak}	R	A_i	Z_{peak}	R	A_i	Z_{peak}	R
95	42.741	1.520	140	60.592	1.277	185	76.904	1.126
96	43.154	1.513	141	60.971	1.273	186	77.250	1.124
97	43.567	1.506	142	61.350	1.269	187	77.594	1.121
98	43.978	1.499	143	61.727	1.265	188	77.937	1.118
99	44.390	1.492	144	62.104	1.261	189	78.280	1.116
100	44.800	1.486	145	62.480	1.257	190	78.622	1.113
101	45.210	1.479	146	62.856	1.253	191	78.963	1.110
102	45.618	1.472	147	63.231	1.249	192	79.304	1.108
103	46.027	1.466	148	63.604	1.245	193	79.643	1.105
104	46.434	1.460	149	63.978	1.242	194	79.982	1.102
105	46.841	1.453	150	64.350	1.238	195	80.320	1.100
106	47.246	1.447	151	64.722	1.234	196	80.658	1.097
107	47.651	1.441	152	65.092	1.230	197	80.995	1.095
108	48.056	1.435	153	65.463	1.227	198	81.330	1.092
109	48.459	1.429	154	65.832	1.223	199	81.666	1.090
110	49.862	1.423	155	66.200	1.220	200	82.000	1.087
111	49.264	1.417	156	66.568	1.216	201	82.334	1.085
112	49.665	1.412	157	66.935	1.213	202	82.666	1.083
113	50.066	1.406	158	67.302	1.209	203	82.999	1.080
114	50.466	1.400	159	67.667	1.206	204	83.330	1.078
115	50.865	1.395	160	68.032	1.202	205	83.660	1.075
116	51.263	1.390	161	68.396	1.199	206	83.990	1.073
117	51.660	1.384	162	68.759	1.196	207	84.319	1.071
118	52.057	1.379	163	69.122	1.192	208	84.648	1.068
119	52.453	1.374	164	69.484	1.189	209	84.975	1.066
120	52.848	1.369	165	69.845	1.186	210	85.302	1.064
121	53.242	1.363	166	70.205	1.183	211	85.628	1.062
122	53.636	1.358	167	70.564	1.179	212	85.953	1.059
123	54.029	1.353	168	70.923	1.176	213	86.278	1.057
124	54.421	1.348	169	71.281	1.173	214	86.602	1.055
125	54.812	1.344	170	71.638	1.170	215	86.925	1.053
126	55.203	1.339	171	71.994	1.167	216	87.247	1.050
127	55.593	1.334	172	72.350	1.164	217	87.568	1.048
128	55.982	1.329	173	72.705	1.161	218	87.889	1.046
129	56.370	1.325	174	73.059	1.158	219	88.209	1.044
130	56.758	1.320	175	73.413	1.155	220	88.528	1.042
131	57.145	1.316	176	73.765	1.152	221	88.846	1.040
132	57.531	1.311	177	74.117	1.149	222	89.164	1.038
133	57.916	1.307	178	74.468	1.146	223	89.481	1.035
134	58.301	1.302	179	74.818	1.143	224	89.797	1.033
135	58.685	1.298	180	75.168	1.140	225	90.112	1.031
136	59.068	1.294	181	75.517	1.137	226	90.427	1.029
137	59.450	1.289	182	75.865	1.135	227	90.741	1.027
138	59.831	1.285	183	76.212	1.132	228	91.054	1.025
139	60.212	1.281	184	76.559	1.129	229	91.366	1.023

TABLE II.2

Values of P for various bombarding energies E.

E (MeV)	P
50	0.984
100	0.577
200	0.338
300	0.248
400	0.198
500	0.167
600	0.145
700	0.129
800	0.116
900	0.106
1000	0.098
1500	0.072
2100	0.055

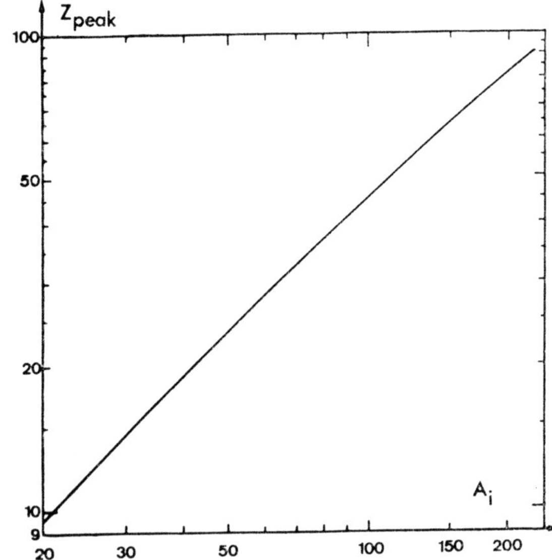

Fig. II.7 The quantity $Z_{\text{peak}} = SA_i - TA_i^2$ plotted as a function of the isotope mass number A_i.

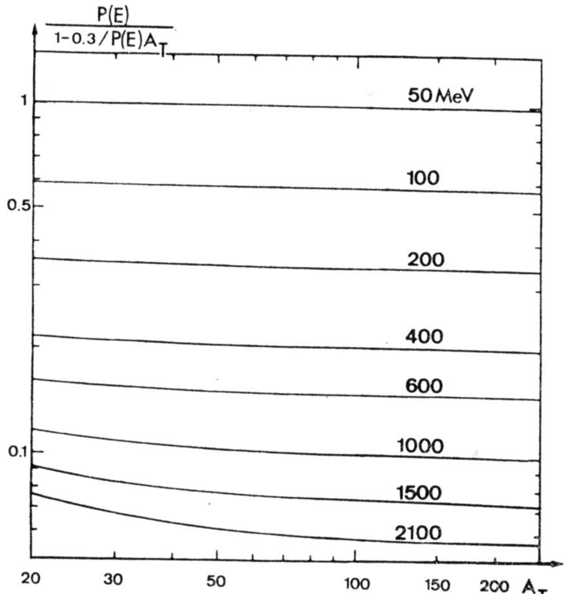

Fig. II.8 A plot of the quantity $P(E)/[1-0.3/P(E)A_T]$ as a function of the target mass number A_T for various bombarding energies E.

from fig. II.7. The expression

$$\frac{P(E)}{1 - 0.3/P(E)A_T}$$

is plotted in fig. II.8 as a function of A_T for various values of E as a parameter. The functions $f_1(A_T)$ and $f_2(E)$ have already been given in figs. II.3 and 4.

3 Activation data for a number of target elements of the naturally occurring isotopic composition

Using a programme made by Rudstam himself according to formula (2.9), the spallation cross-sections were calculated for a number of target elements and for all the radioactive isotopes which could be produced from the target element in each case, except those below $A=22$, as the formula is not expected to be good below $A=20$. To this effect a set of cards was prepared for all radioactive isotopes emitting either a beta particle or a gamma ray from $^{22}_{11}$Na up to $^{208}_{83}$Bi. Those with half-lives below 5 minutes or above $(10^{10}-1)$ days were omitted. These isotope cards included, besides Z, A and the half-life, the energies and emission probabilities of the various gamma rays (one positive electron being counted as 2 quanta of 0.51 MeV), and also the emission probabilities of negative and positive electrons separately. There was only room available for four gamma rays per isotope, so that in cases where more than four gamma rays were emitted, there was some grouping to bring them into four classes, the average values of which were then written down.

If some radioactive nuclides produced by spallation decayed into other radioactive nuclides, the emission of the daughter was added to the emission of the parent in the appropriate manner. When two isomers were produced, they were assumed to be in equal quantities, in agreement with experimental findings. However, if one isomer was short-lived and decayed into a long-lived one, only the long-lived one was assumed to be produced.

A set of cards was also made for the target nuclides, including such data as A, Z, average atomic number of natural isotopic composition, and weight percentage in the produce considered (which could be either the natural isotopic composition or some chemical compound). For each computer run one energy card is added to indicate the bombarding energy of the incident particle, and one time card to indicate the particular irradiation and cooling times.

By these means, an attempt was made to calculate tables which should present in the best possible and exhaustive manner data on radioactivity induced by high energy spallation processes. Thus a very long irradiation time was chosen, i.e. 5000 days, which is practically infinite for our purpose (see distribution of isotopes according to their half-lives in ch. I). The data are computed for typical cooling times: 0, 1, 6 h, 1, 7, 30, 180, 360 d, so that a decay curve can easily be drawn in each case. With the help of the theorem stated in ch. I, section 3, it is then possible to deduce from the decay curve for infinite irradiation the values applying to finite irradiation and cooling times. All data are computed for a flux of 10^6 particles/sec cm^2. The particle energies chosen are 50, 100, 600, 2900 MeV. The data obtained for this last value are then believed to hold for any bombarding energy above 2900 MeV, and could thus be applied to 30 GeV machines and tentatively to 300 GeV ones. The 600 MeV data correspond to cyclotrons and meson-factories, whereas the 50 and 100 MeV data apply to isochronous machines accelerating protons and can be tentatively extended to the case of deuterons and alpha particles by multiplying by the factor 2.4 found from fig. II.5.

The data selected for presenting the best possible information on induced radioactivity list as follows:

a) the gamma dose rate in millirad/hour per gram at 1 cm of a supposedly point source,

b) the gamma danger parameter times the activating flux, in millirad/hour (gamma radiation field in a cavity embedded in a large body of uniformly activated material),

c) the gamma emission rate in quanta per second per gram,

d) the total beta particle emission rate per second per gram (includes positive as well as negative beta particles),

e) the negative beta particle emission rate per second per gram.

All gamma activity includes the annihilation radiation from the positive beta particles, which are supposed to annihilate in the source.

The target elements range from ^{12}Mg to ^{83}Bi. They result from a choice of the most used materials from a structural point of view or otherwise. The target isotopic composition is assumed to be that found in nature. As the properties change rather smoothly with the target atomic number, it is believed that an interpolation between elements presented in the tables would lead to usable values.

In order not to impede the flow of the text, the tables for spallation-induced radioactivity are given in appendix B at the end of the book.

An inspection of these tables shows several general features of the variation of induced activity with the energy of the bombarding particle and the atomic number of the target element. Let us consider, for instance, the first table, with the values at zero cooling time. One notices that there is a variation of activity with bombarding energy. For all targets there is an increase when going from 50 to 100 MeV. Then, for most light targets say up to iron, there is a maximum with increasing energy, lying between 100 and 600 MeV. For medium weight elements, from cobalt to molybdenum, the activity is about the same for energies from 100 to 2900 MeV. For heavier elements, above molybdenum, the tendency is to increase with energy up to 2900 MeV. Let us consider now the variation with atomic number, say at 2900 MeV. Apart from a few exceptions, the tendency for the activity is to increase regularly, up to the zirconium, niobium, molybdenum region. These elements appear really to be the most strongly activated by high energy spallations in the whole table of elements. For heavier elements, the activity decreases a little up to lead and bismuth. The formula of Rudstam is not claimed to be very good for these last heavy elements,

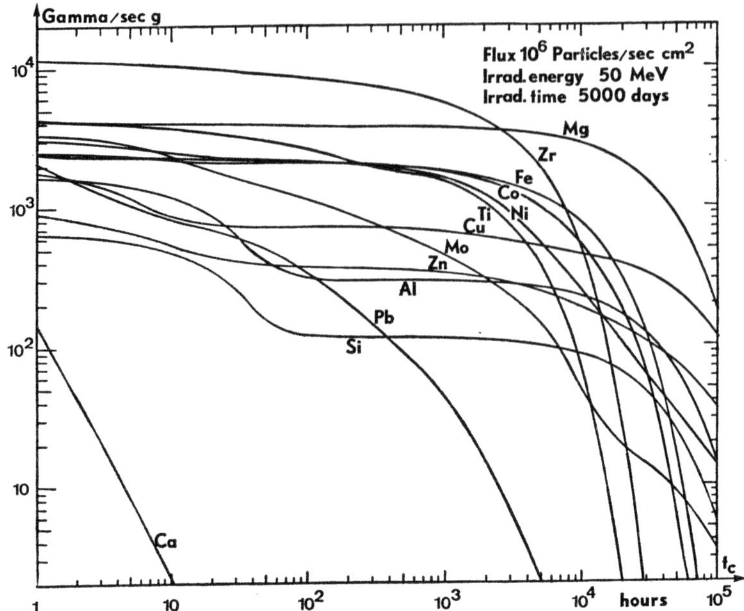

Fig. II.9 Calculated decay of gamma activity of various elements bombarded by 50 MeV protons.

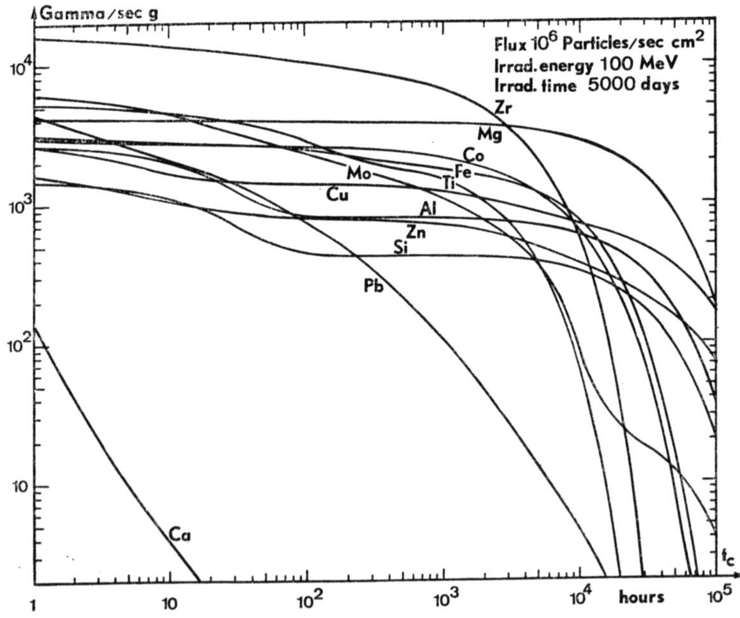

Fig. II.10 Calculated decay of gamma activity of various elements bombarded by 100 MeV protons.

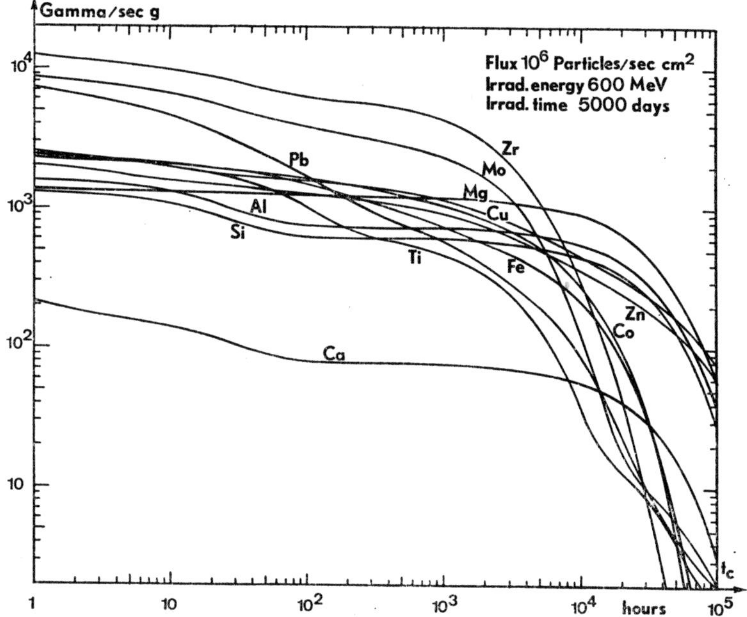

Fig. II.11 Calculated decay of gamma activity of various elements bombarded by 600 MeV protons.

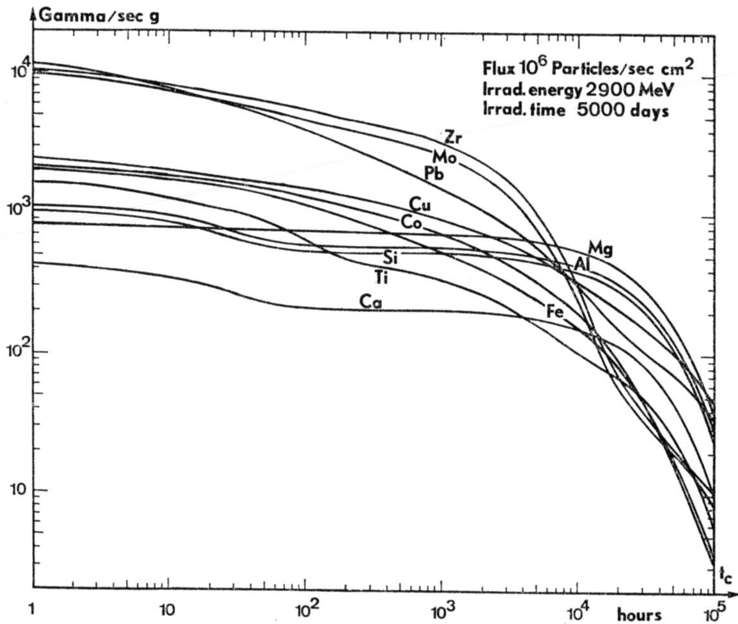

Fig. II.12 Calculated decay of gamma activity of various elements bombarded by 2900 MeV protons.

either because its parameters were derived on the basis of experimental data gained with medium weight elements, or as it does not take fission phenomena into account. Fission is thought to happen to a noticeable extent in bismuth targets only.

We will present now a number of graphs to give an idea of the results obtained. For clarity of reading, these graphs can cover only a small number of target elements. Also to keep the number of figures low, only a few typical cases of irradiation energies and types of activities will be presented.

Figs. II.9 to 12 show the decay curves of the gamma activity in the most used structural materials, to which zirconium has been added, because its activity is the highest. The bombarding energies are 50, 100, 600 and 2900 MeV, the irradiation time is practically infinite. The activity is given in gamma quanta emitted per gram per second. All positrons are supposed to annihilate in the source and to yield two 0.51 MeV quanta, which are included.

What is obvious at first sight is the long cooling time needed in most cases to decay by a factor 10 from the value at 1 hour cooling time. The order of magnitude is one year, this figure being especially correct for the usual structural materials like iron, copper, nickel. Lead decreases faster, light metals more slowly, due to the large contribution of the 2.5 years half-life of ^{22}Na. Calcium appears to have less gamma activity than other materials, at least in the first half year and especially at lower bombarding energies.

Figs. II.13 to 16 show the decay curves for the beta activity, given in beta particles emitted per second and gram. The beta particles of both signs are included here, the positrons being assumed not to have yet annihilated. Beta activity is often neglected as a cause of hazard to personnel in the presence of gamma radiation. If one compares the beta emission rate with the gamma emission rate one finds in fact that the betas are on the average less numerous than the gammas by a factor 3 to 5 for most but not all elements. But this is not all, as the beta energy is usually higher than the gamma energy as is also the biological effect. In view of the damage produced, especially to the eyes, it is as well to be in possession of the relevant data.

We note here also the long decay times, which are similar to those of gamma rays. As in the latter case, the decay curves for different elements lie well apart for low bombarding energies and tend to group themselves in a rather compact bunch for high bombarding energies.

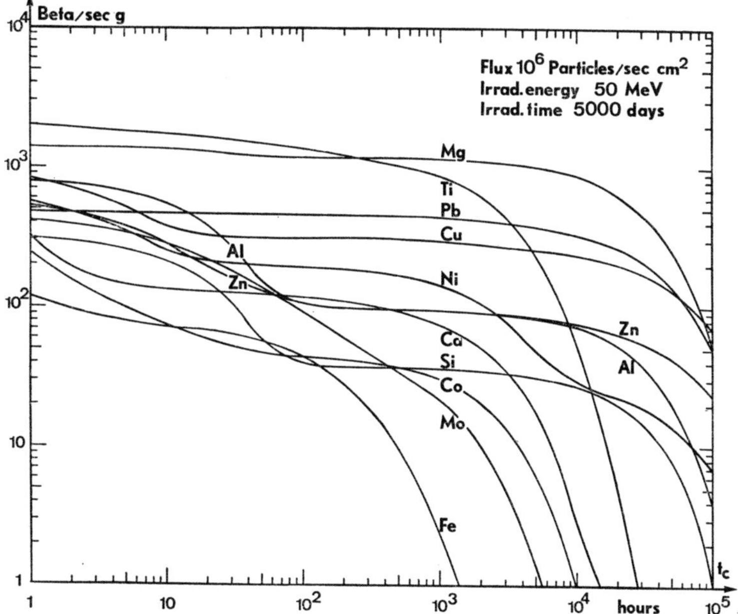

Fig. II.13 Calculated decay of beta activity of various elements bombarded by 50 MeV protons.

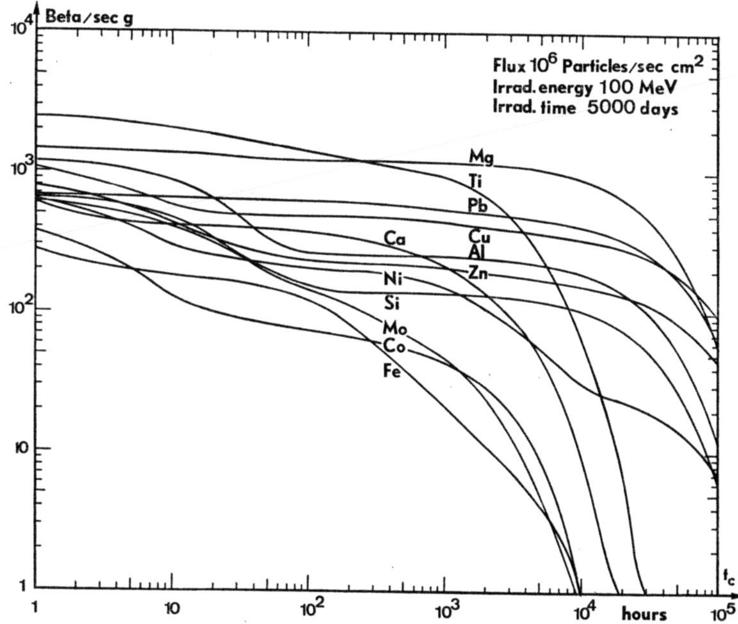

Fig. II.14 Calculated decay of beta activity of various elements bombarded by 100 MeV protons.

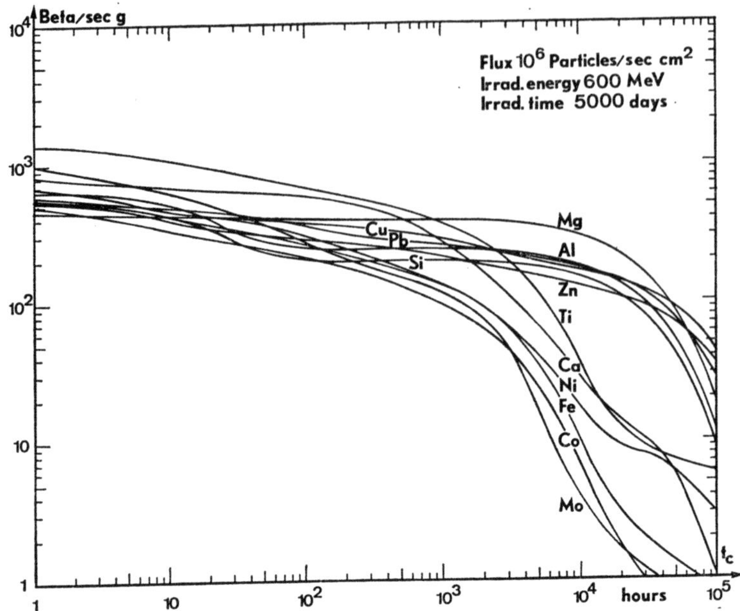

Fig. II.15 Calculated decay of beta activity of various elements bombarded by 600 MeV protons.

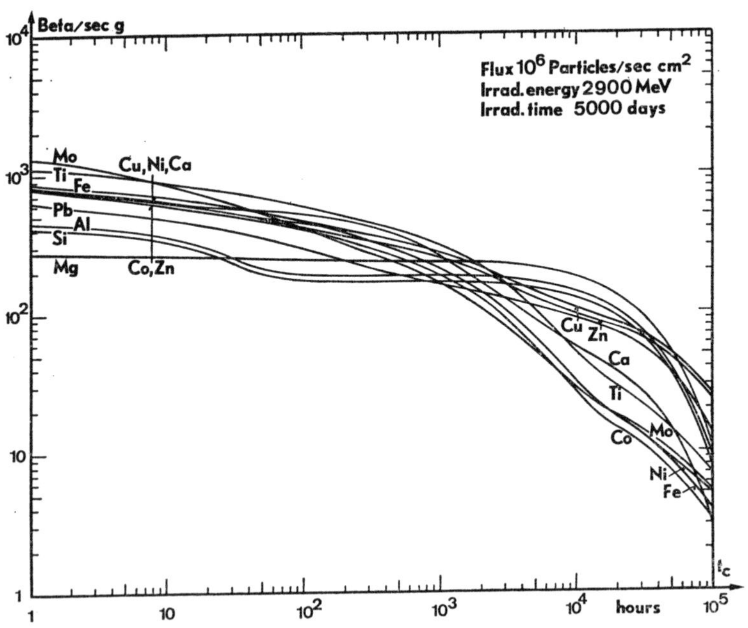

Fig. II.16 Calculated decay of beta activity of various elements bombarded by 2900 MeV protons.

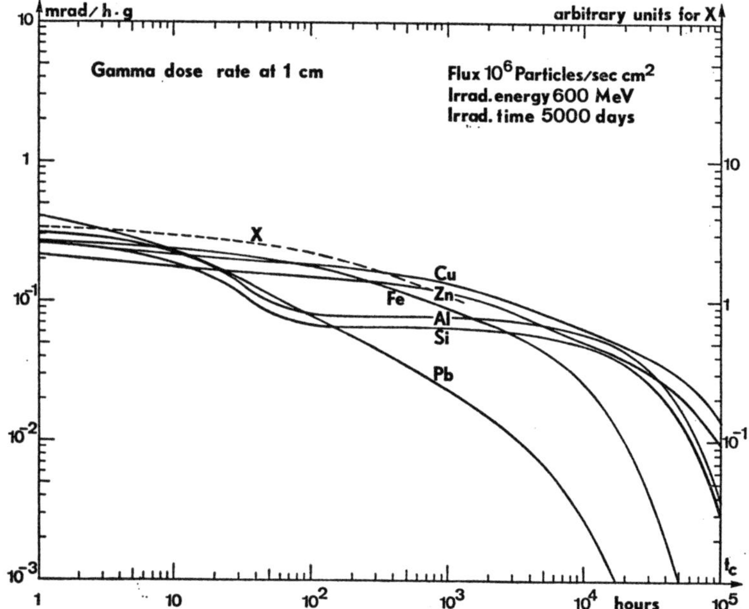

Fig. II.17 Calculated gamma dose rate per gram at distance of 1 cm as a function of cooling time, with experimental dose rate decay in hall of CERN Synchrocyclotron (X-curve).

Another interesting feature to display is the gamma danger parameter, as it can be used to calculate directly the intensity of the radiation field, thanks to the solid angle property. The gamma danger parameter is in fact the radiation field experienced by an observer who sees thick radioactive bodies under a solid angle of 4π, i.e. is completely surrounded by radioactive material. This can happen in a tunnel, or when one crawls into the magnet gap of a cyclotron. Actually it is this danger parameter which is the most used quantity when facing radiation safety problems. This danger parameter follows, of course, the decay and varies also with irradiation time. For the convenience of the user, a large number of decay curves of the danger parameter have been computed and plotted for the elements C, Al, Fe, Ni, Cu, Ag, W, Pb as well as for the compounds found in shielding walls as H_2O, SiO_2, $CaCO_3$ and $BaSO_4$, and are presented in appendix B for various bombarding energies. Here we will content ourselves with presenting the data for Al, Fe, Cu and Pb at 600 MeV in a somewhat different form. The danger parameter will be plotted against

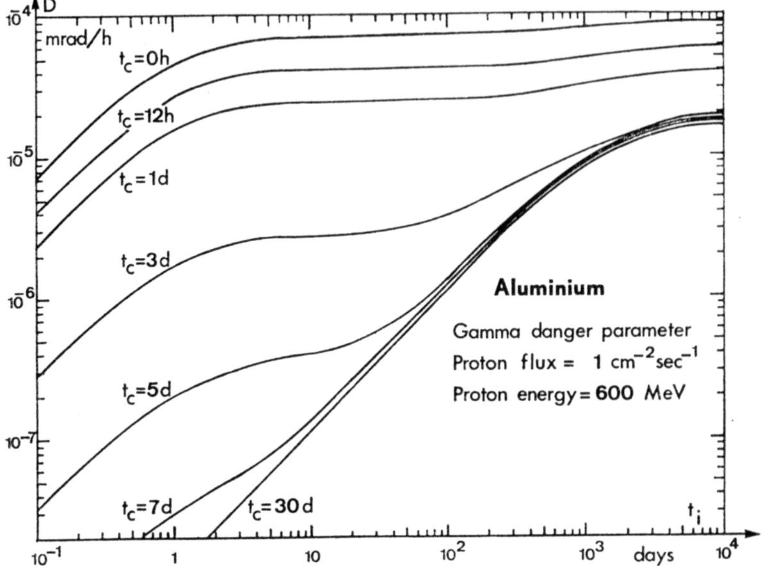

Fig. II.18 Calculated gamma danger parameter for aluminium at 600 MeV.

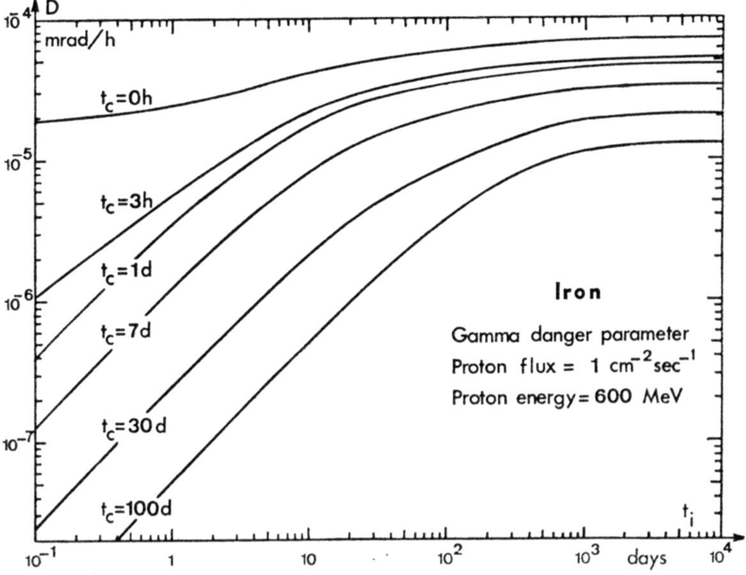

Fig. II.19 Calculated gamma danger parameter for iron at 600 MeV.

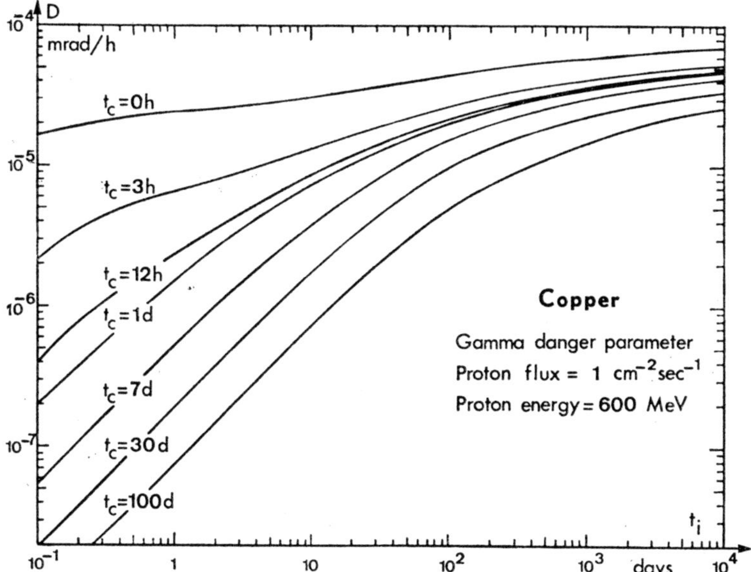

Fig. II.20 Calculated gamma danger parameter for copper at 600 MeV.

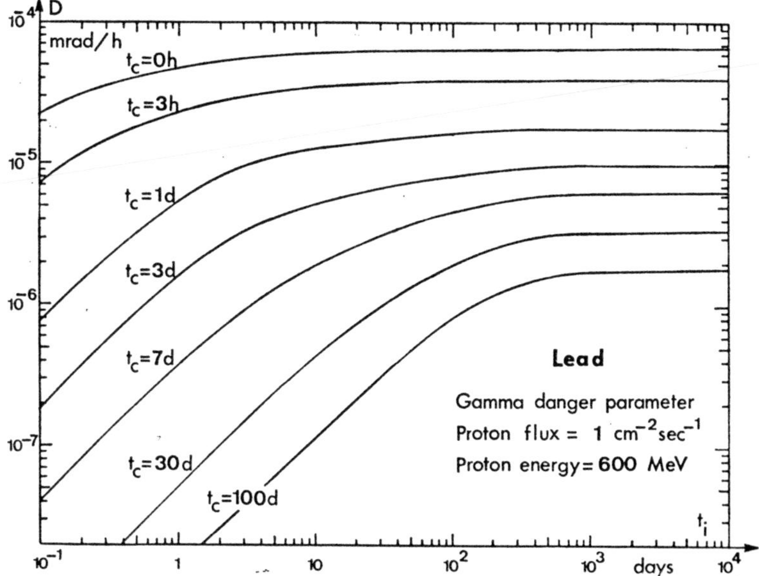

Fig. II.21 Calculated gamma danger parameter for lead at 600 MeV.

irradiation time, so as to show its increase up to the saturation value, for various decay times as parameters to the curves. Figs. II.18 to 21 show how the danger parameter D increases with irradiation time for the above-mentioned metals, and can be used to predict radiation field increases in machines which were made to operate for a relatively short period.

An inspection of these figures shows some differences between them. The activity induced in aluminium has obviously two levels corresponding to one short-lived activity (^{24}Na, 15 h) and one long-lived activity (^{22}Na, 2.5 y). The iron and copper curves rise more smoothly, which shows the presence of a large number of isotopes at the same time. The lead curves do not show any predominant long-lived activity and are typical for a mixture of relatively short-lived nuclides.

4 Some experimental data on activity induced by very high energy particles

Up to this point, we have only presented data computed from Rudstam's formula. It is now appropriate to compare these data with experimental findings. As a matter of fact the cross-section formula itself does not need to be checked, as it is an empirical best fit to numerous existing experimental results, the precision of which is discussed already in Rudstam's paper. However, what needs to be checked as a whole is what was added in the computations, i.e. all the radionuclide data, as probabilities of emission of gamma or beta rays, energies of the gamma quanta, and so on. It is quite a difficult task to select this data among the relevant bibliographical material. Not all the decay schemes of radioactive atoms are known to good accuracy. Much of the information given in the literature is of a qualitative character, and especially the branching ratios for the different gamma rays emitted in the decay of a given atom may differ from author to author. However, it was endeavoured to select the best data, and make reasonable assumptions in the cases where the data available were incomplete, in the hope that errors on particular nuclides would smooth out statistically in the final result. The number set of radionuclide cards used reached in the end 870. It was particularly to check the validity of this stock piling of data that the following experiment was made.

Three elements with sufficiently different atomic numbers were selected in the periodic table in order to give an example of one relatively light, one medium weight and one heavy element. These were nickel ($A=58.3$), niobium ($A=93$) and bismuth ($A=209$). For each element three samples were prepared: a very thin one (a small fraction of a millimetre), a medium one (2 mm thick) and a thick one (5 cm). The purpose was to measure the beta and gamma activity with the foils, the gamma dose rate at a given distance with the medium samples, and to make a direct measurement of the gamma danger parameter with the larger ones following the method given in ch. I, last section.

The diameter of all samples was 2.1 cm. The samples were then grouped by elements in stacks and the three stacks placed beside each other and irradiated in the external proton beam of the CERN 600 MeV Synchrocyclotron. Additional aluminium foils served as monitors to determine the number of protons having passed through each stack by the ^{27}Al(p,3pn)^{24}Na reaction. The irradiation time chosen was in the order of magnitude of one day (in fact 14 h). If one refers to the isotope distribution according to half-lives presented in fig. I.3 one sees that more than one half of the existing radionuclides have a half-life inferior to 14 h, independently of atomic number. We thus get a fair check of our calculations, as the major part of the radionuclides produced will have reached a level of activity near saturation. Also the irradiating beam was pure (only protons) and rather monochromatic ($E_{\mathrm{kin}}=597\pm3$ MeV).

The results obtained are presented in figs. II.22 to 25, reduced to a standard flux of 1 proton/sec cm^2. The full lines refer to the experiment, the dotted ones to the calculation.

The first two figures show the gamma and beta emission rates deduced from the counter measurements by dividing the registered gamma and beta counts by the efficiency of the counting arrangement taken for an energy (beta or gamma) of 0.5 MeV, which is an arbitrary, but seemingly good average value. The counting efficiency had been measured with calibrated sources for the gamma counter, and by comparison of beta and gamma emissions for the beta counter, whereby sources of ^{11}C, ^{18}F, ^{24}Na and ^{40}K were measured on both counters at the same time. For the beta counts, corrections for dead time and sample thickness (always for an energy of 0.5 MeV) were also applied. A description of the exact calibration procedure will be found in Barbier, Hutton, Pasinetti (1966). In view of the systematic error included because of the choice of an average energy, the experimental

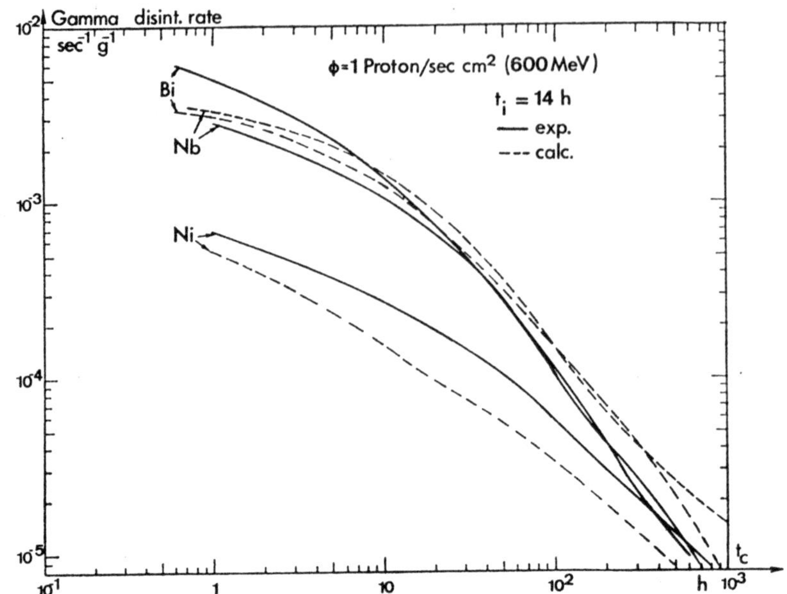

Fig. II.22 Measured and calculated gamma emission rates of Ni, Nb, Bi-samples vs cooling time.

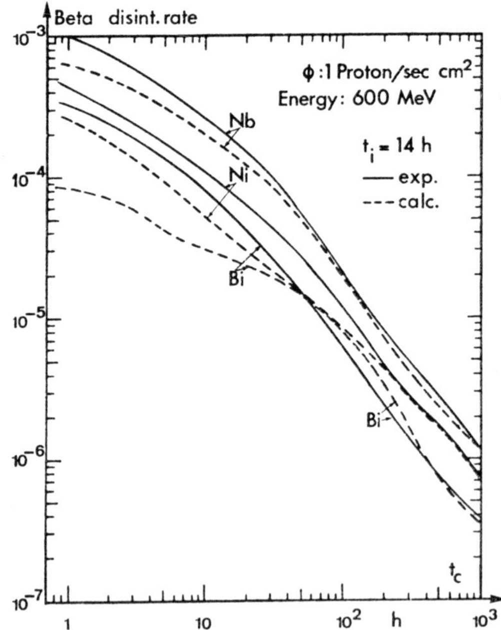

Fig. II.23 Measured and calculated beta emission rates of Ni, Nb, Bi-samples vs cooling time.

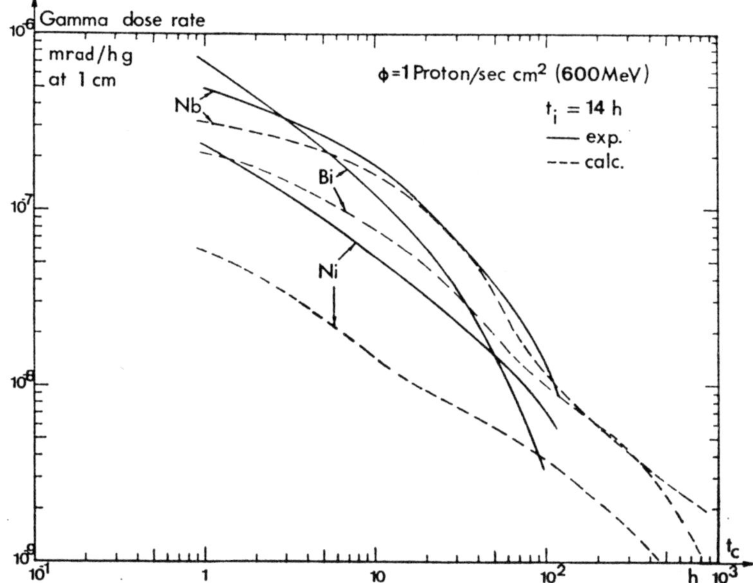

Fig. II.24 Measured and calculated gamma dose rates from Ni, Nb, Bi-samples vs cooling time.

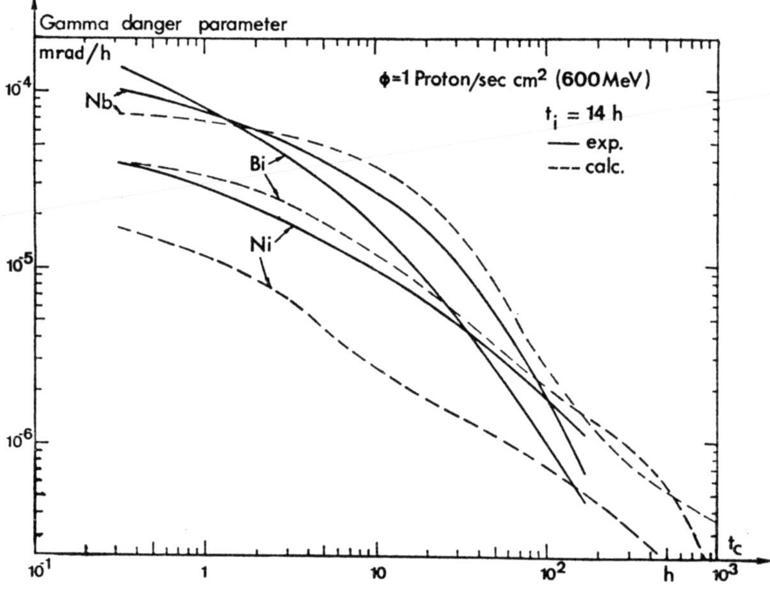

Fig. II.25 Measured and calculated gamma danger parameters from Ni, Nb, Bi-samples vs cooling time.

and calculated curves are not badly matched, although discrepancies up to a factor 2 or 3 appear for certain measurements at certain cooling times. Given the difficulty of doing more precise measurements of this sort, the agreement between calculations and experimental findings is to be considered as satisfactory on the whole, especially if we consider the two figs. II.24 and II.25 as well, in which the discrepancies are not larger.

In these figures the radiation field at a distance from the samples was measured directly with a dose rate meter giving readings in millirad/h, first for a point-like sample in which the self-absorption was negligible, and second on the axis of a very long cylindrical sample, from which the danger parameter was derived with formula (10.2) of ch. I.

Thus we have four measurements by four different methods involving three different sorts of instruments which show an acceptable agreement between the experiment and the activity values calculated with Rudstam's formula and our set of radionuclide cards.

We arrive now at more general, although less clearly defined, experiments, consisting of the irradiation of samples at precise locations inside or around accelerators. The advantage is that one measures the activation which will occur under real conditions of flux, time, intensity etc. The difficulty is that these conditions are not well known: the types, energies and numbers of the incident particles can only be guessed, as the activating fluxes in and around accelerators contain a mixture of all particles and energies. Besides primary protons and protons degraded in energy one can find high energy, evaporation and thermalized neutrons, pions of all energies, deuterons, alpha particles and other ions, fragments and even hyperons, depending on the location inside or outside the accelerator, and on the amount of matter the flux has to pass through from the point it has originated. The relative numbers of all these particles and their emission energy spectra, which depend on angle and bombarding energy, are often unknown at the location of interest. Also frequent changes of target location in the accelerator, as well as shut-downs or breakdowns make it difficult to evaluate the correct exposure time and flux, especially over long irradiation periods. This is what makes it difficult to predict the radiation level, not to mention the usually intricate geometry of the parts struck by the flux.

We shall now present several groups of decay curves taken with various samples irradiated at some locations of interest in or near accelerators and try to outline the interesting features found in each case.

As a first example let us present a 2-years irradiation case of various samples located inside the tank of the CERN 600 MeV Synchrocyclotron, downstream 120° from the target region, 9.5 cm above median plane at a radius equal to the radius where the targets were positioned (see figs. II.26 and 27). At such a location the samples are hit mainly by protons that have been scattered at small angles by the target in the vertical direction.

The distance travelled by the protons from the target to the samples is about 4.5 metres. It is calculated that with the usual 4 cm thick beryllium target the expected proton density is of the order of 10^8 protons/cm^2 with the normal 1 μA beam current. The samples are also exposed to the high energy neutron flux produced not only in the target, where it is estimated that 8% of the beam undergoes nuclear interactions, but also on the pole pieces of the magnet, where the remaining 92% of the beam is lost. The high energy neutrons ($E_n > 100$ MeV) generated in the beryllium target have been measured by Perret at an angle $\theta = 21°$ to the forward direction of the incoming protons. He finds the flux value of 1.2×10^8 neutrons/sec cm^2 at 1 m from the target for 1 μA beam current. Other experiments have shown an angular dependence of the form $e^{-\theta/30°}$ for the high energy particles activating ^{11}C ($E_n > 20$ MeV) so that one is a position to evaluate the neutron flux from the target at the location of the samples for which the emission angle is $\theta = 45°$. The flux found in this way is 4×10^6 neutrons/sec cm^2. To this flux we have, of course, to add the high energy neutrons from the remaining 92% of the beam lost in the pole pieces all over the 360° of the external orbit. This is hard to evaluate; however, we could imagine a further contribution of once or twice the value found for the direct flux from the target.

There are, however, two other sorts of neutron still liable to activate our samples. The first comprises the evaporation neutrons with energies up to 15 MeV, which are generated concurrently with the high energy cascade neutrons. It is fair to estimate them to about the same number as the cascade neutrons. Wallace and Moyer state that at cyclotron energies about one cascade neutron and half an evaporation neutron are produced from one primary proton. So we arrive at a figure of the order of 10^7 neutrons/sec cm^2 for the evaporation particles on the samples.

The second sort of neutrons able to activate our samples consists

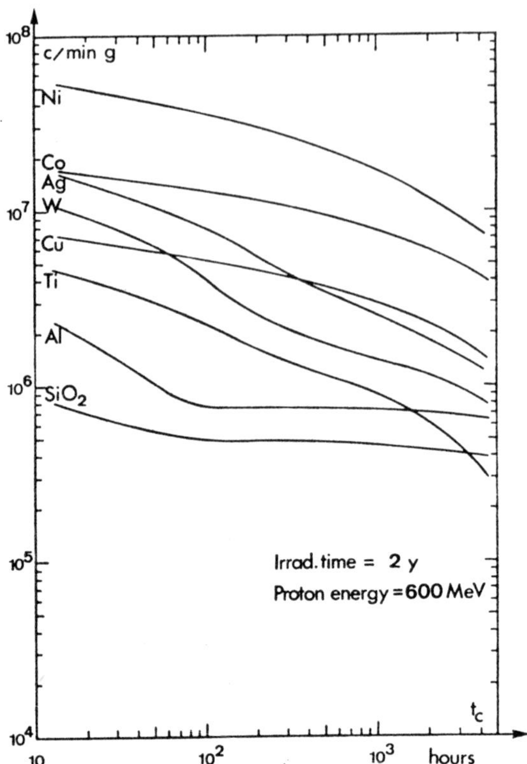

Fig. II.26 Gamma decay of various samples irradiated inside the CERN Synchrocyclotron machine tank, predominantly by protons.

of those which have been thermalized in the pole pieces and make an atmosphere in the machine tank. Other measurements by Perret indicate that this level is about 4×10^6 neutrons/sec cm^2 in the tank.

We see how involved it is to define the irradiating fluxes under real machine conditions. The main component at the location we have chosen purposely for it above the extreme proton orbit is the high energy proton component. At other locations the other fluxes mentioned certainly dominate. Their influence even in our case may not be negligible due to the high cross-sections involved in all cases of activation by thermal neutrons.

The decay curves of the various samples exposed are shown in figs. II.26 and 27. They demonstrate on the whole a fair agreement with the curves in fig. II.11. As an exercise one can use some of the results to calculate the mean value of the irradiating flux received

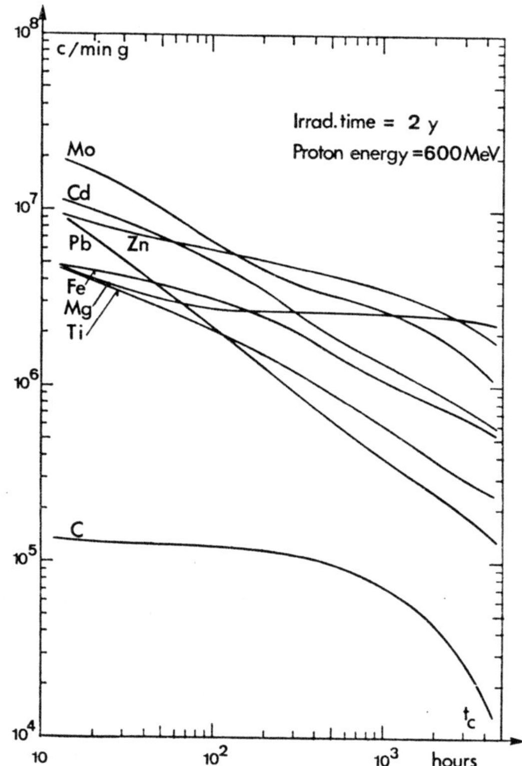

Fig. II.27 Gamma decay of samples irradiated inside the CERN Synchro-cyclotron machine tank at the same time and under the same conditions as those of fig. II.26.

over a long period. We took the ^{7}Be (53.d) level in carbon and the ^{22}Na (2.6 y) level in aluminium and found 1.8×10^8 protons/sec cm^2 with the ^{7}Be and 1.5×10^8 protons/sec cm^2 with the ^{22}Na for the instantaneous flux during machine operation. For the calculation of this instantaneous flux the fact was taken into account that the machine was only in full operation for about two-thirds of the total round-the-clock time. The agreement between these values and the estimated fluxes is satisfactory.

As a second example, we eliminated the protons by placing the samples outside the tank against the 2 mm thick aluminium pion exit window, at a distance of about 2.4 m from the most-used target, and an angle of $\theta = 12°$ with the incident proton direction, that is nearly in the forward direction. All protons scattered from the cy-

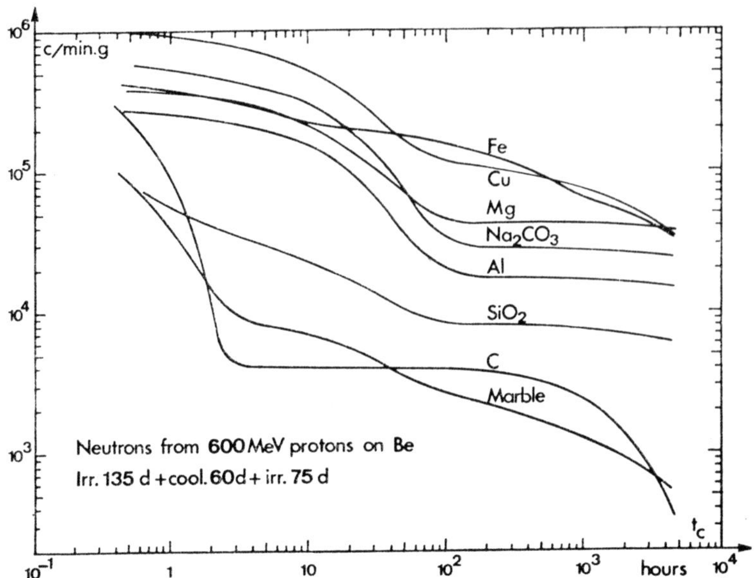

Fig. II.28 Gamma decay of various samples irradiated just outside the CERN Synchrocyclotron machine tank.

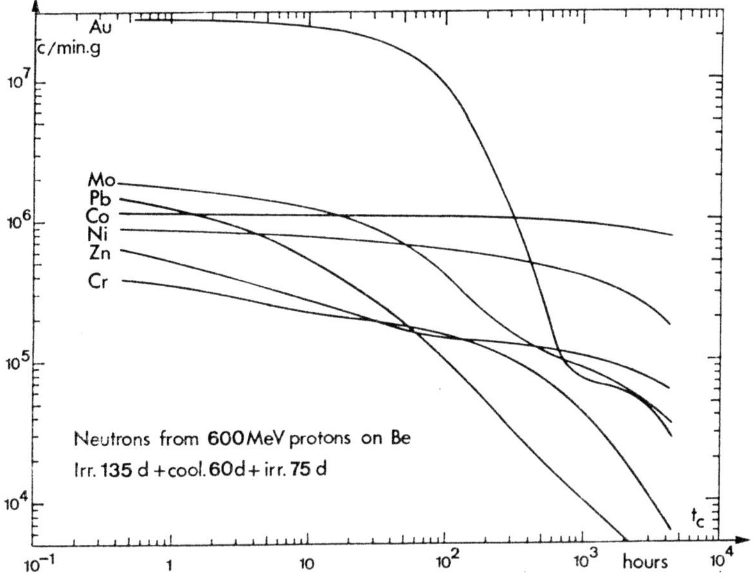

Fig. II.29 Gamma decay of samples irradiated just outside the CERN Synchrocyclotron machine tank at the same time and under the same conditions as those of fig. II.28.

clotron target are deflected away by the magnetic field of the machine. The high energy integrated neutron flux from the target is, after Perret, about 2×10^7 neutrons/sec cm^2 at this location for an internal beam current of 1.0 μA, which was probably the actual machine current during the period considered. The contribution of the beam fraction lost in the pole pieces is probably small against this, due to the eccentric position of the samples with respect to the region where most of the beam is lost (dee region). The contributions of the evaporation and thermal neutrons are probably as in the first example above or smaller, due to the same eccentric sample location.

The experimental results are shown in figs. II.28 and II.29. The overall aspect of the decay curves is the same as for the preceding example. However, signs of thermal neutron activation are present for the curves of the gold and cobalt samples. The very high activity of the 65 h half-life ^{198}Au is apparent. Also the cobalt curve seems more horizontal than before, which indicates production of ^{60}Co.

The long-period averages for the instantaneous bombarding flux found with the ^{7}Be and ^{22}Na plateaus are 5×10^6 and 1.5×10^7 neutrons/sec cm^2, using the proton cross-sections at 600 MeV. These figures are factors 4 and 1.4 times less than what one expects from the direct flux measurements by Perret. Although the disagreement between expected and measured induced activity seems to be systematic in this case, it is not considered to be large, owing to the difficulty in evaluating the exact machine performance including internal beam intensity and switch-on time over a long period, and also due to the lack of information on the exact neutron spectrum and cross-sections. A further illustration of the difficulty in appreciating average irradiating fluxes will be found in the next example.

As a third and last example of activation by very high energy particles, we will examine the decay curves of several similar groups of samples exposed during 5 months from May 26th to October 20th, 1964 in the neighbourhood of four different targets located at units 1, 6, 60, 64 of the CERN Protonsynchrotron, which was usually working at an energy of 19 GeV. The internal beam current intensity of the synchrotron was about 3×10^{11} protons per second during this period and the machine was operated ca. 70% of the time. The distance to the target was between 1.2 and 2.7 metres and the angle to the forward direction between 1 and 2°. The results are presented in the form of decay curves of the materials examined, which included lead, copper, iron, aluminium, quartz, marble and carbon. The decay curves

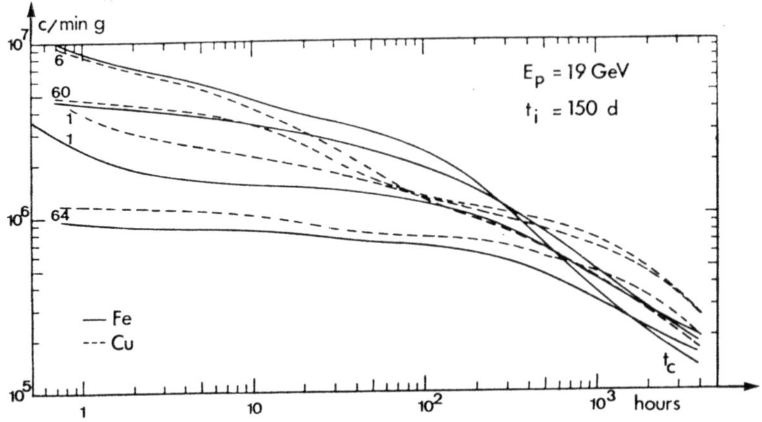

Fig. II.30 Gamma decay of iron and copper samples activated 5 months at the CERN Protonsynchrotron.

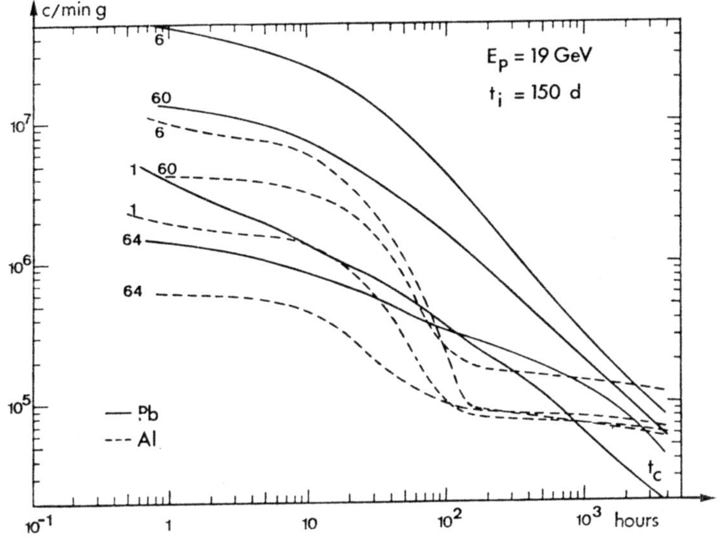

Fig. II.31 Gamma decay of lead and aluminium samples activated 5 months at the CERN Protonsynchrotron.

for the same material at the various locations are presented on the same graph, in order to illustrate the differences in specific activity and slope of decay which can be found in practice and are caused by the irregular operation of each target with respect to irradiation time and fraction of beam allowed to hit the particulier target. Fig. II.30 shows the decay of the target elements Fe, Cu, fig. II.31 that of Pb,

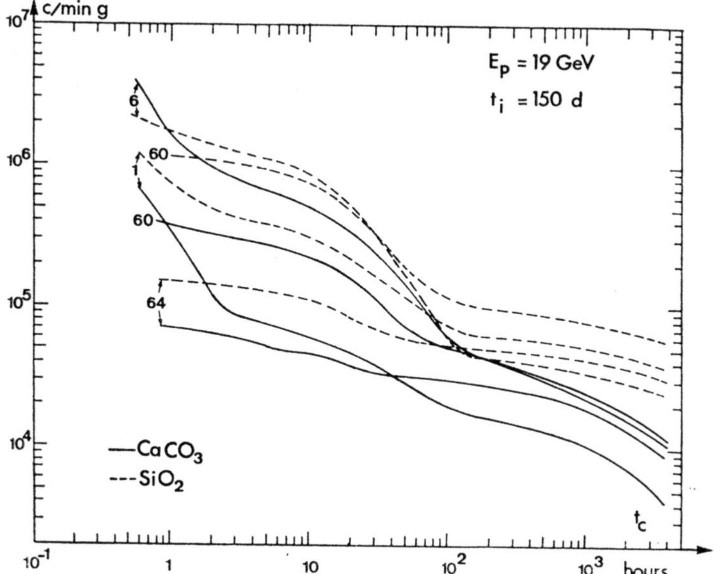

Fig. II.32 Gamma decay of quartz and marble samples irradiated 5 months at the CERN Protonsynchrotron.

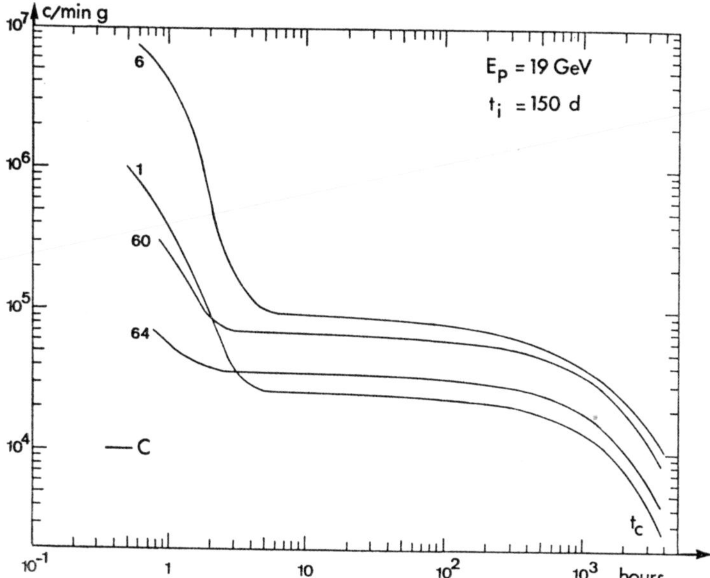

Fig. II.33 Gamma decay of carbon samples irradiated 5 months at the CERN Protonsynchrotron.

Al, fig. II.32 that of SiO_2, $CaCO_3$, and fig. II.33 that of C. One can conclude from an inspection of these figures that differences in the specific activity of the same material can easily cover a factor 10, due only to different targeting conditions in the same machine, although the samples are in all cases at about the same angular position with respect to the target and at comparable distances. The differences are especially apparent during the first days of the decay, reflecting the different targeting conditions. In contrast, the activities of elements with longer half-lives which represent an average of the machine conditions, show less dispersion.

PROBLEMS

CHAPTER II

1. Compare the geometric cross-section σ_g and Ashmore's total proton inelastic cross-section at high energy (eq. 2.8) for ^{9}Be, ^{27}Al, ^{56}Fe, ^{115}In, ^{208}Pb. For the geometric cross-section one can use the expression

$$\sigma_g = \pi r_0^2 A^{\frac{2}{3}} \text{ cm}^2, \qquad r_0 = 1.3 \times 10^{-13} \text{ cm}$$

(1 millibarn $= 10^{-27}$ cm^2, 1 fermi $= 10^{-13}$ cm).

2. Find the production cross-section for ^{24}Na from ^{27}Al, ^{56}Fe, ^{115}In, ^{208}Pb at high energies with Rudstam's formula.

3. Find the production cross-section for ^{24}Na from ^{56}Fe at bombarding energies of 50, 100, 600 MeV and 3 GeV with Rudstam's formula.

4. A 20 GeV proton beam of 10^{12} protons/sec is made to fall on a copper target having a section of 1 cm^2 and a length of 20 cm. Assuming the particle flux density remains constant throughout the target (i.e. neglecting scattering, absorption and build-up) compute the radiation field in rad/h at 1 m from the target for saturation conditions (long irradiation time) at 1 day cooling time without taking into account gamma ray absorption in the target proper.

5. What is the order of magnitude of the number of counts registered by a 3″×3″ NaI(Tl) crystal at a distance of 10 m from the target of problem 4? Assume an overall efficiency of 30% for this crystal and find with the activation the number of photons reaching the surface of the detector which can be taken as a circular disk 3″ in diameter.

6. What is the flux of beta particles of either sign emitted by 1 cm^2 of the target of problem 4, which will reach the eye of an experimenter standing at 1 m? Assume that all the beta particles produced in a layer of 1 mm thickness come out without absorption, and that all others produced at greater depths are fully absorbed.

Assuming that all beta particles have 1 MeV energy, what is the dose rate at 1 m produced by all beta particles, if the surface of the target viewed by the experimenter is 20 cm^2? Compare this figure with the dose rate calculated in problem 4.

7. A tunnel with concrete walls, ceiling and floor runs under the target zone a of high energy accelerator. What is the limit on the high energy particle flux density through this tunnel when the machine is running to give an activation of the tunnel concrete such that the dose rate from the induced activity therein 1 h after shut-down of the machine will not exceed 2.5 mrad/h? Use the danger parameter value for SiO_2 at 500 MeV bombarding energy and for 5000 days irradiation.

CHAPTER III

FISSION PRODUCTS AND ACTIVATION BY THERMAL TO FAST NEUTRONS

Introductory remarks

Neutrons in the energy range 0–15 MeV are quite often the major source of activation in most applications of atomic power nowadays. They also are found in large numbers around high energy accelerators where they are generated in the evaporation process of the target nuclei struck by the high energy particles at the same time and in numbers equivalent to the very high energy (or cascade) neutrons.

It is thus interesting to introduce a chapter which will deal with all the induced activity connected with neutrons in this energy range. We remind the reader that very high energy neutrons have been treated in the preceding chapter on spallation along with protons, pions, deuterons and alpha particles, which also induce this type of reaction at sufficiently high energies.

We shall now examine in the order quoted the activity of fission products from reactors and atomic explosions and the activation of structural materials by fission neutrons. The activation of the same by evaporation neutrons from high energy spallation reactions will also be considered.

1 A few numerical data on fission reactions

We shall begin by recalling a few numerical relationships which are indispensable for figuring out the amount and radioactivity of fission products generated in reactors or in atomic explosions. We can logically proceed as follows.

We start with the number of atoms present in 1 g of fissionable material, say ^{235}U:

$$\frac{N_0}{A} = \frac{6.024 \times 10^{23}}{235} = 2.58 \times 10^{21}/\text{g}. \tag{1.1}$$

This is evidently also the number of fissions per gram of this material. Now the total energy release per fission for various materials can be taken from the following table, which gives also the kinetic energy and the beta and gamma decay energies of the fission products.

TABLE III.1

Energy release per fission (MeV).

Fission element	Fission products kinetic energy	Fission products beta decay	Fission products gamma decay	Fission neutrons kinetic energy	Prompt gamma energy	Total energy release per fission
^{232}Th			10.8			
^{233}U	167.8	8	4.24	5	7	192
^{235}U	178.2	7.8	6.84	4.8	7.5	195
^{238}U			10.9			
^{239}Pu	175	8	6.15	5.8	7	202

The prompt gamma energy is by convention the gamma energy emitted in the fission act itself and up to 1 minute thereafter. The beta and gamma decay energies from the fission products are always taken from 1 minute after the fission onwards.

With these data we can relate the number of fissions or grams of fissionable material burnt up to the reactor power or bomb strength. We continue with the ^{235}U example and write down first the energy release per fission in ergs, calories and tons of TNT (1 ton TNT $\approx 10^9$ cal). So we have

$$\begin{aligned} E/1 \text{ fission } ^{235}\text{U} &= 195 \text{ MeV} = 3.12 \times 10^{-4} \text{ erg} = \\ &= 7.46 \times 10^{-12} \text{ cal} = 7.46 \times 10^{-21} \text{ ton TNT}. \end{aligned} \tag{1.2}$$

The burning-up of 1 g ^{235}U will release

$$\begin{aligned} E/\text{g } ^{235}\text{U} &= 8.05 \times 10^{10} \text{ watt sec} = 1.93 \times 10^{10} \text{ cal} = \\ &= 19.3 \text{ ton TNT}. \end{aligned} \tag{1.3}$$

To arrive at the power generated, in the case of reactors, we have to introduce the burn-up time. It is common practice to express reactor energy release in megawatt days (MWD), the unit being the energy corresponding to a power of 1 megawatt during a day. We calculate now what power P is given by the burning of 1 g ^{235}U in one day. We find

$$P(1 \text{ g } ^{235}\text{U/day}) = 0.93 \times 10^6 \text{ watt} = 0.93 \text{ MW}. \tag{1.4}$$

The fission rate at this level is

$$\frac{dn}{dt} = \frac{2.58 \times 10^{21} \text{ fissions}}{86.400 \text{ sec}} \equiv 2.98 \times 10^{16} \text{ fissions/sec} \tag{1.5}$$

and per MW of operating power it is

$$(dn/dt)_{1 \text{ MW}} = 3.2 \times 10^{16} \text{ fissions/sec} \tag{1.6}$$

whereby the consumption of ^{235}U will be

$$(dm/dt)_{1 \text{ MW}} = 1.075 \text{ g/day}, \tag{1.7}$$

always for a continuous power release of 1 MW.

The fission rate calculated above gives us also the means to calculate the saturation activity of each of the fission products present in the reactor after a long time of operation, because the decaying fission products are immediately replaced in the same numbers by new ones. We have only to multiply this fission rate by the yield for the particular fission product to obtain the saturation activity of this product per MW of operating power of the reactor at the instant of shut-down.

Let Y_ν be the fission yield for the particular fission product ν (defined as the mass of this element produced in the fission of 1 g of the original fissionable material), then the saturation activity A_ν of this particular element obtained per MW of reactor power burning ^{235}U will be

$$A_\nu = Y_\nu A = 3.2 \times 10^{16}\, Y_\nu \text{ dis/sec} = 0.864 \times 10^6\, Y_\nu \text{ curies}, \tag{1.8}$$

where A is the total activity defined in eq. (1.6).

Yield charts will be presented in the next subsection. From this saturation activity it is then possible to calculate the activity at any irradiation and cooling time for this particular isotope as one knows its half-life. One can also find the mass in grams of the isotope. Using its k-factor in rad/h Ci at 1 m given in the appendix, one will arrive at the radiation field involved.

As one knows, every fission gives rise to two products each of which produces on the average about three daughter products before reaching stability. Thus, in a reactor that has been running for a long period, the total disintegration rate of all the fission products will be around 19×10^{16} dis/sec per MW reactor power. However, many of these disintegrations give weak beta or gamma irradiation and thus an overall disintegration rate figure is not very useful for calculating shielding requirements, heating effects or health hazards. To obtain the radiation levels for calculations in such matters it is necessary to consider each fission product separately and add up to find the total effect, and there appears to be no short-cut in these calculations, some of which will be presented in the next subsections.

We come now to the evaluation of the mass of the fission products released in a reactor in a given time interval. The mass of the fission products is in fact practically equal to the mass of the fissionable material burnt up in the time considered. (In the case of ^{239}Pu, for example, there are 0.987 g of fission products made for 1 g of plutonium undergoing fission). Neglecting this mass defect, we arrive at the following relationship between mass of fission products and MWD produced in a reactor for ^{235}U

$$1 \text{ MWD} = 1.075 \text{ g fission products.} \qquad (1.9)$$

In the case of a nuclear explosion burning up ^{235}U, for instance, one can easily calculate from the figures given above the following equivalence

$$1 \text{ kiloton TNT} = 51.8 \text{ g } ^{235}\text{U}. \qquad (1.10)$$

TABLE III.2

Number of curies released per kiloton TNT in a nuclear explosion using ^{235}U.

Time after explosion	1 min	10 min	1 h	8 h	1 d	1 w	1 mo	1 y
Activity of fission products, curies per kiloton TNT	3.9×10^{10}	2.4×10^{9}	3×10^{8}	2.3×10^{7}	6.6×10^{6}	6.4×10^{5}	1.1×10^{5}	$5.5 \times$

An estimate of the number of curies released in the fission products for this explosion force is given in table III.2.

Another useful relationship is the one which connects the gamma ray energy falling on a unit surface to the absorbed dose in rads. Gamma decay curves of fission products are always given in energy units, not in rads per gram at a given distance. To find the latter value, one has to take into account the geometry and calculate the energy flux in watt sec/cm² or MeV/cm² at the location of interest. It is then possible, using the approximate eq. (5.2) of ch. I, to find that

$$1\ \text{MeV/cm}^2 = 2.1 \times 10^{-9}\ \text{rad}; \quad 1\ \text{watt sec/cm}^2 = 1.31 \times 10^4\ \text{rad}. \quad (1.11)$$

These relationships are only valid for gamma quanta above 0.053 MeV energy. However, practically all gamma rays from fission products

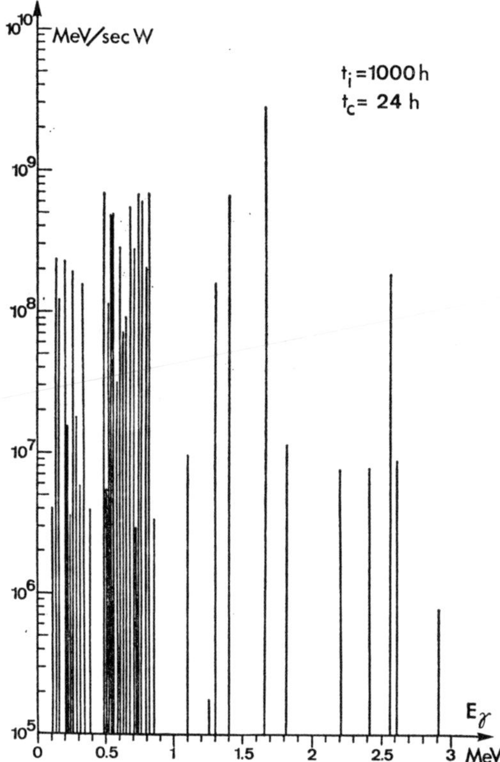

Fig. III.1 Fission products gamma energy spectrum, after 1000 h of operation, 1 day after shut-down.

have energies above this value, as can be seen from the example of fission product spectrum from Moteff presented in fig. III.1. This practically frees us from the necessity of dividing the gamma emission into various groups according to their energy.

2 Various data on fission products from thermal and fast neutron fission

The primary, main source of radioactivity produced in reactors or nuclear explosions is certainly constituted by the fission products. These products have atomic numbers which range between 75 and 160. The relative abundance of fission products depends somewhat on the sort of material that has been fissioned and on the energy spectrum of the neutrons involved. However, as the reader will see from the graphs presented in fig. III.2 for fission by thermal neutrons and in fig. III.16 for fission by a fission spectrum of fast neutrons, the differences are, on the whole, not very large, and the data relevant to thermal reactors will be to a great extent applicable to the fission products of fast reactors and nuclear explosions. We will thus begin with thermal reactors and then give some complements for the other cases just mentioned.

2.1 *Thermal reactors*

Extensive data exist from the literature on the yields and the activity of fission products from thermal reactors. Let us first take the yields. When one refers to a yield, it is necessary to define properly what is meant. In fact, the word has been used to describe three different quantities. The yields given in fig. III.2 are 'chain' or 'mass' yields. They indicate the fraction of the total cross-section devoted to the production of the sum of all isotopes having one and the same mass number and are usually expressed in percent of the total fission cross-section. As an example, it is seen from fig. III.2 that all products of ^{235}U with mass number 91 are formed with 6% of the total cross-section. From the curves in the figures, one can conclude that for various fuels and neutron spectra (thermal or fast) the shape of the 'chain' yield distribution does not vary very much. This shape exhibits clearly two peaks of maximum production from $A=90$ to 100 and from $A=133$ to 145 where the 'chain' yields are 6%. In between there is a valley where the yields decrease by as much as two orders of magnitude for thermal neutrons. The decrease is less pronounced with higher energy neutrons.

If one requires more detailed information the next question will be: 'how much of each particular isotope does one get after fission?'. Here one is obliged to take account of a physical phenomenon. The fission products formed in the fission act are in general not stable and decay by beta emission, a process which does not alter the product

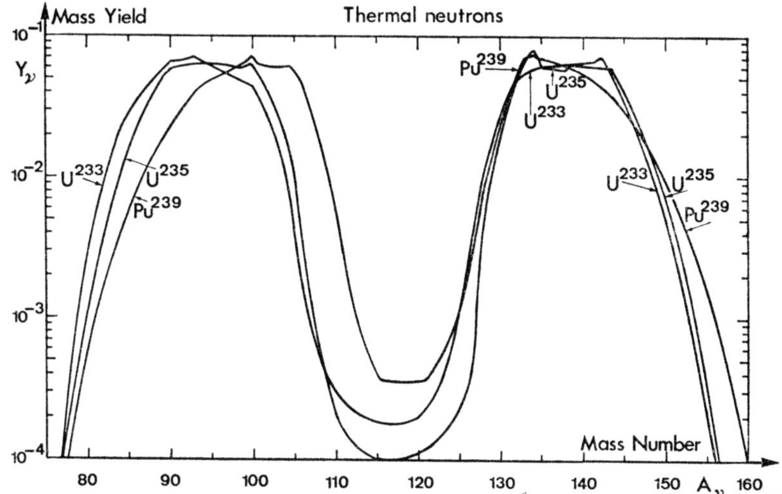

Fig. III.2 Mass yield curves for the thermal neutron fission of ^{233}U, ^{235}U, ^{239}Pu.

mass. The decay times (half-lives) of most fission products are short, so that it is soon irrelevant to know if a certain product has been formed directly in the act of fission or if part of it results from the beta decay of one or several precursors. What one in fact measures after some time when making a chemical analysis are the yields of the products including beta decays of their precursors. These yields are called 'cumulative' yields; charts have been drawn to show how these cumulative yields are distributed throughout the nuclides and are presented in figs. III.3 and 4 for the thermal neutron fission of ^{235}U. The isoyield lines are only approximate and represent a tentative best fit made by hand to present published data. The yields are again expressed in percent of the total fission cross-section. Inspection of these figures shows that the cumulative yield function plotted over a

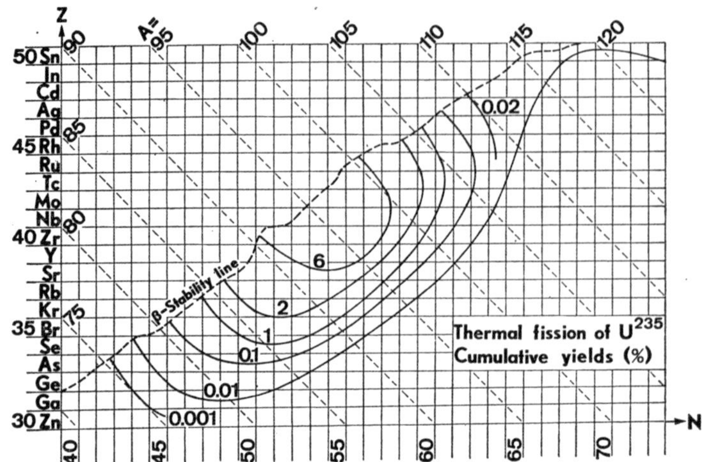

Fig. III.3 Cumulative yields for thermal fission of ^{235}U expressed in percent of the total fission cross-section for $Z=30$ to 50.

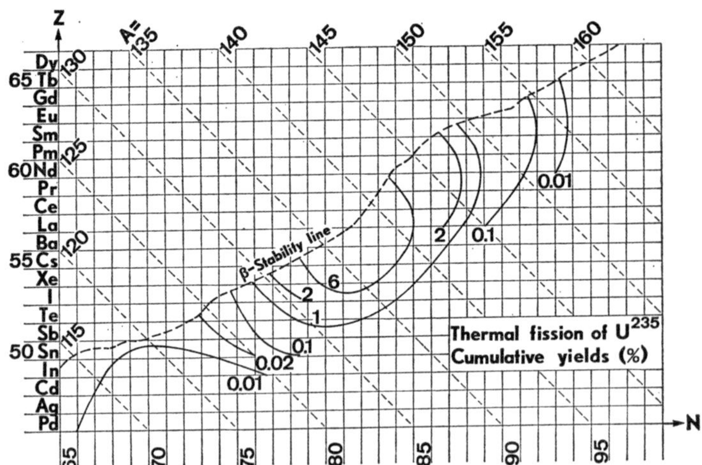

Fig. III.4 Cumulative yields for thermal fission of ^{235}U expressed in percent of the total fission cross-section for $Z=50$ to 65.

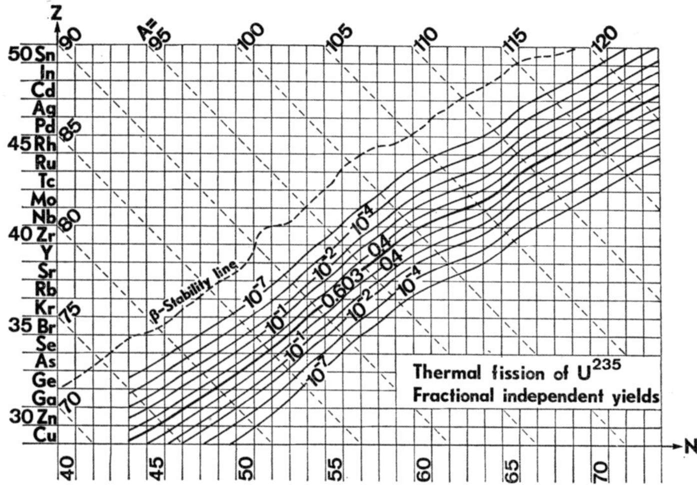

Fig. III.5 Fractional independent yields expressed as the ratio of the independent yield of the particular isotope to the chain yield of the chain to which it belongs, for $Z=30$ to 50.

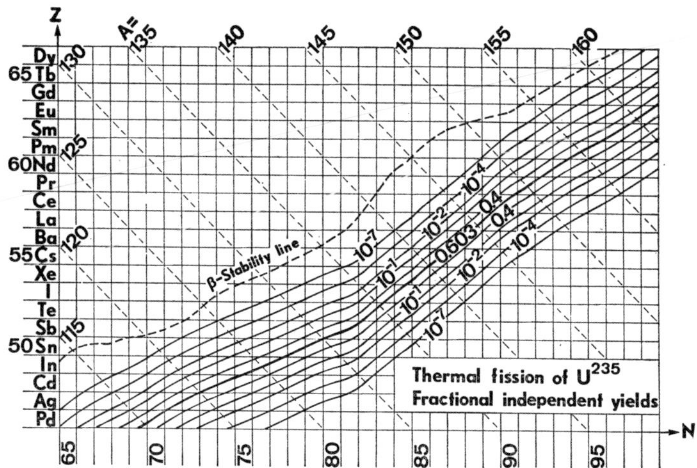

Fig. III.6 Fractional independent yields expressed as the ratio of the independent yield of the particular isotope to the chain yield of the chain to which it belongs, for $Z=50$ to 65.

(N, Z)-nuclides chart has the form of a mountain which would rise towards the beta stability line from the right, whereas on the left of this line there are practically no thermal neutron fission products. The mountain has two peaks at the same atomic numbers as the two maxima of the 'mass' yields distribution curve. The cumulative yields on the beta stability line are in fact the 'mass' or 'chain' yields as beta decay chains stop there.

Still more detailed information on the fission products is gained by measuring the so-called independent isotopic yields, i.e. the yields for the production of the individual nuclides in the fission act itself before the quantity of each nuclide has been increased by the beta decay of its possible precursors. Such measurements are difficult due to the short half-lives. However in some cases the contributions of the various precursors can be separated. In other cases, when the nuclide in question is 'screened' from its precursors by a stable nuclide, a direct measurement is possible. From a number of such individual measurements general laws for the individual isotopic yields distribution over the (N, Z)-chart on the elements have been deduced and are presented in the following figures which were drawn from the results of Wahl and co-workers (private communication). First, the reader is referred to figs. III.5 and 6 which show the distribution of the independent isotopic yields expressed as a fraction of the corresponding chain yields. It has been found that this distribution does not vary with atomic number, and that it has a maximum of about 0.6 of the total chain yield on a line roughly parallel to the beta stability line, but lying to the right of the latter, throughout the region of interest ($Z=30$ to 65). The exact position of this maximum independent yield line when expressed in terms of chain yields can be found in the figures, as well as the steepness of descent of the two side slopes.

From the chain yields of fig. III.2 and the fractional independent yields of figs. III.5 and 6, the individual isotopic yields of each fission product, as formed in the fission act itself before beta decay has set in, can now be calculated by simple multiplication. The result is shown in figs. III.7 and 8, always for thermal neutron fission of ^{235}U. The reader will observe two mountains in the maximum production region, with a steep descent of the yields on both sides, one towards the beta stability line and one away from it. In the two maximum regions the individual cross-sections reach 3.6% of the total thermal neutron fission cross-section of ^{235}U, which lies around 690 barn,

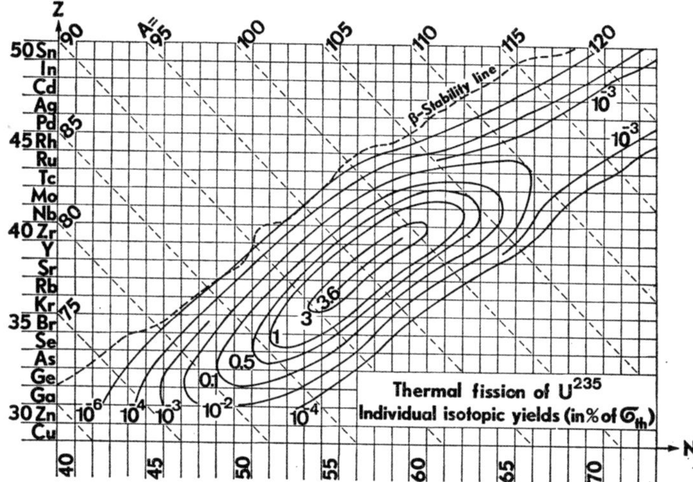

Fig. III.7 Individual isotopic yields of fission products from the thermal fission of ^{235}U, before beta decay of these fission products has set in, expressed in percent of the thermal neutron fission cross-section, for $Z=30$ to 50.

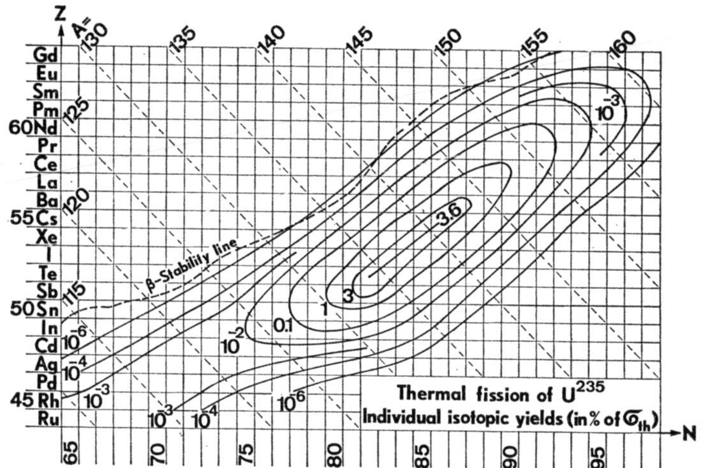

Fig. III.8 Individual isotopic yields of fission products from the thermal fission of ^{235}U, before beta decay of these fission products has set in, expressed in percent of the thermal neutron fission cross-section, for $Z=50$ to 65.

giving 25 barn for the most abundant fission products formed in the fission act itself.

It is often of interest also to calculate the saturation activity of a fission product per gram of fuel element, given the neutron flux the fuel has been subject to. For this we have to relate the power level of the reactor to the neutron flux density reigning therein. The fission rate per gram of fuel element is obviously equal to the number of fuel atoms per gram times the neutron flux times the fission cross-section

$$\mathrm{d}n/\mathrm{d}t = (N_0/A)\Phi\sigma, \tag{2.1}$$

where N_0 is Avogadro's number and A the atomic number of the fuel.

The fission cross-sections generally assumed for various fuel elements are given in table III.3 for three kinds of neutron spectra (thermal spectrum, fast reactor spectrum, ^{235}U fission spectrum) according to Keepin.

TABLE III.3

Fission cross-sections in barns for various fuel elements and several neutron spectra.

	Thermal	Fast reactor	^{235}U fission
^{235}U	582	1.59	1.25
^{239}Pu	748	1.83	1.85
^{233}U	527	2.37	1.90
^{238}U		0.12	0.08
^{232}Th		0.03	0.08

The fission rate permits us now to calculate the power level per gram of fuel element, using the equivalence (1.6). Also, by multiplying the fission rate of the fuel element by the independent mass yield of a particular fission product, we find the production rate of this particular product per gram of fuel element in the flux considered, and hence we are able to calculate the activity of this product after any operation time and at any cooling time after shut-down of the reactor, knowing the half-life of the fission product.

Extensive calculations of the growth and decay of the activity of all individual fission products from thermal reactors can be found in the work of Blomeke and Todd. These authors also apply some corrections as a function of flux intensity which takes care of the neutron poisoning of the fuel by the fission products at very high fluxes. In this chapter we will content ourselves with presenting only global data summed over all fission products to the effect of enabling the reader to foresee radiation fields from these in practical cases.

We begin with the activity of the beta and gamma radiation in curies per megawatt after infinite operation, according to Stehn and Clancy (fig. III.9). Although the curies number is not of very great

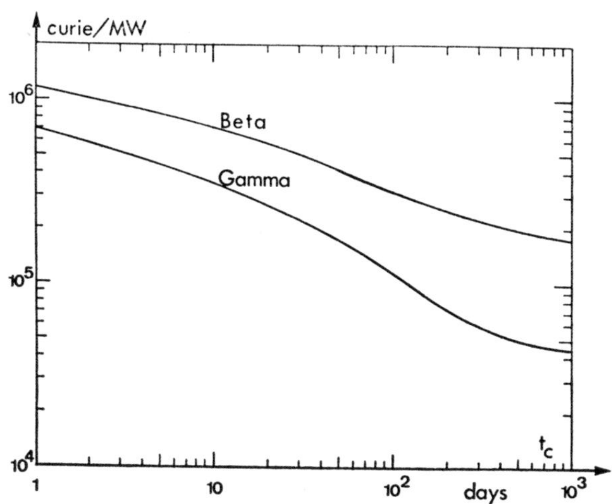

Fig. III.9 Beta and gamma activity of fission products after infinite operating time, in curies per megawatt.

help in calculating radiation fields, it is good to know it as a reference for comparison, and it can be used to calculate average values, such as the mean power or the mean radiation field per decay, etc. We also see from the curves that the decay is slow, roughly a factor of 10 from 1 to 1000 days, and that the beta activity is larger by a factor 2 to 3 than the gamma activity.

The beta and gamma decay powers can be summed up to give the total decay power of the fission products, used, for instance, when calculating the cooling requirements after shut-down. Fig. III.10 presents the data of Stehn and Clancy again on this matter, in watts of decay power per megawatt of operating power after infinite operating time. What can be said about this figure is that the decay is slow. It follows roughly a $t_c^{-\frac{1}{3}}$-law in this case where all the radioactive isotopes produced are saturated. The figure is also interesting in that it extends to a cooling time of 1000 years, showing that the decay between 1 and 1000 years is only a factor of 10.

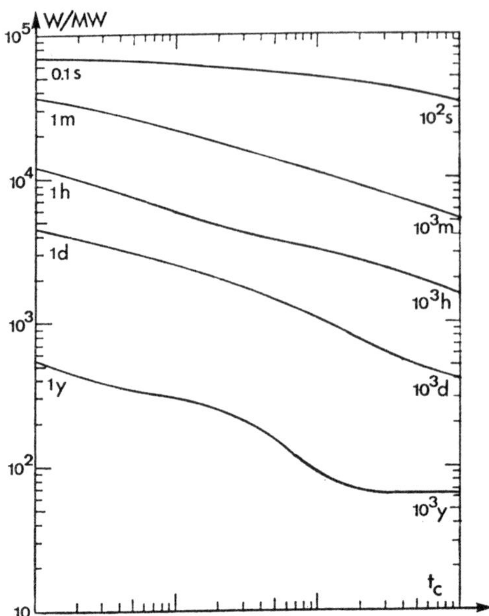

Fig. III.10 Total decay power of fission products after infinite operating time, in watts per megawatt of operating power.

It is appropriate now to give the beta and gamma decay powers separately. For convenience in practical use we have given in figs. III.11 and 12 the decay curves of these powers for various operating times, ranging from 1 hour to infinity, combining the data compiled by Perkins and King (1 to 1000 hours) and by Stehn and Clancy (infinity). The data from the two sources are seen to be compatible. The scale to the right is in watt per megawatt, to the left in MeV/sec watt. It is noticed by comparison of these two figures that the orders of magnitude of beta and gamma powers are the same, throughout the whole cooling times. From the gamma power decay curves one can easily compute the radiation field at a given distance of a point-like source of the given fuel by using the equivalence given in the preceding section, eq. (1.11).

A case of special interest is that of an irradiation burst, when the operating time can be considered to be short in comparison with the half-life of practically all the fission products. According to fig. I.9 of ch. I, which gave the distribution of the half-lives of the thermal fission products of ^{235}U, there are in practice only a few isotopes under 10^{-2} day (14.4 min). So an operating time of a few minutes

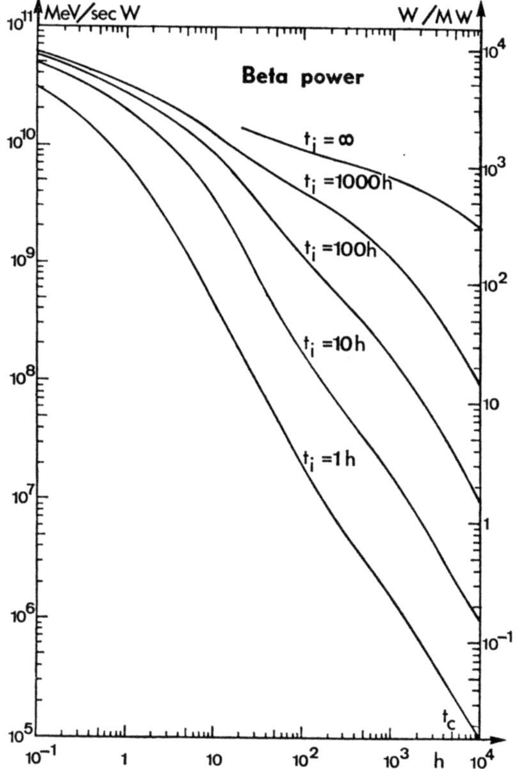

Fig. III.11 Beta decay power of fission products for different operating times.

would come into this category, to which nuclear explosions typically belong. The decay curves for burst irradiation cannot be fitted into the preceding figures as the units to which reference is made are different. It is usual practice not to define a power level during the burst, but to indicate the total energy output (the integral of power over time). This is measured in megawatt days for reactors and in equivalent tons of TNT for nuclear explosions. One can also give the data per single fission. Fig. III.13 shows the beta and gamma activities in curie/MWD and fig. III.14 the total decay powers in watt/MWD following the burst, after Stehn and Clancy. The decay is much faster than with saturated fuel, and follows this time approximately a t_c^{-1}-law, at least up to 100 days, where the beta and gamma decays become faster.

The total decay power is represented in fig. III.14 as a function of time. The time scale extends up to 10^4 years. Two intervals of time are

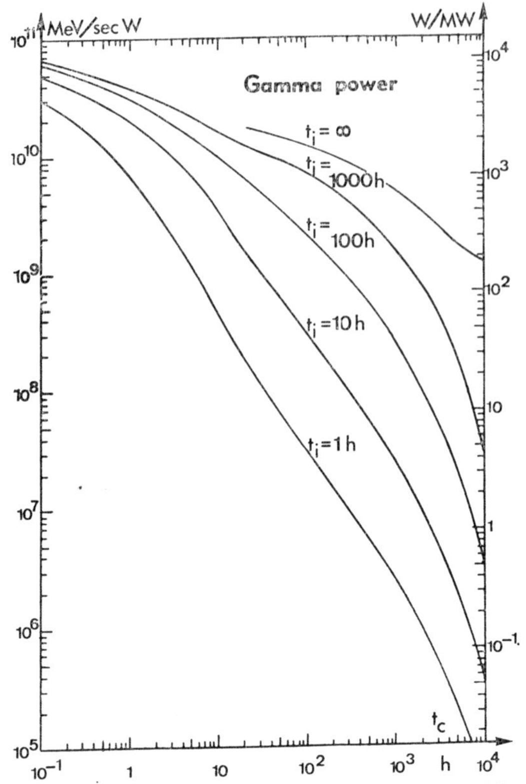

Fig. III.12 Gamma decay power of fission products for different operating times.

observed, one at about 10 years and the other at about 10^3 years cooling time, where the activity is stationary for a while. This is because there are only a few radionuclides left at such long cooling times and because the order of cooling time has not yet been reached after which the activity of the next isotope decreases markedly.

Fig. III.15 shows the beta and gamma decay powers separately (from Perkins and King). The ordinate units to the right are watt/MWD and to the left MeV/sec fission. Both gamma and beta powers are seen to be practically of the same magnitude, as was found in the case of longer irradiations. However the number of curies is greater for beta than for gamma decays, as was also the case before (fig. III.13).

2.2 *Fast reactors and nuclear explosives*

We shall again commence by giving the fission mass yield curves for

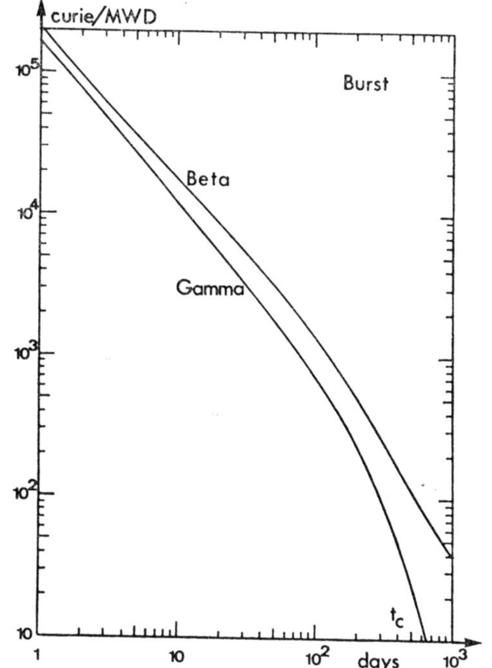

Fig. III.13 Beta and gamma activity of fission products following burst.

the fuel elements most used in fast reactors and for an incident fission neutron spectrum. Fig. III.16 shows the mass yields to be expected from the elements ^{232}Th, ^{235}U and ^{238}U with such a spectrum. The curves have a similar appearance to those of fig. III.2. A slight difference is found in the yields of the fission products with atomic numbers between 105 and 125 which are generally higher with the fission spectrum. However, as the yields are much smaller in this region than in the 85–105 and 125–150 regions, where the yields are larger by two orders of magnitude, the overall effect will not be very pronounced.

In the following figures we find some examples of activity decay from the fast fission of ^{235}U and ^{239}Pu for various operating times, taken from the data of Burries and Dillon. First comes as usual the gross fission product activity, expressed as total curies per watt of reactor power which is presented in fig. III.17.

As we know, the curie is defined as the activity of any radioactive species decaying at a rate of 3.7×10^{10} disintegrations per second. Since most fission products decay first by beta emission, total curies are nearly the same as beta curies. In the following figs. III.18 and 19

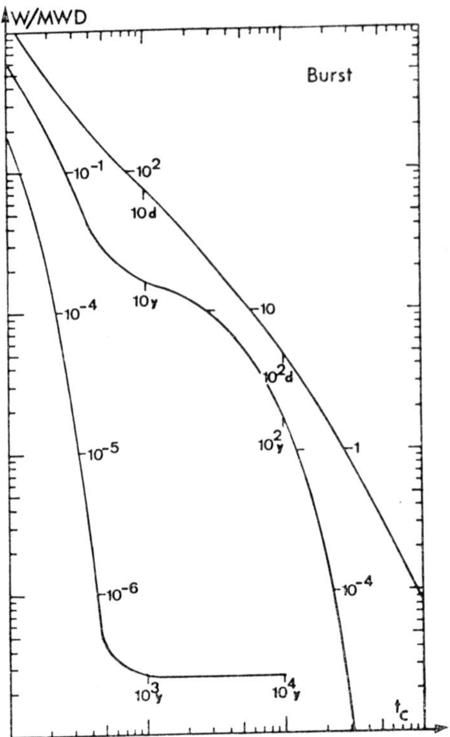

Fig. III.14 Total decay power of fission products following burst.

the beta and gamma energy dissipations are given separately as a function of decay time.

It is interesting to compare the fast fission data to the thermal fission data presented in the preceding subsection. This we have done for the fuel element ^{235}U by reading off a few values in figs. III.9, 11, 12 and in figs. III.17, 18, 19, which we now present in tabular form (table III.4).

As one sees from the table the values for thermal and fast fission are practically identical, at least at 100 days cooling time. At 2 days cooling time the fast fission dissipation values given are roughly twice as large as the thermal fission ones.

The last topic we have to mention in the context of this subsection is the decay of fission poducts from atomic explosions. This has been, of course, measured over and over again. We shall content ourselves with presenting the decay curve of the gamma activity normalized at

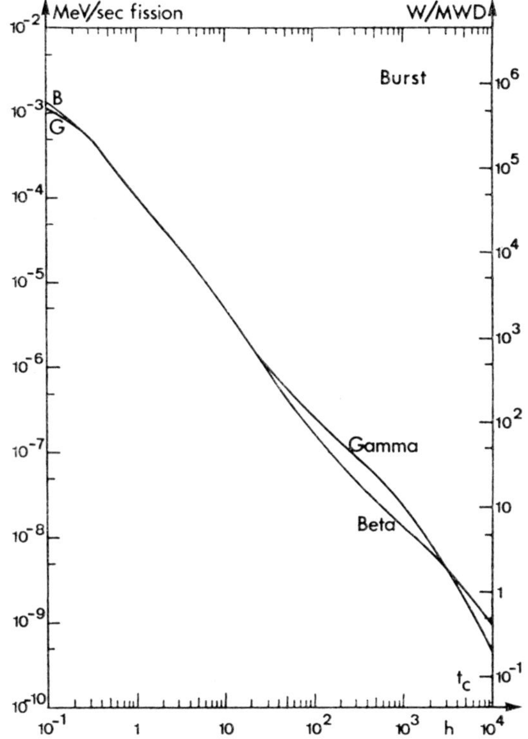

Fig. III.15 Beta and gamma decay power of fission products following burst.

TABLE III.4

Comparison of fission product activity for thermal and fast fission of ^{235}U.

	Cooling time	Thermal fission $t_i = \infty$	Fast fission $t_i = 10^4$ d
Beta activity in curie/MW	2 d	1×10^6	1.6×10^6
	100 d	0.3×10^6	0.3×10^6
Beta dissipation power in W/MW	2 d	1.6×10^3	6×10^3
	100 d	6×10^2	5.5×10^2
Gamma dissipation power in W/MW	2 d	2×10^3	5×10^3
	100 d	4×10^2	4×10^2

one hour after the explosion (fig. III.20). The decrease is seen to follow a $t_c^{-1.2}$-law. The curve is valid up to cooling times of a few thousand

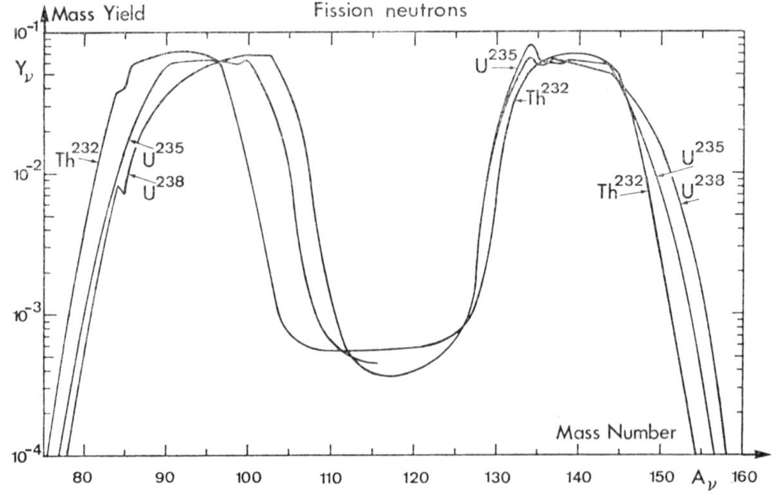

Fig. III.16 Mass yield curves for fission of ^{232}Th, ^{235}U and ^{238}U by a fission neutron spectrum.

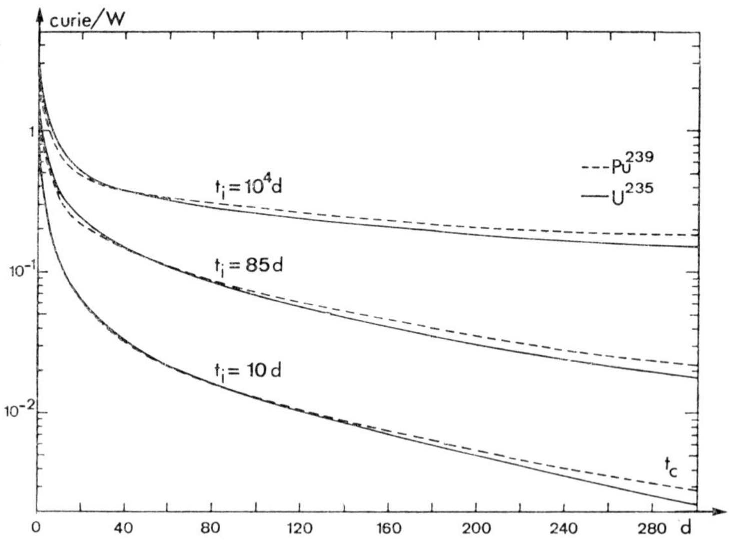

Fig. III.17 Gross fission product activity from fast fission of ^{235}U and ^{239}Pu for various operating times.

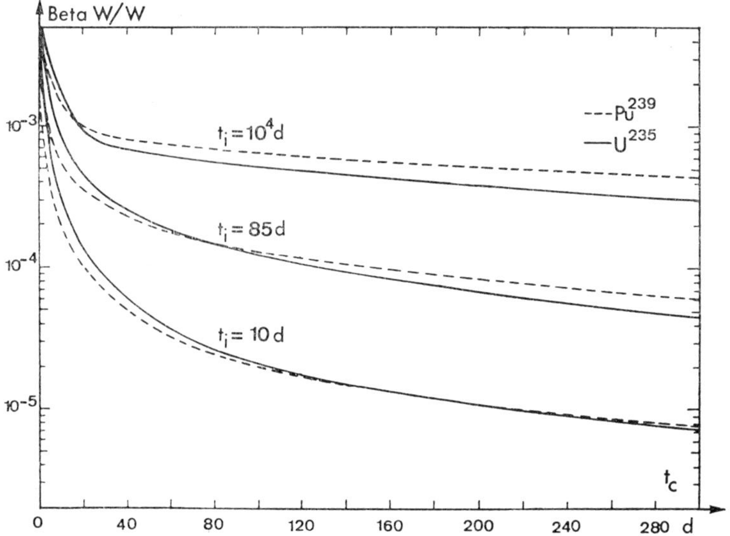

Fig. III.18 Beta energy dissipation by fission products from fast fission of ^{235}U and ^{239}Pu.

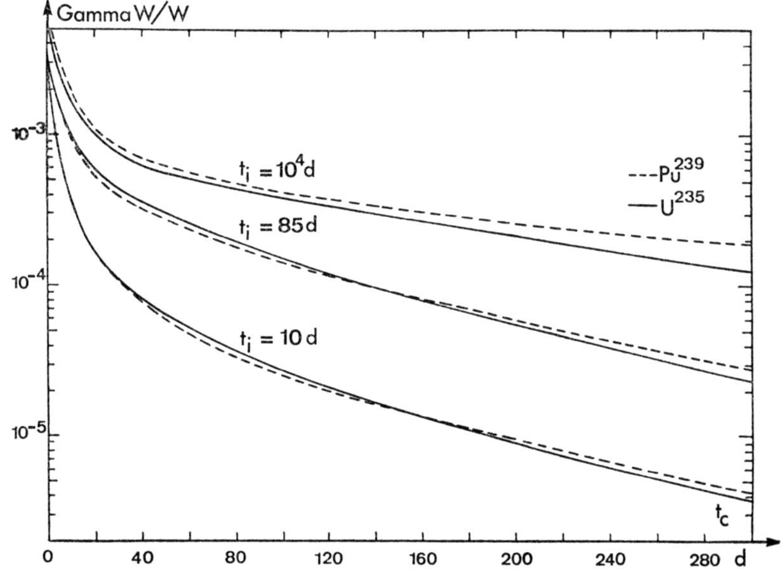

Fig. III.19 Gamma energy dissipation by fission products from fast fission of ^{235}U and ^{239}Pu.

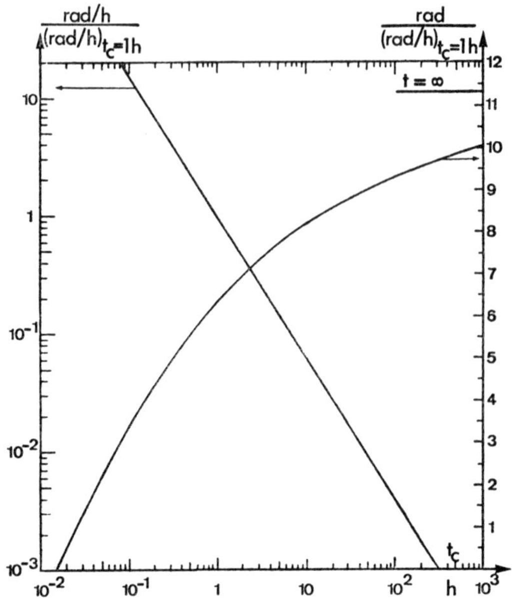

Fig. III.20 Decrease of gamma dose rate from fission products of a nuclear explosion with time (in rad/h, left-hand scale) and rise of the accumulated total dose from 1 minute after the explosion in rads (right-hand scale), both referred to the dose rate in rad/h existing 1 hour after the explosion.

hours (three months). After this time the decay is known to be much faster (compare the decay of a burst irradiation, fig. III.13). The dose rate is given in units of the dose rate existing 1 hour after the explosion as reference dose rate (left-hand scale).

Fig. III.20 also indicates the total dose accumulated at the location of interest as a function of the cooling time, to allow calculation of the doses received by personnel exposed for some period in this cooling time. Rather awkwardly the total dose in rads is referred to the figure of the dose rate in rad/h one hour after the explosion (right-hand scale). As an example the total dose in rads accumulated after an infinite cooling time is seen to be equal to 11.3 times the figure giving the dose rate in rad/h one hour after the explosion at the place of interest.

3 Activation of structural materials by thermal, epithermal and fast neutrons

Besides fission products, activated structural materials are a source of hazard in facilities where a large number of fission or evaporation neutrons are released. This is the case not only with reactors, but also with accelerators, since we find here, as well as very high energy cascade particles generated in the spallation process, an almost comparable number of so-called evaporation neutrons, produced by the unstable nucleus remaining when the cascade particles have left.

We shall consider the activation produced by thermal, epithermal and fission neutron spectra usually found in reactors first, as there is some information available on this matter, and we shall examine the case of evaporation neutrons from compound or excited nuclei, on which much less is known, afterwards.

The problem here is to take the excitation function of the target element for the particular reaction (cross-section as a function of neutron energy), to multiply it by the expected neutron energy spectrum normalized to unit flux and to integrate the product over the relevant energy interval in order to obtain a so-called average cross-section. This can then be multiplied by the total flux in the energy interval considered to find the induced activity with the usual formulae. Such calculations are long and have been made by various authors for specific neutron spectra which they encountered in their work. We shall here use the data of Culp and Page, as reviewed and presented by Layman and Thornton.

The spectrum of reactor neutrons used by these authors represents a close fit to the neutron energy spectrum which had been measured at a given position in a water shield surrounding a particular water moderated reactor (the so-called Ground Test Reactor) by Romanko and Dungan. This spectrum was divided in 3 energy ranges:

a) the thermal region, with

$$0 < E_{\mathrm{n}} < 0.4 \times 10^{-6}\ \mathrm{MeV} \tag{3.1}$$

b) the epithermal region, with

$$0.4 \times 10^{-6}\ \mathrm{MeV} < E_{\mathrm{n}} < 0.1\ \mathrm{MeV} \tag{3.2}$$

c) the fast region, with

$$0.1\ \mathrm{MeV} < E_{\mathrm{n}} < 9\ \mathrm{MeV}. \tag{3.3}$$

In each of these energy ranges the form of the energy spectrum was different.

In the first range (thermal region) the measurements showed that the energy distribution was practically maxwellian. This means that the neutrons are in thermal equilibrium with the surrounding atoms. As a consequence, where a neutron with a given energy leaves a given volume element, it is immediately replaced in steady state conditions by another of the same energy which flies in from the outside of this volume element. The maxwellian or thermal energy distribution gives for the density or number of neutrons with an energy between E and $\mathrm{d}E$ present in some given volume an expression of the form

$$E^{\frac{1}{2}} \exp(-\text{const } E)\mathrm{d}E.$$

What activates is in fact the flux of particles through the volume, which is obtained by multiplying the density of the particles by their velocity, which is proportional to $E^{\frac{1}{2}}$. One arrives thus at the expression for the flux density which was used for the calculations and reads as follows

$$\varphi(E) = 1.46\times 10^{15} E \exp(-3.82\times 10^{7} E) \frac{\text{neutrons}}{\text{cm}^2 \text{ sec MeV}} \tag{3.4}$$

for the energy interval (3.1), where $\varphi(E)$ has been normalized so that the integral of $\varphi(E)$ between 0 and 0.4×10^{-6} MeV will give 1. The energy is always given in MeV.

In the second region (the so-called epithermal neutrons), the conditions are different. The fission neutrons are being slowed down, mainly by repeated collisions with the atoms of the moderator. One finds experimentally that the neutron flux has roughly a $1/E$ dependence upon energy. Such a dependence is often encountered when fast neutrons are being slowed down and when there is enough hydrogen around, as this element is known to change the neutron energy in collisions by the largest amounts possible due to the equality of its mass with that of the neutron. The flux density used in this region is

$$\varphi(E) = \frac{1}{12.4E} \frac{\text{neutrons}}{\text{cm}^2 \text{ sec MeV}} \tag{3.5}$$

for the energy interval (3.2).

This expression, integrated between 0.4×10^{-6} and 10^{-1} MeV, gives again 1.

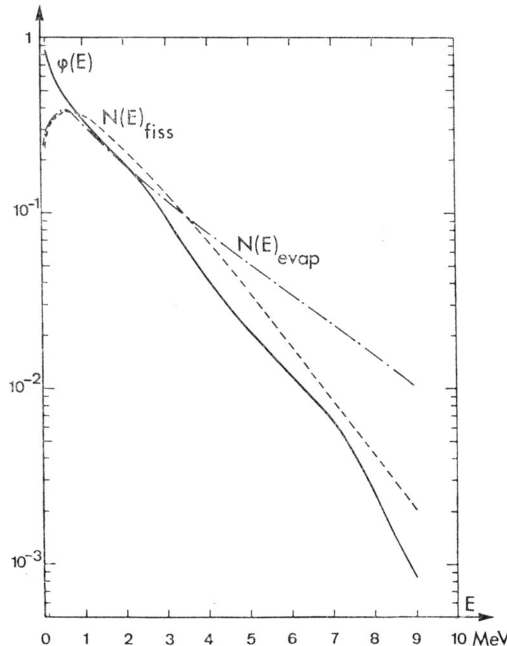

Fig. III.21 Fast neutron flux energy distribution $\varphi(E)$ in the ground test reactor water shield at a position $3\frac{1}{2}$ inch from the face of the core, according to Layman and Thornton (full line). The neutron energy spectrum $N(E)_{\text{fiss}}$ from the thermal fission of ^{235}U after Grundl (dashed line) and the evaporation neutron spectrum $N(E)_{\text{evap}}$ from silver bombarded by 190 MeV protons after Gross (line with dots and dashes) are shown for comparison.

In the third region (the fast neutrons) the flux measurements showed a rather smooth exponential decrease with energy. The exponential decay resembles the tail of the fission neutron distribution as measured at a distance from a small quantity of fissionable material, the fission of which occurs either spontaneously or by bombardment with a jet of neutrons from an external source (Bowman et al., Leachman). Note that here there is no equilibrium with the surroundings, so theat the neutron energy distribution function gives also the flux at a distance, taking into account the appropriate geometrical factors. The flux energy distribution of the neutrons used for the calculation of the average cross-section in this range (3.3) cannot be expressed by a simple formula and is taken from the graphical representation as shown in fig. III.21. For comparison the neutron spectrum arising from the thermal fission of ^{235}U according to Grundl is also shown in this figure.

Once these spectral flux distributions were defined, the averaged cross-sections of the nuclear reactions expected in the energy interval were calculated from the formula

$$\bar{\sigma} = \int_{E_{n1}}^{E_{n2}} \varphi(E)\sigma(E)\mathrm{d}E \tag{3.6}$$

E_{n1}, E_{n2} designating the limits of the neutron energy interval considered.

The nuclear reactions taken into account were the (n,γ) reaction in the thermal and epithermal ranges, and the (n,p) and (n,α) reactions in the fast neutrons range. This was done for each of the naturally occurring isotopes of the target element considered, and the corresponding averaged cross-sections were added in proportion of the natural abundances of the particular isotopes. The results for a number of selected structural materials can be found in a table presented in appendix D of Layman and Thornton's book, and were used by the present author to compute the gamma dose rate decay curves of these materials for an infinitely long irradiation time. These curves, which represent the gamma radiation field in rad/h at 1 metre from a quantity of 1 gram of the particular element irradiated by a flux of 1 neutron/sec cm^2 in the energy range indicated for an infinite time, are given in figs. III.22 to 24. For a finite irradiation time the reader will, as usual, subtract the activity values at the times t_i+t_c and t_c according to the well-known theorem. In cases where this procedure leads to inaccurate readings, especially for short activation times, the reader is referred to Layman and Thornton's book, which gives graphs showing the build-up of radioactivity with irradiation time from 10 to 1000 hours, at various cooling times (1, 10, 100, 1000 hours). The units used by these authors are roentgen/h lb at 1 foot distance. The equivalence in our units is

$$1 \text{ roentgen/h lb at 1 foot} = 2\times 10^{-4} \text{ rad/h g at 1 m} \tag{3.7}$$

the factor for converting from roentgen to rad(tissue) being 0.965, from pound to gram 1/454, from feet to metres distance $(0.305)^2$.

Also a table of the saturation values for each isotope formed in each neutron energy region is given in appendix C to enable the reader to compute himself the decay curve with accuracy in the case of irradiation times other than infinite. Only isotopes of half-life longer than 10 minutes were considered throughout all these calculations.

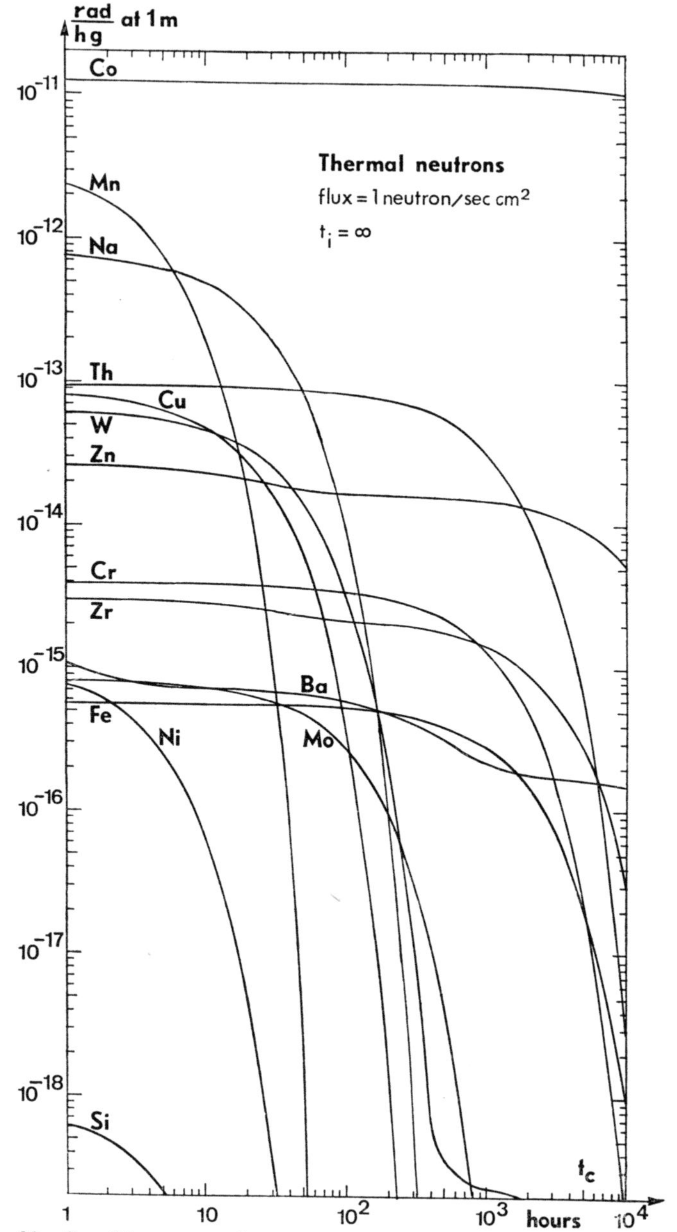

Fig. III.22 Specific gamma dose rate decay curves for selected target elements, thermal neutron activation, unit flux, irradiation to saturation.

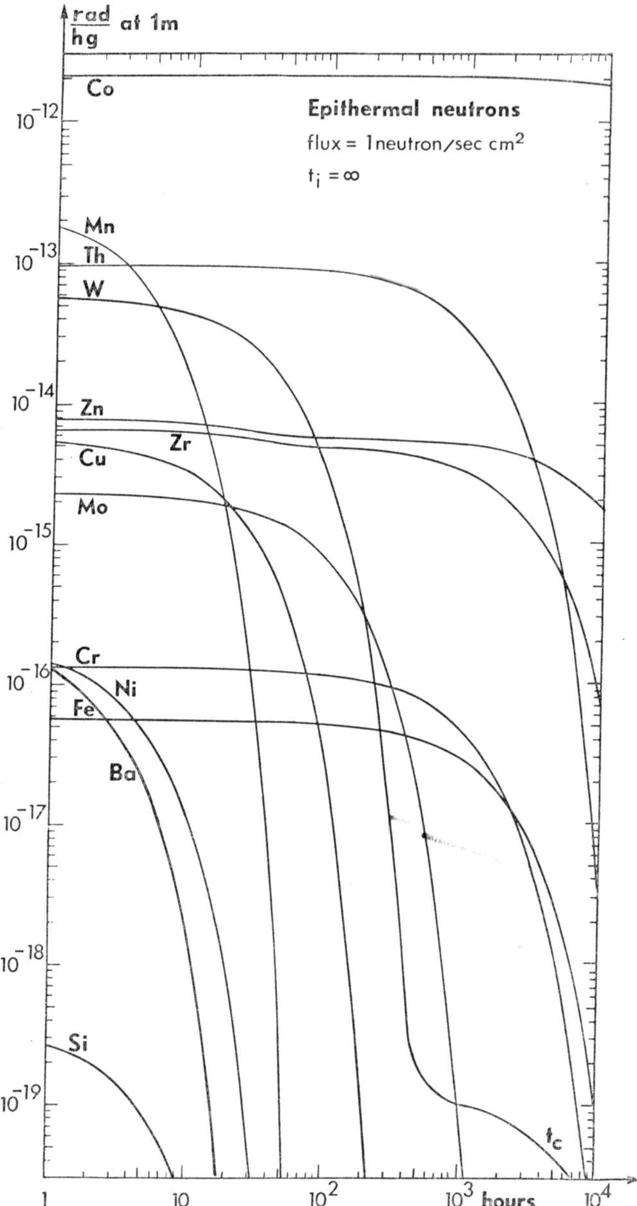

Fig. III.23 Specific gamma dose rate decay curves for selected target elements, epithermal neutron activation, unit flux, irradiation to saturation.

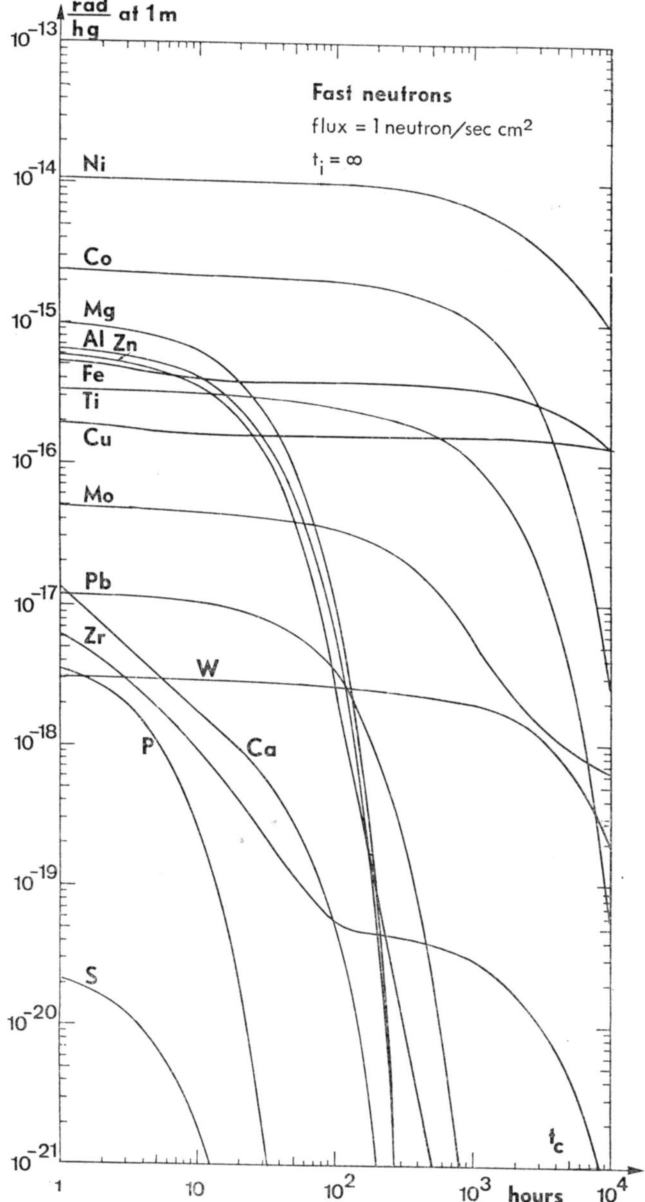

Fig. III.24 Specific gamma dose rate decay curves for selected target elements, fast neutron activation, unit flux, irradiation to saturation.

Inspection of the decay curves presented in figs. III.22 to 24 shows several features of interest.

First, in contrast to the decay curves found in ch. II for high energy spallation reactions, which lay nearly all in the same order of magnitude (cf. figs. II.9 to 12 and II.17), the activities induced by reactor neutron fluxes are considerably dependent on the target element. As an example, for thermal neutron activation, the saturation values at 1 metre lie between 10^{-11} rad/hg for cobalt to 10^{-18} rad/hg for silicon, so that 7 orders of magnitude are involved. The same is true for epithermal and fast neutron activation.

Another characteristic of the decay curves is their form, which is the same on log-log paper in nearly all the cases, one curve being easily deducible from another by horizontal and vertical translation. This is because there is usually only one radioisotope which is responsible for the decay. In this case a horizontal translation on a log-log plot means going over to another half-life and a vertical translation going over to another saturation activity. In some cases two or more radioisotopes are activated. This is seen immediately from the form of the decay curve which shows a step, or a departure from the usual form.

Let us examine now the specific gamma decay curves for each type of neutron activation separately. We begin with fig. III.22, which shows the thermal region. Most activated are the elements Na, Co, M, Th, Cu, W, Zn. The longest half-lives are found with Zn and Zr. Copper is much more activated than iron, but decays much faster, the activities of both being equivalent after 100 h. The iron activity has a half-life of about 1000 h. W, Zn and Zr give birth to two radioisotopes. Some materials are absent from this graph, such as Al, Mg, S, P, Ti, V, not because they have not been examined, but because they are not activated by thermal (and epithermal) neutrons producing the (n,γ) reaction.

Fig. III.23 shows the activities produced by the neutrons in the epithermal region, i.e. also via the (n,γ) reaction. The cross-sections being smaller than from those in the thermal region, the saturation activities found are usually lower, although the integration interval of neutron energies is much larger than in the preceding case. For Zn and Mo only one finds higher dose rates than in the preceding case. The form of each curve is normally the same as in the preceding case, as we have here the same nuclear reaction and as the radioisotopes produced for each target are the same as before. Some isotopes, however, may be missing owing to lack of data.

Fig. III.24 shows the activities induced by fast neutrons. The reactions relevant in this energy region are the (n, p) and (n, α) actions, which lead to other radiosotopes than those produced with the (n, γ) reaction. The cross-sections for these reactions are much smaller than for the (n, γ) reaction considered previously and as a result there is a net shift of the group of curves towards lower activities, although the integration interval of energies is still larger than before. Only Ni and Fe targets are found to yield higher gamma radiation fields than when activated by epithermal neutrons. Some elements activated by thermal and epithermal neutrons are found not to be activated any more by (n, p) and (n, α) reactions. These are Si, Cr, Mn, Sn. In contrast the elements quoted before as not being activated by (n, γ) reactions, i.e. Al, Mg, S, P, Ti, V are now showing up. Elements most activated by fast neutrons are Ni, Co, Mg, Al, Zn, Fe, Ti, Cu. Copper is now somewhat less activated than iron and aluminium but shows the longest half-life with production of ^{60}Co (5.27 years).

All these curves can serve as a guide when selecting structural materials for construction of reactors or accelerators, and be used to get a feeling of the radiation fields to be expected in given neutron fluxes.

An example will now be given of an evaporation neutron spectrum from a target bombarded by high energy protons to show that it is reasonable to apply as a first approximation the fast neutron data given above to accelerator parts subjected to evaporation neutrons arising besides the cascade of very high energy particles from targets or other parts struck directly by an accelerated beam. We have taken the data measured by Gross with silver atoms struck by 190 MeV protons, for which theoretical formulae have also been given by Mitler. Other experimental data have been given by Skyrme. The evaporation neutron spectrum $N(E)_{\mathrm{evap}}$ of Gross, which is the isotropic part of the total neutron spectrum emitted, the forward-peaked part being the cascade neutrons, is represented as a function of neutron energy in fig. III.21 (line with dots and dashes). It is seen to resemble the fission neutron spectrum, although it contains more high energy neutrons.

Examples of energy spectra of stray neutrons around accelerators can be found in the work of Lehman and Fekula, who have used emulsions around the betatron and synchrocyclotron on the Lawrence radiation laboratory site at Berkeley. All exhibit a roughly exponential decrease with energy between 0.5 and 12 MeV.

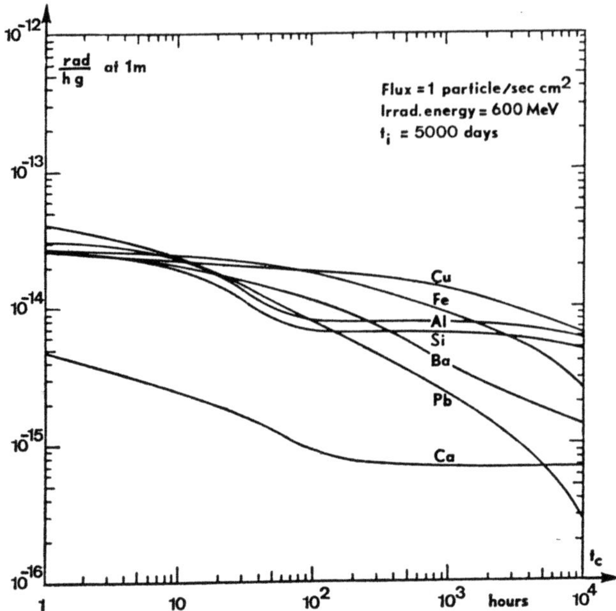

Fig. III.25 Specific gamma dose rate decay curves for selected target elements activated by very high energy particles inducing spallation (unit flux, irradiation to saturation).

In practice, so many factors can influence the neutron spectrum one is faced with, such as the kind of nuclear reaction by which it is generated, the amount and kind of shielding it has to traverse, the reflecting properties of the enclosure etc., that a thorough investigation of the exact form of the spectrum by the usual methods (photographic emulsions, threshold detectors, at higher energies proton recoil counter measurements) is recommended.

The question which the reader will naturally ask himself after having read this chapter on activation by reactor neutrons and the preceding chapter on spallation is which kind of flux activates more. To answer this we have plotted some results of ch. II in the same form as figs. III.22 to 24. The next fig. III.25 shows the radiation field in rad/hg at 1 m to be expected after infinite irradiation by a unit flux of 600 MeV protons. Decay curves are presented for the elements Al, Si, Ca, Fe, Cu, Ba, Pb as examples. This permits a direct comparison with the radiation fields due to thermal, epithermal and fast neutrons for the elements mentioned. One sees that with very high energy particles all curves lie rather near each other, except calcium. The radiation field obtained with very high energy particles are of the order of 10^{-14} rad/hg at 1 m. This lies right in the middle of the activation values expected with thermal and epithermal neutrons, and somewhat above those expected with fast neutrons. The reader can now make the comparison of the curves of the individual elements for himself.

PROBLEMS

CHAPTER III

1. What is the mass of ^{235}U which has to undergo fission in a bomb in order to have an equivalent explosive force of a) 20 kilotons TNT, b) 1 megaton TNT?

2. What is the quantity of ^{235}U consumed per year in a plant of 1000 MW thermal power? To how many megaton bombs is this equivalent (assume all ^{235}U in the bomb is actually burnt up in the explosion)?

3. What is the saturation activity in curies of the barium isotopes ^{139}Ba, ^{140}Ba, ^{141}Ba produced per MW power in a reactor burning ^{235}U? Take 582 barn as fission cross-section for ^{235}U and 30, 190, 450 mb as independent production cross-sections for the three barium isotopes mentioned. To which barium isotopes weights does this amount? The half-lives of the three elements are 83 m, 12.8 d, 18 m.

4. What is the saturation activity of all the fission products in curies per megawatt of reactor power in a reactor that has been running for a long period, assuming that each fission gives rise to two products, each of which produces on the average three daughter products, before stability is reached?

5. What is the total weight and the total activity in curies of the fission products one minute after a bomb explosion of a) 20 kilotons TNT, b) 1 megaton TNT?

6. A reactor delivering 1000 MW thermal power operates with a flux of 10^{14} thermal neutrons per cm^2 per sec. What is the total weight of uranium it needs (assume ^{235}U thermal fission only and fuel enriched to 20% in the isotope ^{235}U)? What time does it take to burn up a quantity of fuel equal to that just calculated at this power level (use answer of problem 2)?

7. Consider a rod containing 1 kg of ^{235}U. What is the total decay power 1 h after shut-down, if it has been fissioned with the fission rate calculated in problem 6? What is the gamma power at the same time? What is the gamma radiation field in rad/h at 10 m distance from this rod considered as a point source?

8. A one megaton bomb has been exploded. Express the energy released in MWD. Assume that all fission products are evenly distributed in a sphere with a radius of 5 km, 10 minutes after the burst. Calculate the gamma radiation field existing at the centre of this sphere at this time.

9. What is the radiation field at 1 m from a piece of 1 cm^3 of copper when it is taken out of a reactor, where it has been irradiated for a time of several hundred hours? Assume thermal neutrons alone, with a flux of 10^{13} neutrons/sec cm^2. The same question is asked for a piece of 1 cm^3 of iron. The densities of copper and iron are 8.9 and 7.8.

10. What is the radiation field at 1 m from a piece of 1 cm^3 of cobalt when it is taken out of a reactor where it has been irradiated for 300 h? Assume a thermal neutron flux of 10^{13} neutrons/sec cm^2. What is the number of curies produced? The density of cobalt is 8.8, the *k*-factor for ^{60}Co is 1.26 rad/hCi at 1 m and the half-life of this isotope 5.26 years.

11. Take 1 kg blocks of aluminium, iron, copper and lead and expose them for several years to equal fluxes (10^7 cm^{-2} sec^{-1}) of thermal, epithermal, fast, and high energy (600 MeV) neutrons. These kinds of fluxes are often found in equal numbers around accelerators. Calculate the radiation field at 1 m from the block considered as a point source (neglect self-absorption) for each kind of flux and material, and zero cooling time.

CHAPTER IV

COMPOUND NUCLEUS REACTIONS AND THRESHOLD DETECTORS

Introductory remarks

Some introductory remarks to this chapter will help the understanding of its scope and structure. In chs. II and III of this book, we have examined the rather clear-cut questions of the radioactivity induced by spallation at very high energies and by neutron fluxes at reactor energies (evaporation or thermal neutrons). Charged particles do not penetrate the Coulomb barrier of the nucleus at these low energies. In contrast, at the high energies where spallation takes place, the charge difference between protons and neutrons, for instance, is not of importance any more.

Between those two 'extreme' cases, there lies a wide energy range for the kinetic energy of the incident particle, extending from a few MeV to a few tens of MeV, say even to 100 MeV, where other phenomena than either neutron capture or complete destruction of the target nucleus by spallation will occur. As we have said above, the neutral particles are the first to enter the nucleus, due to the absence for them of the Coulomb barrier, which usually repels charged particles. As the energy increases, the excitation energy of the nucleus which has absorbed the neutron will increase and it will soon be in a position to expel (or evaporate) one or more neutral particles. In the same energy range and as the energy increases still further, one or more charged particles can be expelled in their turn, so that a new evaporation mechanism (or channel) appears, which begins to compete with the preceding one. Concurrently, charged and uncharged particles can be evaporated together, either independently by emission of a neutron plus a proton, for instance, or bound, as is the case when a deuteron or an alpha particle is emitted.

All these reactions, which are characterized first by the absorption of the incoming particle by the target nucleus, and second by the subsequent emission of one or several particles by the same nucleus in an excited state, are known under the generic name of 'compound nucleus' reactions, a term introduced in 1936 by Niels Bohr. For ease of writing, each of these reactions is labelled by an expression in brackets, where the incoming particle appears first, and the outgoing one or ones second. Thus one speaks of a (n, p) reaction or a (p, pn) reaction when one is concerned with an incoming neutron and an outgoing proton in the first case or with an incoming proton and with a proton and a neutron coming out, either free or bound together, in the second case.

As the kinetic energy of the incoming particle is increased further, still more particles can leave the nucleus after the impact. Thus one goes gradually into the spallation region and there is a vast energy region where rather 'simple' compound nucleus reactions, involving the evaporation of only one or two particles or of a deuteron or an alpha particle, are possible at the same time as more involved 'spallation' reactions, with the expulsion of many more nucleons or heavier fragments of the nucleus. This region can be considered to extend from a few tens to a few hundreds of MeV kinetic energy of the incident particle. In this energy range, and when the target nucleus is very heavy, fission can also take place as an alternative to spallation.

It is to this whole realm of closely related phenomena that the present chapter will be devoted. It will, however, be clear to the reader that a complete presentation of all the possible cases, with their relative probability as a function of energy is beyond the scope of this presentation. Also complete knowledge for all nuclei does not exist yet. It is to be hoped that with the appearance of a generation of new accelerators with higher intensities than before in the energy region mentioned (the so-called isochronous cyclotrons) the existing data will be somewhat improved and made more complete. We will thus have to offer only a small compilation of the more simple cases of compound nucleus reactions, which is considered as an example to illustrate the orders of magnitude of the effects to be expected. This will be done for 4 types of incoming particles: protons, neutrons, deuterons and alpha particles.

After having thus given in a first section a review of the simplest among the compound nucleus reactions, a second section will deal with the beginning of the spallation reactions in the same energy range. Some examples, limited to protons, will be given of the excitation functions for the production of several radionuclides in a few structural materials in the energy range from threshold to 50 or 100 MeV.

The two next sections of the chapter deal with monitor reactions and threshold detectors. Both the compound nucleus reactions and the initial spallation reactions can be used, and some are to a large extent, as threshold detectors and monitor reactions for the detection and measurement of particle fluxes. It is thus appropriate to present here the accurately measured excitation functions of the more widely used detectors responding to a number of elementary particles, over the whole energy range from threshold to several tens of GeV. This is done in section 3.

A fourth section is, for the sake of completeness, dedicated to the particular neutron threshold detectors that are convenient for neutron spectra measurements in reactors and for similar work with evaporation neutrons from nuclei bombarded by high energy particles.

Finally section 5 deals with the fission of heavy nuclei by protons, which begins in the energy range considered here.

1 Excitation functions for simple reactions induced by protons, neutrons, deuterons and alpha particles

What is of interest here in the compound nucleus reactions from the point of view of induced radioactivity is the total cross-section for production of the product considered from the target nucleus. Differential cross-sections, angular dependences, branching ratios for going to excited states of the same product nucleus, are not really relevant to our point and will be omitted. In contrast, the dependence of the cross-sections on the kinetic energy of the incoming particle, and especially the value of the cross-sections in the region where they are greatest, are of special interest, as they show in what energy range and to what extent the particular product is likely to be formed. This information should be given for the largest possible number of target nuclei. Thus, for each type of reaction, we would come to a set of curves giving the cross-section σ of the reaction considered as a function of the particle energy, each curve applying to one particular target nucleus. This amount of knowledge, if it existed for each nucleus and particular reaction, would constitute in itself a large reference work of several volumes. However, it exists so far only for a very small number of reactions and particularly of target nuclei. One reason is the amount of work involved in making the cross-section measurements. Another reason is the limited number of accelerators and their limited intensity. Still another reason is that many products have too long or too short half-lives or are not radioactive and hence difficult to measure directly. As a consequence, up to now, we have found only a few excitation functions for only a few reactions and had to be content with that. So we will present here data on the following types of reactions only:

(p,n), (p,2n), (p,pn), (p,α)
(n,p), (n,2n), (n,α)
(d,n), (d,2n), (d,p), (d,α)
(α,n), (α,2n), (α,p), (α,pn).

Furthermore, we have had to abandon the idea of giving the excitation curves for each reaction and target nucleus for which data are available. The reason is that such a presentation does not give a clear picture and does not lend itself well to interpolation or extrapolation. The numerical values lie widely scattered due to differences between authors, and also intrinsically because nuclei with neighbouring charge or mass numbers may have very different responses to the reaction, especially if one takes isotopes lying away from the stability line.

Attempts to bring some order into the data, in order to make predictions possible for target nuclei other than those measured or simply to try to find some general behaviour, have been made by various authors, at least for some types of reactions or at certain energies. Chatterjee studied shell effects in 14 MeV (n,p) reactions. Grover and Caretto made a compilation of results on (nucleon, two-nucleon)-reactions above 100 MeV and interpreted some theoretically. Gardner gave an analytical expression for the cross-section of (n,p) reactions induced by 14 MeV neutrons, which shows well the behaviour of this cross-section when the mass number A of the target is varied with the atomic number Z remaining constant.

In the present work we are faced with the need to give the maximum possible information in the minimum number of pages and we shall follow a suggestion made by G. Rudstam, according to which one should plot the cross-section values as a function both of the particle energy and the mass number of the target nucleus. In this two-dimensional reference system, points with equal cross-sections should be joined by lines and a relief chart of the cross-section constructed resembling very much a geographical map with contours of equal altitude. Because of the large scattering of the results already mentioned, due to variations of the cross-section with the particular target isotope, it is clear that a single-valued cross-section function $\sigma(E, A_T)$ of both E, the energy of the incoming particle, and the mass number A of the target (or even its charge number Z), simply does not exist physically. However, such plots, with the original data limited to isotopes near the stability line, and with smoothing out of the differences between neighbouring target nuclei, can be useful for rough estimates of the order of magnitude of a required cross-section, and of the energy region where it can have its maximum. Also plotting of all the data in this manner is instructive in eliminating erroneous results and in making it possible to smooth out the information in the way best suited to subsequent interpolations; although, and we would like to stress it again, a smooth single-valued $\sigma(E, A_T)$ or $\sigma(E, Z_T)$ function does not exist physically. The purpose here is only to give a rough idea of a smoothed mean value or of a superior boundary value for the phenomena which could be expected.

The σ=const. curves were drawn on the basis of at least 10 measured excitation functions, and even 20 for many reactions, as well as of many isolated cross-section values. The threshold of the reaction is usually well known, as well as the maximum of the cross-section The positions of these maxima which lie on an apparently smooth line have remarkably little scatter. Little is known on neutron cross-sections above 20 MeV. For very light nuclei ($A<20$) it can happen that either little is known or the values scatter very much. In the cases where the evidence was poor, arbitrary prolongation of the σ=const. lines into the range of E or A values in question was, of course, omitted. For deuteron and alpha-particle induced reactions, less data are available than for proton and neutron reactions, and large (E, A) regions have been left blank on the graphs. It is hoped that the charts presented, though very imperfect, will help in the making of predictions and also show the work which remains to be done. We shall now proceed towards examining them individually.

We commence with the proton reactions (p, n), (p, 2n), (p, pn) and (p, α) for the cross-sections of which smoothed approximations are given in figs. IV.1 to 4. For comparison the total inelastic proton cross-

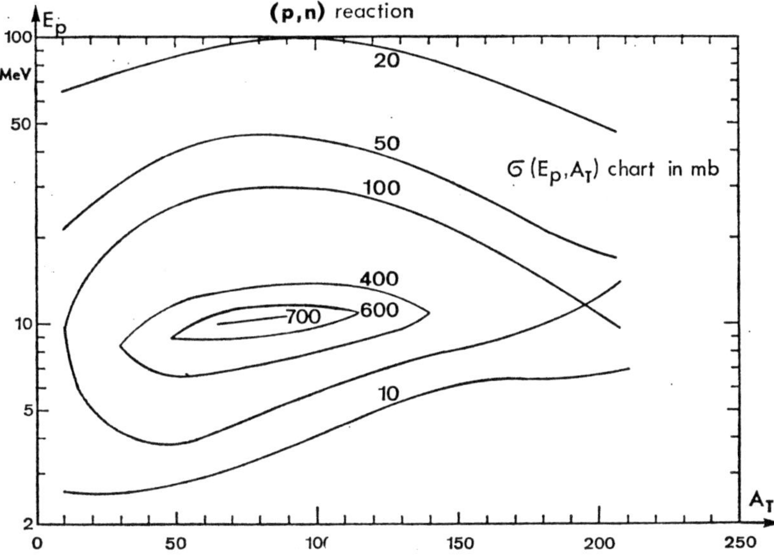

Fig. IV.1 Smoothed approximation of (p, n) reaction cross-section.

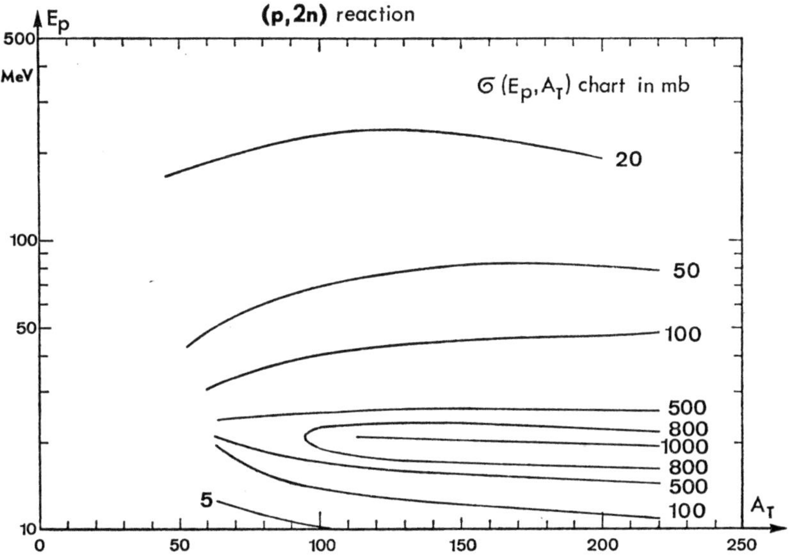

Fig. IV.2 Smoothed approximation of (p, 2n) reaction cross-section.

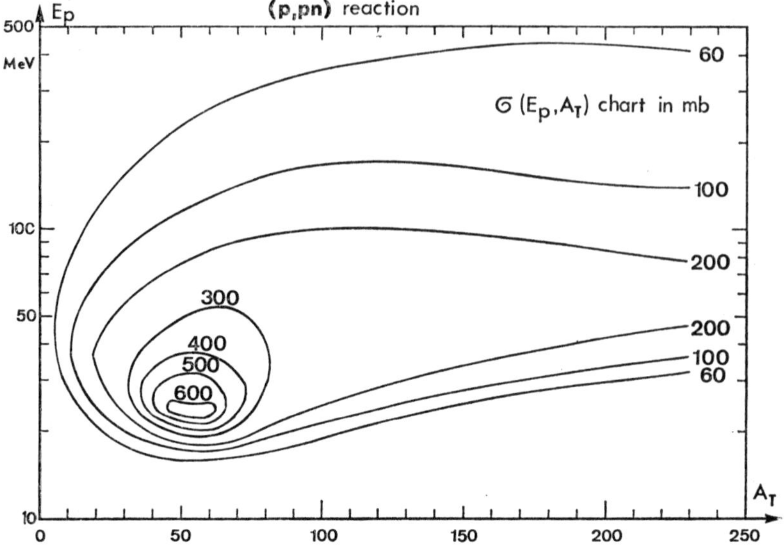

Fig. IV.3 Smoothed approximation of (p, pn) reaction cross-section.

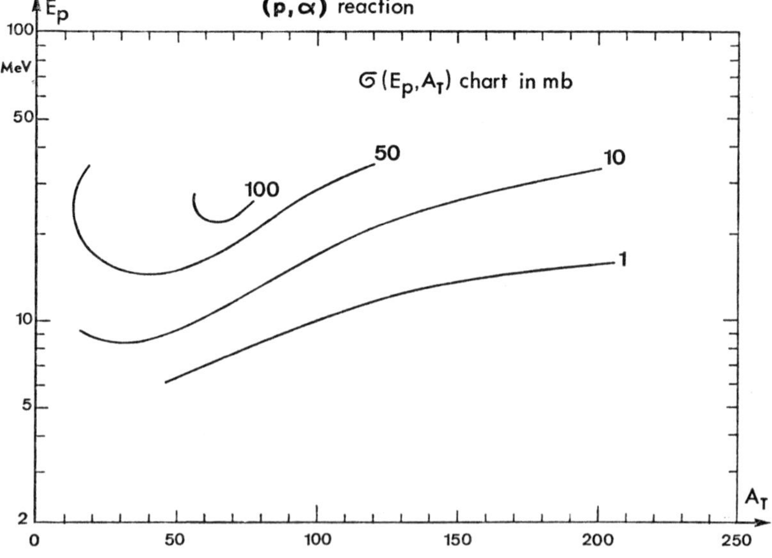

Fig. IV.4 Smoothed approximation of (p, α) reaction cross-section.

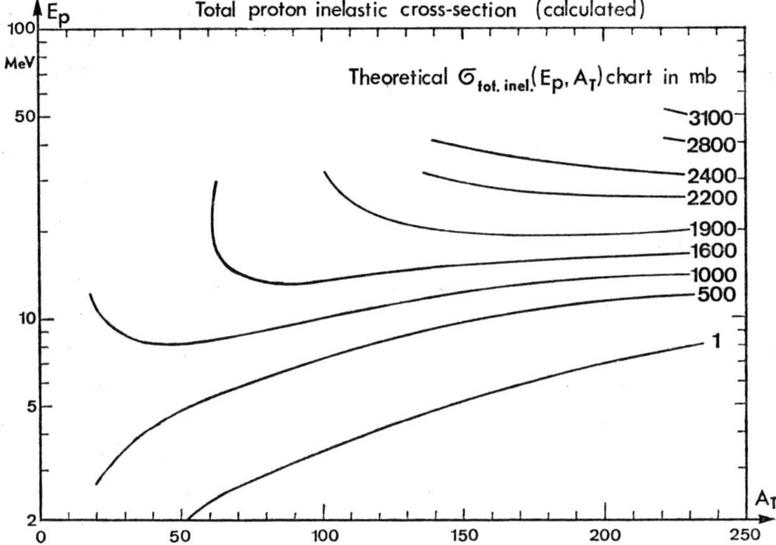

Fig. IV.5 Theoretically calculated total proton inelastic cross-section.

section, which can be calculated theoretically, is given in fig. IV.5.

As a function of E and A, the cross-section of the (p, n) reaction, which is one of the best studied reactions, has the general aspect of an elongated hill, the crest of which lies about parellel to the A-axis, corresponding to an energy value of about 10 MeV for maximum cross-section. On the low energy side the slope is steep, on the high energy side it descends more gradually. Maximum peak values of the order of 700 mb are attained in the mass number range of 50 to 100, which corresponds almost exactly to the total proton inelastic cross-section which is calculated at 10 MeV in this mass number range (see fig. IV.5). The total inelastic cross-section represents, of course, the upper boundary of the sum total of the cross-sections of all the proton reactions. Thus, in these mass number limits and at this energy, the (p, n) reaction seems to be the dominant one. The total inelastic proton cross-section values have been taken from P. Lefort's work 'La chimie nucléaire'. The values for $A < 50$ are only approximate. If we now go back to fig. IV.1 we can observe further that the peak values along the crest of the hill fall towards higher A. This is obviously because of the increasing potential barrier which prevents the proton from reaching the nucleus at this energy and reduces the cross-section. More energy to increase the cross-section at high target mass numbers will not be useful, as the (p, 2n) reaction begins to show up, due to the higher excitation energy of the nucleus, and to compete with the (p, n) reaction, the cross-section of which then begins to decrease.

The next fig. IV.2 shows the smooth approximation to the (p, 2n) reaction cross-section. It has the same overall appearance as the preceding one. However, the crest lies now at 20 MeV proton energy because one needs more excitation energy by the nucleus to get rid of two neutrons. The peak values on the crest tend this time to increase slowly with target mass number and approach or exceed the 1000 millibarn value above $A=100$, i.e. a large amount of the total inelastic cross-section at 20 MeV (which lies around 1600 mb and increases as $A^{\frac{2}{3}}$). This shows that the larger part of the interaction goes via the (p, 2n) channel at this energy above $A=100$.

Fig. IV.3 shows the smooth approximation to the (p, pn) reaction, an extensively studied nuclear reaction. The general features resemble those of the (p, n) reaction. The peak cross-section values lie between 20 and 50 MeV, i.e. at higher energies than in the (p, n) case because it needs energy to evaporate the supplementary proton. Also the peak values, which are a maximum in the range $40<A<60$, decrease towards higher E and A as there must be increasing competition with the (p, 2n) process.

Last, fig. IV.4 gives some data on (p, α) reactions. As a large excitation energy is needed in the nucleus to evaporate the doubly-charged alpha particle, the crest of the hill is pushed towards still higher energies (30 MeV) and the decrease of the peak values with increasing A is still more apparent, because of the competition of other processes.

The next graphs are devoted to neutron-induced reactions of the (n, p), (n, 2n) and (n, α) types. The theoretically calculated total neutron inelastic cross-sections according to a private communication by Lefort are given also for comparison in the energy range 0.2 to 10 MeV. The values are only approximate.

The (n, p) cross-section, shown in fig. IV.6, appears not to be very large, having a maximum zone of about 150 millibarns at 13 MeV in the range $30<A<40$. It falls when one goes to higher target mass numbers because of the increasing potential barrier opposing the evaporation of a proton.

In contrast, the (n, 2n) reaction (fig. IV.7) attains very high values, peaking at 1700 mb in the range $A>100$ at about 15 MeV neutron

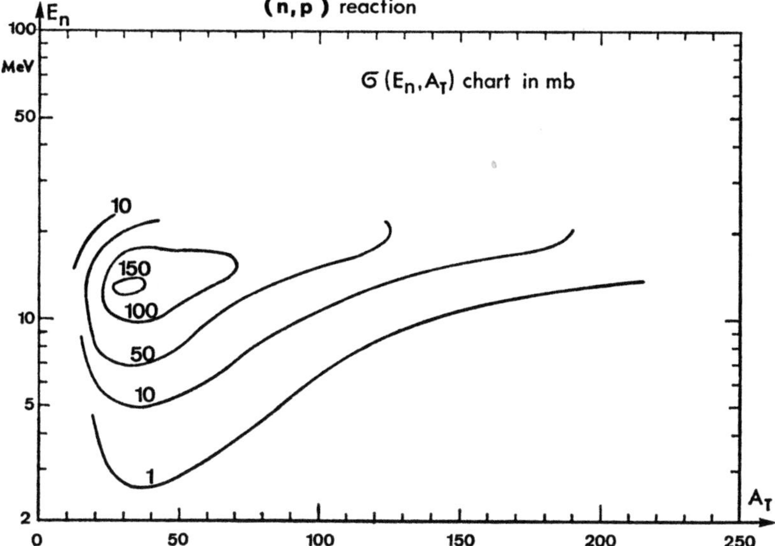

Fig. IV.6 Smoothed approximation of (n, p) cross-section.

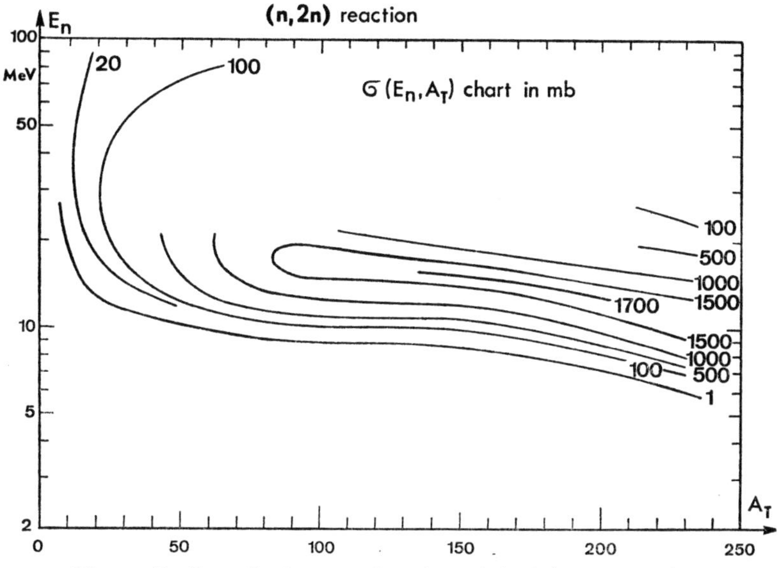

Fig. IV.7 Smoothed approximation of (n, 2n) cross-section.

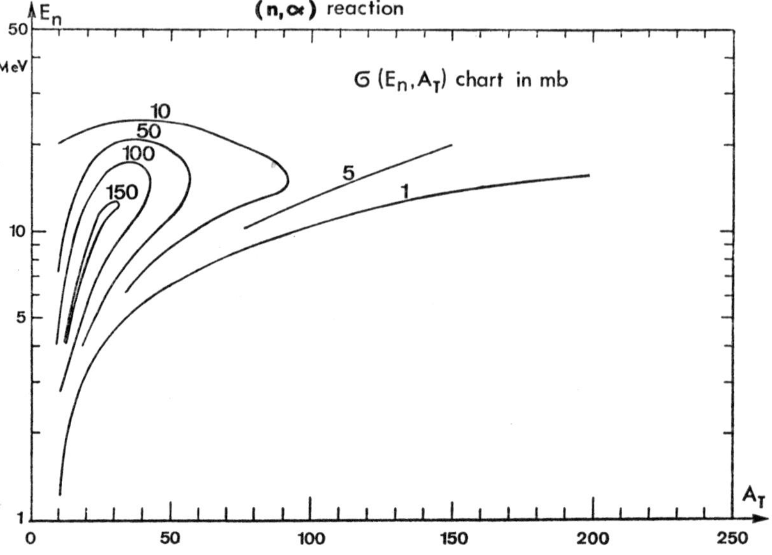

Fig. IV.8 Smoothed approximation of (n, α) cross-section.

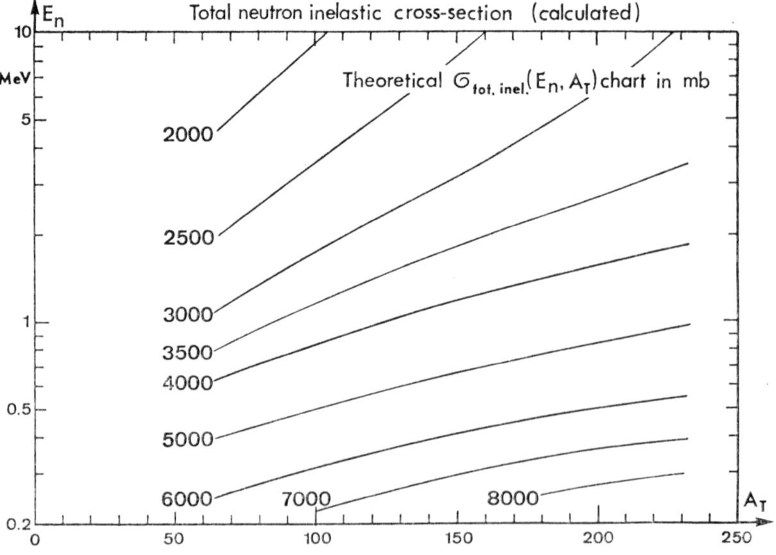

Fig. IV.9 Theoretically calculated total neutron inelastic cross-section.

energy. At this energy the cross-section values are seen to increase gradually with mass number, much like the geometrical cross-section, which increases as $A^{\frac{2}{3}}$. The similarity of the general aspect with the (p, 2n) reaction chart presented above in fig. IV.2 is evident.

Finally the (n, α) reaction chart is presented in fig. IV.8. The cross-sections again are small, with maxima of 150 millibarns for low mass numbers $A<25$. The decrease towards high mass numbers is very pronounced. Again some similarity of behaviour exists with the (p, α) reaction of fig. IV.4.

Because the interest of this presentation was pointed out to the author by several colleagues and in order to facilitate the rapidly expanding work on activation analysis with charged particles (see the paper by R. S. Tilbury), the compilation was extended to some deuteron and alpha particle reactions. As these reactions have been studied to a lesser extent than proton and neutron reactions, we will present in the following what are only preliminary and coarse cross-section charts relevant to these particles.

The first deuteron reaction, the (d, n) reaction (fig. IV.10), has a cross-section configuration somewhat similar in shape to the (p, n) reaction, with the peak values at the same energy around 10 MeV and a decrease towards higher mass numbers. However, these values are much smaller with deuterons (150 vs. 700 millibarn at most).

The (d, 2n) reaction (fig. IV.11) resembles even closer the (p, n) reaction. The crest of the rise has the same value (around 600 mb), lies however in the same mass range ($50<A<150$) at 13 instead of 10 MeV energy of the incoming particle i.e. somewhat higher.

The (d, p) reaction (fig. IV.12) has an appreciable cross-section (up to 500 millibarns) in the 5 to 10 MeV kinetic energy region up to $A=80$. It is the important deuteron reaction besides the (d, 2n) reaction.

The (d, α) reaction (fig. IV.13) is more similar in cross-section shape to the (n, α) than to the (p, α) reaction and has the same order of magnitude (150 mb in the 10 MeV region for light nuclei with $A<30$).

We now come to study the compound nucleus reactions with alpha particles and will present some results on (α, n), (α, 2n), (α, p), (α, pn) reactions, together with the total alpha inelastic cross-section, as calculated by Huizenga and Igo.

The alpha particle data are interesting in themselves, and also because they lend themselves to comparison with proton data. This is because, if the same compound nucleus is formed in two different ways,

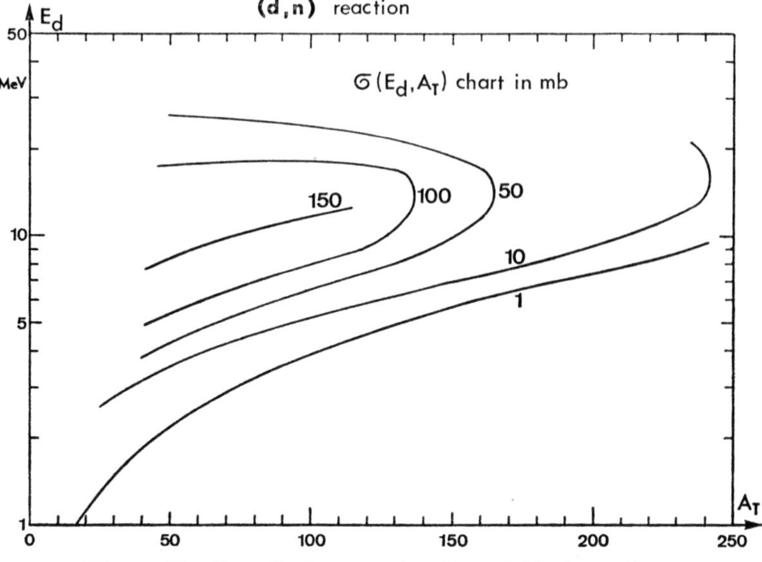

Fig. IV.10 Smoothed approximation of (d, n) reaction.

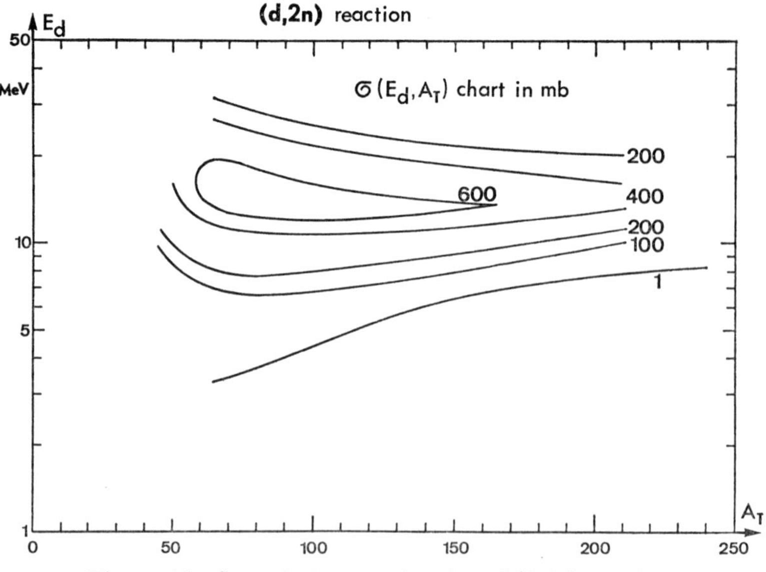

Fig. IV.11 Smoothed approximation of (d, 2n) reaction.

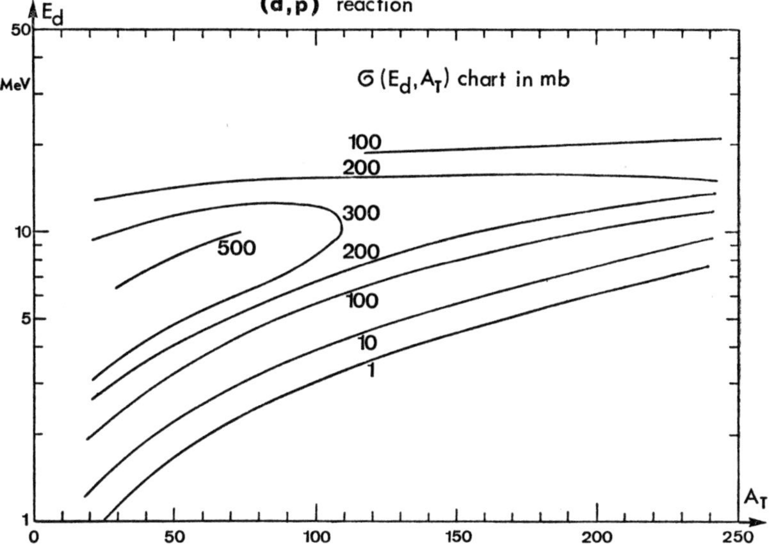

Fig. IV.12 Smoothed approximation of (d, p) reaction.

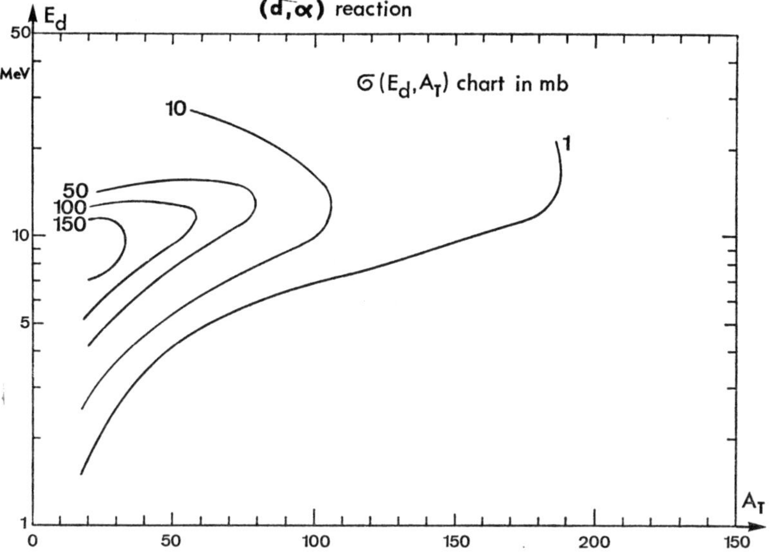

Fig. IV.13 Smoothed approximation of (d, α) reaction.

i.e. by proton bombardment of a given nucleus and by alpha particle bombardment of another nucleus, the decay of the two identical compound nuclei is the same, irrespective of the way they have been formed, if they have been provided with the same excitation energy. Goshal bombarded ^{60}Ni with alpha particles and ^{63}Cu with protons. He could show that the excitation curves for the (α,n) and (p,n) reactions, for the $(\alpha,2n)$ and $(p,2n)$ reactions and also for the (α,pn) and (p,pn) reactions were the same, except a shift in the kinetic energy of the bombarding particle by an amount of about 7 MeV, the binding energy of the proton in the compound nucleus being larger by this amount than that of the alpha particle. Or in other words, the energy difference corresponds to the difference between the masses of the ^{63}Cu+^{1}H and ^{60}Ni+^{4}He compound nuclei.

Fig. IV.14 shows the smoothed approximation for the (α,n) reaction. This chart and the following ones were, of course, drawn quite independently of the charts for the proton reactions. The peak cross-section values are around 500 mb in the $50<A<80$ range, the overall shape is similar, except the descending slope towards higher energies is much steeper. This fact indicates that one cannot continue for any length of time expelling only one neutron, and that other reactions soon become dominant as one increases the kinetic energy of the incoming alpha particle.

Fig. IV.15 shows the $(\alpha,2n)$ cross-section chart. The similarity with the $(p,2n)$ chart is striking, with peak values around 1000 mb and steep slopes in both cases. As one goes towards higher mass numbers the position of the peak is shifted somewhat to higher energy values with alpha particles, whereas it was shifted to lower energy values with protons. This is probably due to the double charge of the alpha particle which makes the penetration of the Coulomb barrier more difficult at higher mass numbers.

The next reaction is the (α,p) reaction shown in fig. IV.16. Logically it could be compared with the (p,α) reaction. In fact it resembles this reaction topologically (see fig. IV.4) with a peak between $A=50$ and $A=100$. However, the maximum cross-section value with alpha particles (500 mb) is much larger than with protons (100 mb) as it is easier to expel a proton than an alpha particle from the excited nucleus.

Last comes the (α,pn) reaction (fig. IV.17) on which data are scarce. A peak similar to the (p,pn) reaction one is, however, observed in the $50<A<70$ range, with the usual 7 MeV shift in energy.

The calculated total alpha inelastic cross-section is finally given as a

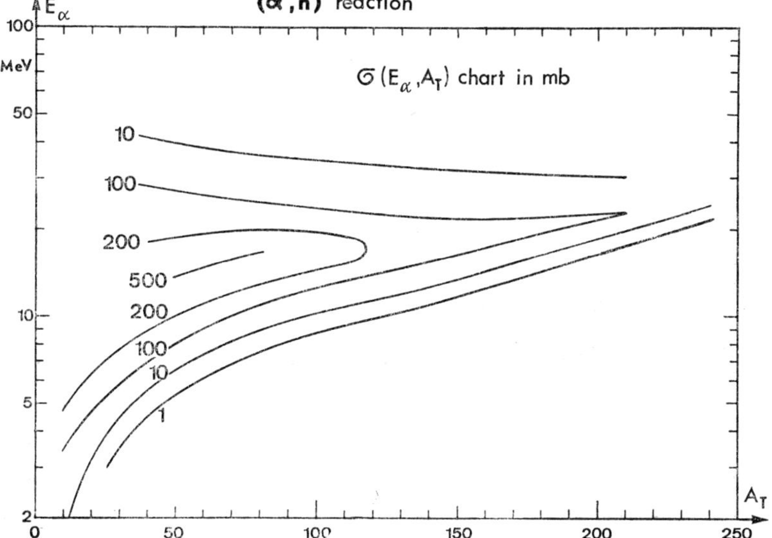

Fig. IV.14 Smoothed approximation of (α, n) cross-section.

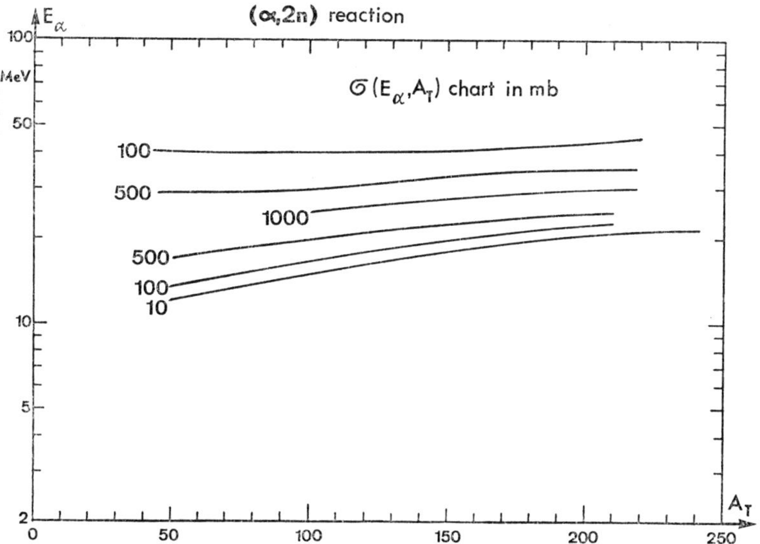

Fig. IV.15 Smoothed approximation of (α, 2n) cross-section.

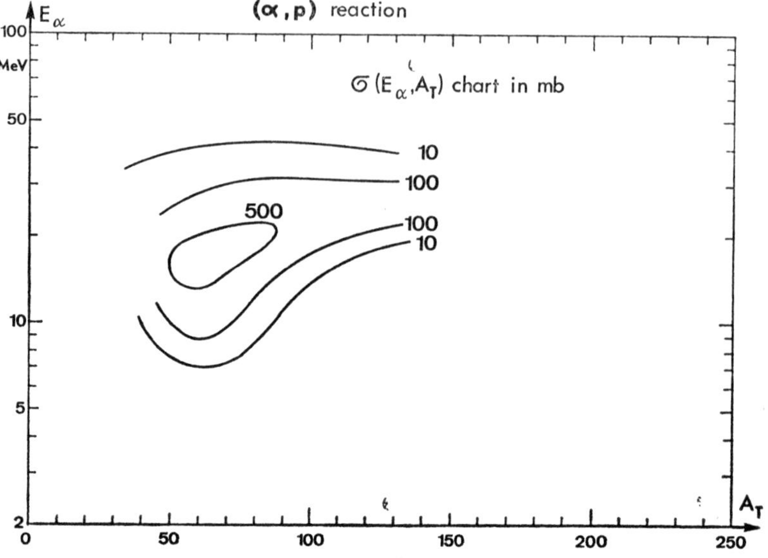

Fig. IV.16 Smoothed approximation of (α, p) cross-section.

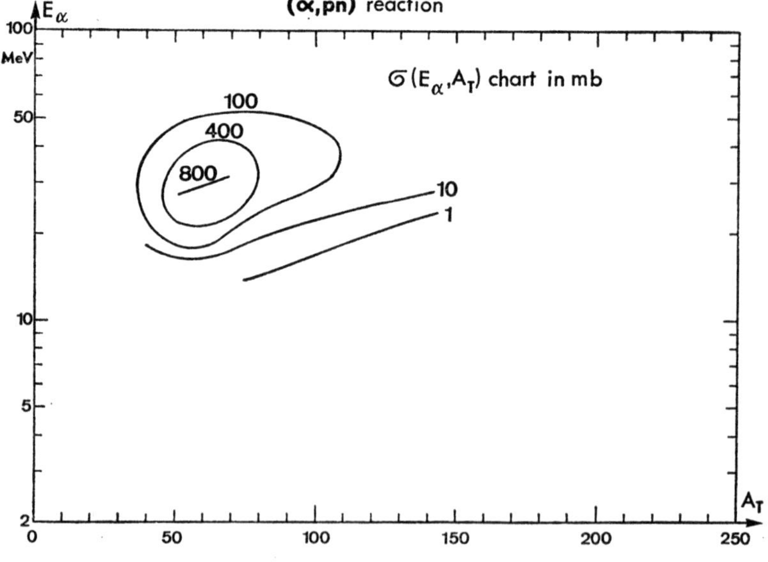

Fig. IV.17 Smoothed approximation of (α, pn) cross-section.

reference (fig. IV.18). One will notice that, when shifted down in energy, it resembles closely the total proton inelastic cross-section (fig. IV.5).

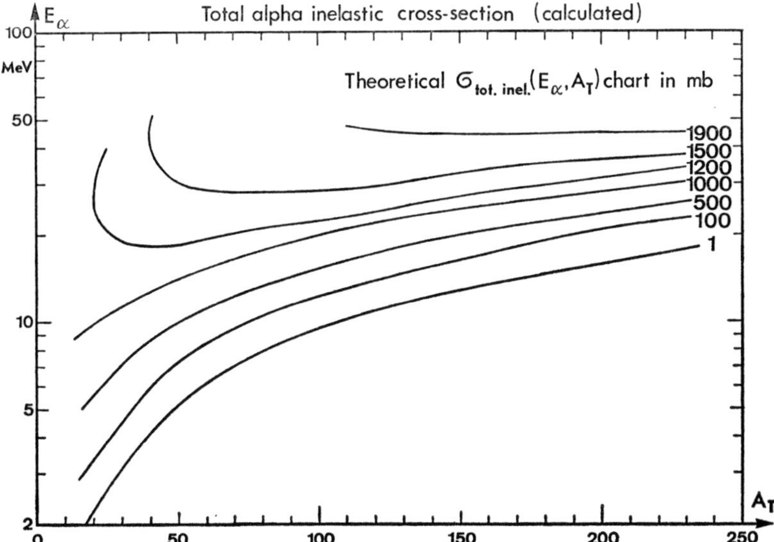

Fig. IV.18 Theoretically calculated total alpha particle inelastic cross-section.

The review which has just been made should not convey to the reader the impression that he is in possession of the cross-sections for the reactions listed and in the energy and mass number ranges for which information is available from the graphs. These were prepared from rather incomplete data only to give an impression, and possibly some approximative numerical values, to serve as a guide in further work. One should remember that such a thing as a cross-section surface over kinetic energy and atomic number does not exist. To present this smoothed approximation, factors as large as 2 or 3 in the peak region and as large as 10 elsewhere had to be neglected. Also cross-sections for target elements far from the stability line can be vastly different from the smoothed approximation which was drawn on consideration of target isotopes in the immediate neighbourhood of this line only. The only real answer to the seeker of information is the direct compilation of excitation functions for all nuclides and projectiles known up to date, such as that which the charged cross-section data centre in Oak Ridge is beginning to edit (see literature under McGowan, Milner, Kim).

2 Excitation functions for the production of a few radionuclides in C, N, O, Al, Fe, Cu, Ag, Au by proton bombardment up to the levelling off

In the preceding section, a number of 'simple' reactions of the types (p,xpyn), (n,xpyn) etc. with $x+y \leq 4$ have been discussed with the help of charts giving the smoothed approximation of the cross-section against energy and atomic number.

As the bombarding energy of the incident particle increases above the threshold of these simple reactions, yet more particles of the compound nucleus get away due to its higher excitation energy. The nucleus remaining after full de-excitation has taken place is thus more and more distant from the original one in terms of charge and mass numbers. Nuclides further and further away from the target begin to be formed. The excitation functions (cross-section versus energy of the incident particle) thus rise from a certain threshold. It is found that after some energy interval, perhaps after having gone through a maximum, the cross-sections arrive at some level where they are practically constant, having only a slight variation with energy. It is in this region that Rudstam's formula can be considered to apply rather well. However, there is up to now no good theory to describe quantitatively the fast rise of the cross-section of nuclear reactions of the spallation type above threshold. One is thus left with the experimental investigations alone. Unfortunately, these are at present rather scarce in this energy region (20 to 100 MeV) but it must be hoped that with the appearance of high-intensity sector-focused cyclotrons just in the right energy region the necessary cross-section work will advance.

For the time being we shall have to be content to present a few well-studied examples of the phenomena described to give the reader an impression of the evolution of the cross-section with target number and energy.

One of the reactions of this type which has been somewhat investigated and is moreover not covered by Rudstam's formula is the production of ^{7}Be in various nuclei ranging from carbon to gold. It is shown in fig. IV.19 where one can see how the thresholds rise with increasing atomic number of the target. It is also apparent that at very high energies the cross-section for production of ^{7}Be is not very much dependent on target mass number.

Audouze, Epherre and Reeves have reviewed the experimental

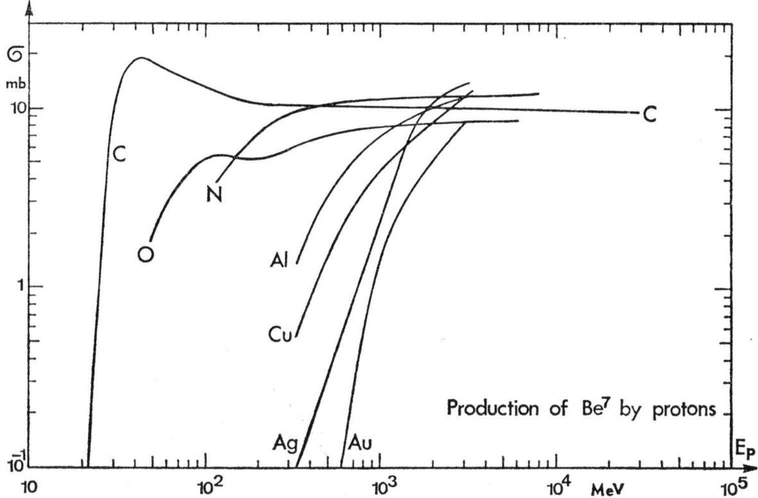

Fig. IV.19 Excitation functions for the production of ^7Be in C, N, O, Al, Cu, Ag, Au by protons.

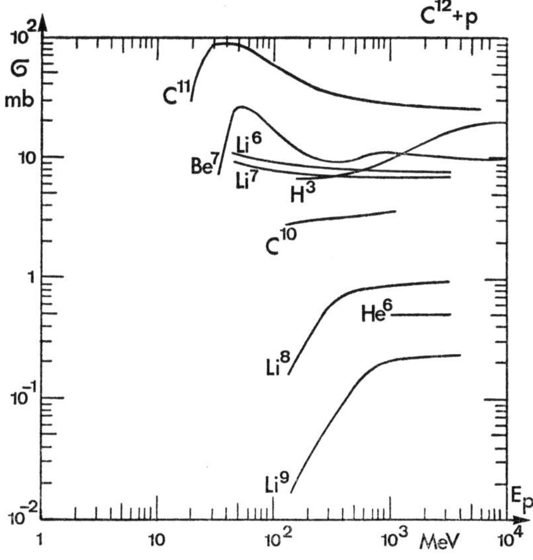

Fig. IV.20 Production cross-section of various isotopes in carbon by proton bombardment.

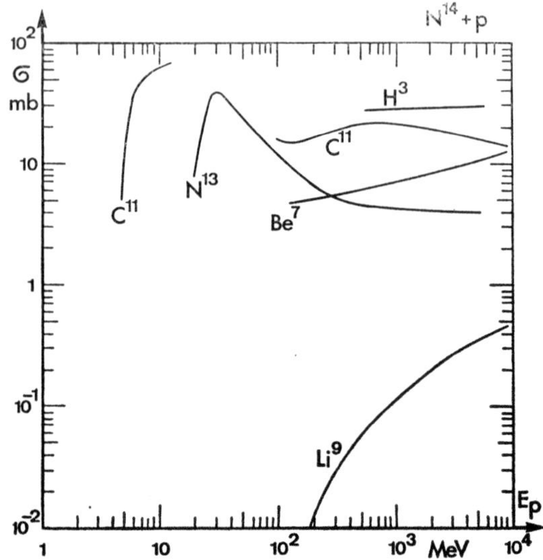

Fig. IV.21 Production cross-section of various isotopes in nitrogen by proton bombardment.

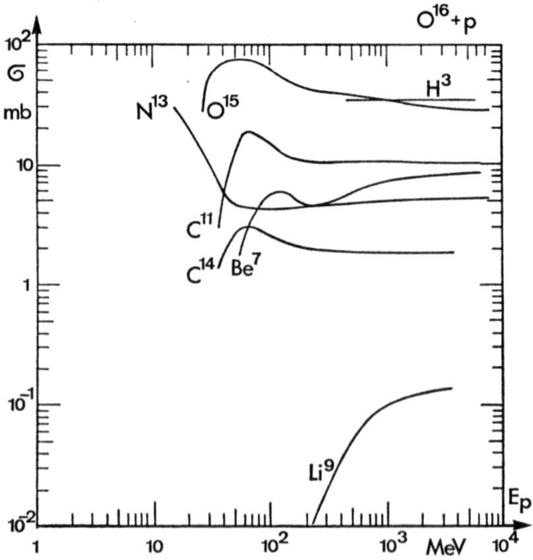

Fig. IV.22 Production cross-section of various isotopes in oxygen by proton bombardment.

results published on the formation of spallation products by proton bombardment of the light elements ^{4}He, ^{12}C, ^{14}N, ^{16}O, such as ^{3}He, ^{6}Li, ^{7}Be, ^{11}C, ^{13}N, ^{15}O and others. They give excitation functions for all these reactions which are worthwhile mentioning, as the various thresholds and the levelling off at higher energies are well apparent (see figs. IV.20 to 22).

Other good examples of excitation curves at the beginning of the spallation process are given in the work of Williams and Fulmer, who have investigated the radionuclides produced in natural iron and copper by protons in the energy range up to 70 MeV. Cross-sections are given in fig. IV.23 for the production of ^{48}V, ^{51}Cr, ^{52}Fe, ^{52}Mn, ^{54}Mn, ^{55}Co, ^{56}Co, ^{58}Co in natural iron. All show the expected rise and levelling off with a smooth slope. Fig. IV.24 shows the production of ^{57}Ni, ^{56}Co, ^{57}Co, ^{60}Co, ^{61}Cu, ^{62}Zn, ^{65}Zn in natural copper, all of which except ^{65}Zn again display the same features of a steep rise and a

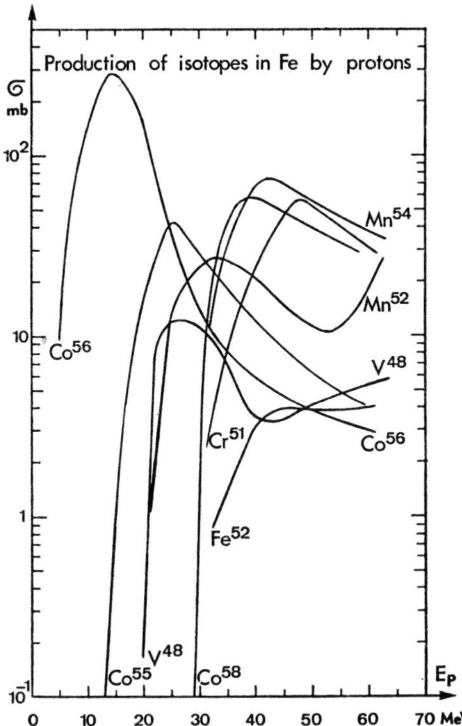

Fig. IV.23 Excitation functions for radioactive isotopes produced by protons of less than 60 MeV in natural iron.

smooth descent. The ^{65}Zn case can be explained by the large cross-section of the ^{65}Cu(p, n)^{65}Zn reaction in the 10 MeV range (700 mb according to fig. IV.1), the threshold being at about 2 MeV.

Williams and Fulmer have applied these data to residual radiation studies for medium energy proton accelerators, an energy range in which predictions on induced radioactivity are impossible by other means. In particular they have computed residual radiation dose rates outside slabs of material irradiated by collimated beams of particles. The excitation function and gamma branching ratio for each of the radionuclides constitute the basis of the calculation. The slab thickness is divided into small slices and the beam energy in the middle of each slice is evaluated. The corresponding cross-section is then used for each radionuclide, the absorption of the gamma ray in the material is considered and the total dose rate is obtained by summing the contributions from all slices of the slab and all the radio-

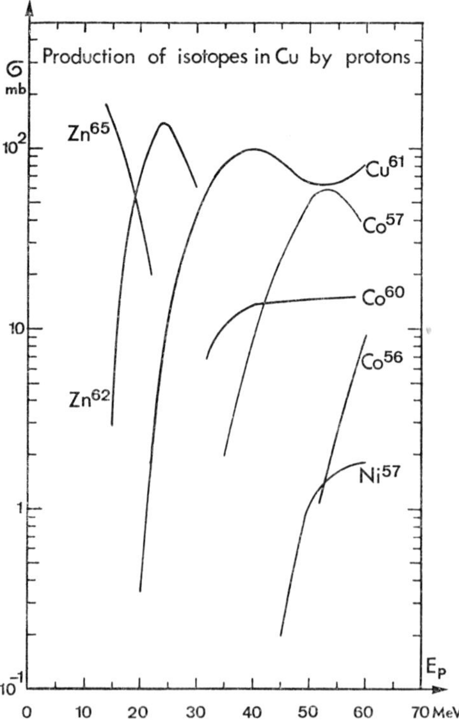

Fig. IV.24 Excitation functions for radioactive isotopes produced by protons of less than 60 MeV in natural copper.

nuclides. It is interesting to give an example of their results to know the order of magnitude of the radiation fields to be expected in front of proton beam stoppers in that energy range. Fig. IV.25 gives dose rate decay curves for thick targets of C, Al, Fe, Cu for a bombardment time of 5 years (practically saturation). These curves are for 50 MeV protons incident on stacks of material sufficient to stop the beam. The ordinate scale is the dose rate at 1 metre for a collimated beam of 1 mA. It is also the dose rate outside a large slab uniformly irradiated with a flux of 10^{11} particles/cm^{2}sec.

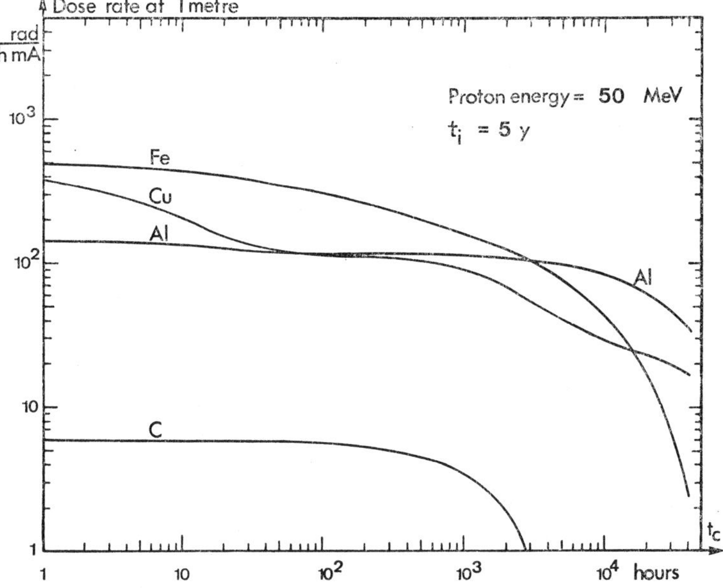

Fig. IV.25 Dose rate decay curves for thick targets of C, Al, Fe, Cu after bombardment with 50 MeV protons for 5 years (Fulmer, private communication).

3 A few monitor reactions for various particles of very high energy

Due to the existence of a threshold as outlined in the preceding sections, activation detectors, using the induced activity to measure a flux of known particles, have often been called threshold detectors, although they are also currently used in an energy region far away from the reaction threshold. Activation detectors prove particularly useful as they can be utilized as monitors with high precision, provided one is dealing with a pure, monoenergetic beam, i.e. if the beam has only one sort of particle and if the energy shows little dispersion. However, if the energy spread is large in a region where the cross-section varies, or if one has a mixed flux of various types of particles having widely different cross-sections, measurement with one detector only is only useful as an indication of the presence of radiation. If several detectors with different known responses are used and if the flux includes only one type of particle, iterative mathematical procedures can give quantitative results of some precision, as is current procedure for the measurement of the energy spectrum of reactor neutrons (see next section 4). Before passing on to this major if particular application, we shall be content in this section to give a number of cross-section vs energy curves for reactions which are in common use for monitoring pure high energy charged particle beams produced by high energy accelerators. The most frequently used are carbon and aluminium detectors. However this list is not comprehensive and any target producing a radionuclide, whose excitation function for the particle considered is known, can in principle be utilized as monitor, provided that the decay activity of the radionuclide selected can be measured with ease and precision. For good sensitivity a relatively high cross-section in the energy region of interest is also desirable.

Carbon is used either pure, or in the form of polythene or of scintillators. It is usually the ^{11}C beta-plus decay which is used. However the ^{7}Be decay which has a much longer half-life is also used sometimes. A plastic scintillator has the advantage of converting the positive electron into light in the detector itself without necessity of putting the detector on a crystal and gives thus higher sensitivity. Fluxes of the order of 1 to 10 particles per second per cm^2 could be detected by this method. However, as plastic scintillators include atoms other than C, such as N, O, etc., a particular calibration is necessary against pure carbon. This calibration is, of course, only valid for one type of flux, or particle, and one energy. Fig. IV.26 gives the excitation functions for the production of ^{11}C in ^{12}C by protons, neutrons, negative pions, deuterons, ^{3}He atoms and alpha particles. Though incomplete, the curves show that the thresholds are different for the various particles, and that the cross-sections tend to level off after a few hundred MeV bombarding energy at values which are all within a factor 2 or 3 of each other.

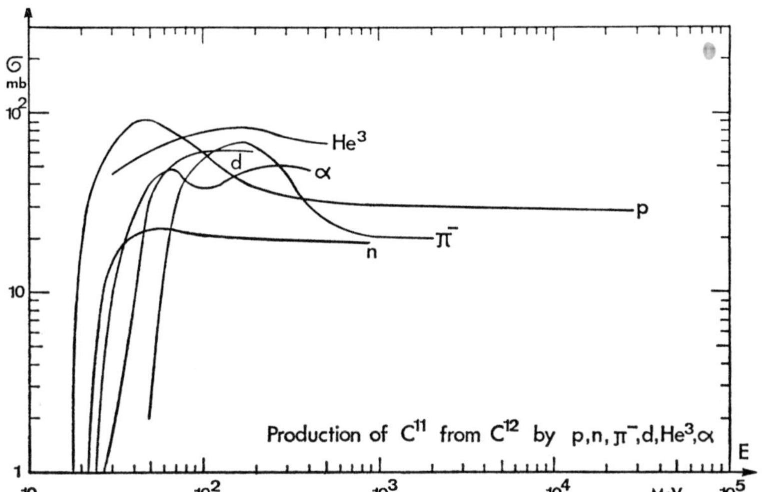

Fig. IV.26 Excitation functions for the production of ^{11}C from ^{12}C by protons, neutrons, negative pions, deuterons, ^{3}He atoms and alpha particles.

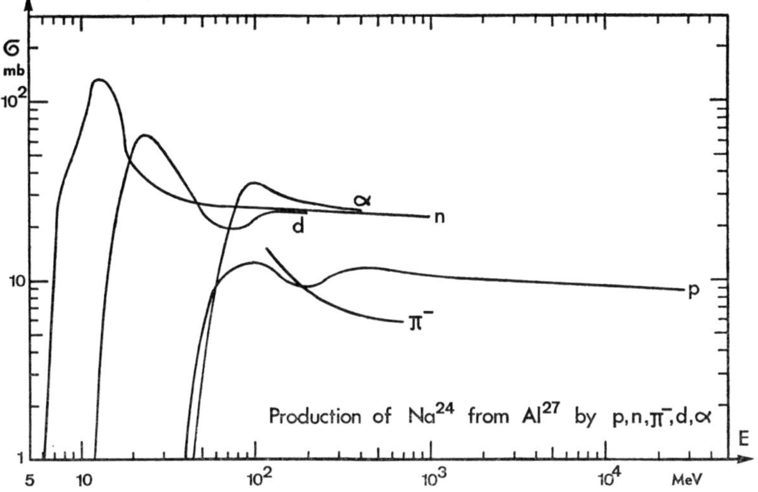

Fig. IV.27 Excitation functions for the production of ^{24}Na from ^{27}Al by protons, neutrons, negative pions, deuterons and alpha particles.

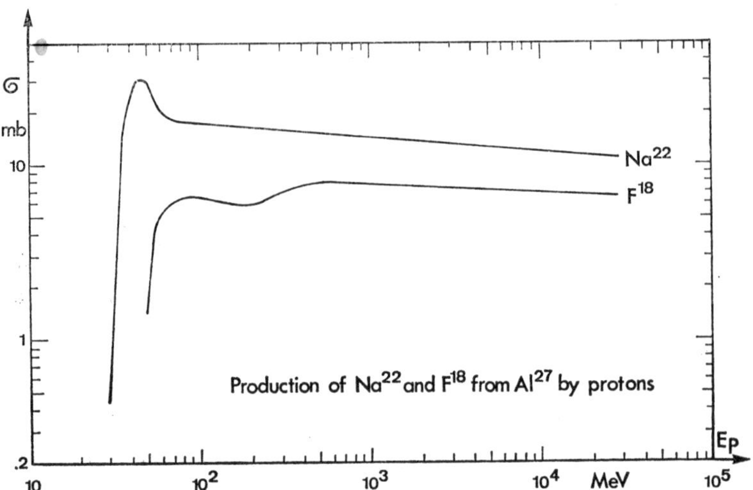

Fig. IV.28 Excitation functions for producing ^{22}Na and ^{18}F by protons from ^{27}Al.

The interest of the carbon and aluminium detectors is that they can be used up to very high energies with an almost constant sensitivity. Above 1 GeV, the cross-sections are practically constant so that the detector measures the number of high energy particles irrespective of their energy.

Besides carbon, aluminium is widely used as threshold, or better, activation detector. It can be prepared in thin foils, does not burn, and gives rise to several radionuclides which can well be measured independently, due to their difference in half-lives and decay radiation: ^{24}Na, ^{22}Na and ^{18}F. Fig. IV.27 gives the cross-sections for the production of ^{24}Na from ^{27}Al by protons, neutrons, negative pions, deuterons and alpha particles. One sees again the various thresholds. All curves level off towards higher energies. Fig. IV.28 shows the production of the isotopes ^{22}Na and ^{18}F in aluminium by protons up to 30 GeV, which is the present limit of experimentation.

Whereas the ^{11}C and ^{24}Na isotopes give information on the average flux received in time spans of the order of 1 h and 1 d before irradiation ceased, in contrast ^{22}Na with its long half-life of 2.5 years can be used to record average flux levels over periods of months.

Besides the C and Al detectors, many other reactions have been used in the high energy region, depending on the kind of flux and the scope of the experiment. As examples we shall mention tritium and ^{37}A production in meteorites, bismuth fission (starting at 80 MeV),

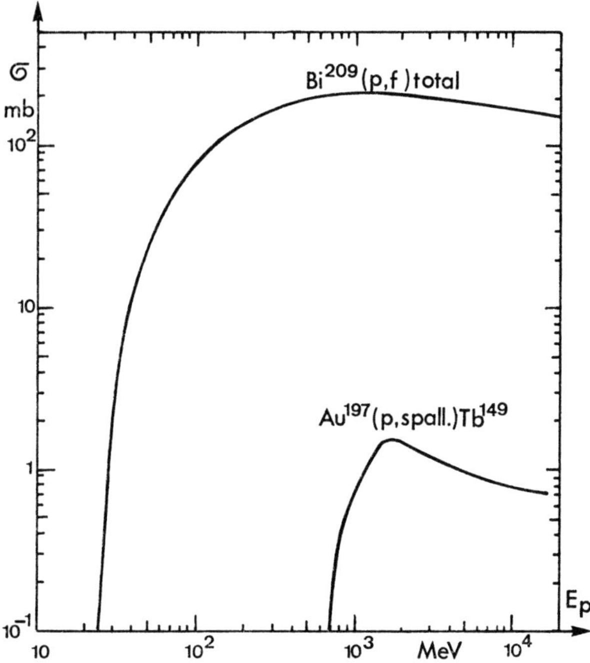

Fig. IV.29 Total cross-section for fission of bismuth by protons and partial cross-section for the production of the alpha-emitting branch of ^{149}Tb from gold by protons.

production of ^{149}Tb by spallation from gold (threshold at 600 MeV), etc. (see fig. IV.29). For the bismuth fission, the cross-sections for the production of a number of radionuclides can be found in the paper of Jodra and Sugarman.

The experimenter has in each case to find the detector or monitor which is most suited to his needs. In this respect the smooth approximation charts of the 'simple' reactions presented in section 1 of this chapter will be useful to suggest detectors hitherto not thought of and presenting a maximum response in the energy region and for the particle required.

4 Activation detectors for reactor and high energy neutrons

Reactor neutron fluxes, from thermal to fast, have been extensively investigated with activation detectors. The same begins to apply to the neutron atmosphere around high energy proton or electron accelerators. It is thus appropriate to present by way of examples some information on this subject although the question is far-reaching and has been extensively studied. As an example the International Atomic Energy Agency in Vienna has to date devoted two symposia to this matter, one on neutron dosimetry (Harwell 1961) and one on neutron monitoring (Vienna 1966), at which a number of contributions were presented on activation detectors.

The first materials to be selected for neutron flux measurements were those which exhibited a particularly high cross-section in the thermal region as indium, cobalt, gold, for instance. It was soon discovered that in mixed radiation fields including higher energy neutrons, it was advisable to introduce a cover or a shield which would absorb the thermal neutrons (such as cadmium) making it possible to discard the high energy background by taking the difference between the base and the covered detector. This led to the Cd-In-Cd-In sandwich and to many others. Another way of using shields around activation detectors was to degrade the energy of the faster neutrons down to thermal energies where the detectors were highly sensitive. With this in view, boron was used (Davis) or materials with hydrogen content such as paraffin, polythene, etc. giving the so-called Bonner spheres (Bramblett et al., Burrus, Hankins). Foil activation detectors were also used bare in the fast neutron region using prominently (n, p) reactions (Cross, Barral and McElroy). These detectors were then also applied to dose determination. Fission reactions of all kinds, from bismuth to plutonium, and neutron spallation at high energies (for instance that of mercury leading to ^{149}Tb) have also been used.

Recently the introduction of fission fragment detectors using etching techniques (Widdell, Becker) has further enhanced the use of activation detectors, as recoil protons or alpha particles from simpler (n, p) or (n, α) reactions can also be detected by the same procedure.

We shall now give some examples of how activation detectors have been used to determine neutron energy spectra over a wide range of energy. Of course, one detector is insufficient: several detectors with different threshold and excitation functions are required. The results of the measurements are then subject to a mathematical iteration process which makes it possible to arrive at an approximation of the spectrum. Such experiments and mathematical procedures, applied to the determination of the neutron energy distribution from various fission materials and reactors, have been described. As an example we give here the excitation functions of the various detectors used by

Grundl (see also the work of Ringle, or Kohler). These included the following reactions: ^{27}Al(n, p and n, α), ^{31}P(n, p), ^{56}Fe(n, p), ^{235}U(n, f), ^{238}U(n, f), ^{237}Np(n, f) and are presented in fig. IV.30. It can be seen that the various thresholds are quite different, and that after some increase with energy, the cross-section becomes practically constant in all the cases.

Another example of interest is the neutron detectors investigated by Heertje in the energy range up to 25 MeV with accelerator-produced neutrons. Heertje used for this purpose a 26 MeV deuteron beam

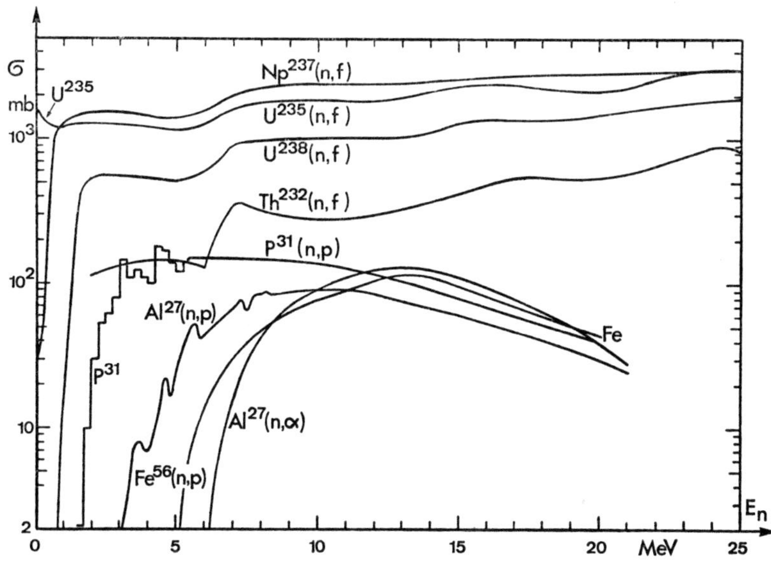

Fig. IV.30 Fundamental detector cross-sections used by Grundl. The excitation functions have been extended towards higher energies using the data in BNL 325. The ^{232}Th(n, f) curve has been added.

extracted from a cyclotron and falling on a beryllium target. He first determined his neutron energy spectrum with known activation detectors, and could then determine a few more including the reactions ^{14}N(n, 2n), ^{16}O(n, 2n), ^{19}F(n, 2n), ^{24}Mg(n, p), ^{35}Cl(n, p), ^{58}Ni(n, 2n), ^{115}In(n, γ)^{116}In, ^{127}I(n, 2n). The corresponding excitation functions, together with those used for his preliminary spectrum determination, are presented in fig. IV.31. Both threshold and absolute value of the cross-section maximum vary considerably among the cases considered.

An interesting application of the neutron reactions studied in this chapter is activation analysis using 14 MeV neutrons. Neutron generators for this particular energy are readily constructed. The chart presented in fig. IV.32 (from Tokyo Shibaura Electric Co.) shows the product $\sigma\theta/A$ at 14 MeV for the reaction used, which can be (n, p), (n, d), (n, α), (n, 2n) or (n, fission) according to the material looked for. The half-life of the radionuclide produced is taken as abcissa. The type of the reaction is indicated by a particular symbol. As one can see from the chart, most of the interesting elements found in structural materials can be detected and their concentration measured by this method.

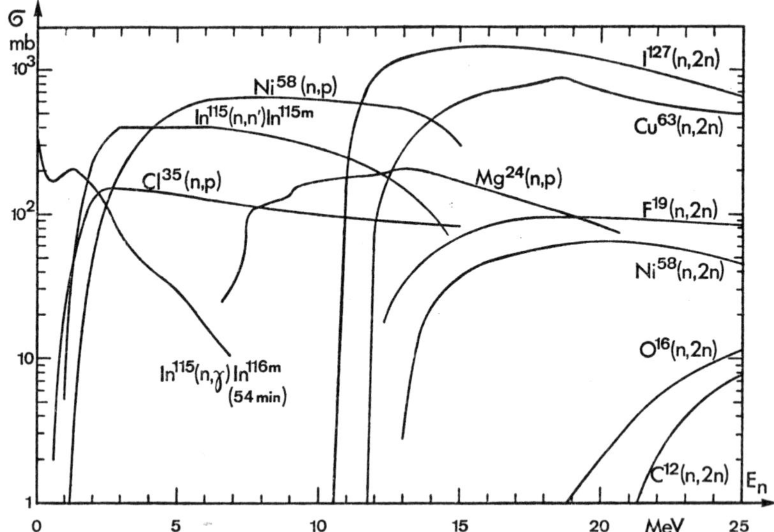

Fig. IV.31 Cross-section vs energy curves for various detectors investigated by Heertje. The ^{16}O(n, 2n) and ^{19}F(n, 2n) curves of Heertje have been slightly corrected using the data in BNL 325.

We shall not give data on thermal cross-sections, as these have been exhaustively studied for activation analysis among other purposes and published in data sheets (Brookhaven National Laboratory report BNL 325). We will simply recall to the reader the capture or total cross-sections for some elements often used as covers for or as activation detectors in the lowest energy region. Fig. IV.33 gives the shape and position of the largest absorption resonance peak for indium, cadmium and gold, as well as the hydrogen, gadolinium and

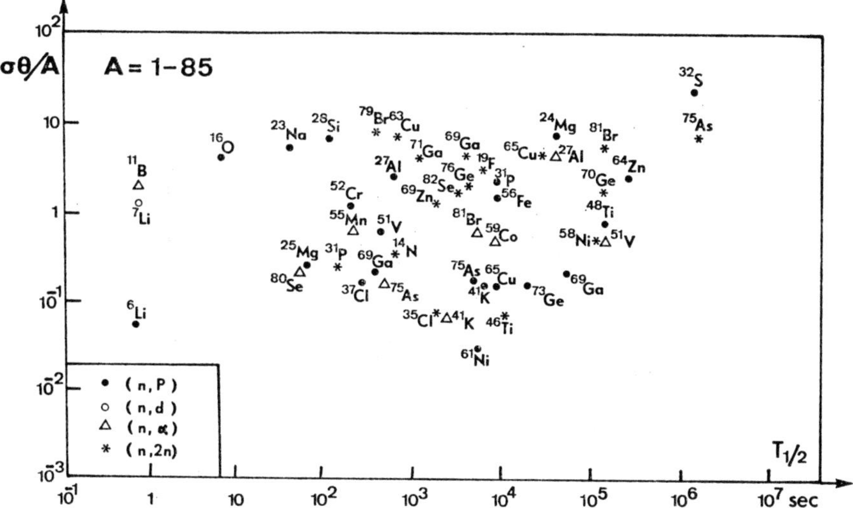

Fig. IV.32a Chart of activation of elements with $1<A<85$ by 14 MeV neutrons (A: atomic weight, θ: abundance ratio, σ: cross-section in millibarns).

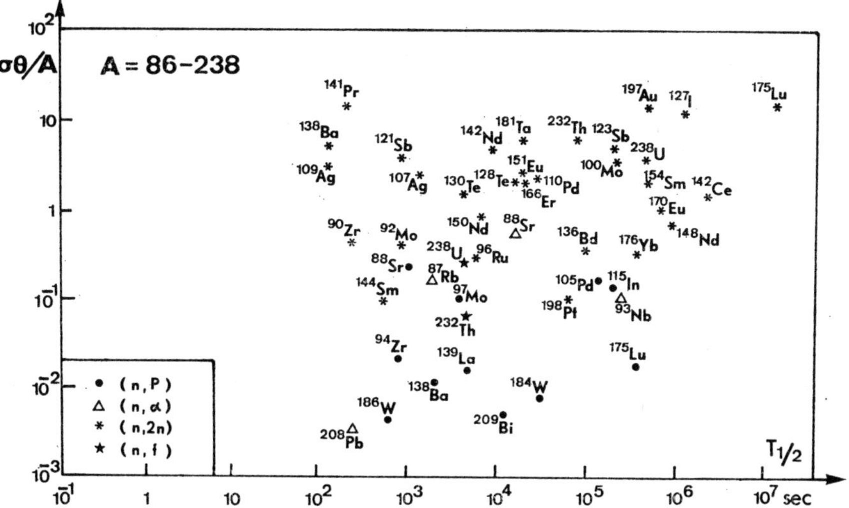

Fig. IV.32b Chart of activation of elements with $86<A<238$ by 14 MeV neutrons (A: atomic weight, θ: abundance ratio, σ: cross-section in millibarns).

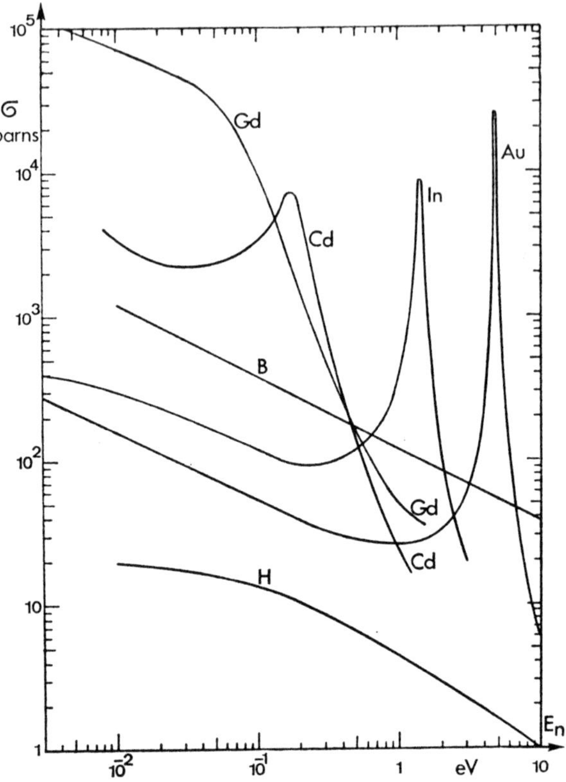

Fig. IV.33 Capture cross-sections for indium, cadmium, gold and total cross-sections for hydrogen, natural boron and gadolinium at low neutron energies.

boron total cross-sections in the energy range 10^{-3} to 10 eV, as it is often convenient to find these curves at hand.

5 Fission by high energy protons

Another phenomenon which it is appropriate to include in this chapter is fission by high energy protons. This kind of fission starts at a proton energy which is sufficient to overcome the Coulomb barrier of the nucleus, that is around 10 MeV. The nuclei undergoing fission by proton bombardment are, of course, to be found only towards very high atomic numbers, say from tungsten upwards. Such a heavy nucleus is then capable of fission, in addition to spallation which has already been considered, whereby various nucleons or clusters thereof are successively knocked out before the final spallation product

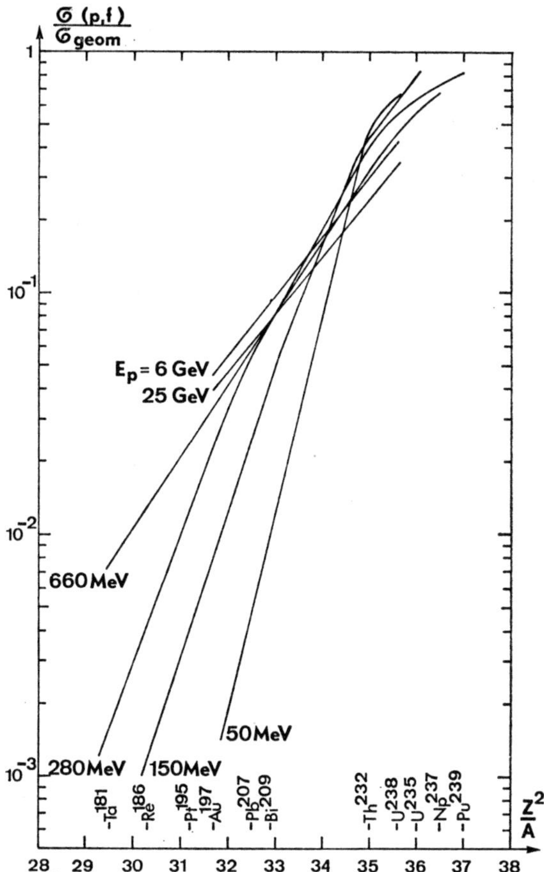

Fig. IV.34 Ratio of proton-induced fission cross-section to geometrical cross-section for various proton energies as a function of target Z^2/A.

remains. This spallation product is generally not too far from the original nucleus for high-Z target elements. In the case of fission, however, the target nucleus divides into two parts more or less comparable in weight, the sum of which amounts roughly to the weight of the original nucleus.

Of course both kinds of disintegration of a heavy target nucleus under high energy proton bombardment are present at the same time and compete. Experiments show that the amount of fission relative to the total interaction or the geometric cross-section increases as the target atomic number increases. Fig. IV.34 shows this clearly. The ratio $\sigma(\mathrm{p,f})/\sigma_{\mathrm{geom}}$ is plotted against the target parameter Z^2/A, where

Z is the atomic and A the mass number of the target, for various incident proton energies. One sees that for all energies the fission part is smaller than 10% for all target elements under thorium. For target elements from thorium upwards and for all energies, the fission part is more than half the geometrical cross-section. Thus the case is really clear-cut, allowing us in a first approximation to neglect fission for all elements up to and including bismuth, and to neglect spallation for all elements from thorium upwards, when bombarded by protons, and if we have to calculate only the gross activity as a whole. In fact since the fission and spallation products are very numerous and widely distributed over a great range of atomic numbers in these cases, their overall effects on the residual radioactivity will be roughly of the same order of magnitude, so that we can neglect spallation versus fission or the contrary as soon as the cross-section for one of these effects is appreciably smaller than the cross-section for the other effect.

We come now to examine in more detail the excitation curves for the

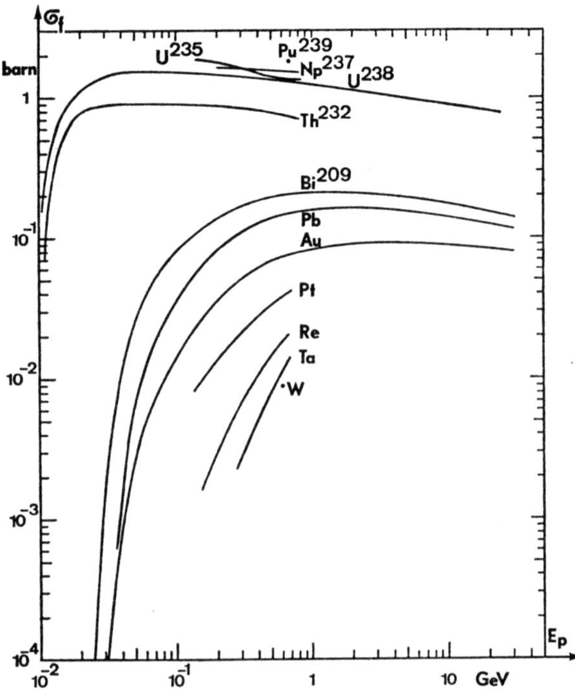

Fig. IV.35 Proton-induced fission cross-section for various target elements as a function of proton energy.

fission processes induced by high energy protons in different materials. Fig. IV.35 shows the cross-section for fission for various target elements ranging from tungsten to plutonium, with peak values starting around 10 millibarns for tungsten and going up to 2 barns for ^{235}U and ^{239}Pu. The curves exhibit an early and steep rise, especially for uranium and other neutron-fissionable elements, a broad maximum in the 600 MeV proton energy region and a very slow descent of not more than a factor 2 down to 30 GeV, the highest energy where measurements are available now for some of the higher-Z materials.

As usual when investigating fission phenomena, the next information which is useful when we have an idea of the fission cross-section is the chain or mass yields of the fission products, a quantity giving the sum of all the yields of the products having the same mass number (isobars). It will be remembered that for all neutron-fissionable

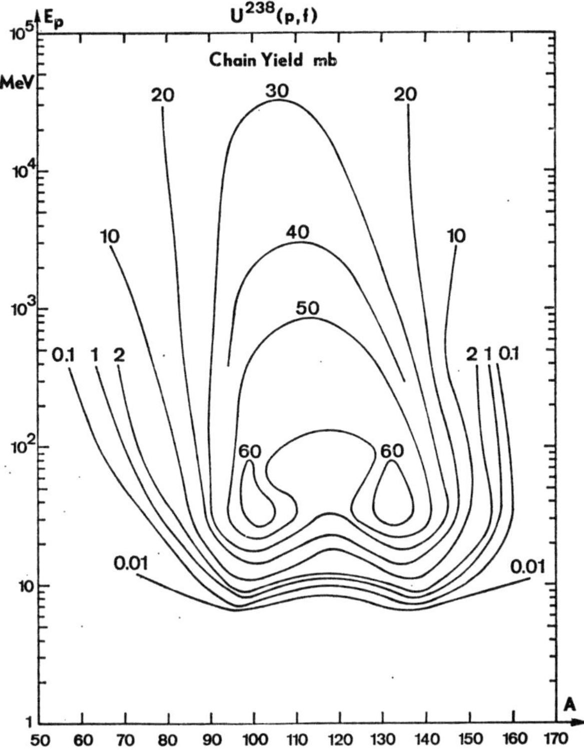

Fig. IV.36 Fission products chain yields of the proton-induced fission of ^{238}U as a function of chain mass number and proton kinetic energy.

elements a curve with two humps, the well-known 'camel curve' was found for the dependence of the chain yields on the mass numbers of the fission products. From fig. IV.36 we see that this is also the case for protons at lower bombarding energies, up to about 100 MeV, whereas at energies higher than this, the two maxima merge together giving a mass yield curve with one broad top, showing that the mechanism of fission changes somewhat when the energy imparted to the target nucleus by the incident proton increases very much. Thus the distribution of fission products varies considerably with the energy of the incoming proton, in a manner which can be deduced from the figure.

This is reflected also in the behaviour of the individual independent yields with energy. Few measurements of independent isotopic yields exist at low bombarding energies; however, the curves in figs. IV.37 and 38 which are drawn for proton energies from 100 MeV upwards will indicate the trend rather well. The independent yields for various isotopes of the iodine and caesium fission products families have been drawn as a function of the energy of the incident proton. The cross-sections for production of the neutron-deficient isotopes are seen to increase with energy, those for neutron-abundant isotopes to decrease, until a rather equal value for both is attained, which remains nearly constant up to very high energies.

By analysing and putting together various data available at the moment for independent isotopic yields of some fission products of ^{238}U by protons in the energy range 170 to 680 MeV (where the mass yield distribution seems to remain somewhat stationary according to fig. IV.36) we have tried to draw an isoyield chart for proton-induced fission of this material throughout the whole system of elements. This chart is presented in figs. IV.39 and 40. The iso-yield curves have been drawn according to a best fit found by the eye. The position of the peak line going through the maximua of the charge distributions of the independent yields had been partly indicated already by various authors (Hagebø, Pappas, Friedlaender and others). The independent isotopic yield values on this peak line have been found either by extrapolation of yields at nearby locations or by similarity to the chain yield curve versus fission product mass number elsewhere. The agreement of the two procedures was satisfactory. Of course the chart presented is to be considered only as a drawing exercise giving a rough idea and approximative values of the cross-sections involved owing to the still insufficient number of exact determinations of independent isotopic yields over all mass numbers. The only fission

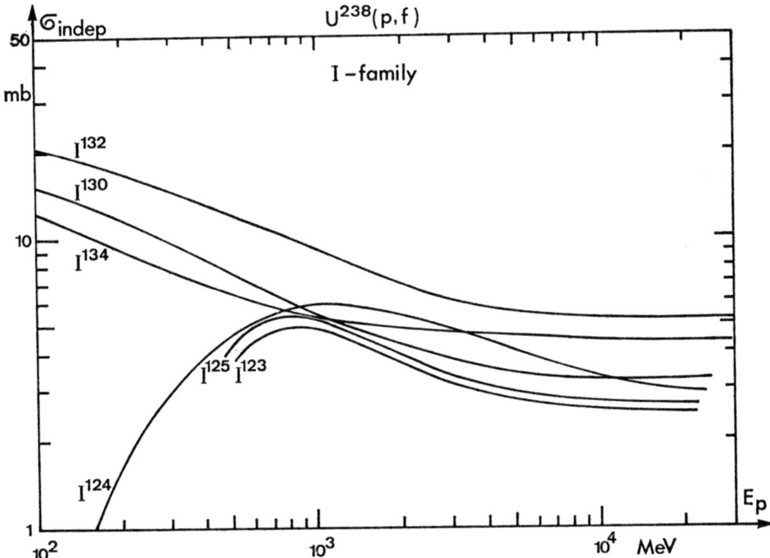

Fig. IV.37 Independent yields of various iodine isotopes for proton-induced fission of ^{238}U as a function of proton energy.

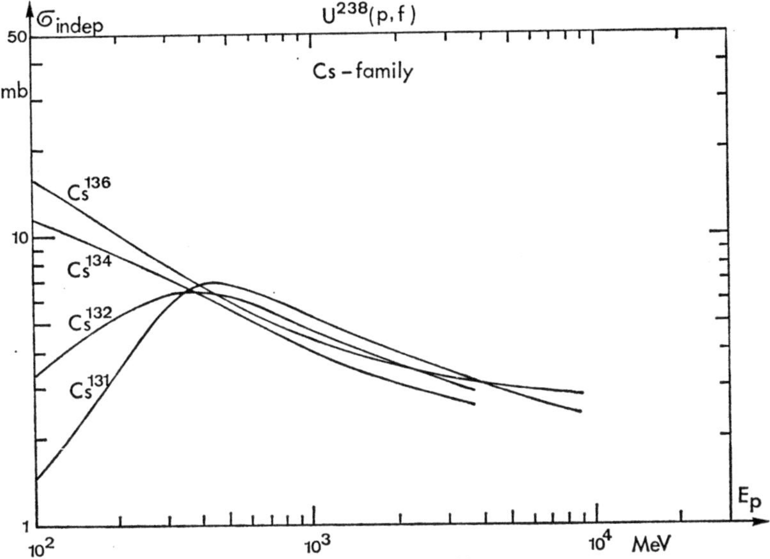

Fig. IV.38 Independent yields of various caesium isotopes for proton-induced fission of ^{238}U as a function of proton energy.

products region which can be considered as correctly covered in this diagram is the region with masses between $A=127$ and $A=134$ (Pappas and Hagebø). Data for attempting to make a similar drawing for proton energies in the 6 to 30 GeV region have been really insufficient up to now, so that we must content ourselves with examining in some detail the isoyields chart for the energy region covered here to get acquainted with the principal features of high energy proton induced fission.

It is clearly seen from figs. IV.39 and 40 that the independent yield function forms an elongated hill, the fission products ranging from Mn($Z=25$) to Tb($Z=65$). The height of this hill (maximum independent yields) is 15 millibarns, a value which remains constant from Zr($Z=40$) to I($Z=53$). The maximum yields do not lie on the beta stability line, but on a parallel line on the right, corresponding to a neutron excess of about 6 units with respect to the stable natural isotopes. For a constant A value the charge distribution curves are symmetrical and parabolic, and extend to the left further than the beta stability line. Thus it is possible by high energy protons induced fission to produce neutron-deficient isotopes having a beta-plus decay mode. In particulier fig. IV.40 shows that the effect is enhanced when passing to higher energies. The gain in the independent isotopic yields can be roughly a factor 10 in the region Ag($Z=47$) to Sb($Z=51$) when going from 170 to 680 MeV incident proton energy. The yields on the other, neutron-abundant, side of the charge distribution do not seem to be affected by a change in the bombarding energy and are the same throughout the whole energy region considered.

We shall now direct our attention towards the other reaction occurring in parallel with fission when bombarding high-Z nuclei with high energy protons, the spallation process already mentioned. It is probable that spallation products also fall into the range where we have found the fission products. Of course, once the product is found in this range, it is impossible to determine whether it has been formed via fission or via spallation. However, we have good reasons to consider that the spallation contribution in the fission products range is not appreciable at 600 MeV and for ^{238}U. This is because one finds another range, which lies in fact very close to the original target nucleus, where products are formed essentially only by spallation, as their mass difference from this original nucleus is very small. This zone has been investigated and fig. IV.41 shows the independent yields found there, i.e. between Pt($Z=78$) and U($Z=92$), in the case of an energy

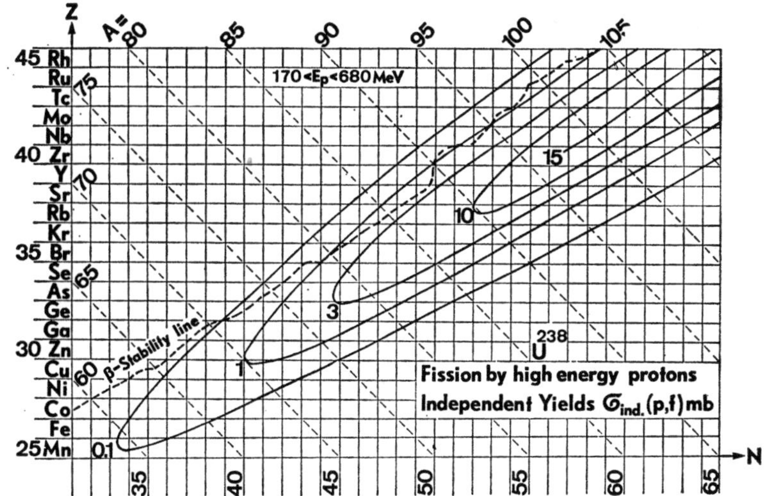

Fig. IV.39 ^{238}U fission by protons with energies between 170 and 680 MeV: independent yields in millibarns of fission products from $Z=25$ to $Z=45$.

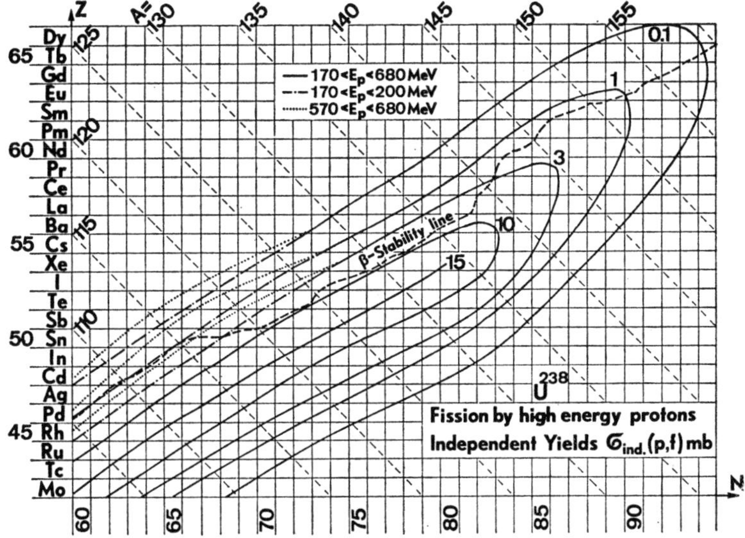

Fig. IV.40 ^{238}U fission by protons with energies between 170 and 680 MeV: independent yields in millibarns for fission products from $Z=45$ to $Z=65$.

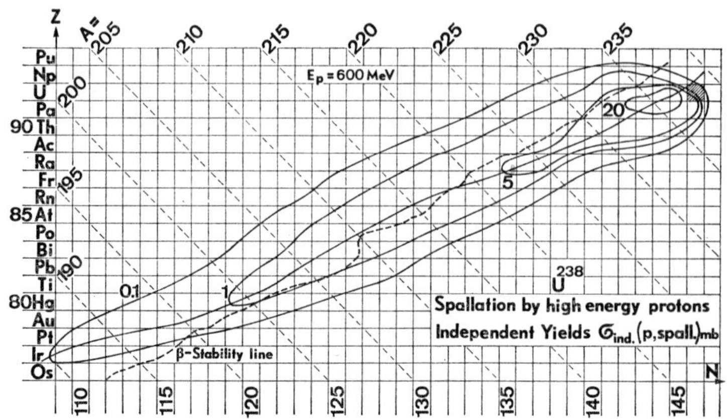

Fig. IV.41 ^{238}U spallation by 600 MeV protons: independent yields in mb.

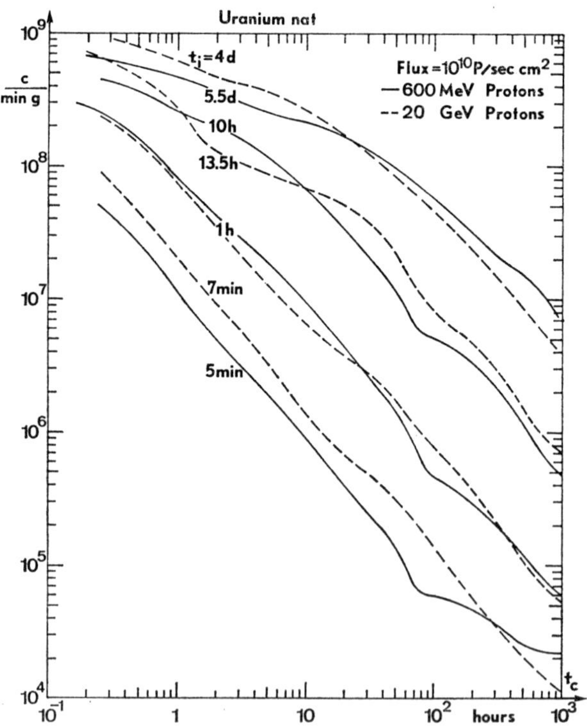

Fig. IV.42 Gross activity decay of natural uranium bombarded by 600 MeV and 20 GeV protons for various irradiation times in counts per minute per gram registered with sample in contact with a 3″ × 3″ NaI crystal (incident flux 10^{10} protons/sec cm^2).

of 600 MeV for the bombarding proton. In contrast with the fission results, the spallation products distribution appears to be fairly symmetrical with respect to the beta stability line, neutron-abundant and -deficient isotopes being formed in equal numbers by this spallation process. The independent isotopic yields are of the order of 20 mb in the immediate vicinity of the target nucleus, lie above 5 mb between Ra($Z=88$) and Th($Z=90$) and above 1 mb between Hg($Z=80$) and Fr($Z=87$).

In view of the increased use of uranium and other heavy materials in accelerator techniques, mainly as analysing slits, beam collimators and stops, it has appeared necessary to make some direct measurements of the total gamma-induced activity of natural uranium under high energy proton bombardment in order to determine the radiation hazard of such objects. The results are presented in figs. IV.42, 43 and 44 for bombardment energies of 20 GeV, 600 and 50 MeV. At the two

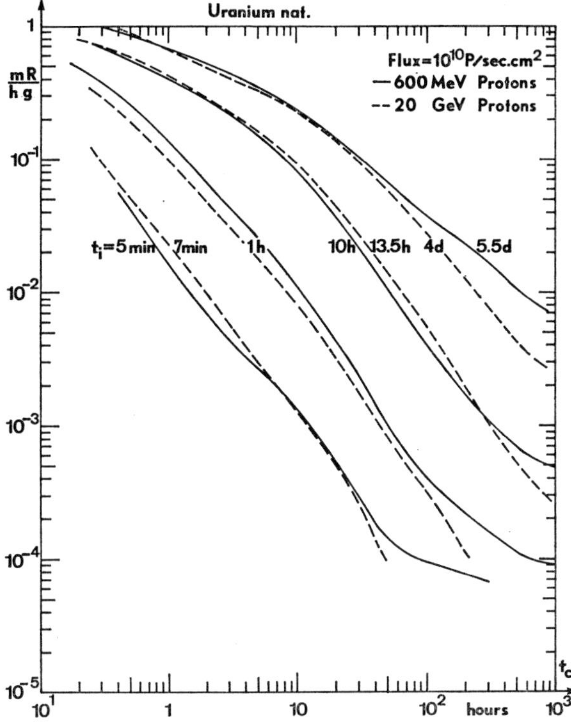

Fig. IV.43 Decay of the radiation field around natural uranium bombarded by 600 MeV and 20 GeV protons for various irradiation times in millirad/hour per gram at 1 metre distance (incident flux 10^{10} protons/sec cm^2).

energies first mentioned, the ranges of the proton in the material are large compared with the sample thickness and it is useful to indicate the specific activity or hazard per gram of target material at these energies. This is done in the two figs. IV.42 and 43 which give the number of counts per minute per gram registered on 3″×3″ NaI crystal in contact, and the radiation field in millirad/hour per gram at 1 metre for a series of bombardment times as a function of cooling times. The data were reduced to an incident flux value of 10^{10} protons/sec cm^2. It is seen that there are no apparent differences between the 600 MeV and 20 GeV cases. The experimental data gained at 50 MeV

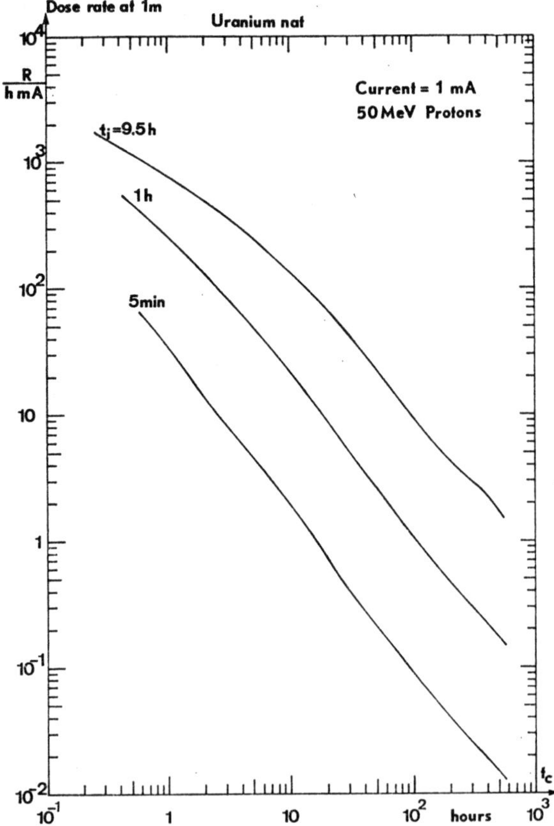

Fig. IV.44 Decay of the radiation field of a natural uranium target which has fully absorbed a 50 MeV proton beam with a intensity of 1 mA for various irradiation times, measured in rad/h at 1 metre distance in front of the target.

are somewhat different in principle. Because of the low bombarding energy, the uranium samples used absorbed the beam completely. Fig. IV.44 gives then the radiation field decay curves in rad/hour in front of a beam stop having absorbed the beam completely. The data have been reduced to 1 mA beam intensity. This information can then be compared with that presented already in fig. IV.25 for C, Al, Fe, Cu targets and the same proton energy, but for very long irradition time (5 years). It is seen that at 1 hour cooling time, uranium stops which have been irradiated 9.5 hours are as active as iron stops which have been bombarded 5 years by the same 50 MeV 1 mA proton beam.

PROBLEMS

CHAPTER IV

1. At what bombarding energies can one expect the (p, n) reaction cross-section with ^{89}Y as target to be approximately 10, 20, 50, 100, 400, 600, 700 mb?

2. How does the cross-section of the (p, pn) reaction at a bombarding energy of 25 MeV vary with target mass number? Give approximate cross-section values for mass numbers $A=20$, 50, 100, 150, 200.

3. Take a mass number $A=80$ and a bombarding energy of 20 MeV. What are the expected cross-sections for the (p, n), (p, 2n), (p, pn), (p, α) reactions? Add them up and compare with the theoretical total proton inelastic cross-section at this energy and this mass number.

4. What are the energies at which the cross-section values for the ^{127}I(n, 2n) reaction are approximately expected to be equal to 1, 100, 500, 1000, 1500 mb according to the smooth approximation? Compare with the curve given in fig. IV.31.

5. Favourite activation detectors are elements which have only one stable isotope in nature. They are cheaper than artificially separated stable isotopes and the radioisotopes which they produce can be ascribed to one nuclear reaction only, given the bombarding particle. These elements are listed below:

^{9}Be, ^{19}F, ^{23}Na, ^{27}Al, ^{31}P, ^{45}Sc, ^{55}Mn, ^{59}Co, ^{75}As, ^{89}Y, ^{93}Nb, ^{103}Rh, ^{127}I, ^{133}Cs, ^{141}Pr, ^{159}Tb, ^{165}Ho, ^{169}Tm, ^{197}Au, ^{209}Bi.

With these, proton reactions of the type (p, n) or (p, pn) or both always lead to radioactive products. In which mass number and energy ranges does one find the highest cross-sections for these two reactions and what are their orders of magnitude?

The same question is asked for neutron detectors, based on the (n, 2n) reaction, which is the most widely used, since the (n, p) and (n, α) reactions have decidedly much smaller cross-sections.

The same question is asked for deuteron detectors based on the (d, 2n) and (d, p) reactions.

The same question is asked for alpha particle detectors based on the (α, 2n) reaction.

CHAPTER V

RADIOACTIVITY INDUCED BY HIGH ENERGY ELECTRONS AND PHOTONS

1 The electromagnetic cascade and photonuclear reactions

When a high energy electron enters a material, the well-known phenomenon of the electromagnetic cascade occurs, in which a shower of electrons, positrons, and photons develops. The development of the shower continues until the energy of the photons is small enough for the attenuation by Compton scattering and photoemission of the atomic electrons to become dominant. In this work we are interested in the high energy photons ($>$10 MeV) in the shower before attenuation by atomic electrons begins, and in particular we are interested in the interaction of these photons with the nucleus.

When a photon interacts with a nucleus, the photon is absorbed and the excited nucleus emits one or more nucleons, thereby changing to a different element or to a different isotope of the same element. This new nucleus may well be radioactive, and it is the build-up of such nuclei which causes the residual radioactivity in an electron accelerator.

There is an additional process which also contributes to the activity of a machine. Many of the photonuclear interactions will give off high energy particles, which will in turn interact with other nuclei to produce new isotopes, and very high energy particles may even initiate a nuclear cascade in addition to the initial electromagnetic shower.

2 The total number of photonuclear interactions in a cascade initiated by one electron

The energy and density of photons in a cascade will vary with the distance along the cascade, and we may define $\gamma(k,t)\,\mathrm{d}k$ to be the average number of photons with energies between k and $k+\mathrm{d}k$ at a depth t in the cascade. We may also define the total track length (B. Rossi, High Energy Particles, New York, 1952, p. 218):

$$g_0(k) = \int_0^\infty \gamma(k,t)\,\mathrm{d}t \tag{2.1}$$

where $g_0(k)\,\mathrm{d}k$ is the total distance travelled by all the photons in the shower while their energy lies between k and $k+\mathrm{d}k$.

This track length has been computed (Rossi, p. 271), and for a cascade initiated by an electron of energy E_0 we know:

$$g_0(k)\,\mathrm{d}k = 0.57\,\frac{E_0}{k^2}\,X_0\,\mathrm{d}k\ \ \mathrm{g/cm^2} \tag{2.2}$$

where X_0 is the radiation length in g/cm^2 in the material containing the cascade.

Now, if the total cross-section for the interaction of a photon with one of the nuclei of the material is $\sigma(k)$ cm^2, then the photons will interact with all the nuclei in an amount $\sigma(k)g_0(k)\,\mathrm{d}k$ grams of the material. The number of nuclei per gram is N_0/A, where N_0 is Avogadro's number (6×10^{23} mole^{-1}) and A is the atomic weight of the material. Hence the number of photonuclear reactions produced by the photons of energy k to $k+\mathrm{d}k$ in the cascade will be, per incident electron initiating a cascade,

$$\begin{aligned}\mathrm{d}Y &= (N_0/A)\sigma(k)\,g_0(k)\,\mathrm{d}k\\ &= 0.57\,\frac{E_0X_0N_0}{A}\,\frac{\sigma(k)\,\mathrm{d}k}{k^2}.\end{aligned} \tag{2.3}$$

The total number of photonuclear interactions in the cascade will be obtained by integrating this over the energy of the photons in the cascade. That is, the yield or total number of photonuclear reactions in a cascade initiated by one electron of energy E_0(MeV) is

$$Y = 0.57\,\frac{E_0X_0N_0}{A}\int_0^{E_0}\frac{\sigma(k)\,\mathrm{d}k}{k^2}. \tag{2.4}$$

Substitution of the value of Avogadro's number gives:

$$Y = 3.42 \times 10^{23} \frac{E_0 X_0}{A} \int_0^{E_0} \frac{\sigma(k)\,\mathrm{d}k}{k^2} \tag{2.5}$$

interactions per incident electron, where energies are measured in MeV and $\int(\sigma/k^2)\,\mathrm{d}k$ is in cm^2/MeV. The problem is now reduced to that of evaluating the integral in eq. (2.5) for the material concerned, and for this we may divide the problem into several energy ranges, and study each one in turn.

For gamma ray energies up to about 50 MeV, the cross-sections have been studied carefully by many people, and it is found that the reactions are dominated by the well-known 'giant resonance' phenomenon occurring at about 20 MeV. Between 80 and 300 MeV there is a process for which the photodisintegration of a pseudo-deuteron in the nucleus appears to be a useful model. In this model it is assumed that the photon is absorbed by a deuteron (proton–neutron pair) in the nucleus, and that the deuteron then disintegrates. It will be shown later that the total number of such deuteron disintegrations in the shower is about one thousand times less than the number of giant resonance reactions, but the process is important because it may give rise to high energy neutrons which can themselves activate further nuclei. Finally, above 300 MeV meson photoproduction starts, and is of the same order as the deuteron photodisintegration process.

3 The giant resonance photonuclear reactions

The excitation functions of all photonuclear reactions exhibit a large peak at photon energies ranging from 25 MeV for light target nuclei to 14 MeV for heavy target nuclei. This peak is commonly called the 'giant resonance'. It is explained by the fact that the electromagnetic wave associated with the incident photon induces a collective oscillation of all the protons in the nucleus as a whole against the neutrons, which happens to be in resonance with the photon wave at this photon energy. This motion can be accompanied by the knocking out of one or several charged and uncharged nucleons from the target nucleus. When the target element is sufficiently heavy, fission has even been observed by this mechanism, also with a sharp cross-section peak at giant resonance energies.

Since most of the interactions occur around the giant resonance energy k_0, we may make a simplifying approximation for the integral

$$\int_0^{50\text{ MeV}} \frac{\sigma(k)\,\mathrm{d}k}{k^2} \cong \frac{\int_0^{50} \sigma(k)\,\mathrm{d}k}{k_0^2}. \tag{2.6}$$

The integrated cross-section $\int \sigma(k)\,\mathrm{d}k$ has been measured for many elements and reactions in this range. It should be noted that at the giant resonance there are many possible reactions, typically (γ, n), $(\gamma, 2\mathrm{n})$, (γ, p), (γ, np), (γ, α) etc. The available data on these reactions for various elements together with the measured excitation functions, integrated cross-sections and giant resonance energies, have been summed up in the form of graphs presented in the following figures.

First, photonuclear cross-section charts giving the cross-section as a function of photon energy and target mass number for the reactions just mentioned have been drawn in much the same manner as the charts drawn in ch. IV for compound nucleus reactions. As before only a smooth approximation is drawn freely through the points given by various authors for a number of target nuclei. The data available were too scarce to make it possible to discover or retain a fine structure in the graph, which has simply the appearance of a hump extending over a well-defined energy region and through all mass numbers. When going from one type of reaction to another, the energy range in

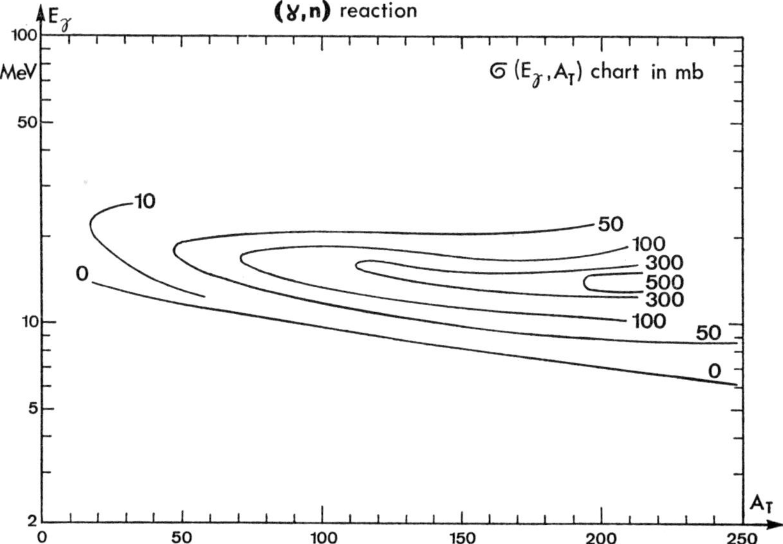

Fig. v.1 Smoothed approximation of the (γ, n) reaction cross-section.

which the reaction takes place may shift, and the magnitude of the maximum cross-section values found changes markedly.

The most prominent giant resonance reaction is the (γ,n) reaction (fig. v.1). It is the first to appear, with thresholds in the 10 MeV photon energy region, and the one with the highest maximum cross-section, increasing smoothly from 50 to 500 mb for mass numbers between 50 and 200, at photon energies of 15 to 18 MeV.

The next reaction is the $(\gamma,2n)$ reaction. The thresholds are somewhat higher, lying around 15 MeV, and the maximum cross-section values increase from 20 to 200 mb in the mass number range extending from 50 to 200 at photon energies of 18 to 20 MeV (fig. v.2).

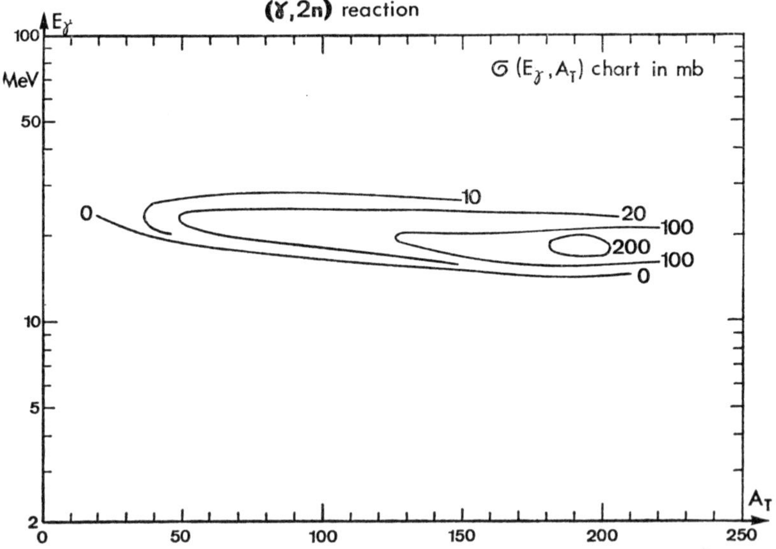

Fig. v.2 Smoothed approximation of the $(\gamma, 2n)$ reaction cross-section.

The photonuclear reactions in which charged particles are produced have somewhat different behaviour. There is a definite mass number range, in the neighbourhood of $A = 50$, in which the peak cross-section values of the giant resonance are the highest. These peak values decrease then considerably with increasing target mass number, due to the difficulty of expelling a charged particle from a high-Z nucleus, an effect which was also seen in compound nucleus reactions.

The (γ,p) reaction has highest peak cross-section values of about 50 mb for target mass numbers A, lying between 30 and 40, and photon energies of 20 MeV (fig. v.3).

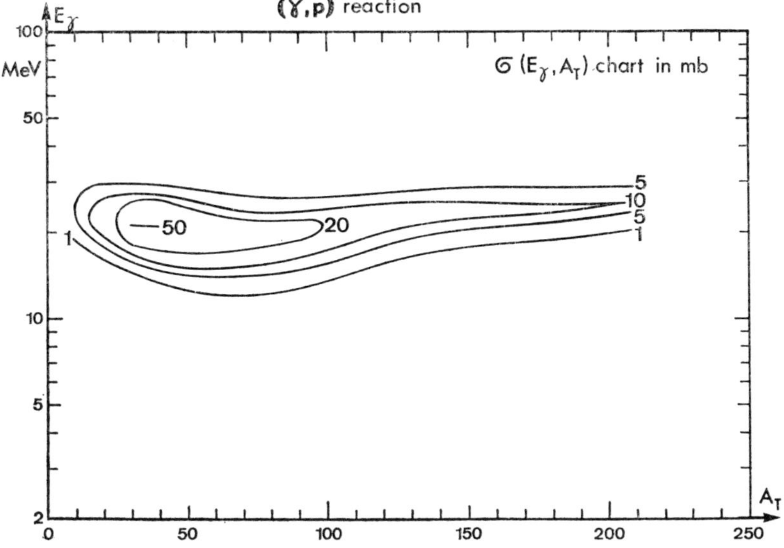

Fig. v.3 Smoothed approximation of the (γ, p) reaction cross-section.

The (γ, np) reaction starts later, with thresholds at 20 MeV and maximum peak values of 10 mb around $A=50$ for 30 MeV photon energy (fig. v.4).

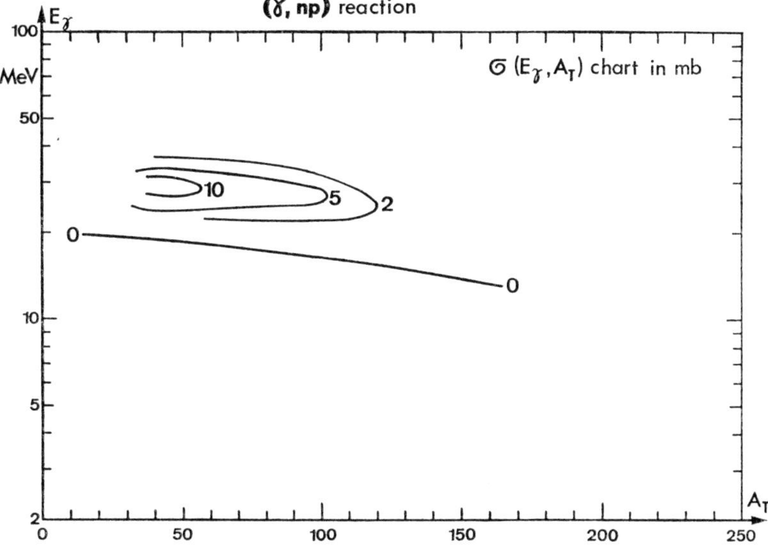

Fig. v.4 Smoothed approximation of the (γ, np) reaction cross-section.

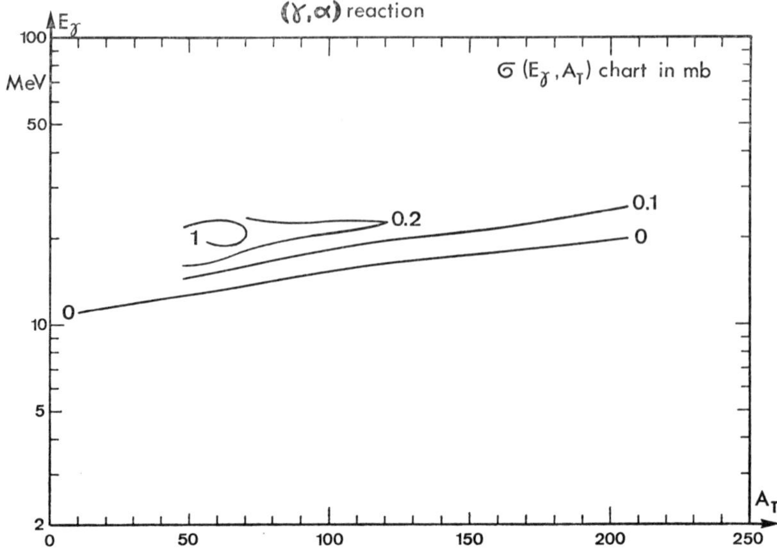

Fig. v.5 Smoothed approximation of the (γ, α) reaction cross-section.

The (γ,α) reaction has much lower maximum cross-sections (1 mb only) for $60<A<70$ and 20 MeV photon energy (fig. v.5).

The excitation functions of a few particular reactions which are sometimes used as monitors are given as examples in the following fig. v.6. They include the well-known $^{12}C(\gamma,n)^{11}C$ reaction as well as the reactions leading to ^{24}Na from ^{31}P, ^{27}Al and to ^{56}Mn from ^{59}Co. For the three latter reactions, the cross-sections are, of course, much smaller (about 100 times) and the peak is less marked than for the first reaction, indicating that spallation occurs in a rather constant manner above the giant resonance energy. However, we will not continue with the study of this new phenomenon now but will return to the giant resonance reactions with the aim of extracting from this compilation of cross-sections all that is required to arrive at precise values of induced radioactivity for a series of common materials.

We have seen from eqs. (2.4) to (2.6) that what we need in order to calculate the yields are the integrated cross-sections and the integrated cross-sections divided by k_0^2, k_0 being the photon energy at which the resonance cross-section peak occurs depending on the type of reaction and target mass number. As an example the giant resonance photon energies $k_0(\gamma,n)$ for the (γ,n) reaction indicated by various experimenters (fig. v.7) for a number of targets are plotted as a function of A_T. Some authors draw the straight line indicated as a best fit. A fine

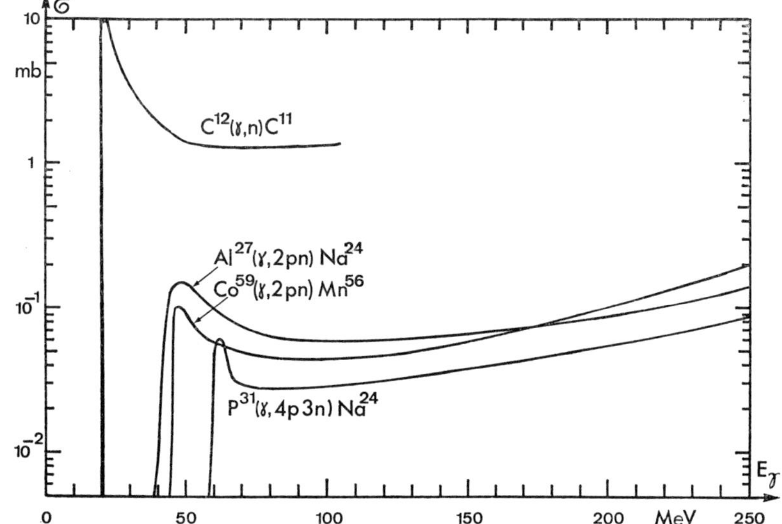

Fig. v.6 Excitation functions for some photonuclear monitor reactions.

structure of $k_0(\gamma, n)$ as a function of A_T is not recognizable with any certainty.

The integration of the cross-section curves over the photon energy, to which we now proceed, could have been made by using the charts

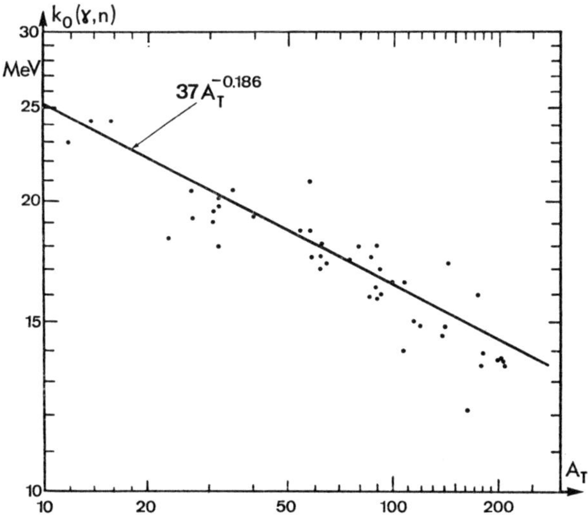

Fig. v.7 Giant resonance photon energy for the (γ, n) reaction as a function of target mass number.

just presented. However, the values found by the authors who have measured the resonance curves were used directly, plotted as a function of A_T and a smooth curve was drawn freely through the admittedly scattered points. From figs. v.8 and 9, which show this data for the five photonuclear reactions investigated, the reader will see that the curve is seldom more than a factor of 2 off the authors' values, and that no regular structure departing from this smooth curve can be reliably admitted, a finding which justifies our procedure. We have

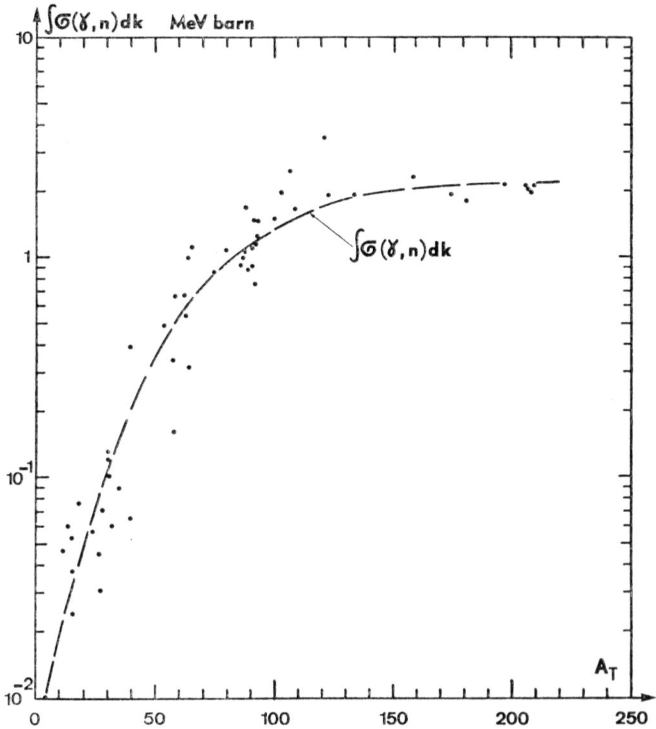

Fig. v.8 Integrated cross-section of the (γ, n) reaction in the giant resonance region plotted against target mass number.

thus at least gained values which represent well the order of magnitude of the integrated cross-sections of the five photonuclear reactions examined in the giant energy region and for all target mass numbers.

Further work has been done in dividing the integrated cross-section values by the squared resonance energy values k_0^2 in each case. The integrated cross-section values were this time taken from the smooth curves of figs. v.8 and v.9, whereas the k_0-values were still the indi-

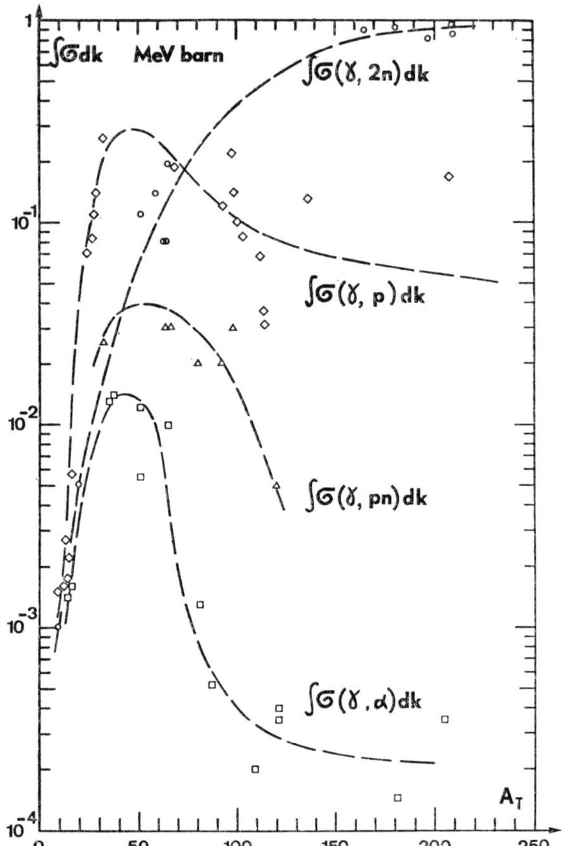

Fig. v.9 Integrated cross-section of the $(\gamma, 2n)$, (γ, p), (γ, pn) and (γ, α) reactions in the giant resonance region plotted against target mass number.

vidual values mentioned by the various authors without any attempt to smooth them out. The results of the integrated cross-sections are presented in figs. v.10 and 11. Only for the (γ, n) reaction are the points rather numerous. The other reactions are less well known and our curves are drawn through a smaller number of points, so that larger errors are to be expected here.

We now proceed to the calculation of the yield, i.e. the total number of photonuclear reactions of a given type in a cascade initiated by one electron, according to eq. (2.4). As this yield Y is directly proportional to the initial electron energy E_0, it is appropriate to calculate the yield per unit energy of the incident electron Y/E_0, a quantity which depends then only on the material and the nuclear reaction considered.

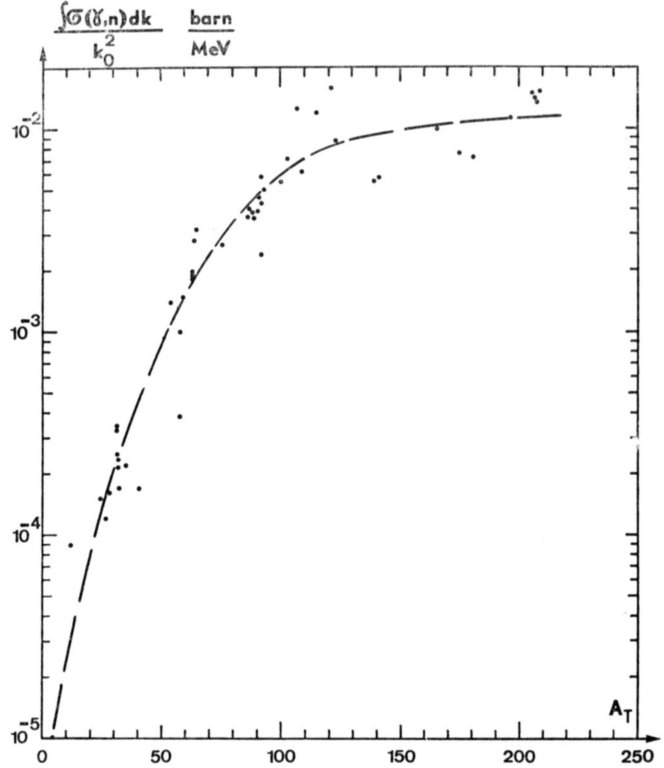

Fig. v.10 The integrated cross-section divided by the square of the resonance energy for the (γ, n) reaction plotted against target mass number.

The quantity Y/E_0 is the number of nuclear interactions per incident electron per MeV. A condition is, of course, that the electron energy E_0 be very large in comparison with the photon giant resonance energy k_0. As only giant resonance reactions are considered here, without regard to pseudo-deuteron disintegration and meson production effects, we shall from now on call the quantity Y/E_0 computed the 'giant resonance yield' per electron per MeV. As we have seen, it is enough to multiply by a factor $3.4 \times 10^{23}\, X_0/A$ the integrated cross-sections divided by k_0^2, X_0 being the radiation length of the electromagnetic cascade in the target material. For this radiation length a well-established formula exists (Rossi, High Energy particles, p. 220). As an example we give in the following table v.1 the values of X_0 obtained for a series of common substances as a function of their mass number A.

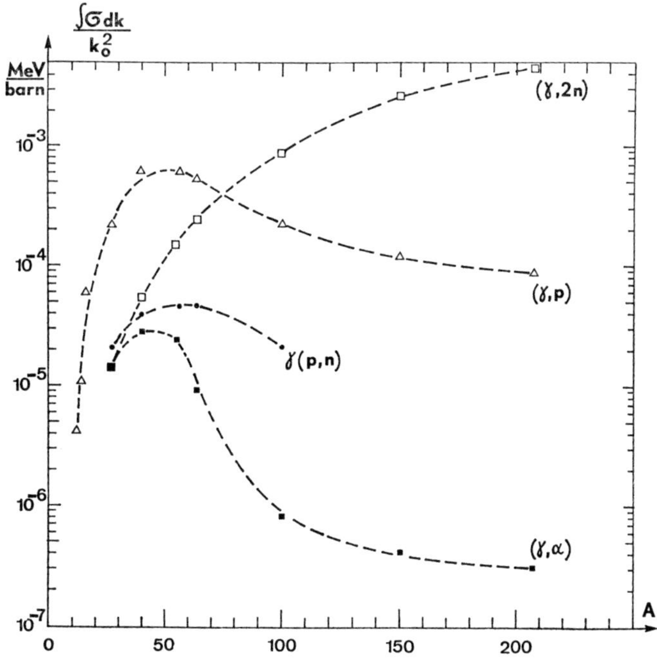

Fig. v.11 The integrated cross-section divided by the square of the resonance energy for the $(\gamma, 2n)$, (γ, p), (γ, pn), (γ, α) reactions plotted against target mass number.

TABLE v.1

Radiation length X_0 in various substances.

Substance	A	X_0 (g cm^{-2})
Carbon	12	44.6
Nitrogen	14	39.4
Air	14.78	37.7
Water	14.3	37.1
Oxygen	16	35.3
Aluminium	27	24.5
Argon	39.9	19.8
Iron	55.84	14.1
Copper	63.57	13.1
Lead	207.2	6.5

With these data it is possible to evaluate the 'giant resonance yields' per incident electron per MeV for all the elements and photonuclear reactions on which we have information. The results are presented in graphical form again in figs. v.12, 13 and 14, where the values of Y/E_0 obtained from the curves drawn in the preceding graphs have been computed for the (γ,n), $(\gamma,2n)$, (γ,p), (γ,pn) and (γ,α) reactions as usual.

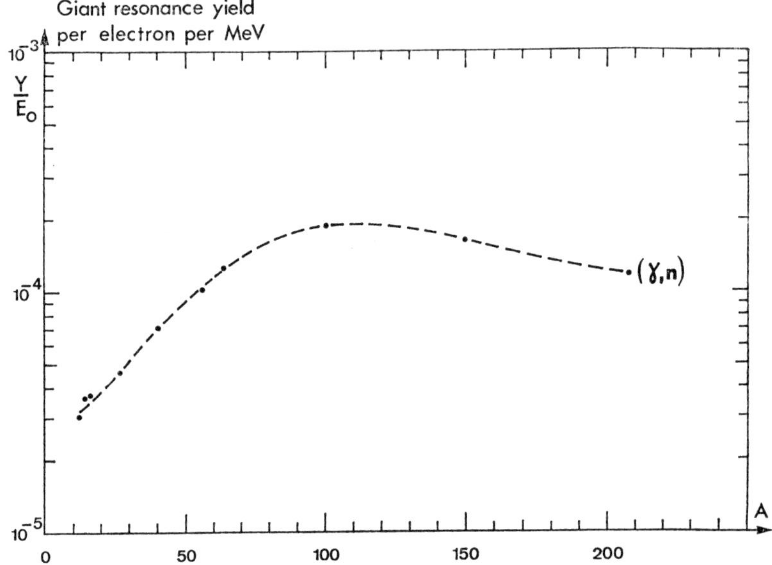

Fig. v.12 Giant resonance yield per electron per MeV for the (γ,n) reaction as a function of target mass number.

The variation of Y/E_0 with target mass number A is as follows. For the (γ,n) and $(\gamma,2n)$ reactions the yields increase steadily by an order of magnitude when going from $A=20$ to $A=100$, above which they tend towards a practically constant value of 10^{-4} for the first and 5×10^{-5} for the second reaction mentioned. For the other photonuclear reactions listed a maximum appears in the $A=40$ region for all curves.

The use of these curves is simple. To obtain the number of nuclear reactions of a given type taking place in a cascade initiated by one electron of energy E_0, one reads off the value of Y/E_0 and multiplies by the electron energy, provided this energy is much larger than the giant resonance photon energies k_0.

The absolute yields having been determined in this way, one now has to examine, for each reaction and target nucleus, what is the product of the reaction, whether it is radioactive or not, what is its probability of emitting gamma rays, and what is its k-factor (rad/h Ci at 1 m). One is then in a position to calculate the number of radioactive nuclei produced by all the nuclear reactions of each particular type induced in the whole cascade initiated by one electron and the resulting radiation field.

For this one has to assume a regular beam of electrons of given numbers falling per unit time on the target, in order to calculate the total induced activity produced by the whole cascades generated by these electrons in a target sufficiently large to absorb the beam completely. The induced activity is always defined as a decay rate and necessitates the introduction of a regular current of activating particles in our case where the incident flux per unit surface cannot be used, due to the subsequent development of the electromagnetic cascade. Also the dependence of the decay rate on activation and cooling times becomes apparent. For one single electron initiating the cascade, one can only calculate the number of the various kinds of radioactive nuclei produced, but there is no decay rate, as there is only one radio-

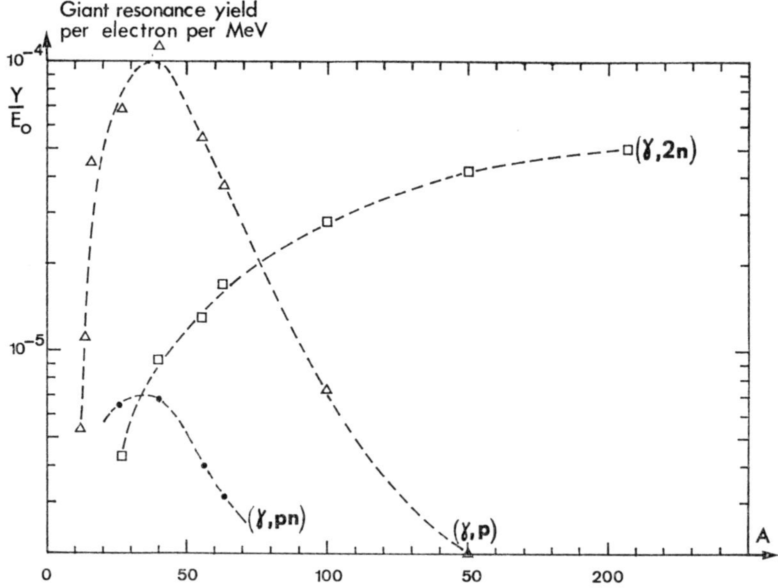

Fig. v.13 Giant resonance yield per electron per MeV for the $(\gamma, 2n)$, (γ, p), (γ, pn) reactions as a function of target mass number.

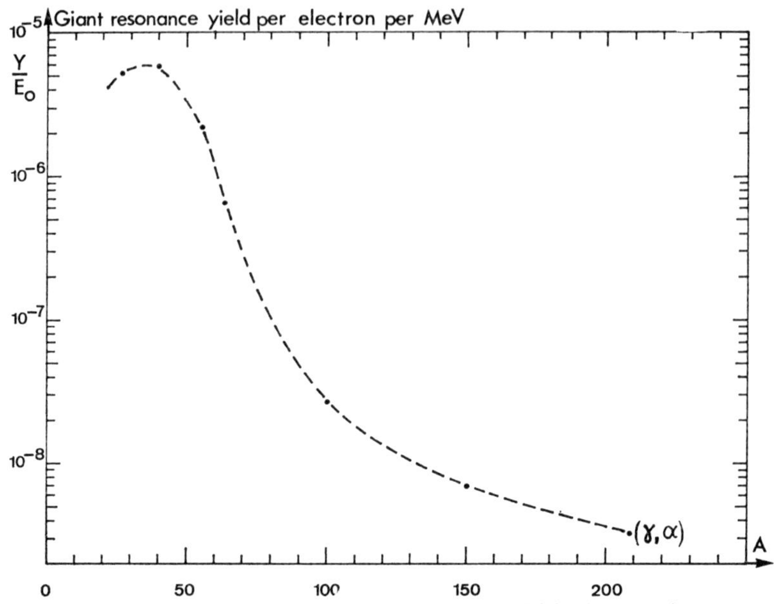

Fig. v.14 Giant resonance yield per electron per MeV for the (γ, α) reaction as a function of target mass number.

active nucleus decaying for each species. In contrast, as soon as one has introduced an electron current and found the decay rates, it is also possible to draw the decay curves of each process expected from a given target nucleus and sum up to obtain the magnitude and form of the decay for a volume of this material that has absorbed the showers produced by the electron current completely. The calculation is carried out as usual and is now briefly outlined.

4 Calculation of direct activation in the giant resonance region

In this section we shall show how to calculate the gamma disintegration rates, the beta disintegration rates, the radiation field in mrad/h at 1 cm from a point source containing all the radioactive nuclei produced. Each of these will be calculated assuming a flux of 1 electron/sec per MeV kinetic energy for a given irradiation time t_{irr} and a given cooling time t_{cool}.

Let us assume we know the yield of each particular isotope that can be made by giant resonance reactions in the cascade initiated by one electron of kinetic energy E_0 incident on a thick target containing 100% of the parent isotope, for each of the parent isotopes for which we have experimental data.

Let Y_ν be this yield, i.e. the number of radioactive isotopes of kind ν produced by one electron of kinetic energy E_0 incident on a thick target of pure parent isotope. We take a regular flux of 1 electron per unit time (second) striking the target for a time t_{irr} (irradiation time), and the problem is to find the activity of the target at a time t_{cool} (cooling time) after the electron flux has stopped.

The decay rate, normalized to one radioactive nucleus, is of course

$$-\frac{\mathrm{d}n_\nu}{\mathrm{d}t}(t_{\text{cool}}) = \frac{1}{t_\nu}\exp[-t_{\text{cool}}/t_\nu] \tag{4.1}$$

where t_ν is the time constant of the decay ($t_\nu = 1.442 t_{\nu,\frac{1}{2}}$). Hence if we have radioactive nuclei created at a constant rate which is equal to Y_ν, because we have assumed a current of 1 electron/sec, for a time t_{irr}, the decay rate at a time t_{cool} after the end of irradiation will be

$$-\frac{\mathrm{d}n_\nu(t_{\text{irr}}, t_{\text{cool}})}{\mathrm{d}t} = \int_{\tau=0}^{t_{\text{irr}}} Y_\nu \frac{1}{t_\nu}\exp[-(t_{\text{cool}} + t_{\text{irr}} - \tau)/t_\nu]\,\mathrm{d}\tau$$

$$= Y_\nu \exp[-t_{\text{cool}}/t_\nu](1 - \exp[-t_{\text{irr}}/t_\nu])$$

disintegrations per unit time of the isotope ν. (4.2)

Now, let $\varepsilon_{\nu\varkappa}$ be the number of γ-photons with energy $E_\varkappa$ emitted by each disintegration of the isotope ν. The gamma disintegration rate for ν is then given by

$$n_{\nu\gamma} = \sum_\varkappa Y_\nu \varepsilon_{\nu\varkappa} \exp[-t_{\text{cool}}/t_\nu](1 - \exp[-t_{\text{irr}}/t_\nu]) \tag{4.3}$$

photons per unit time at t_{cool} after irradiation for t_{irr} by a beam of 1 electron/sec.

Assuming that all the nuclei of type ν produced during t_{irr} are concentrated in a point, we may easily obtain the γ-photon flux at unit distance from this point source, i.e. the flux at 1 cm

$$\Phi_\gamma = \frac{1}{4\pi}\sum_\varkappa Y_\nu \varepsilon_{\nu\varkappa} \exp[-t_{\text{cool}}/t_\nu](1 - \exp[-t_{\text{irr}}/t_\nu]) \tag{4.4}$$

photons/cm^2 sec.

To convert this flux to a dose rate we use fig. I.10, in which is given the number $f_\varkappa$ of photons/cm² of energy $E_\varkappa$ needed to produce a dose of 1 rad. Thus we get the dose rate at 1 cm

$$R_\nu = \sum_\varkappa \frac{3.6\times 10^6}{4\pi} Y_\nu \frac{\varepsilon_{\nu\varkappa}}{f_\varkappa} \exp[-t_{\text{cool}}/t_\nu](1 - \exp[-t_{\text{irr}}/t_\nu]) \tag{4.5}$$

mrad/h for unit flux of 1 GeV electron.

The numerical factor 3.6×10^6 is for conversion from rad/sec to millirad/h. In practice we will not have a target of pure isotope, the target will contain several isotopes ν each of percentage p_ν. The dose rate at 1 cm due to this mixture is given by

$$R = \sum_{\nu,\varkappa} \frac{3.6\times 10^6}{4\pi} Y_\nu \frac{p_\nu \varepsilon_{\nu\varkappa}}{f_\varkappa} \exp[-t_{\text{cool}}/t_\nu](1 - \exp[-t_{\text{irr}}/t_\nu]) \text{ mrad/h}$$

$$= \sum_\nu p_\nu R_\nu \text{ mrad/h}. \tag{4.6}$$

The beta disintegration rate will be given by a similar formula:

$$n_\beta = \sum_{\nu,\mu} p_\nu Y_\nu \varepsilon'_{\nu\mu} \exp[-t_{\text{cool}}/t_\nu](1 - \exp[-t_{\text{irr}}/t_\nu]) \tag{4.7}$$

disintegrations per sec, where $\varepsilon'_{\nu\mu}$ is the number of beta particles of energy E_μ emitted by each disintegration of the isotope ν.

The decay curves calculated in this manner are presented for various materials in fig. v.15 which shows the number of photons emitted per second from radioactive nuclei in the target absorbing the cascade completely, for an electron current of 1 electron/sec and infinite irradiation time, per MeV energy of the electron beam.

The various materials are seem to differ enormously in their response to the activation in the electromagnetic cascade. The activities can differ by as much as two orders of magnitude at the beginning of the cooling period. Apart from this effect, one can distinguish two groups of target materials, those whose main activities have a half-life not extending above 1 or 2 days (Al, Cu, Ca, Pb), and those whose main activities have a much larger half-life and remain apparent for cooling times above 100 days (Fe, Ni, Co, Zn, W, Au). After 10 days cooling, for instance, only the activities of the second group remain, and at near maximum strength.

The curve for each target material is evidently the sum of the various exponential decay curves valid for each species of radioactive nuclides formed by each particular photonuclear reaction. For the interest and convenience of the reader, we give in appendix D for each of the target materials investigated tables and curves giving the decomposition of the total decay curve in its various constituents, indicating for each of these the particular isotope produced and the type of photonuclear reaction which has led to its production. With these tables and curves the reader can immediately deduce which are the main nuclear reactions responsible for the activity with each target

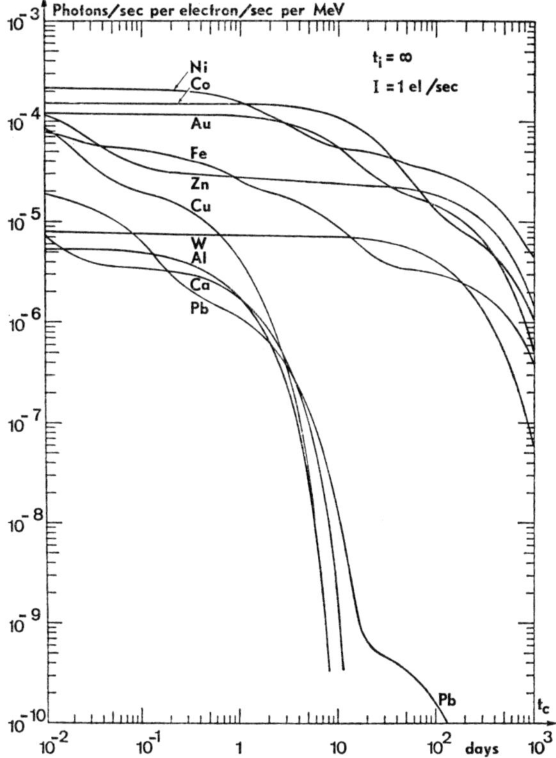

Fig. v.15 Total photon emission rate from radioactive nuclei in large targets of various materials irradiated by an electron current of 1 electron/sec, per MeV incident electron energy.

material, which are the radionuclides of interest, and what are their half-lives. The tables in the appendix thus make it possible to calculate quickly the decay curves for irradiation times other than long ones.

Another physical quantity that can be calculated with our procedure as shown above is the radiation field in rad/h at 1 metre from a supposed point source containing all the radioactive nuclei created in the electromagnetic cascade resulting from a steady current of 1 electron/second. Although the large target does not constitute exactly a point source, such numerical values can be used with convenience to compute the radiation field levels at distances large compared with the dimensions of the target block, provided self-absorption of the emitted gamma rays from the radioactive nuclei in the target is neglected. Ultimately corrections for this self-absorption can be applied, depending on the lateral extension of the target block with respect to the axis of the beam. Such data have been computed and are presented in fig. v.16, which shows the radiation fields at 1 metre from a point source containing all the radioactive nuclides produced in the cascade by a 1 electron/sec current, calculated per MeV incident electron energy and drawn as a function of cooling time. The two same groups of fast and slowly decaying target materials are also found here.

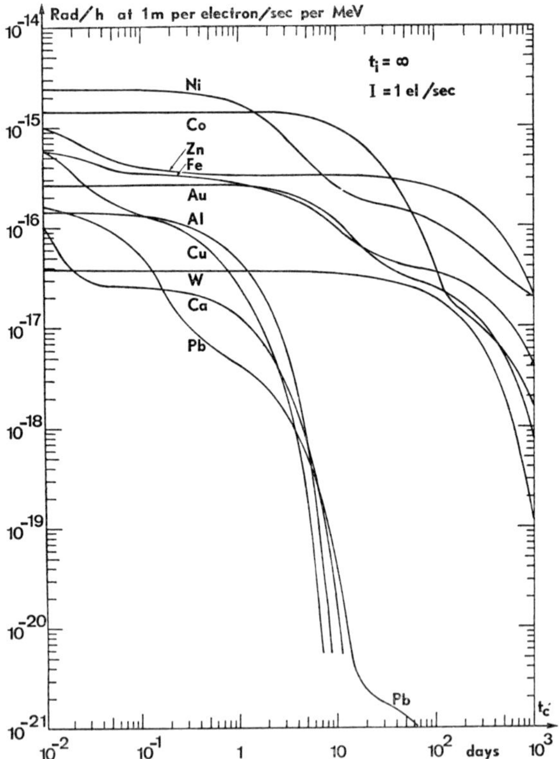

Fig. v.16 Radiation fields at 1 metre from a point source containing all the radioactive nuclei produced in a target by the electromagnetic cascade from a 1 electron/sec beam per MeV beam energy.

In the appendix the reader will also find for the radiation levels the full decomposition of the various decay curves into their respective constituents and be able to trace the important reactions and calculate decay curves for cooling times other than infinite.

5 Photofission

A particular type of photonuclear reaction which takes place with heavy target elements is photofission. The excitation curves found for the photofission of uranium and transuranic elements as americium, neptunium and plutonium exhibit a sharp and important peak at giant resonance energies between 10 and 20 MeV for the incident photon (fig. v.17). At higher photon energies the photofission cross-

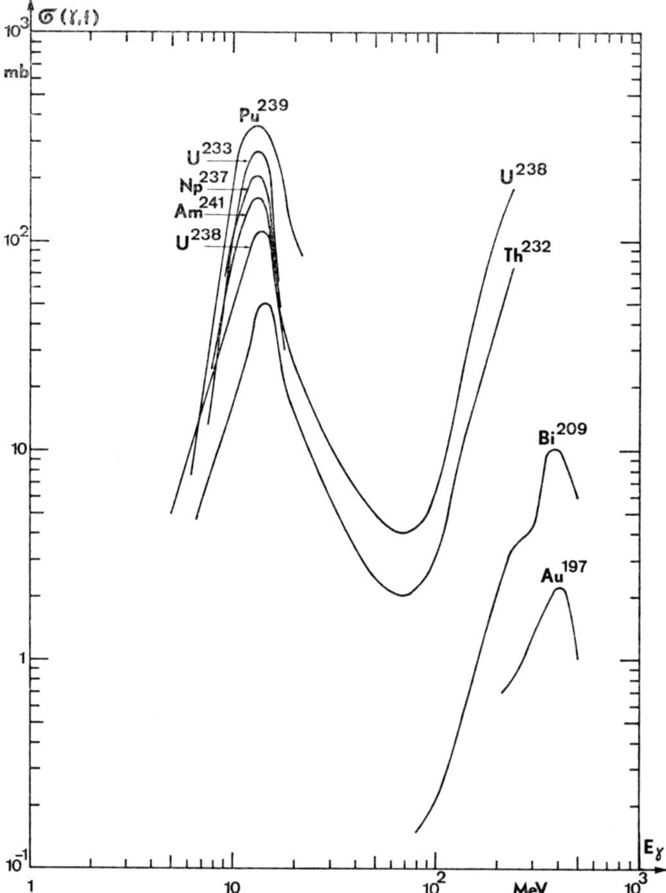

Fig. v.17 Excitation curves for the photofission reaction on heavy elements.

section decreases by two orders of magnitude up to 100 MeV. Above this value, it increases again sharply due to a new mechanism of fission which one could ascribe either to the neutron field in the pseudo-deuteron disintegration mechanism or to the incipient production of photopions in the nucleus or to both. For elements lighter than uranium, such as bismuth and gold for example, the giant resonance photofission has not been found. The photofission processes start above thresholds in the vicinity of 100 MeV photon energy and exhibit a maximum at about 400 MeV before decreasing again.

The distribution of the fission products from the photofission process has been studied as a function of the energy of the incident photon. For uranium for example the chain yields are given in the graph of fig. v.18 as a function of both the atomic number of the fission product

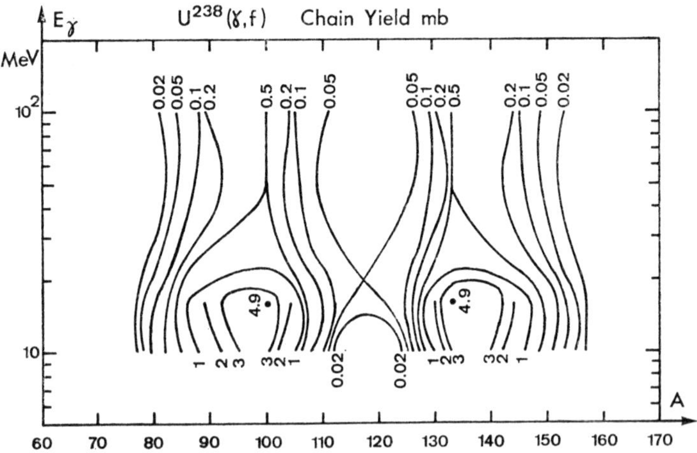

Fig. v.18 Chain yields for the photofission of ^{238}U, drawn as a function of the mass number of the fission products chain and the energy of the incident photon.

and the incident photon energy. At all photon energies between 10 and 100 MeV the fission product distribution is seen to have two well marked humps, in contrast with the chain yield distributions found for fission with high energy protons, where the two-hump distribution went gradually over to a one-hump distribution with increasing proton energy. Here the two-hump structure remains regardless of the energy. Only the magnitude of the chain yields is strongly dependent on energy, having a sharp maximum with peak values of 4.9 mb at 15 MeV photon energy. At 100 MeV photon energy the peak chain yields are only 0.5 mb.

6 Energy spectrum of photoneutrons and photoprotons from giant resonance reactions

Before we bring to an end the section on giant resonance reactions, it is appropriate to indicate what are the energies of the secondary particles produced in these reactions, such as neutrons in the (γ, n), $(\gamma, 2n)$, (γ, pn) reactions and protons in the (γ, p), (γ, pn) reactions, for instance. This information is of interest, as these particles activate further and in some cases one might wish to apply a correction. In fact, activation by secondary particles is found on the whole to be an effect of much smaller magnitude than activation by the direct electromagnetic cascade. Fig. v.19 shows a few typical neutron spectra as measured by G. Cortini et al. These authors have done extensive measurements in the energy region above 2 MeV neutron kinetic energy. The curves drawn here in fig. v.19 are theirs, extended to the left over energies lower than 2 MeV by using the expression

$$\frac{dN}{dE} = \frac{1}{E_{\max}}\left(\frac{E}{E_{\max}}\right)\exp[-E/E_{\max}]\,\text{MeV}^{-1} \tag{6.1}$$

where dN/dE is the number of neutrons per MeV kinetic energy interval and $E_{\max}$ the energy found from the slope of the exponential curves fitting the measured values best in the range 2 to 5 MeV.

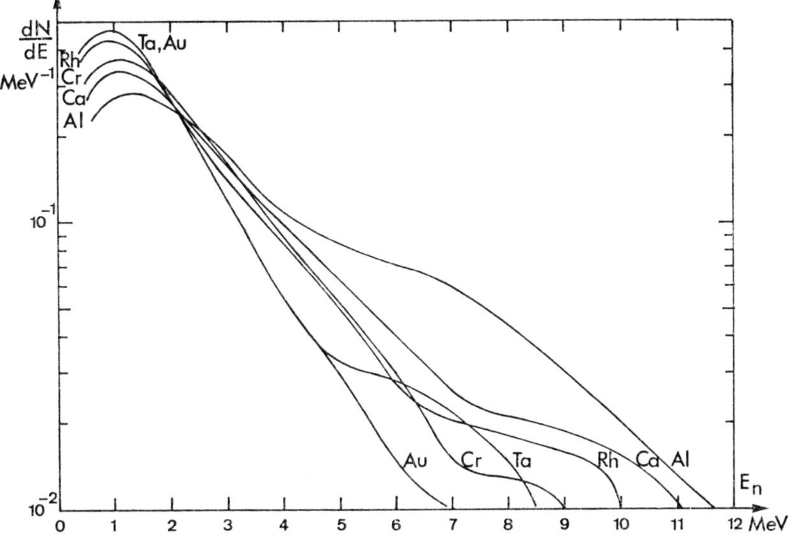

Fig. v.19 Neutron energy spectrum from giant resonance reactions.

On the whole, except at the tail at the high energy end, the bulk of these neutrons are seen to be evaporation neutrons. This spectrum is similar to that of reactor neutrons and the procedure outlined for reactor neutrons in ch. III should be applied when calculating the radioactivity these neutrons could induce in target or other materials.

The next fig. v.20 shows proton energy spectra from various authors found when bombarding different targets at different photon energies. The energy values and target materials are listed alongside the curves in each case. The curves are drawn in arbitrary units following the data given by the various authors, without any attempt to normalize them with respect to a well-defined incident flux power. The conclusion one can draw from inspection of fig. v.20 is that the protons emitted in photonuclear reactions have an energy range similar to that of the photoneutrons, say 0–15 MeV. The peaks of the distributions are somewhat higher, ranging from 1 to 10 MeV.

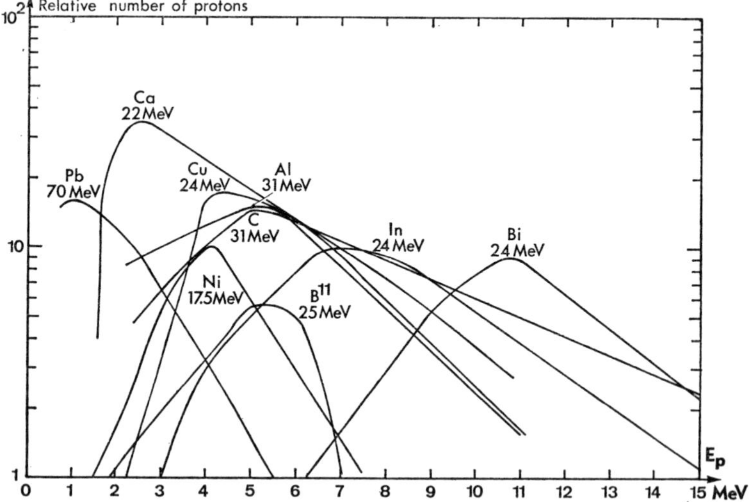

Fig. v.20 Energy spectra of photoprotons from giant resonance reactions.

However, activation of surrounding materials by photoprotons can be neglected in most cases, as one knows from compound nucleus data presented in ch. IV that the activation thresholds for protons usually lie above 10 MeV.

7 The pseudo-deuteron photodisintegration process

When a photon hits a nucleus, and the photon energy is above the giant resonance energy but below the meson production threshold, it appears that the photon interacts as if it 'saw' only two nucleons in the nucleus: either a proton–neutron or a proton–proton pair. Of course, one of the two must be charged, i.e. be a proton. In the case where there is a proton close to a proton interacting with the photon, electric quadrupole disintegration occurs. In the case where there is a proton close to a neutron, forming a so-called 'pseudo-deuteron', photodisintegration of this deuteron occurs. Let us examine this latter process, since the former is of less importance.

For every gamma ray absorption, leading to the photodisintegration of the 'pseudo-deuteron', a neutron and a proton will be emitted in opposite directions in the centre-of-mass system of the pseudo-deuteron. One of these, or both, or none may be subsequently reabsorbed in the nuclear matter before escaping from the nucleus. The momentum distribution affects the energies of the outgoing proton or neutron which would otherwise each have half of the gamma ray energy. Various authors (Levinger, Stein, Dedrick, Wilson, Panofsky, de Staebler) have investigated the photodisintegration of the deuteron, either free or in the nucleus. They have arrived at the following expression for the apparent pseudo-deuteron cross-section for photodisintegration σ_A in a nucleus of mass number A, neutron number N and charge number Z in terms of the photodisintegration cross-section σ_{D} of the free deuteron

$$\sigma_A(k) = \alpha \frac{NZ}{A} \sigma_{\mathrm{D}}(k), \tag{7.1}$$

k being the kinetic energy of the incident photon of which these cross-sections are a function. In this expression α is a coefficient determined experimentally, whose value was given as 1.5 in the earlier works (Panofsky, Dedrick) and is now taken as

$$\alpha = 4 \tag{7.2}$$

in the latest paper of de Staebler (SLAC 9). As regards the variation of the free deuteron photodisintegration cross-section $\sigma_{\mathrm{D}}(k)$ with the photon energy k, it is usually taken as nearly constant up to k=300 MeV and then varying as the inverse square of the photon energy (R. Wilson):

$$\sigma_{\mathrm{D}} = 7 \times 10^{-29}\ \mathrm{cm}^2 \quad \text{for } 80 < k < 300\ \mathrm{MeV} \tag{7.3}$$

and

$$\sigma_{\mathrm{D}} = \frac{6}{k^2} \times 10^{-24}\ \mathrm{cm}^2 \quad \text{for } k > 300\ \mathrm{MeV} \tag{7.4}$$

where k should be expressed in MeV and 1 barn $= 10^{-24}$ cm^2.

Hence we get for the deuteron photodisintegration within a nucleus

$$\sigma_A = 2.8 \times 10^{-28} \frac{NZ}{A}\ \mathrm{cm}^2 \quad \text{for } 80 < k < 300\ \mathrm{MeV} \tag{7.5}$$

and

$$\sigma_A = \frac{2.4}{k^2} \times 10^{-23} \frac{NZ}{A}\ \mathrm{cm}^2 \quad \text{for } k > 300\ \mathrm{MeV}. \tag{7.6}$$

Evaluation of the integral in fig. v.5 is now trivial. We have to take it between the limits 80 MeV and the maximum photon energy, which is equal to the maximum electron energy E_0.

$$\int_{80}^{E_0} \frac{\sigma_A \mathrm{d}k}{k^2} = 2.8 \times 10^{-28} \frac{NZ}{A} \int_{80}^{300} \frac{\mathrm{d}k}{k^2} + 2.4 \times 10^{-23} \frac{NZ}{A} \int_{300}^{E_0} \frac{\mathrm{d}k}{k^4}$$

$$= \frac{NZ}{A} \left[3.1 \times 10^{-6} - \frac{8}{E_0^3} \right] 10^{-24}\ \mathrm{cm}^2\ \mathrm{MeV}^{-1}. \tag{7.7}$$

For energies E_0 in excess of 200 MeV one can ignore the second term in brackets and obtain

$$\int \frac{\sigma_A \mathrm{d}k}{k^2} \approx 3 \times 10^{-30} \frac{NZ}{A}\ \mathrm{cm}^2\ \mathrm{MeV}^{-1}. \tag{7.8}$$

Substitution in eq. (2.5) gives for the number of interactions per electron of E_0 MeV which proceed by the deuteron photodisintegration mechanism

$$Y_{\mathrm{dp}} = 3.42 \times 10^{23} \frac{E_0 X_0}{A} \int_0^{E_0} \frac{\sigma(k)\,\mathrm{d}k}{k^2} = 10^{-6} \frac{NZ}{A^2} E_0 X_0. \tag{7.9}$$

The deuteron photodisintegration yield per incident electron per MeV would then be

$$\frac{Y_{\mathrm{dp}}}{E_0} = 10^{-6} \frac{NZ}{A^2} X_0. \tag{7.10}$$

For the radiation length X_0 one often uses the approximation

$$X_0 = 225 A^{-\frac{2}{3}} \,\mathrm{g\,cm^{-2}}. \tag{7.11}$$

The NZ/A curve itself is practically proportional to A:

$$NZ/A \approx 0.25A. \tag{7.12}$$

Thus in the end we have

$$\frac{Y_{\mathrm{dp}}}{E_0} = 5.6 \times 10^{-5} A^{-\frac{2}{3}} \tag{7.13}$$

pseudo-deuteron interactions per incident electron per MeV, a function which is shown in fig. v.21 as a function of atomic weight A. It can be seen from this figure that the total number of interactions by the

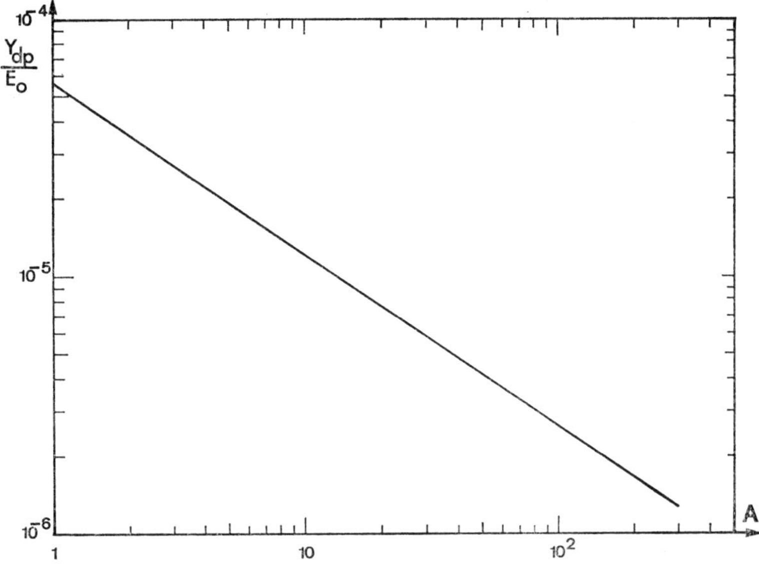

Fig. v.21 Yields per incident electron per MeV for the pseudo-deuteron photodisintegration process.

photodisintegration of the pseudo-deuteron is generally at least an order of magnitude less than in the giant resonance process (compare with the Y/E_0 curves in figs. v.12, 13 and 14). Especially in the $A=60$ region, where we have most of the structural metals, the ratios of Y_{dp}/E_0 to $Y_{(\gamma,\mathrm{n})}/E_0$, $Y_{(\gamma,\mathrm{p})}/E_0$ and $Y_{(\gamma,\mathrm{pn})}/E_0$ are 3.25×10^{-2}, 6.5×10^{-2} and 0.8. This last figure calls for a further remark. The photodisintegration mechanism of the pseudo-deuteron, the global yield of which has just been calculated, is a process which (seen from the outside of the nucleus) is similar to a (γ,n) or a (γ,p) or a (γ,pn) reaction according to whether the proton remains absorbed in the nucleus, or the neutron, or none of both. The probability that both proton and neutron can escape from the nucleus without either one being absorbed in the nuclear matter has been calculated by Stein. This probability-of-escape for both nucleons at the same time is a function of the target nucleus diameter and can be written as follows:

$$P(x) = \frac{3}{x^3}\left[2 - \mathrm{e}^{-x}(x^2 + 2x + 2)\right] \tag{7.14}$$

where

$$x = 2R/\lambda \tag{7.15}$$

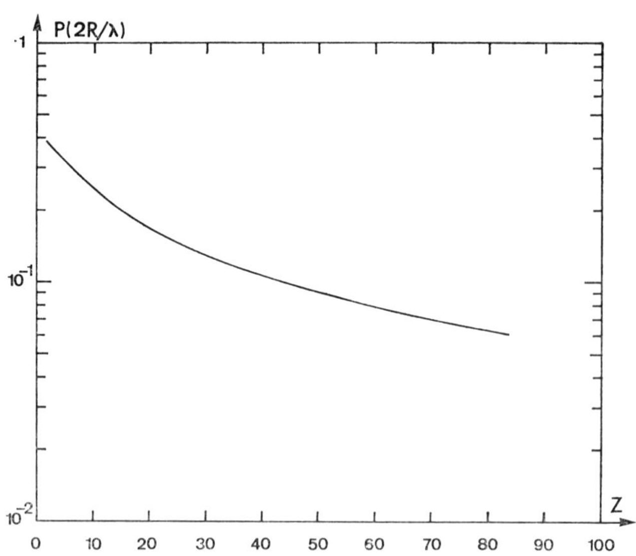

Fig. v.22 Probability of escape at the same time of both nucleons emitted in the pseudo-deuteron photodisintegration mechanism, as a function of target atomic number.

is the ratio of the nuclear diameter $2R$ to the mean free path λ of the nucleon in nuclear matter. Fig. v.22 gives a graph of this function. It is seen that for $A=60$ the probability that both p and n escape is 0.08 which makes this process less probable by this factor than the (γ, pn) process by the giant resonance mechanism.

Thus, as a whole, we can neglect the induced activity of the daughter nuclei produced in the cascade from the pseudo-deuteron disintegration process compared with that of the daughter nuclei produced in the same cascade via the giant resonance reactions. However, the importance of these reactions lies in the fact that they produce high energy nucleons, which can in turn activate as spallation-inducing particles, because of their high kinetic energy gained from the photon. The induced activity can then be computed by the methods given in ch. II. This pseudo-deuteron disintegration mechanism is of importance also for shielding calculations.

8 Nuclear reactions induced by very high energy photons

With the recent commissioning of the new very high energy electron accelerators the study of the photonuclear reactions above the meson production threshold (100–200 MeV) is progressing rapidly. However, the results are still preliminary and incomplete so that a complete review of the field is not really feasible for the moment. For this reason we will limit ourselves to present a few examples, which will enable the reader to get an idea of some of the phenomena and orders of magnitude involved.

We begin with a few excitation functions for the photoproduction of π-mesons from light nuclei (fig. v.23). The curves exhibit a more or less pronounced maximum. The order of magnitude of the cross-sections found for photoproduction of charged pions for low-Z nuclei is about 10^{-30} cm^2 per nucleon. The pion yields are found to increase with $A^{\frac{2}{3}}$. The probability of these reactions per event is thus about a thousand times less than in the giant resonance peak. However, as machines with electron energies of tens of GeV become available, the feeble cross-section can be balanced by the multiplicity of photons available in the shower after degradation of energy has occurred, and it will have to be decided in each particular case to what extent these events do contribute to the total activation.

Precise cross-section determination of photonuclear reactions is in practice difficult. When the high energy electrons impinge on the target

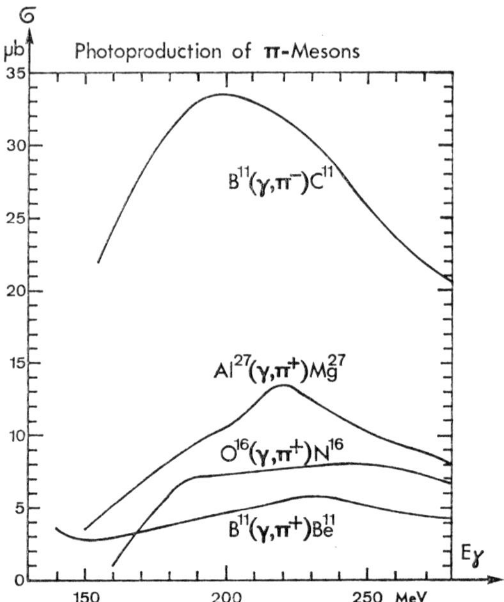

Fig. v.23 Excitation curves for the photoproduction of π-mesons from light nuclei.

(often referred to as 'radiator') which is used to convert them into photons through the 'bremsstrahlung' mechanism, the photon spectrum produced is a continuum, where the low energies are favoured and which goes towards zero spectral density after reaching the energy of the incident electrons. It is this spectrum which is then used for the investigation. What one actually measures are the so-called 'yields'. The yield is the production rate of a given radionuclide per g/cm^2 of target element when a standard amount of photon energy is being passed through the target per unit time. The target referred to here is, of course, the experimental target exposed to the photon beam, after clearing the beam coming out of the radiator target of its charged particles (electrons, pions, protons etc.) by a suitable sweeping magnet. The yields are usually expressed as a cross-section per equivalent quantum. The definition of the equivalent quantum is based on the form of the bremsstrahlung spectrum. Let $n(k, k_{\max})$ be the spectral density distribution of this spectrum, where k is the particular photon energy and $k_{\max}$ the maximum photon energy. The spectral density distribution is in fact a function of both. The energy content in the incident photon beam is

$$V = \int_0^{k_{max}} n(k, k_{max}) k \, dk. \tag{8.1}$$

If we divide this by the maximum energy k_{max} of the bremsstrahlung spectrum, we obtain a number of photons which is the number we would have if all the incident energy were equally distributed between fictitious photons of the maximum energy k_{max}

$$Q = k_{max}^{-1} \int_0^{k_{max}} n(k, k_{max}) k \, dk. \tag{8.2}$$

These fictitious photons are called 'equivalent quanta'. As one must have a number of photons to define a cross-section one uses this number to define it and so one arrives at the 'cross-section per equivalent quantum'

$$\sigma_{equiv.\ quant.} = Q^{-1} \int_0^{k_{max}} \sigma(k) \, n(k, k_{max}) dk, \tag{8.3}$$

where $\sigma(k)$ is the photonuclear reaction cross-section at the photon energy k. The incident photon energy per square centimetre can be measured by physical means, for instance with an instrument of the 'quantameter' type as described by Wilson. The numerator is obtained by activation measurements.

At high energies one can show that the spectral distribution $n(k, k_{max})$ can be approximated as Q/k where Q is the number of equivalent quanta.

$$n(k, k_{max}) \approx Q/k. \tag{8.4}$$

One can also show that in this case the number of quanta in a logarithmic interval equal to unity is exactly Q (see problem 4). Thus we get

$$\sigma_{equiv.\ quant.} = \int_0^{k_{max}} \sigma(k) \frac{dk}{k} = \int_0^{k_{max}} \sigma(k) \, d(\ln k) \tag{8.5}$$

and by taking the derivative

$$\frac{d\sigma_{equiv.\ quant.}}{d(\ln k)} = \sigma(k). \tag{8.6}$$

In this manner, by a differentiation procedure applied to the yield curves (the cross-sections per equivalent quantum as a function of energy), one arrives at the order of magnitude of the photonuclear reaction cross-sections. In practice such differentiation procedures can give rise to large errors. This is why the experimental data are usually presented by the authors in the form of cross-section curves per equivalent quantum given as a function of k_{max}, the maximum photon energy in the spectrum, which is also the electron or machine

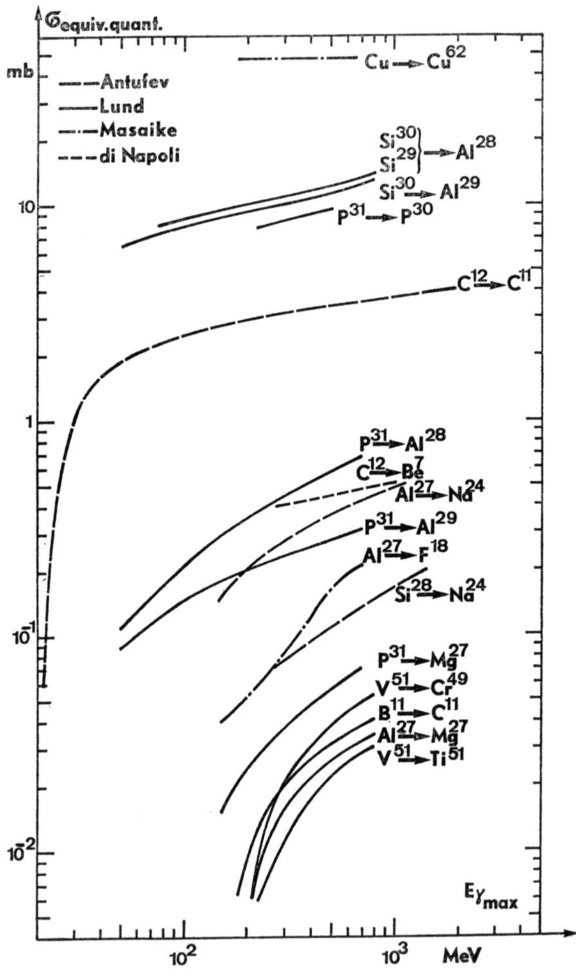

Fig. v.24 Yields of various photonuclear reactions at high energies as a function of maximum 'bremsstrahlung' energy.

energy used at the time of the experiment. The equation derived above provides also a good way of checking the yields found experimentally at high energies. This is to integrate the cross-section divided by k, when it is known at low energies and to see if one finds the expected values.

In the adjacent fig. v.24 we have given a few examples of nuclear photoreactions represented by their yield curves as reported by various authors. Several of them are especially important, as they are often used as monitor reactions, in particular, the reactions $^{12}C(\gamma,n)^{11}C$ and $^{27}Al(\gamma,2pn)^{24}Na$. Due to calibration difficulties there may exist systematic errors of the order of 10% between different laboratories. The sources of the data are the articles by Masaike, Antufev et al., di Napoli et al., and private communications of Lund university's photonuclear group. Recent measurements of (γ,n) yields up to 6 GeV have been made by de Carvalho and co-workers (private communication) and are shown in fig. v.25. The ^{12}C curve agrees with the data in the preceding figure.

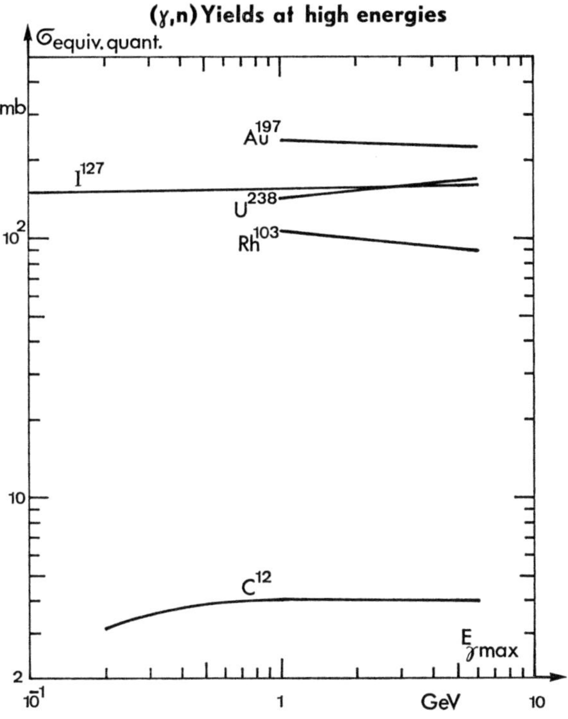

Fig. v.25 Yields of (γ,n) reactions at high energies.

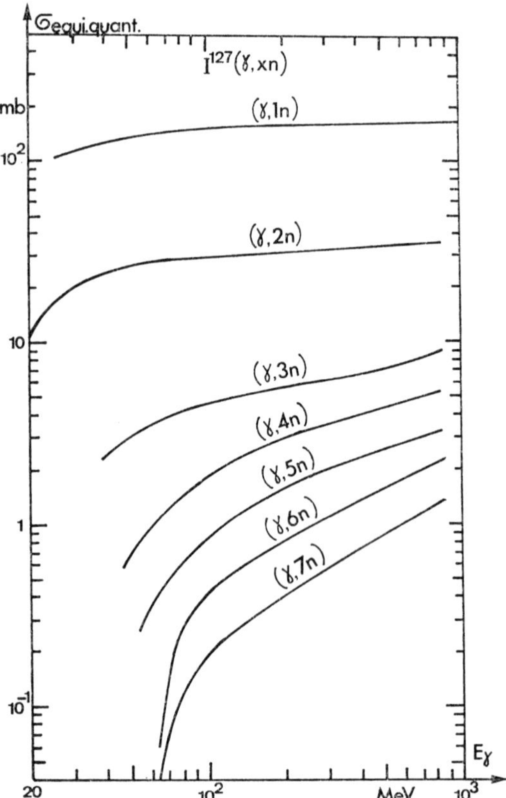

Fig. v.26 Yields of the (γ,xn) reactions on ^{127}I as a function of maximum 'bremsstrahlung' energy.

As a further particular example, we show in fig. v.26 the (γ,xn) reactions yields measured in a systematic way by Jonsson et al. One sees how the yields decrease with the complexity of the reaction, i.e. with increasing numbers x of neutrons expelled from the target nucleus, and how furthermore these complex reactions are favoured towards higher energies as appears from the increasing slopes of the curves with larger x.

9 Irradiation experiment with high energy photons on various materials

It is difficult to design irradiation experiments with high energy electrons and photons to check the calculations we have made, in particular the tables and graphs for the giant resonance reactions presented in the appendix. From the data given in sections 7 and 8 of this chapter and pertaining to the photonuclear reactions above the mesonic threshold, it appears that the cross-sections are very much smaller at high energies than in the giant resonance region. From the activation point of view the high energy of the incident electron is in fact used more to produce a large number of photons in the electromagnetic cascade it generates, than to produce a nuclear reaction.

It is thus appropriate to irradiate the material foils, not in the direct electron beam, but after the so-called 'radiator', in which an electromagnetic cascade is allowed to develop and which consists of a high atomic number material. The thickness of this radiator is made in such a way that it extends up to the point where the shower has its maximum density of electrons and photons. At this point the energy of the photons is also well degraded, so that one can expect large numbers of them to fall within the giant resonance region.

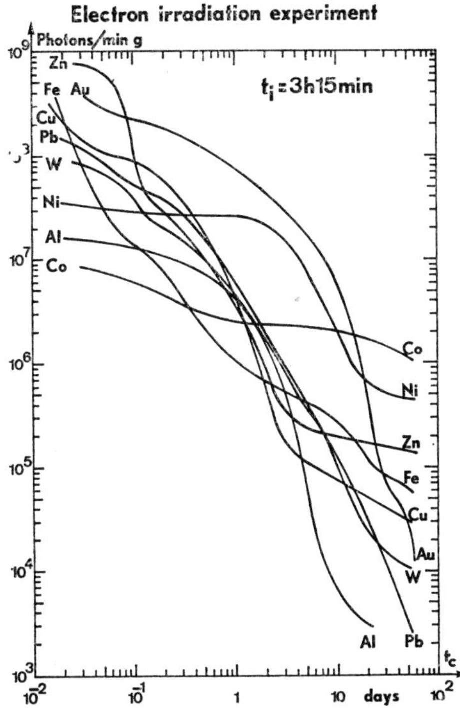

Fig. v.27 Decay curves of foils exposed 3 h 15 min behind 5 radiation lengths of lead to an electromagnetic shower generated by 6 GeV electrons.

In this experiment we would like to mention, the 6 GeV external electron beam of the DESY accelerator in Hamburg was made to fall on a 25 mm thick lead plate, representing 5 radiation lengths. The beam cross-section was 1 cm² in the front of the radiator block and smaller than the foils at the exit (2×2 cm²). The mean electron current was 0.57×10^{11} electrons/sec, the irradiation time 3 h 15 m and the total charge delivered by the beam 1.07×10^{-4} Coulomb. The foils were counted with a NaI crystal at a distance of 3 cm, the counter being set with a threshold of 75 keV gamma energy. The counts registered with this arrangement were then normalized to 4π solid angle and divided by the foil weights, so as to obtain the total number of photons emitted in the whole space per minute per gram of material. These results are shown in fig. v.27 (Bathow, Freytag, Tesch, private communication).

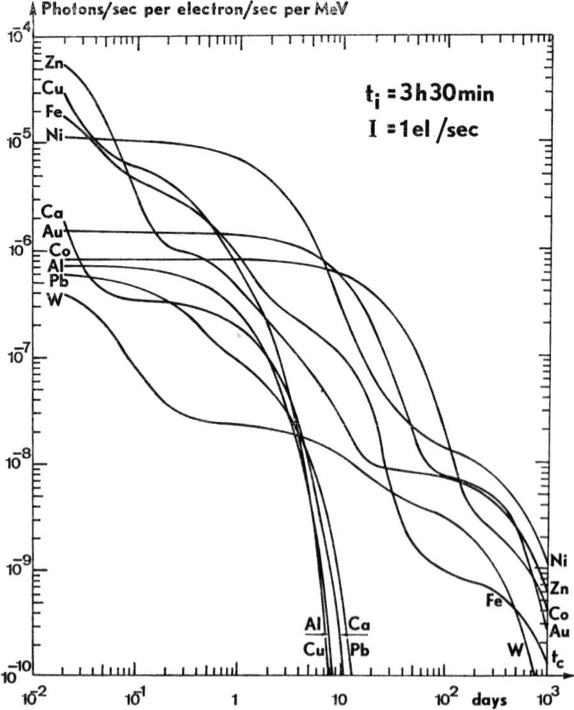

Fig. v.28 Total photon emission rate from all radioactive nuclei produced in large targets of various materials irradiated during 3 h 30 min by an electron current of 1 electron per second per MeV incident electron energy.

For the purpose of comparison, decay curves of the same materials and 3 h 30 min irradiation time have been calculated on the basis of the theory developed in sections 1 to 4 of this chapter and are presented in fig. v.28.

It is to be stressed here that what was actually calculated, i.e. the total radioactivity generated by the shower in a target block absorbing it completely, is not exactly comparable to what was measured experimentally, i.e. the local activation at the maximum of development of the cascade. However, the curves compare favourably, in particular a number of the main radioactive isotopes calculated are recognizable in the experimental curves.

PROBLEMS

CHAPTER V

1. Give the maximum cross-section, the approximate energy at which it occurs and the product of the following reactions:

(γ, n) on ^{19}F, ^{23}Na, ^{55}Mn, ^{59}Co, ^{115}In, ^{197}Au;
$(\gamma, 2n)$ on ^{63}Cu, ^{197}Au;
(γ, p) on ^{40}A;
(γ, np) on ^{24}Mg, ^{40}A;
(γ, α) on ^{65}Cu.

2. Calculate the total photons/sec per electron/sec per MeV for a large aluminium target having absorbed the beam completely for an irradiation time of only 3 hours 30 minutes.

3. The energy of a beam of photons is found with a quantameter to be 1 Wsec, the maximum photon energy being 6.25 GeV. What is the number Q of equivalent quanta?

4. Assume the bremsstrahlung spectrum is C/k from $k=0$ on, where C is a constant. Calculate the number Q of equivalent quanta and show that it is equal to C. Show that Q is also equal to the number of real quanta present in a logarithmic interval equal to 1.

CHAPTER VI

SOME ASPECTS OF RADIOACTIVITY INDUCED BY HEAVY IONS

1 Introductory remarks

Research on nuclear forces and radiochemistry carried out with heavy ion accelerators acquires more and more importance as the time goes on. The number of installations delivering heavy ion beams increases steadily. Three types of machines have broken through for this kind of work: Van de Graaff electrostatic generators, linear accelerators and cyclotrons. The recent appearance of sector-focused isochronous cyclotrons has allowed a considerable increase in available beam current, resulting in a renewed interest of the scientific community in the whole field.

The radioactivity induced by heavy ions at the energies available nowadays (10 MeV per nucleon) is due to many different types of nuclear reactions, only a few of which have been investigated in any detail. This is due to the complexity of the projectile, which is now a nucleus, in contrast with the cases hitherto examined, where the projectiles in question were protons, neutrons, photons and possibly deuterons and alpha particles. The heavy ion, being an assembly of protons and neutrons bound by nuclear forces, is able to cause many new modes of interactions compared with a single particle. It is of interest to stress a few of these differences. First, the size of the projectile being larger, the total interaction cross-sections will be higher. Due to the charge of the heavy ions, the Coulomb barrier effects will be here more pronounced. Also stripping effects are to be expected in the case of grazing collisions, one or a few nucleons being transferred from the target to the projectile or vice versa. Compound nucleus reactions with complete fusion of both interacting nuclei will lead to nuclides of atomic weight much higher than that of the target. These will be in a high state of nuclear excitation and in particular have very high angular momenta, due to the large angular momentum imparted at the moment of collision by the heavy incoming projectile. In this high excited state, the easiest way to cool down is to boil off one or several neutrons. These so-called (HI,xn) reactions, where HI stands for heavy ion and x for the number of neutrons evaporated by the compound nucleus, have been studied extensively, but respresent only a small part of what an excited compound nucleus can do. More complex modes of disintegration of the compound nucleus are found in great numbers, with subsequent ejection of several charged as well as uncharged nucleons or clusters thereof. As the bombarding energy increases, the incident ion penetrates more and more the target nucleus in a manner similar to that described for high energy single nucleons, and effects leading to complete spallation of the target are observed, with a wide variety of possible products.

Other reactions of interest produced more easily with heavy ions than with single nucleons include, for instance, fission of heavy elements. Due to the weight of the projectile which combines with the target nucleus, fission phenomena are observed with lighter targets than in the case of high energy protons. In the cases where the nucleus so formed does not fission, but de-excites by boiling off a few neutrons or by some kind of spallation, it has become possible to produce a quantity of hitherto unknown neutron-deficient isotopes in the top region of the nuclides chart, say between gold and uranium, to the left of the beta stability line. Even nuclides which decay by proton emission have also been found in this region (Flerov, 1964). Another use of heavy ion accelerators is the production of new elements in the region above uranium by the bombardment of natural uranium or still heavier reactor-produced elements as targets. The heaviest presently known transuranics have been prepared in this unique manner, sometimes in only minute quantities.

It would be a little far-bringing to give a detailed account of the knowledge available to date on heavy ion reactions, which is dispersed in the literature. Instead we shall present on each type of reaction we have mentioned a few selected examples of data suitable for giving an impression of the features of the phenomenon considered and an idea of the order of magnitude of the cross-sections which have been measured. We should stress that the field of heavy-ion induced reactions is far from being investigated on a large scale, and that the information at hand is clearly insufficient to cover by interpolation a large range of targets and projectiles, even for one selected type of interaction. As projectiles, carbon, oxygen, nitrogen and neon are practically the only elements that have been used on a large scale to the present date. Also the target elements chosen by the investigators are usually light elements, seldom going above $A=50$. In a few cases, elements ranging from gold to plutonium have been used. Also the energies at which the experiments have been made are strictly limited to the range of the machine of the laboratory where the investigation was made. Thus the reader has to consider the compilation presented in this chapter more as a series of particular examples than an attempt to present a formal and complete picture of the nuclear reactions induced by heavy ions. It is hoped that the present work will help those interested in nuclear reactions to select target, projectile and bombarding energy best suited to the needs of their particular experiment.

As to the question of the radiation fields from induced activity around heavy ion accelerators, attention must be drawn to a particular feature of the heavy ions, their very small range in matter, due to their high electric charge and strong ionizing properties. In fact, the only components activated directly by the heavy ions are the targets and the parts of the accelerating system which the ions hit. The radioactive layers are thus very thin, much thinner than the gamma radiation attenuation lengths, so that one can assume in this case that the radioactivity is deposited at the surface of these components and that no self-absorption of the gamma rays occur. Apart from these parts, which are activated primarily and must be looked after with extreme care, all the surrounding structure of the accelerator and the associated equipment is also liable to become radioactive. This is because of the large quantity of neutrons liberated in any heavy-ion induced reaction. These neutrons are of the evaporation type and have spectra resembling reactor neutrons, with maximum energies of about 10 to 15 MeV; they traverse large amounts of matter and can activate all around. They can be dealt with in the manner indicated in ch. III.

We shall now go through the various types of heavy-ion induced nuclear reactions on which we have information and we will terminate giving some values of the gross numbers of secondary neutrons produced by heavy ions stopped in matter.

2 Total reaction cross-sections for heavy ions

The first kind of data which is basic for our study if the total nuclear reaction cross-section when two nuclei collide. The situation is as follows. There is a geometrical cross-section, which is calculated in the assumption that each nucleus is a sphere with a radius

$$R_0 = r_0 A^{\frac{1}{3}} \tag{2.1}$$

where A is the mass number of the nucleus and r_0 a radius parameter, usually taken as

$$r_0 = 1.5 \times 10^{-13} \text{ cm.} \tag{2.2}$$

The geometrical collision cross-section is then equal to the surface of a disc whose radius is the sum of the radii of the two colliding particles:

$$\sigma_{\text{geom}} = \pi r_0^2 (A_1^{\frac{1}{3}} + A_2^{\frac{1}{3}})^2 \tag{2.3}$$

the indices 1 and 2 referring to the particles considered.

The geometrical cross-section is, of course, independent of energy. It is useful to have a chart of the geometrical cross-sections at hand, to figure out an upper limit for total nuclear reaction cross-sections. Fig. VI.1 shows such a chart, where lines of equal geometrical cross-

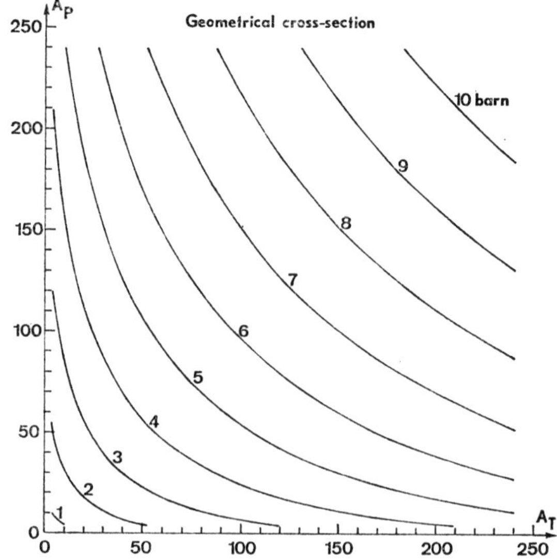

Fig. VI.1 Geometrical cross-sections for nucleus–nucleus collisions. A_T: target mass number, A_P: projectile mass number.

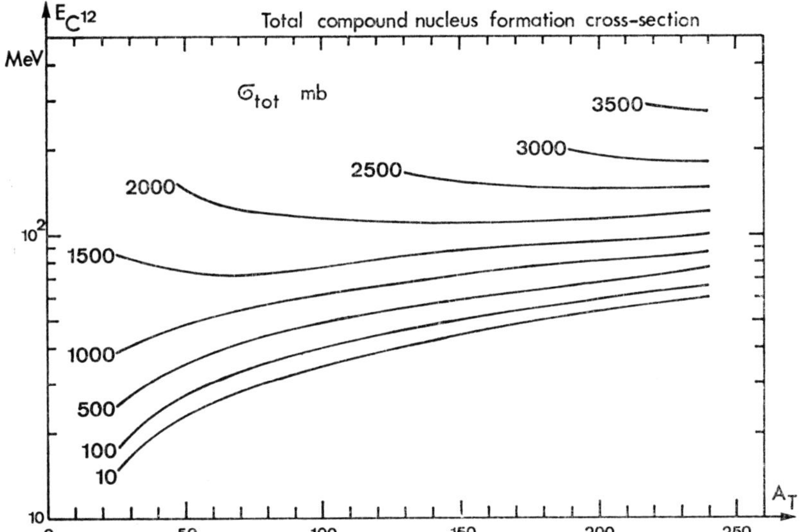

Fig. VI.2 Compound nucleus formation cross-sections for ^{12}C ions as a function of bombarding energy and target mass number.

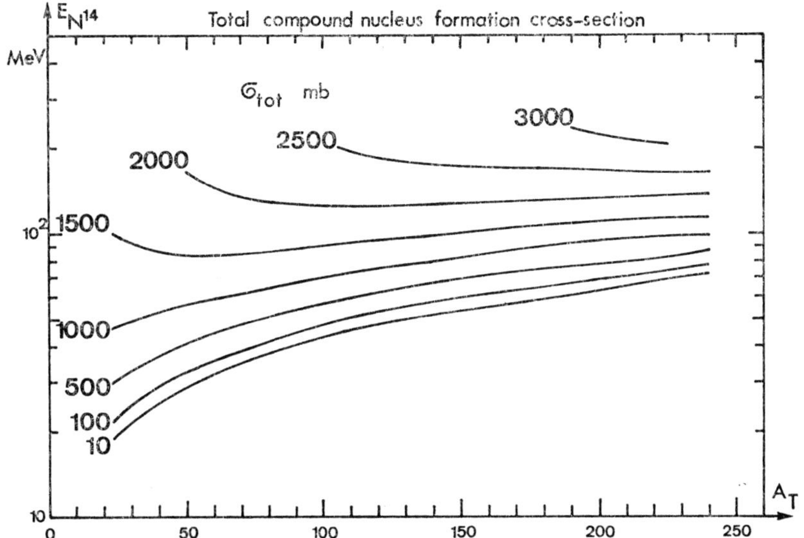

Fig. VI.3 Compound nucleus formation cross-sections for ^{14}N ions as a function of bombarding energy and target mass number.

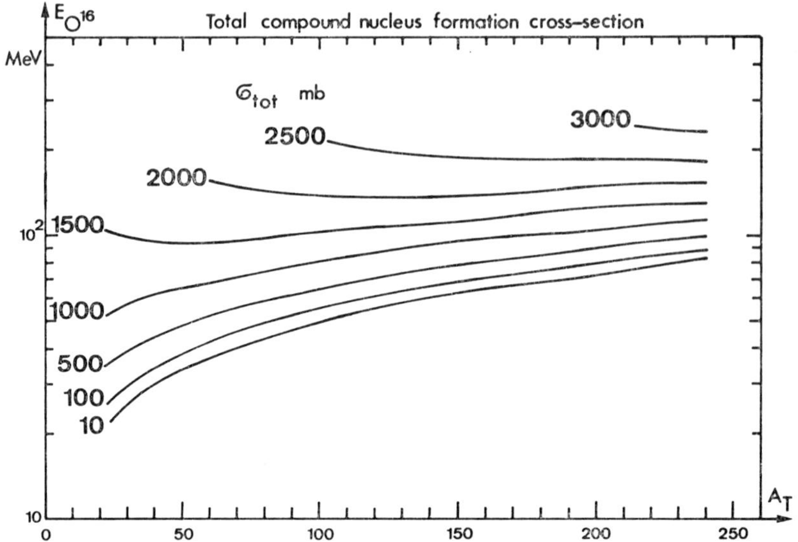

Fig. VI.4 Compound nucleus formation cross-sections for ^{16}O ions as a function of bombarding energy and target mass number.

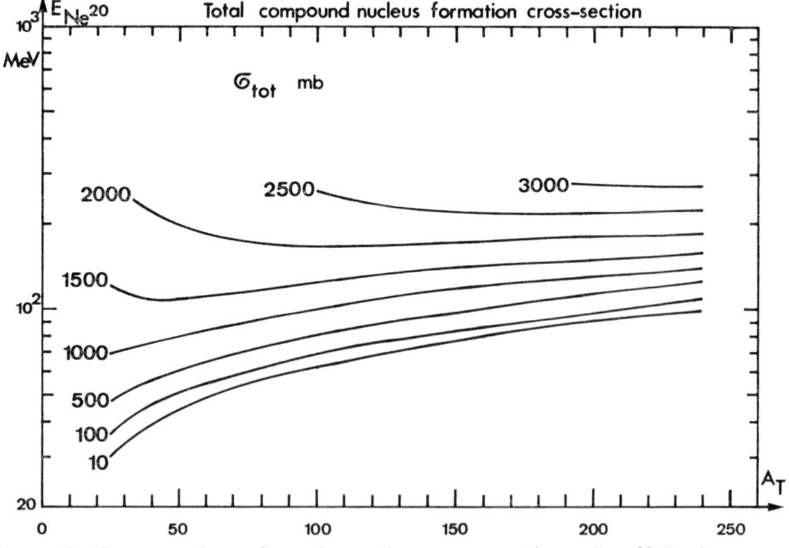

Fig. VI.5 Compound nucleus formation cross-sections for ^{20}Ne ions as a function of bombarding energy and target mass number.

sections are drawn as a function of the projectile and target atomic numbers.

Geometrical cross-sections range from 0.7 barn for the alpha–alpha case to 10 barns for the uranium–uranium case.

Another typical cross-section is often used for reference besides the geometrical one. Heavy-ion induced nuclear reactions proceed in most cases by the compound nucleus mechanism whereby the projectile and target nuclei form after the impact one so-called compound nucleus, which can then disintegrate in various ways. The usual postulate in this theory (Blatt and Weisskopf) is that there is no nuclear interaction when the closest distance of approach of the nuclei is greater than R_1+R_2 (the sum of the radii of the nuclei considered). When this distance is smaller than R_1+R_2 there is a probability of nuclear interaction which can be calculated using various models of nuclear forces and taking also into account the potential energy of the system before the collision, i.e. the Coulomb repulsion and the angular momentum involved. With these, the compound nucleus formation cross-section can be found. The cross-sections one gets are smaller than the geometrical cross-sections and tend towards them at high bombarding energies. Numerical computations of this kind have been made by T.D. Thomas for carbon, oxygen, nitrogen and neon projectiles and various targets, permitting graphical interpolation for other targets than those taken for the computation. Fig. VI.2 to 5 show graphs of the compound nucleus formation cross-section drawn for these projectiles as a function of target mass number and ion bombarding energy. They are similar in shape, although they have higher absolute values, to the total nuclear cross-section graphs for protons, deuterons and alpha particles given in ch. IV. Because the overwhelming majority of nuclear reactions proceed via the mechanisms taken into account for the calculation of the compound nucleus formation cross-section, this cross-section is often referred to in the literature as the total nuclear reaction cross-section. An exception are the nucleon transfer reactions, which only happen in grazing collisions and are significant only at energies lower than or approaching the Coulomb barrier, where the compound nucleus cannot yet be fully formed. We will examine them to begin with in the next section.

3 One or several nucleons transfer reactions

When examining in turn the main kinds of nuclear reactions induced by heavy ions, it is appropriate to begin with the transfer reactions. These reactions, also called pick-up or stripping reactions, happen when the projectile and target nuclei experience a grazing collision, i.e. a collision where they hardly come in contact with each other, in contrast with the case, which will be examined next, where one has complete fusion of the two nuclei into a compound nucleus. In grazing collisions, the principal effect is that one or several nucleons are

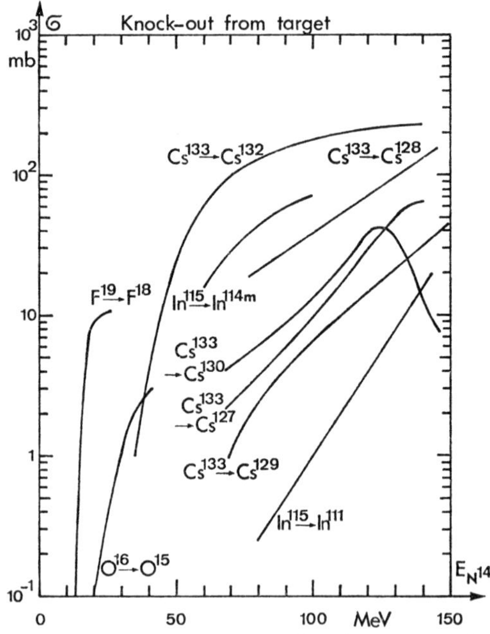

Fig. VI.6 Transfer reactions induced by ^{14}N ions on various targets. Case where the incident ion knocks out one or several nucleons from the target.

simply transferred either from the projectile to the target, or vice-versa. In some cases, up to 6 neutrons have been observed to pass from one nucleus to the other. Charged particles or charged and uncharged particles at the same time can be also exchanged, as one will see from the graphs presented in fig. VI.7. We have limited ourselves to one single projectile, the ^{14}N ion, and have shown the excitation functions for various reactions on various target nuclei. In fig. VI.6 are grouped neutron pick-up reactions by the projectile with the elements ^{16}O, ^{19}F, ^{115}In and ^{133}Cs as targets. With the heavier targets the cross-section values are seen to reach a few hundred millibarns, which is already an appreciable fraction of the compound nucleus formation cross-sections, which amount to 1500 mb in the 100 MeV region. The next fig. VI.7 shows the reactions in which it is the target that picks up nucleons from the projectile. Besides the one-neutron transfer, one finds also cases where the nucleon groups pn, 2pn, 2p2n and 2p have been transferred to the target. The cross-sections appear to be lower than in the preceding fig. VI.6.

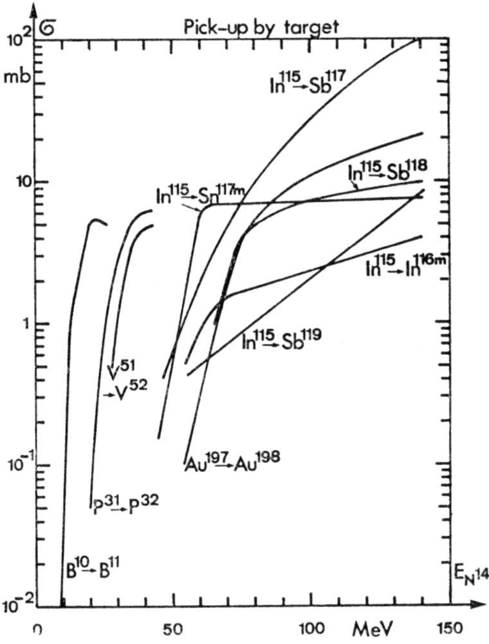

Fig. VI.7 Transfer reactions induced by ^{14}N ions on various targets. Case where the target picks up one or several nucleons from the incident ion.

4 Neutron evaporation from the compound nucleus

By far the larger proportion of events when two complex nuclei interact lead to the formation of a compound nucleus by complete fusion of two colliding bodies. This compound nucleus is always in a high excited state due to the high kinetic energy brought about by the incident nucleus. The whole kinetic energy, however, is not available as excitation energy because the compound nucleus obtains a high rotational energy from the angular momentum received at the moment of the impact, when the trajectory of the incident ion does not pass exactly

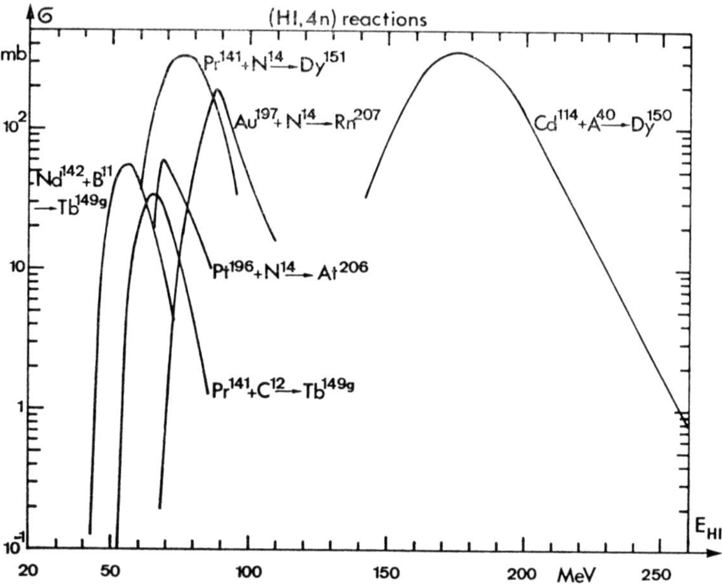

Fig. VI.8 Excitation functions of the (HI, 4n) reaction for various targets and projectiles.

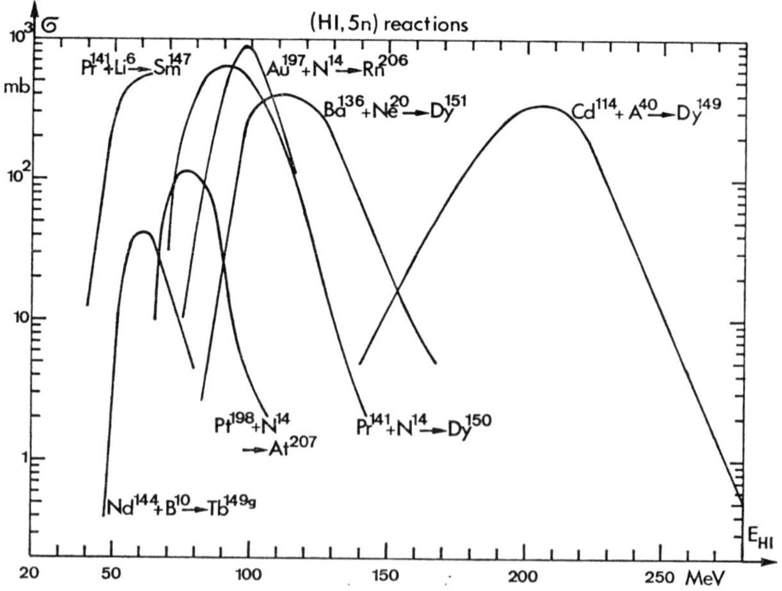

Fig. VI.9 Excitation functions of the (HI, 5n) reaction for various targets and projectiles.

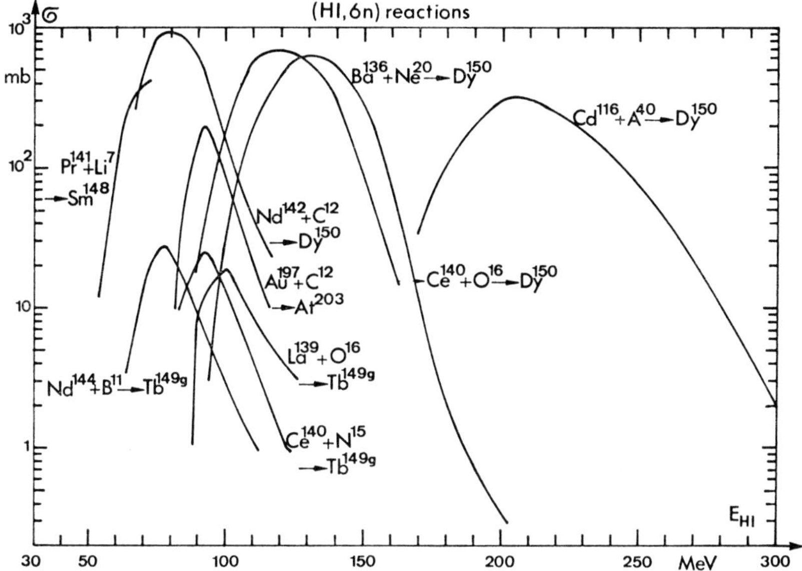

Fig. VI.10 Excitation functions of the (HI, 6n) reaction for various targets and projectiles.

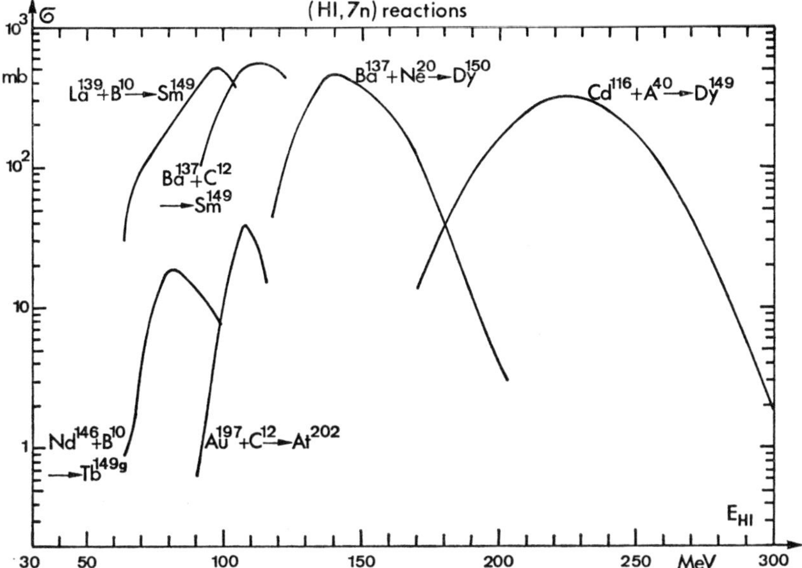

Fig. VI.11 Excitation functions of the (HI, 7n) reaction for various targets and projectiles.

through the centre of the target nucleus. Thus large variations in excitation energies are to be expected in the compound nucleus. There are many ways in which the excited compound nucleus can come to a stable state. Either it is stable and then it de-excites by gamma emission, or it is unstable and then it can de-excite by beta or alpha decay, or by evaporation of a few neutrons, or by emission of a number of charged and uncharged particles (spallation) or by fission into two or more large nuclei. Some neutron-deficient compound nuclei produced with the help of heavy ions have even been found to display proton decay.

We will leave aside the simple cases where the compound nucleus is stable, or is known to decay by beta or alpha emission, as they appear to be relatively rare. Usually the compound nucleus does not remain in the form in which it has been produced owing to the fact that it lies far away from the beta stability line, generally to the left, and has high excitation energy. We shall examine in turn the cases where it boils off neutrons, where it experiences spallation, and where it fissions, which have been studied to some extent. The rest of the present section will be devoted to the neutron boil-off reactions from the compound nucleus, usually called (HI,xn) reactions, x being the number of neutrons given off by the compound nucleus.

Little data is available on (HI,n), (HI,2n), (HI,3n) reactions, because they are either rare or difficult to observe. In contrast a number of investigators have gathered information on the reactions (HI,4n) up to (HI,8n). Fig. VI.8 shows the cross-sections of the first of these reactions measured for a variety of targets and projectiles as a function of bombarding energy. The heavy ions used include ^{11}B, ^{12}C, ^{14}N, ^{40}A. One observes the typical form of the excitation function for a compound nucleus reaction, i.e. a sharp rise, a narrow maximum region, and a sharp fall. The lightest projectiles have the lower thresholds (about 40 MeV for the boron ion), the excitation peaks shifting slowly towards higher bombarding energies as the atomic number of the projectile increases. The maximum values of the cross-sections can reach in some cases several hundred millibarn, or be ten times smaller in other cases. They can decrease by as much as three orders of magnitude when the energy of the incident ion increases due to competition with many other nuclear reactions, an effect which is never seen with protons as incident particles. The width of the excitation peak, which was very small with low-Z projectiles, increases also when going towards heavier incident ions. This is also due to the increasing number of possible nuclear reactions which compete with the particular reactions (HI, 4n) under consideration when the complexity of the projectile nucleus increases.

Fig. VI.9 shows similar data for the (HI, 5n) reaction. As before, only a few characteristic examples could be taken and plotted in the figure, out of the large number of excitation functions available in the literature, in order not to overcrowd the drawing. The reader can easily find more if he requires by consulting the bibliography given at the end of the book. As before we find a shift towards higher bombarding energies and a broadening of the peaks when going to higher-Z projectiles. In contrast with the (HI, 4n) reaction reviewed earlier, the maximum cross-sections found are markedly higher, reaching the value of 1 barn. The same applies to the excitation functions of the (HI, 6n) reactions found in the next fig. VI.10, the peak values of which also approach 1 barn.

When going to still higher numbers of boiled-off neutrons (seven, see fig. VI.11 and eight, see fig. VI.12) the peak values commence to decrease again slowly, reaching $\frac{1}{2}$ barn at most for the (HI, 8n) reaction. This means that nature slightly favours the evaporation of 5 or 6 neutrons from the complete fusion compound nucleus with the pro-

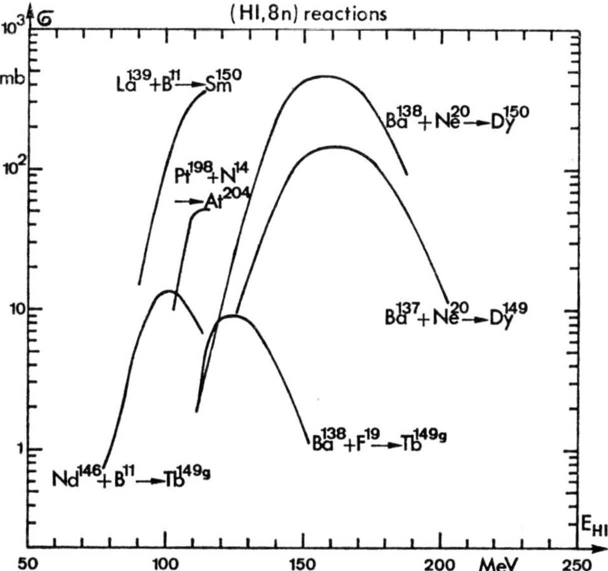

Fig. VI.12 Excitation functions of the (HI, 8n) reaction for various targets and projectiles.

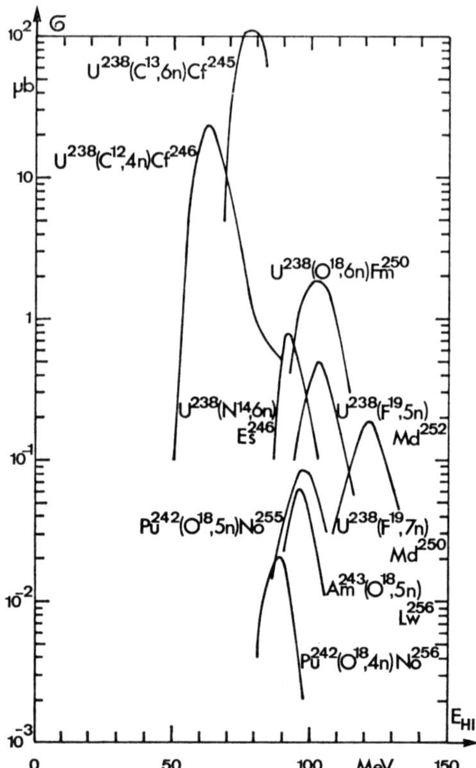

Fig. VI.13 Excitation functions of several (HI,xn) reactions used to produce transuranics.

jectiles used (up to ^{40}A). More energy goes apparently into the other modes of de-excitation. Another effect noticeable when going to higher numbers of evaporated neutrons is the broadening of all the excitation functions. This is again because of the higher number of competing reactions. In fact, all the evaporation reactions with fewer neutrons are competitive, and their numbers go, of course, up with x.

A field of activity where (HI,xn) reactions have found a particularly useful application in nuclear chemistry is the production of very neutron-deficient heavy isotopes in the lead to uranium region and also the production of transuranics. The next fig. VI.13 is devoted to this latter case and shows a number of (HI,xn) excitation functions corresponding to the processes which were used to prepare the higher transuranics californium, einsteinium, fermium, mendelevium, nobelium and lawrencium with atomic numbers 98 to 103. The projectiles used were all of not very different mass number, ranging from ^{12}C to ^{19}F. This is probably why all the excitation functions are found to lie within the bombarding energy region from 50 to 150 MeV. All the curves are sharply peaked, displaying the typical character of a compound nucleus excitation function, rising and falling sharply under the influence of numerous competing reactions. In the cases drawn x has values ranging from 4 to 8. The peak values of the cross-sections measured vary enormously from 10^2 to 10^{-2} millibarn, apparently decreasing when the mass of the transuranic formed increases. Fig. VI.13 does not contain all the excitation functions known to this date for this type of work, but only a few selected examples, for the sake of clarity of the graph. As before, others will be found in the literature referred to later.

5 Spallation reactions induced by heavy ions

We have seen in the preceding sections the case of transfer reactions, where one or a small number of nucleons is believed to be exchanged between the interacting nuclei, and the case of neutron evaporation from a compound nucleus resulting from the complete fusion of target nucleus and incident ion. These (HI,xn) reactions are really a clear-cut case where one knows that a compound nucleus with an atomic number Z equal to the sum of the atomic numbers of the interacting ions is actually formed, as this Z is preserved after the neutrons have been boiled off.

The majority of nuclear reactions induced by heavy ions are, however, more complicated. One finds in most cases that the product of the nuclear reaction is a nucleus vastly different from either the target or the projectile. It has often an atomic number Z and a mass number A which are less than those of the target, indicating spallation of the target induced by the incident ion. It can also have Z and A numbers which are superior to those of the target, indicating some complex mechanism of pick-up of various charged and uncharged nuclei by the target from the projectile. One way to explain such mechanisms, which has been suggested by various authors, is to assume as a first step a complete fusion of target and projectile, followed in a second step by a spallation of the compound nucleus so formed. The phenomenon of spallation would then concern the hypothetic compound nucleus. It is difficult to assert in most cases at the present time if such a compound nucleus exists at an intermediary stage, if it results from a complete fusion of the interacting nuclei, or if some nucleons are already lost before this fusion takes place. We shall include in this subsection a number of examples of reactions measured experimentally which do not fall into the simple cases of nucleon transfer reactions or (HI,xn) reactions, and include a large number of obvious spallation reactions, either of the target nucleus or of a hypothetic compound nucleus. Some transfer reactions involving transfer of more or less elaborate groups of nucleons will, of course, also be present, as one cannot differentiate them easily from the other kinds of reactions mentioned earlier.

We shall concentrate on three examples, for which we have found extensive data: lithium, carbon and nitrogen ions taken as projectiles and interacting with various targets.

In the lithium case, the ^{6}Li ion was accelerated in all the experiments which will be mentioned. Targets include ^{27}Al, ^{54}Fe, ^{197}Au. Due to its small number of nucleons, the lithium-induced reactions are typical of the not very clear-cut cases on the border between transfer reactions, compound nucleus formation, evaporation and spallation. Thus the excitation functions presented in fig. VI.14 for lithium bombardment constitute a sort of review of the various cases possible and can serve as an appropriate introduction to this section. Thresholds for the interaction of the lithium atom with ^{27}Al are seen to lie as low as 3 MeV, with ^{197}Au the thresholds can be estimated to be

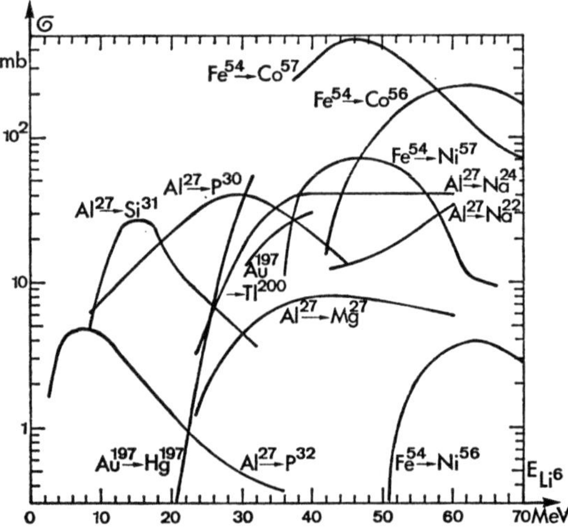

Fig. VI.14 Excitation functions for various nuclear reactions induced by ^{6}Li bombardment of Al, Fe, Au targets.

between 15 and 20 MeV. As with the transfer reactions and unlike the sharp and typical peaks of the compound nucleus evaporation reactions, the excitation curves are generally broad here. They either flatten out or show a decrease when another excitation function starts, i.e. the excitation function for the ^{54}Fe(^{6}Li, 2pn) reaction goes down when the ^{54}Fe(^{6}Li, 2p2n) reaction cross-section increases. This means that the latter channel is preferred as soon as there is enough energy available to expell the 2p2n group instead of the 2pn group of nucleons. On the whole the cross-sections rarely exceed 100 mb, except for the reactions just mentioned with ^{54}Fe, for which peak values of 500 mb are found.

The next case examined is that of ^{12}C ions as bombarding particle, for which some cross-sections are presented in fig. VI.15. As targets we have only carbon and aluminium. The only reaction indicated with the carbon target is the particularly interesting disintegration of the two impinging carbon nuclei into six alpha particles, with a cross-section steadily increasing from 1 to 150 mb in the energy range 40 to 105 MeV. This is, of course, truly a spallation reaction, both for the target and as for the incident nucleus. With the aluminium target, we find reactions where the atomic mass of the target nucleus is increased, as well as reactions where it is decreased. The former kinds for which

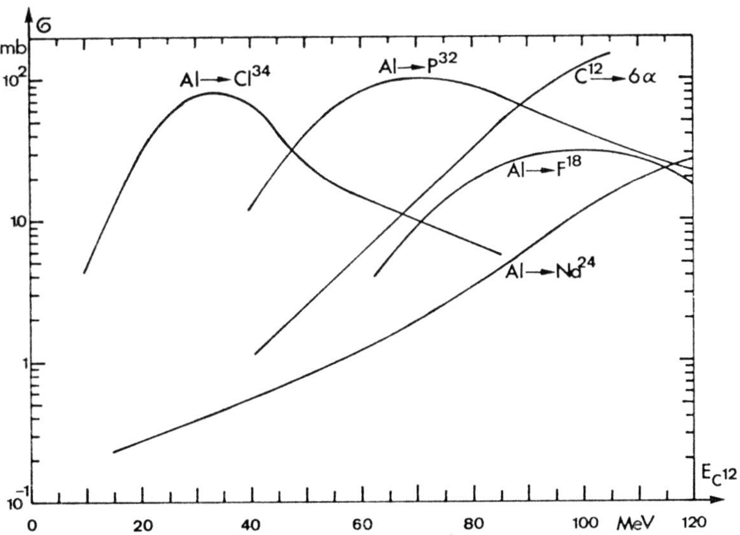

Fig. VI.15 Spallation reactions of compound nuclei formed by ^{12}C bombardment of carbon and aluminium.

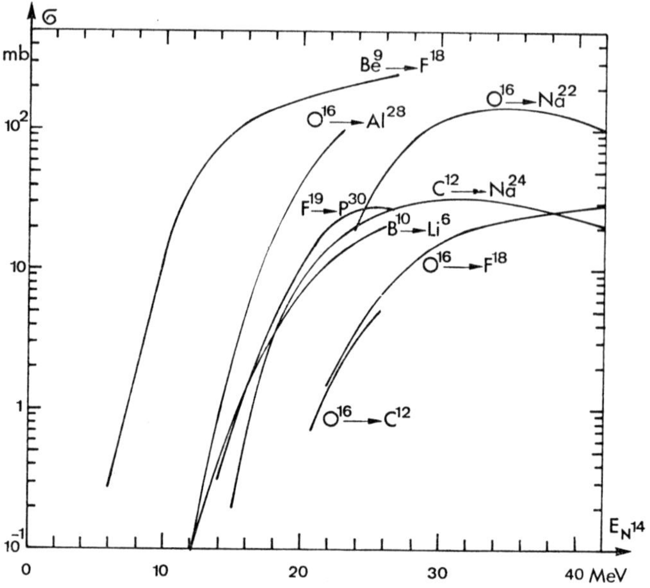

Fig. VI.16 Spallation reactions of compound nuclei formed by ^{14}N bombardment of various light targets from beryllium to fluorine.

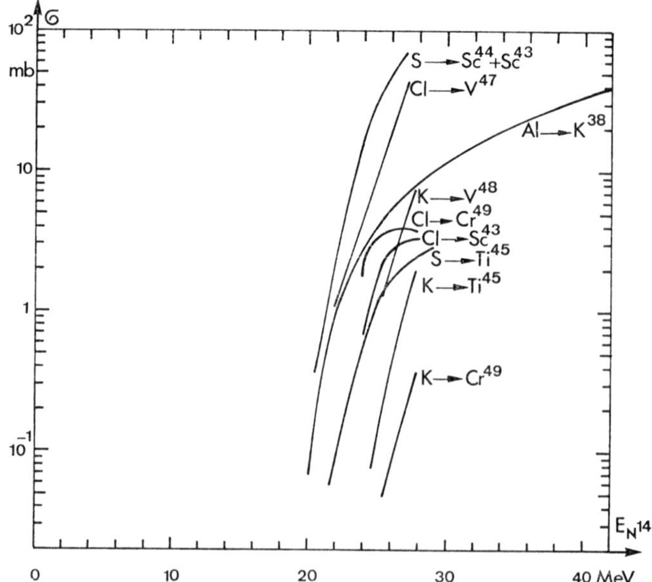

Fig. VI.17 Spallation reactions of compound nuclei formed by ^{14}N bombardment of aluminium, sulphur, chlorine and potassium.

the product is nearest to a hypothetic compound nucleus have the lower thresholds and the steeper rise. So, for instance, the cross-sections for production of ^{34}Cl and ^{32}P start at about 5 and 25 MeV, display maximum values of about 100 mb at 30 and 70 MeV and fall gradually thereafter. In contrast, the ^{24}Na production starts surprisingly early and increases steadily over two orders of magnitude to reach 30 mb at 120 MeV. Further down the mass number scale of the products, ^{18}F production starts later and displays a well-marked maximum of 30 mb at 100 MeV. The last two reactions leading to the production of ^{24}Na and ^{18}F are unquestionably true spallation reactions, whereas those leading to ^{34}Cl and ^{32}P which happen at a lower energy could still be interpreted as more or less direct channels for the de-excitation of the compound nucleus ^{39}K.

The next and last case presented here is that in which ^{14}N ions are used as projectiles. Since many excitation functions have been measured experimentally with this ion and are available from the literature, it is interesting to present a number of them for various targets in ascending order of atomic mass. For the sake of clarity two figures will be needed. Most of the reactions listed lead to a product which has a mass higher than that of the target, so they may be thought to proceed through the formation of a compound nucleus and spallation thereof. In fig. VI.16 which we consider first, the targets bombarded by the ^{14}N ion are ^{9}Be, ^{10}B, ^{12}C, ^{16}O, ^{19}F. The thresholds lie under 5 MeV for the ^{9}Be target, and above 10 MeV for the others. The general trend of all the curves is a sharp rise followed by a smooth maximum or a plateau, at cross-section values ranging from 200 mb for the $^{9}Be(^{14}N,\alpha n)^{18}F$ reaction to 5 mb for the $^{16}O(^{14}N,9p9n)^{12}C$ reaction. Unfortunately data above 40 MeV bombarding energy are lacking.

The next fig. VI.17 shows a few spallation reactions of compound nuclei formed by ^{14}N bombardment of aluminium, sulphur, chlorine and potassium. For these heavier elements only a small part of the excitation functions is available, showing the thresholds, which increase with target weight, and lie now in the energy region 20 to 25 MeV. It is likely that the plateaus will lie between 10 and 100 mb and even higher.

6 Fission induced by heavy ions

The main features of heavy-ion induced fission will be derived with the help of the following figures, which will permit us to make some comparisons with the cases of fission we have examined already, i.e.

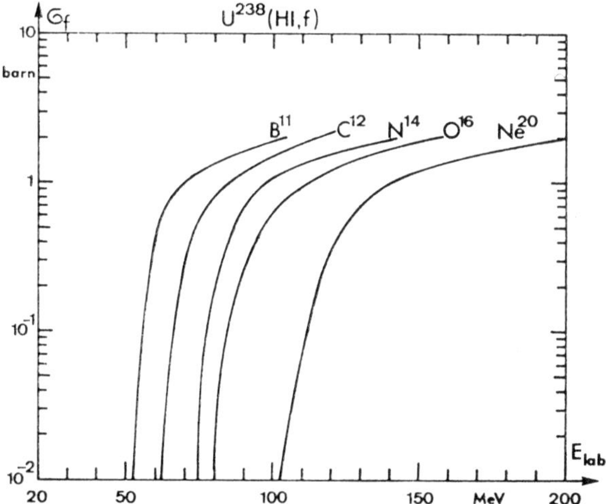

Fig. VI.18 Cross-section for the fission of ^{238}U by various projectiles as a function of bombarding energy.

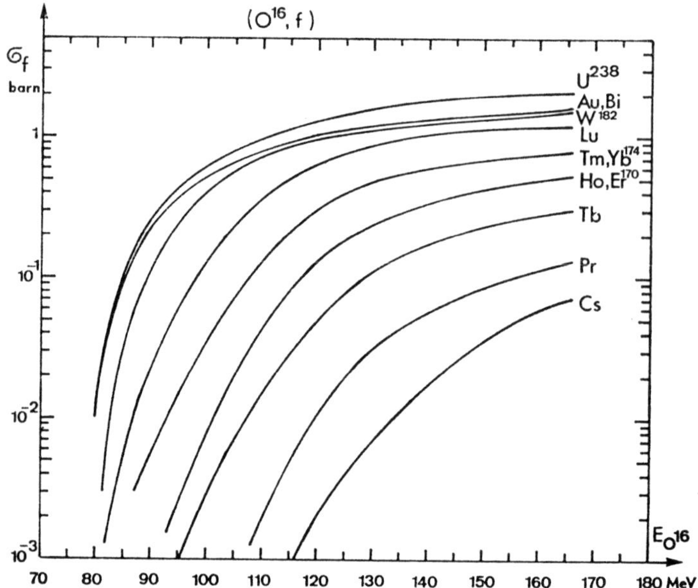

Fig. VI.19 Cross-sections for the fission of various target elements by ^{16}O ions as a function of bombarding energy.

fission induced by neutrons, high energy protons and photons. It is appropriate to start from ^{238}U as target, as it is the material that fissions best. Also this case is usually well studied and can serve as a bridge to link the vastly different kinds of fission just mentioned. For heavy ions particularly, it can be used to make comparisons between various projectiles and bombarding energies. Fig. VI.18 shows the excitation functions for fission of ^{238}U by ^{11}B, ^{12}C, ^{14}N, ^{16}O and ^{20}Ne nuclei as a function of the bombarding energy in the laboratory system. The steep rise of the curves at threshold, obviously due to the Coulomb barrier, increases regularly with the atomic number of the incident particle. The shape of the curves is the same as that of the theoretical total reaction cross-sections as calculated by T. D. Thomas. As a matter of fact, Viola and Sikkeland could prove quite accurately that for ^{238}U the fission cross-section is practically the total reaction cross-section for all energies and projectiles.

For target materials of lower mass, the measurements show that the fraction of the total reaction cross-section which goes into fission decreases steadily as the target mass number and the bombarding energy decrease. This fraction is still about 1 for bismuth and even gold at the higher energies. However, it goes down to about 0.1 for cesium, for instance, at the highest bombarding energies used, as can be computed. Fig. VI.19 gives an account of this phenomenon. It shows the fission cross-sections measured as a function of bombarding energy for various targets with ^{16}O as a projectile. With cesium, for instance, the fission cross-section hardly reaches 0.1 barn. However, it is quite a feat to be able to fission nuclei as low down in the periodic system as cesium, which one cannot bring to fission with neutrons, protons or photons.

At the high energies necessitated by the Coulomb barrier to induce fission with heavy ions, the fission is usually symmetrical, as is evident from the mass yields curve presented in fig. VI.20 for ^{238}U target and 147 MeV ^{22}Ne incident ions. The maximum chain yields are of the order of 3.5%, i.e. lower than the two maxima of the 'camel' curve for thermal fission. The region of the maximum chain yields is around $A=125$, i.e. half the value of the mass number of a compound nucleus obtained by complete fusion of target and projectile followed by evaporation of about 10 nucleons.

To conclude, we would like to present an example for which a number of independent isotopic yields have been measured with an accuracy sufficient to draw a chart with isoyield lines. It is the case where gold

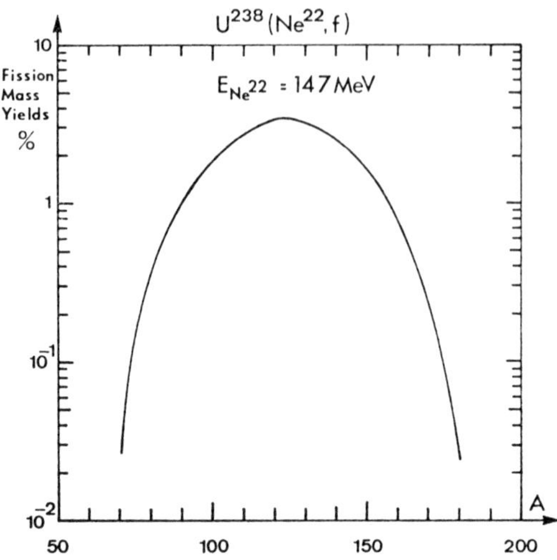

Fig. VI.20 Mass yield curve for the fission of ^{238}U by ^{22}Ne ions of 147 MeV kinetic energy, in percentage of the total weight.

is fissioned by carbon ions of 115 MeV kinetic energy. Fig. VI.21 shows the independent isotopic yield distribution drawn over the atomic and neutron numbers of the fission products as ordinate and abscissa. Again the fission appears to be symmetrical, as the distri-

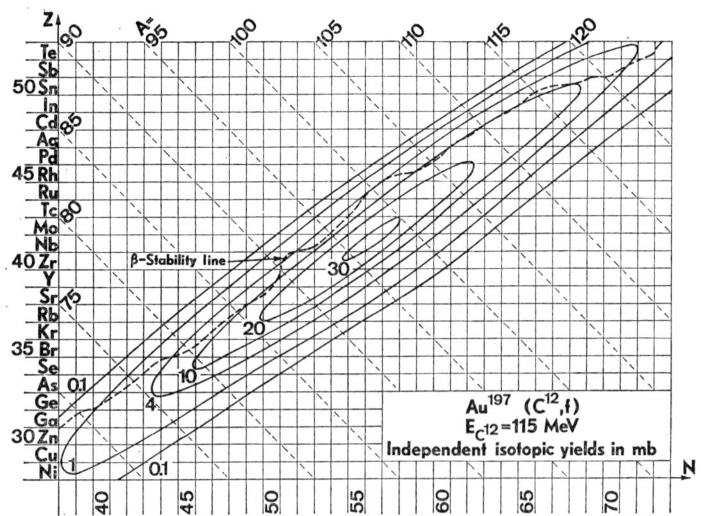

Fig. VI.21 Independent isotopic yields in millibarn for the fission of ^{197}Au by ^{12}C ions of 115 MeV kinetic energy.

bution has the form of a hill with only one peak, instead of the two found in thermal fission of ^{238}U. The maximum independent isotopic yields have the order of 35 mb and are found in the mass number range 95 to 100, the average of which is again half of the mass number of a compound nucleus obtained by complete fusion of target and projectile, followed by evaporation of roughly 10 nucleons. The position of the crest line of the whole distribution relative to the beta stability line is seen to be very near, much nearer than for thermal fission. The consequence is that a number of isotopes lying to the right of the beta stability line, i.e. neutron-deficient isotopes, are produced as fission products by heavy ion bombardment. This fission of heavy elements is another possibility of producing neutron-deficient isotopes with heavy ions.

7 Production of transuranics by reactions other than (HI, xn)

In section 4 of the present chapter, we have given examples of the production of transuranics by means of compound nucleus reactions involving evaporation of several neutrons. These reactions are important, as they make it possible to form new elements with the largest possible electrical charge. Other types of reactions have also been investigated for the production of transuranics and we shall give here a few of them for the sake of completeness.

When uranium and thorium are bombarded with very light elements, such as lithium or boron, excitation functions for some products have been found which do not resemble the typical curves for compound nucleus reactions with their sharp maximum. As the next fig. VI.22 shows, in many cases the excitation functions display a prolonged steady increase or a substantial plateau. This is similar to deuteron-induced reactions and it is felt by the investigators (Fleury, Mivielle, Simonoff, 1967) that one must have a dissociation of the light projectile, which emits one deuteron or one tritium atom at the moment of the impact. It is this deuteron which penetrates the target and triggers the interaction.

This is particularly the case for the ^{232}Th and ^{6}Li reaction, which resembles the (d, 2n) reaction on ^{238}U measured by Wing. For the ^{232}Th and ^{7}Li reaction, the agent is believed to be a tritium atom. In the cases of the boron-induced reactions leading to the production of plutonium the same mechanism applies, with a deuteron as intermediary, leading to the formation of ^{238}Np, which experiences beta-minus decay. The other curves leading to curium are in contrast typical of a compound nucleus reaction, involving subsequent evaporation of a proton and several neutrons.

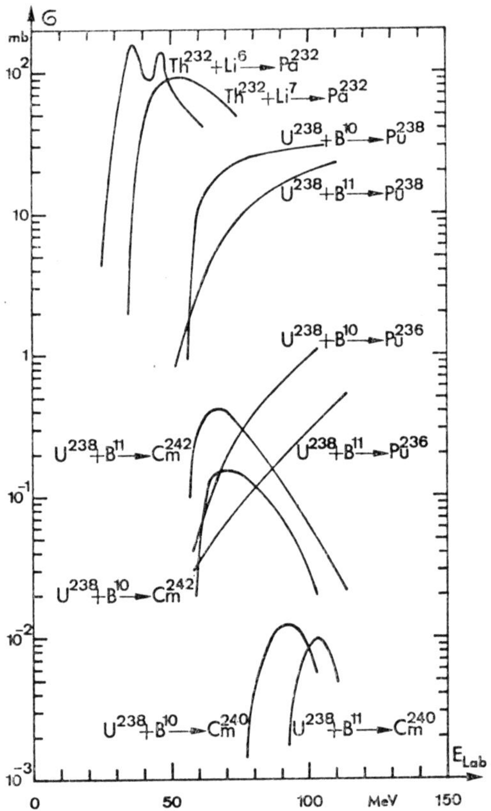

Fig. VI.22 Production of transuranics by bombardment of uranium and thorium with light nuclei.

8 Neutron production by heavy ion bombardment

Clearly, from the information gathered in the seven preceding sections, which is valuable for predicting the cross-sections of some typical interactions, it is not possible to calculate the total activity induced in some material bombarded by a given ion. However, this could be measured directly. Gross-activity measurements of this kind performed in practical cases are not available, however, at the moment to the best knowledge of the author. They should be advocated as the first thing to do when setting up heavy ion accelerators in order to get a feeling of the radiation fields which will emanate from the targets.

It is of interest here to remind the reader of a particular feature of

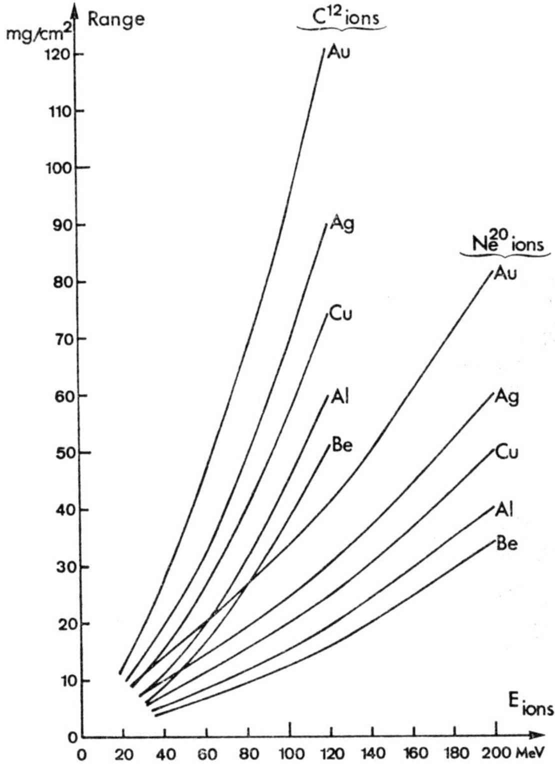

Fig. VI.23 Range of ^{12}C and ^{20}Ne ions in various materials as a function of bombarding energy.

heavy ions which will be determining from the point of view of the induced activity of the whole accelerator. This feature is the very small range of heavy ions in matter, which is due to their high electric charge. Ranges of the order of 50 milligram/cm² are found, for instance, with ^{20}Ne ions of 200 MeV kinetic energy in copper. Fig. VI.23 gives a few range curves in several materials as a reference basis. The consequence is that only a surface layer of the targets and of those parts of the machine structure which the ion beam hits directly will be activated via heavy-ion induced reactions. The rest of the targets or the machine structure will, also, however, be found to be radioactive, because of the large neutron fluxes generated by the heavy ions when they undergo nuclear interactions. These neutrons will propagate easily from the spots where beam is lost to other places of the structure and activate for themselves the materials they traverse. It is thus necessary to have some information on neutron production by heavy ion beams, a subject which fortunately has been investigated by several workers. We are thus in a position to foresee the order of magnitude of the neutron fluxes to be expected around targets and hot spots and to compute the activity induced by the neutrons in the manner which has been outlined in ch. III for the various neutron energies. With the ion energies obtained to this date it will appear that there are no high energy cascade neutrons inducing spallation and that the neutrons found are mainly evaporation neutrons with an energy spectrum which resembles that of reactor neutrons.

The first figure to be investigated is, of course, the total number of neutrons produced by an ion beam in a target thicker than the range of the ions in the material. The curves in fig. VI.24 give this information for ^{12}C, ^{14}N, ^{20}Ne ions of about 10 MeV per nucleon kinetic energy and a variety of targets, according to the measurements of Hubbard, Main and Pyle (1960) with the $MnSO_4$ bath. One sees that the numbers of neutrons produced per incident ion range from 5×10^{-4} to 25×10^{-4} when going from carbon to uranium as a target material. This order of magnitude can seem to be low, but it must

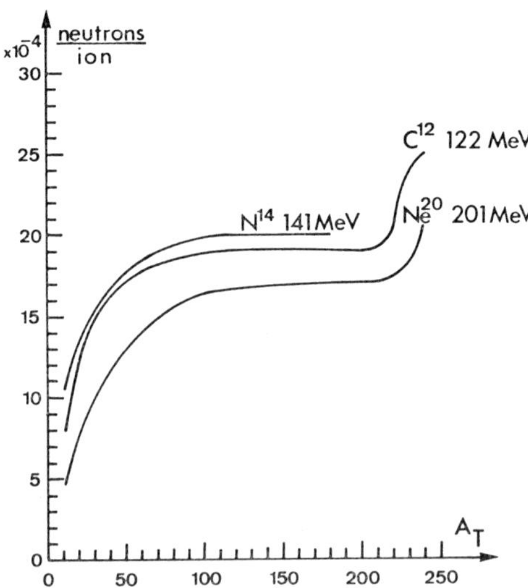

Fig. VI.24 Neutron yields from thick target bombardment by heavy ions of approximately 10 MeV/nucleon as a function of target mass number.

be kept in mind that most of the incident particles are stopped within a very short range by Coulomb scattering due to their high charge.

The same investigators have also measured the effective cross-section σ_{1n} for producing one neutron, using thin targets this time. The values found range from 2 barns for aluminium to 20 barns for lead, are similar for the three ion sorts investigated and are also roughly equal to the values found for the one-neutron production cross-sections with deuterons and protons, as will be seen from fig. VI.25.

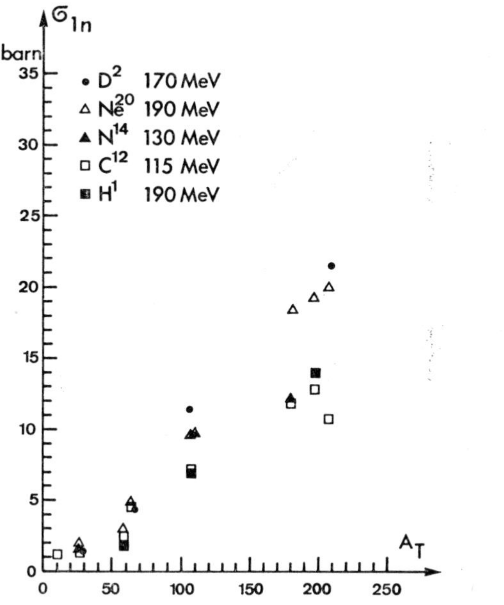

Fig. VI.25 Effective cross-sections for producing one neutron by proton, deuteron, ^{12}C, ^{14}N, ^{20}Ne bombardment, as a function of target mass number.

These effective cross-sections for producing one neutron are, of course, only symbolic equivalent quantities appropriate for the calculations. In fact, several neutrons are emitted on the average at each nuclear interaction of a heavy ion. This is already apparent if one considers that the σ_{1n} values mentioned above are much larger than the geometrical or the total nuclear reaction cross-sections σ_{tot} mentioned at the beginning of this chapter. One can easily calculate the average number of neutrons emitted by the excited compound nucleus by determining the ratio of σ_{1n} to σ_{tot} in each case, and numbers are found lying between 1 and 7.

Another interesting characteristic of the neutrons produced is their energy spectrum. Broeck (1961) has measured neutron spectra for Al, Ni, Cu, Au targets bombarded by ^{16}O ions of 10.5 MeV/nucleon energy. Fig. VI.26 shows the differential cross-sections found per steradian and MeV neutron energy interval, as a function of neutron energy in the centre-of-mass system. Maximum values at zero neutron energy range from 25 to 600 mb/MeV sterad when going from aluminium to gold.

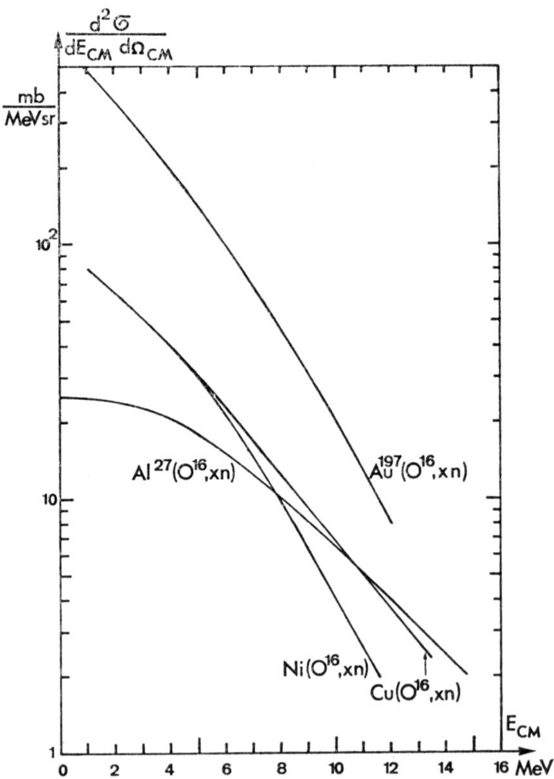

Fig. VI.26 Experimental neutron energy spectra in the centre-of-mass system averaged over solid angle.

In the laboratory system various energy spectra have also been measured by the same investigator. As an example the spectra from the Cu(^{16}O,xn) reaction in the laboratory system taken at a variety of angles to the forward direction are given in fig. VI.27. The usual enhancement of the particle production in directions nearer to the forward direction is clearly visible. This must be taken into account in practical calculations of distribution of radioactivity in the neighbourhood of targets struck by heavy ion beams.

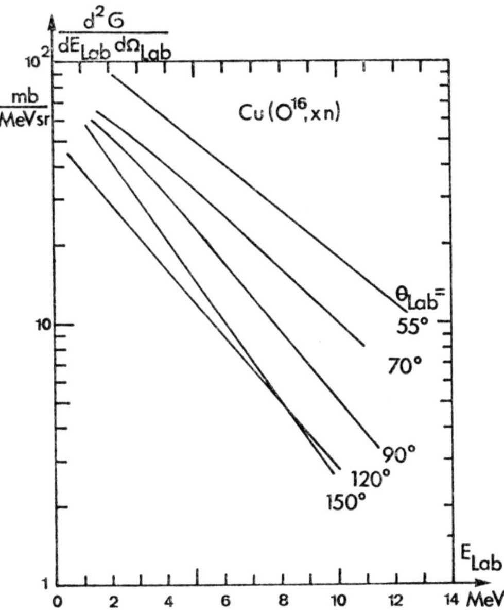

Fig. VI.27 Neutron energy spectra from the Cu(^{16}O,xn) reaction in the laboratory system.

PROBLEMS

CHAPTER VI

1. Give approximately the geometrical cross-section and the total compound nucleus formation cross-section at 100 MeV of neon atoms on neon, iron, silver, lead targets.

2. Find the production cross-sections of the isotopes 132, 130, 129, 128, 127 of cesium from ^{133}Cs by the neutron knock-out mechanism with nitrogen ions of 75 MeV.

3. Find the cross-sections for the ^{197}Au(^{14}N, 4n) and ^{197}Au(^{14}N, 5n) reactions at 90 MeV. To what fraction of the total compound nucleus reaction cross-section does the sum of the two cross-sections mentioned above amount?

4. What is the bombarding energy to be chosen if one wants to make ^{246}Cf by ^{12}C bombardment of ^{238}U? What is the maximum cross-section to be expected?

5. What is the cross-section to adopt above 40 MeV when using the $^{27}\text{Al}\rightarrow{}^{24}\text{Na}$ reaction to monitor a beam of ^{6}Li atoms?

6. What is the cross-section to adopt at 120 MeV when using the $^{27}\text{Al}\rightarrow{}^{24}\text{Na}$ reaction to monitor a beam of ^{12}C atoms?

7. What is the cross-section for obtaining ^{238}Pu when bombarding ^{238}U with ^{10}B atoms above 120 MeV?

8. Let us assume we bombard a layer of gold 20 mg/cm^2 thick with 10^{16} carbon atoms of 115 MeV energy. What is the number of yttrium atoms which have been formed by fission after the radioactive isotopes in the neighbourhood of ^{89}Y, which decay into this element, have actually decayed? Add the independent yields of ^{89}Zr, ^{89}Y, ^{89}Sr.

CHAPTER VII

RADIOACTIVITY INDUCED IN TISSUES

1 Clinical and biological aspects of induced activity in tissues

The study of radioactivity induced in tissues is of great interest for several reasons. First, and sadly enough, it is often the only means to assess the dose received by a person in a radiation accident if this person has not worn personal monitors. Even if such monitors were worn, one can have recourse to induced activity measurements on the person to find whether the dose distribution over the body was uniform or not, and in some cases to determine the average dose by measuring the blood activity for instance. Another case of accidental exposure is unnoticed irradiation of long duration at a low level of dose, as can happen with insufficient shielding. Here also induced activity of long life-time in the body can help figuring out a posteriori the dose received. In all cases a rapid and correct estimate of the doses is required by the physicians in order that they may choose correctly the therapy which has to be applied to the patient who has been unintentionally exposed.

Another advantage of studying induced activity in tissues is that it gives clues with respect to biological changes or damage that has occurred because of the irradiation. As one knows, the dose deposited in any material is due partly to ionization losses, which affect the electron shells around the atoms and can destroy chemical bonds, and partly to nuclear interactions, which modify the nature of the nucleus. As the energy of the incoming radiation increases, these nuclear interactions gain in number and importance and account for an increasing fraction of the dose received. As a rather well-determined part of these nuclear interactions gives birth to radioactive nuclei, a good way of measuring them is to use the induced activity of some convenient radionuclides produced in the sample and to calculate back via the total cross-sections what is the total number of interactions having taken place. Such procedures will not only be useful for purely physical dosimetric purposes. They are also required for the assessment of the biological damage inflicted to the cells. It is precisely the nuclear interactions which make the difference between the cases of an irradiation by X-rays and an irradiation by high energy particles and gamma rays. Whereas a part of the damage produced by ionization can be generally recuperated, changes of nuclei are irreversible. This is especially important if one considers the genetic action of the radiation.

For obvious reasons we shall concentrate in the following on human tissues.

2 Chemical composition of various tissues and occurrence of their elements in the average human adult

For any investigation of the kinds described in the preceding section an exact knowledge of the chemical composition of the tissues involved is of primary necessity. Thus it is good to give a detailed analysis of the various sorts of human tissues, bearing in mind that the values given are only accepted averages and that in certain cases direct analysis of the sample examined can be further required. Generally speaking the tissues in the human body can be divided into three classes: fat, bone and parenchymatous tissue (muscle, liver, kidney, etc.). Table VII.1 shows the occurrence in percentage of weight of the main kinds of tissue in the average human adult. One will observe that fat and bone represent only one quarter of the body weight, the remainder being taken by parenchymatous tissue, which is in turn highly diversified in muscle, blood, liver, kidney, skin etc.

A detailed chemical analysis of the constituents of these various tissues will be found in the next table VII.2. The values given are the percentages in weight. They have to be multiplied by the weight of the subject to find the probable total content of the element desired in the body or in a particular organ. One should of course not forget that variations from individual to individual can be large.

One figure deserves special attention in the next table. Recent

TABLE VII.1

Percentages by weight of various tissues in an average human adult.

Bone without marrow	10
Fat	14
Muscle	43
Blood	7.7
Kidneys	0.43
Liver	2.4
Skin and subcutaneous tissue	8.7

experiments in vivo by Anderson and others have shown that the sodium content of the body was smaller than previously believed and that a value of 0.1% body weight was more accurate.

Blood and urine are especially interesting, since they are easy to obtain from irradiated persons and are often used for activity measurements. Table VII.3 gives average values of the constituents of blood and urine; it must be stressed that precisely here variations between

TABLE VII.2

Chemical composition of tissues in the human adult in percent of weight.

Element	Whole body	Bone	Fat	Muscle	Kidney	Liver	Hair	Skin
Oxygen	65.0	40	11	71	71.6	72.3	23.6	66.7
Carbon	18.0	7.7	77	5.8	4	15.5	49.2	11.6
Hydrogen	10.0	5	12	9.6	9.4	11.2	7.6	9.8
Nitrogen	3.0						16	
Calcium	1.5	11.0		0.007	0.02	0.012		0.020
Phosphorus	1.0	5.05		0.22	0.14	0.21	<1	0.065
Potassium	0.2	0.061		0.36	0.175	0.215		0.107
Sulphur	0.25	–		0.25	–	0.19	3.6	
Sodium	0.15	0.18		0.072	0.175	0.19		0.16
Chlorine	0.15	0.19		0.066	0.22	0.16		0.30
Magnesium	0.05	0.105		0.023	0.021	0.022		0.014
Iron	0.006							
Manganese	0.00003	0.3 –0.17		0.05	0.06	0.205		–
Copper	0.0002	1.19–0.41		0.125	0.166	0.71		–
Iodine	0.00004	0.03		–	–	–		–
Tin		0.08–0.05		0.011	0.02	0.06		0.5–1.0
Aluminium		0.5 –0.24		0.015	0.042	0.16		–

TABLE VII.3

Chemical composition of urine and blood.

Elements	Urine	Blood
C		50–55% weight
N		15–17.5 ,,
O		19–24 ,,
H		6.5–7 ,,
P	92 mg/100 cc	28–48 mg/100 cc
S	183 ,,	
Si		0.83 ,,
Cl	405 ,,	273–321 ,,
Na	385 ,,	196 ,,
Mg	8.3 ,,	1.92 ,,
K	228 ,,	168–370 ,,
Ca	20.8 ,,	6 ,,
Fe		44–56 ,,
Al		0.013 ,,
Br		3.72 ,,
Cu		0.098 ,,
Sn		0.012 ,,
F		0.028 ,,
I		0.0043 ,,
Pb		0.015 ,,
Zn		0.88 ,,

individuals, and even in one individual with time, can be considerable. This is also why the C, N, H, O contents of urine are not indicated separately in weight. Rather the more accurate value of total proteins eliminated per 24 hours in the urine is given at the foot of the left column in the following table VII.4, which lists a few global features for urine and blood.

TABLE VII.4

Other features of urine and blood.

	Urine	Blood
Total proteins	55–70 mg/24 h	6.55 g/100 cc
Carbon hydrates		82–92 mg/100 cc
Volume	600–1600 cc/24 h	65.6–71.4 cc/kg body weight
Weight		77 g/kg body weight
Dry residue	55–70 g/24 h	

3 The flux-to-dose conversion factors for various particles

The aim of the work outlined in this chapter is in fact to find ways to assess the flux of particles received by a person by measuring his own radioactivity resulting from the irradiation. It is thus appropriate to examine the flux-to-dose conversion factors to apply in order to find the dose received by the person exposed, assuming that the flux measurement via induced activity in the body could be made reliably.

First, one has to consider the physical dose, i.e. the energy deposited by the radiation in a cubic centimetre of tissue. This dose can vary with the location in the body. When a flux of particles enters the body one can have, depending on the kind of particle and on the energy, an increase or a decrease of energy deposition with depth inside the body. With high energy particles one can have a build-up of secondary radiation. Also the ionization will change with the energy of the particle as it is slowed down. Charged particles exhibit a maximum of energy deposition per unit volume when they are slowed down to low energies, the so-called 'Bragg peak' in the dose versus depth curve. It is thus interesting to know the physical dose both at the surface of the body and at the location inside the body where it is a maximum. The physical dose is usually expressed in rads, one rad corresponding to an energy deposition of 100 erg per cc, as already mentioned in ch. I.

Second, one has to consider the so-called dose equivalent, describing the biological damage to tissue. This damage is proportional to the physical dose, but the proportionality factor is dependent on the nature of the energy deposition, i.e. on the nature and the speed of the particles. This factor is called the radiobiological effectiveness and is determined experimentally by various biological methods in each case. The unit used for the dose equivalent is the rem. The figure for the dose equivalent in rems is then simply derived from the physical dose in rads by multiplying the latter by the radiobiological effectiveness. As in the preceding case the rem dose, as well as the radiobiological effectiveness, can vary throughout the body when the radiation is penetrating.

In the next graphs we will give the flux-to-dose conversion factors in $rad^{-1}cm^{-2}$ and $rem^{-1}cm^{-2}$, i.e. we will give the number of particles per square centimetre necessary to produce 1 rad and 1 rem in soft tissue. Only normal parallel incident fluxes will be considered. Wherever possible the values will be given separately for the surface of the body and for the locations where the dose and dose equivalent are at a maximum. For this purpose the body will be considered to be 30 cm thick and the maximum values referred to are those found within these 30 cm, following a generally agreed procedure.

Fig. VII.1 shows the flux-to-dose conversion factors for neutrons as a function of energy, both at the surface of the body and at this maximum position in $rad^{-1}cm^{-2}$ and $rem^{-1}cm^{-2}$. The curves are based on those published by Hine and Brownell in the thermal and epithermal region, by Blatz in the epithermal and fast neutron region, by Irving, Alsmiller, Moran up to 60 MeV and by Neufeld and Snyder from 60 to 400 MeV. Slight adjustments have been made among the values given by these different authors in order to obtain a continuous curve.

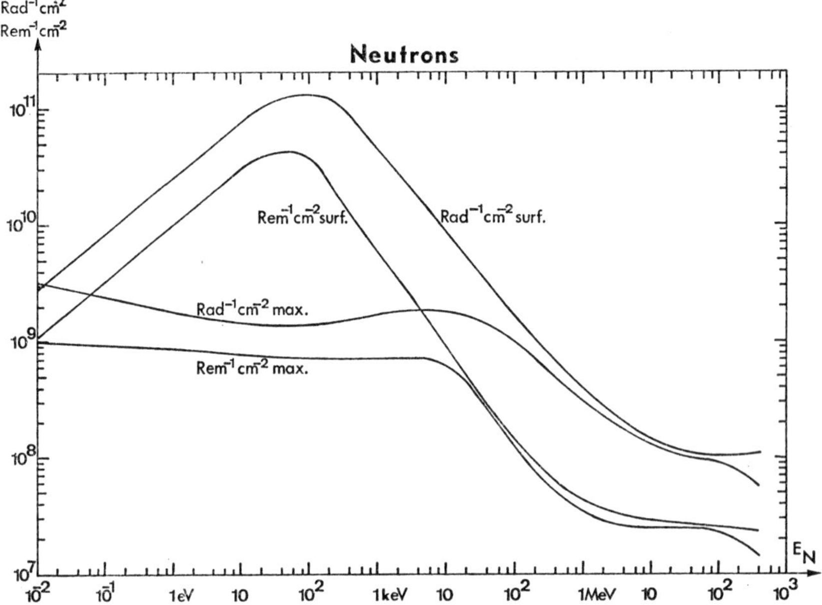

Fig. VII.1 Normal fluxes per cm² to produce 1 rad and 1 rem in tissue at the surface of the body and at the location of the maximum dose for neutrons from thermal energies to 400 MeV.

Fig. VII.2 shows similar data wherever available for some charged elementary particles i.e. pions, kaons and protons. For pions and kaons only the surface physical dose is given, based on the ionization loss curves calculated by Serre. For protons, besides this source, more detailed information exists. The curves are based on the work of

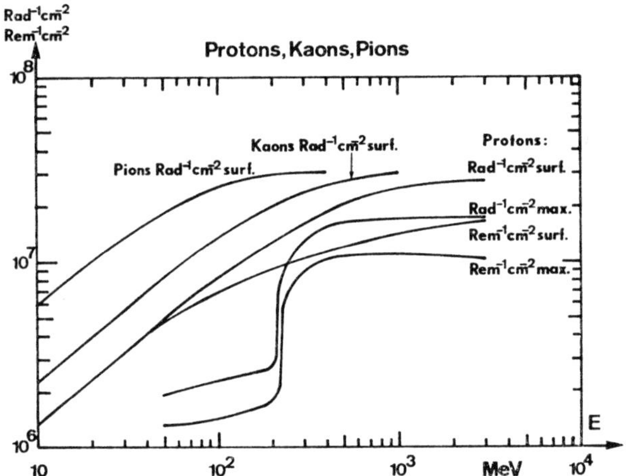

Fig. VII.2 Normal fluxes per cm² to produce 1 rad and 1 rem in tissue at the surface of the body and at the location of the maximum dose for protons, kaons, pions.

Neufeld and Snyder again up to 400 MeV and continued smoothly up to the data obtained by Sklavenitis at 3 GeV.

Fig. VII.3 shows flux to physical dose conversion factors for weak interacting particles, photons, electrons and muons. The curves for photons and electrons are based on those found in Price, Horton, Spinney and in Hine and Brownell in the low energy region. The curve for muons is drawn after Serre. In the high energy region, commencing between 10 and 100 MeV, the data given for electrons and photons both at the surface and at the position of maximum dose within the body are based on the work of Alsmiller, which agrees with the experimental results of Tesch. The lower flux-to-dose conversion factor values found for photons and electrons at a certain depth in the body are due to the development of the electromagnetic shower generated by the incoming photons and electrons. Photons, electrons and muons are considered to have radiobiological efficiencies equal to 1. The contribution to the dose due to neutrons and protons produced in photonuclear reactions seems to be small even at the high end of the energy range considered so that the radiobiological efficiency can be taken as 1 throughout. Thus the curves presented in fig. VII.3 also indicate the rem^{-1}cm^{-2} values for the weak interacting particles mentioned.

As a general comment, one should mention that, especially in the high energy end of the range, all the curves are subject to caution. This is already true for the physical dose curves, which rely partly on theoretical computation, and even more for the dose equivalent, as the radiobiological efficiency factors to be used according to the nature of the secondary particles depositing the energy are still a matter of controversy.

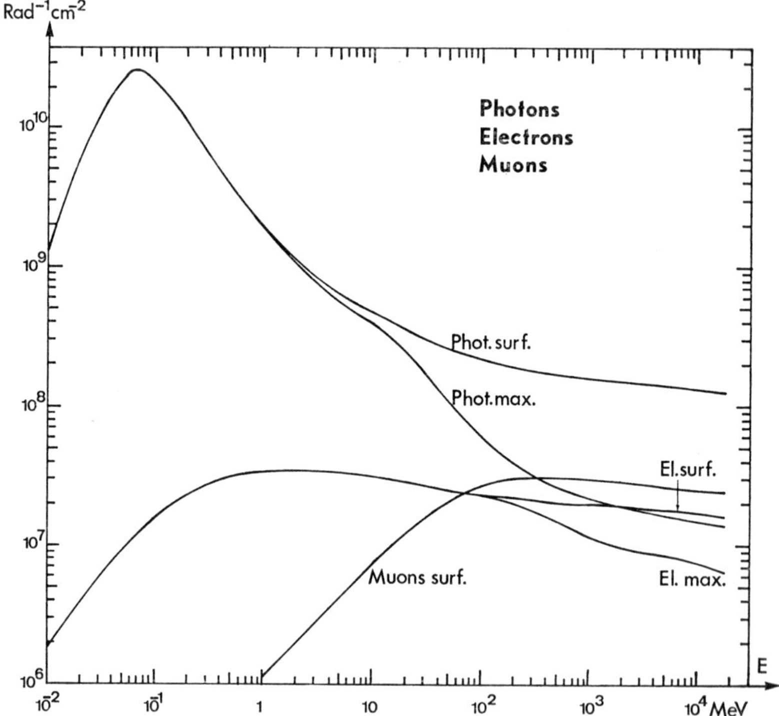

Fig. VII.3 Normal fluxes per cm^2 to produce 1 rad in tissue at the surface of the body and at the location of the maximum dose for photons, electrons and muons.

4 Expected cross-sections for various reactions relevant to body activation

At this stage and before dealing with specific cases it is appropriate to present some data gathered by various authors on the effect of the two kinds of fluxes which will interest us most: reactor neutrons and high energy protons.

As one knows, neutron leakage fluxes from reactors have a continuous energy spectrum. It is thus good to have integrated cross-sections over a given energy range with some appropriate assumption on the form of the spectrum which has some chance of applying in practical cases.

The first case is that of thermal neutrons. They are in thermal equilibrium and have a maxwellian distribution around 0.025 eV. For this kind of spectrum the so-called thermal cross-sections for the (n,γ) process have been carefully measured and are easily found in the literature. A number of them are listed in the second column of table VII.5.

A second case of interest is the $1/E$ neutron spectrum. In the region above 0.4 eV the neutron spectrum is often distributed as the reciprocal of the energy E. It is convenient to define the so-called resonance integral in the way given by Barrall and Mac Elroy

$$I = \int_{4\times 10^{-7}\,\mathrm{MeV}}^{18\,\mathrm{MeV}} \sigma(\mathrm{n},\gamma)\,\frac{\mathrm{d}E}{E} \tag{4.1}$$

where $\sigma(\mathrm{n},\gamma)$ is the microscopic neutron cross-section for the (n,γ) reaction resulting in the particular product nuclide indicated for each reaction in table VII.5. This integral is called the resonance integral because it is taken over the energy region where the cross-sections of most (n,γ) reactions exhibit one or several resonance peaks. The values of this resonance integral recommended by Barrall and Mac Elroy are listed in the third column of table VII.5.

A third and last case of interest for neutrons is of course the fission spectrum, for which one simple form which fits the data well below 9 MeV is

$$\varphi(E) = 0.77E^{\frac{1}{2}}\exp(-0.776E). \tag{4.2}$$

For this expression Barrall and Mac Elroy have computed average cross-sections. The results are listed in column 4 of table VII.5 for a large number of reactions including not only radiative capture but also the types (n,p), (n,2n), (n,α), (n,f). All these data can be of interest when a rapid estimate is required for a spectrum whose form can be approximated by one of those mentioned.

It is interesting to give here also the full curve for the energy dependence of the cross-section of the radiative capture in ^{23}Na, with reference to sodium activation of blood by neutron bursts. The curve

TABLE VII.5

Integrated neutron cross-sections in barns.

Reaction	Thermal	Resonance	Fission	
$^{23}Na(n,\gamma)^{24}Na$	0.535	0.28	0.00026	
$^{55}Mn(n,\gamma)^{56}Mn$	13	15.9		
$^{59}Co(n,\gamma)^{60}Co$	37	74.5	0.00313	(under Cd)
$^{63}Cu(n,\gamma)^{64}Cu$		5.1		
$^{115}In(n,\gamma)^{116m}In$	157	2622		
$^{197}Au(n,\gamma)^{198}Au$	98.8	1535	0.102	(under Cd)
$^{232}Th(n,\gamma)^{233}Th$	7.4	85.4	0.102	(under Cd)
$^{238}U(n,\gamma)^{239}U$	2.73	282	0.078	(under ^{10}B)
$^{115}In(n,n')^{115m}In$			0.17	
$^{64}Zn(n,p)^{64}Cu$			0.0252	
$^{54}Fe(n,p)^{54}Mn$			0.092	
$^{31}P(n,p)^{31}Si$			0.0325	
$^{32}S(n,p)^{32}P$			0.0573	
$^{56}Fe(n,p)^{56}Mn$			0.000977	
$^{34}S(n,\alpha)^{31}Si$			0.00217	
$^{24}Mg(n,p)^{24}Na$			0.00151	
$^{27}Al(n,\alpha)^{24}Na$			0.000705	
$^{127}I(n,2n)^{126}I$			0.00083	

generally used is given in fig. VII.4 and shows, superimposed on the $1/E$ dependence, a typical resonance in the keV region.

Besides the reactors, other sources of danger are high energy accelerators. Above a certain energy, of the order of several hundred MeV, the activation cross-sections are usually flat. At these energies neutrons and pions activate roughly as protons. Thus a collection of data for protons of 600 MeV is a good illustrating example. The production cross-sections for various isotopes in light weight target elements for protons of kinetic energy 600 MeV or near this value are presented in table VII.6. The values in brackets are calculated by means of Rudstam's formula and represent only approximations. In contrast with the values listed in the preceding table, which were given in barns, one sees that here the values are quoted in millibarns. However one has to remember that charged particle beams are intense and well focussed and represent in most cases, with present day intensities, lethal dangers even for very short irradiation times.

TABLE VII.6

Production cross-sections for protons with kinetic energy around 600 MeV (millibarns).

Target \ Isotope	^{7}Be	^{11}C	^{13}N	^{18}F	^{22}Na	^{24}Na	^{27}Mg	^{28}Mg	^{31}Si	^{32}P	^{33}P	^{35}S	^{38}S	^{43}K
^{12}C	10	30.7	5											
^{14}N	11.3	20	4.75											
^{16}O	10.7	9	6.4											
^{23}Na	13													
^{24}Mg														
^{27}Al	4.6	3.3	0.86	(7.7)	19.2	11								
^{31}P				(4)	(8.1)	(13)	(5.7)	(1.03)						
^{32}S					(7)	(10)								
$^{35.5}$Cl				2.2	(4.3)	7.1		0.2		37		6.3		
^{39}K					(2.7)	(4.3)								
^{40}Ca				(1.15)	(3.84)	1.72	(1.7)	(0.29)	(5.4)	(23.2)	(9.4)	(16)	(0.007)	
^{44}Ca														25
^{54}Fe					(0.29)	(0.46)								
^{55}Mn					(0.33)	(0.53)								
$^{63.5}$Cu					0.04	0.03								

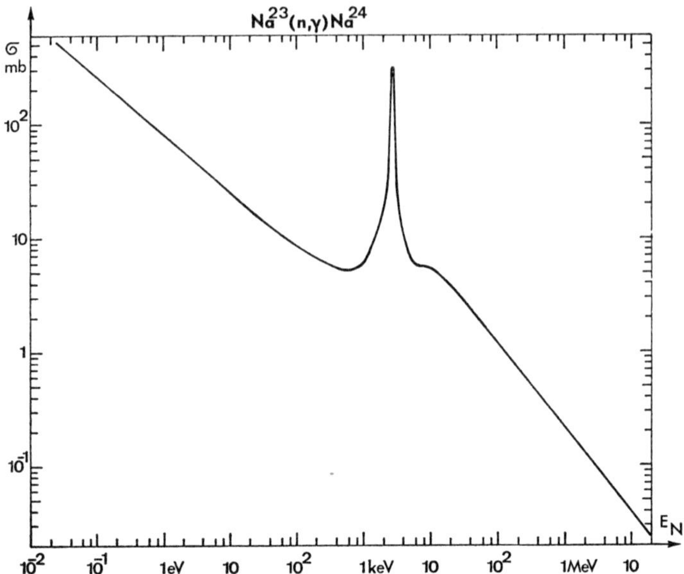

Fig. VII.4 Cross-section of the $^{23}Na(n,\gamma)^{24}Na$ reaction as a function of neutron energy.

5 Activation of blood and hair in the body by reactor or evaporation neutrons

Sodium activation in blood has been used to determine the received dose after reactor accidents. In such accidents resulting from unwanted bursts of criticality, the activating flux released is generally a fast neutron spectrum. These neutrons are also accompanied by a flux of gamma rays which of course do not activate but add their share to the dose already received by the exposed person due to the neutron flux. It has been current practice in the past to determine a posteriori the ratio of gamma to fast-neutron produced doses by a mock-up experiment reproducing in a controllable manner the conditions of the accident. This mock-up experiment also serves to establish the form of the energy spectrum of the neutrons released. We shall here be concerned only with the determination of the neutron dose by body sodium activation and leave the appreciation of the accompanying gamma dose to the experimentalist.

The measurement of the activated sodium in the body after a criticality accident is a very indirect method of measuring the incident flux consisting mainly of fast neutrons. In fact the reaction used is the

$^{23}Na(n,\gamma)^{24}Na$ thermal neutron radiative capture reaction. From the table giving some integrated cross-sections in section 4 of this chapter, it is seen that the integrated neutron cross-section for ^{24}Na formation is 0.535 barn for a thermal spectrum, 0.285 barn in the case of the resonance integral and 0.00026 barn for a fission spectrum. Thus fission neutrons activate sodium very little in comparison with low energy and thermal neutrons. It is not the value 0.00026 barn which is used in this kind of work. This is because the fast neutrons are slowed down and partly thermalized when entering the human body due to the large quantity of water and other light elements they pass through. A theoretical analysis of this effect is difficult but has been done by various authors involved in the dosimetry work made after several reactor accidents (Hurst, Snyder, Jones etc.). Also live animals and phantoms filled with NaCl in water solutions have been irradiated on various occasions with reactor fluxes of several kinds to check the results of these calculations by direct measurements of the specific ^{24}Na activity produced, either in the animal's blood or in the saline solution (Harris, Smith).

It is already apparent from what we just said that there is no general recipe for sodium activation in man, owing to the extreme sensitivity of the activation mechanism to neutron energy and spectrum. What we shall endeavour to do in the following is only to give a brief account of the figures established thus far, in order to be able to calculate the expected sodium activity per unit incident flux as soon as one has an idea of the spectrum of the incident neutrons.

First, to the question of sodium content in man, we have already mentioned that recent investigations (for instance Anderson et al.) have shown that the figure of 0.15% of body weight given in table VII.1 according to the ICRP recommendations on the standard man is in error by excess. The total sodium in the body if about 0.1% of the body weight. The standard 70 kg man thus carries about 70 g of ^{23}Na. It was also found that there is no difference to be made between total sodium and exchangeable sodium, as practically all the sodium passes through the blood. The blood gives thus a good picture of the general sodium in the body. Sodium content in the blood was found to be 1.906 mg ^{23}Na/ml and in blood serum to be 3.17 mg ^{23}Na/ml. In the case of urine it appears possible to correlate ^{24}Na activity with neutron dose, particularly if the first void after exposure is not used and if samples are collected before new sodium is given to the exposed individual.

The next question is related to the probability for the fast neutrons entering the body to be thermalized and captured by the sodium. Smith, as well as Sanders and Auxier and others have examined this matter experimentally as a function of neutron energy or for typical criticality accident fluxes. For the exploitation of the experimental results and for calculating later on the expected activity, a quantity known as probability of capture ξ has been defined by the authors mentioned above. It is assumed that in fast neutron irradiation of a body containing sodium the produced ^{24}Na activity is directly related to the total number of neutrons captured by the ratio Σ_{Na}/Σ_T where Σ_{Na} is the macroscopic absorption cross-section at thermal energies of ^{23}Na in the body and Σ_T is the total absorption cross-section for the thermalized neutrons. The relationship between this macroscopic Σ and the microscopic σ of the radiative capture by ^{23}Na or the total absorption by all other atoms is as follows

$$\Sigma = \sigma n \tag{5.1}$$

n being the number of atoms present per cm^3 and participating in the reaction of which σ is the microscopic cross-section. As the medium in which the fast neutrons propagate is composed of various elements, the resulting Σ will be the sum of such contributions from all the elements constituting the body.

For the ^{23}Na(n,γ)^{24}Na reaction in a thermal spectrum the recommended cross-section is 0.536 barn. With a 0.1% concentration in the whole body one finds

$$\Sigma_{Na} = 1.4 \times 10^{-5}\ \text{cm}^{-1}. \tag{5.2}$$

For the absorption of the thermalized neutrons by all other atoms present in the body, Hurst, Ritchie and Emerson (1959) calculate with the elemental composition of tissue a value

$$\Sigma_T = 0.0234\ \text{cm}^{-1}. \tag{5.3}$$

Table VII.7 shows with which cross-sections the value of Σ_T has been calculated.

As a result the ratio of the two macroscopic cross-sections takes the value

$$\Sigma_{Na}/\Sigma_T = 6 \times 10^{-4}. \tag{5.4}$$

TABLE VII.7

Calculation of the macroscopic total neutron absorption cross-section Σ_T.

Fraction of body weight f (g/cm³)	Element	Mass number A	Thermal neutron cross-section σ (10^{-27} cm²)	Reaction type	Number of atoms per gram $6\times10^{23}/A$ (g⁻¹)	Number of atoms per cm³ $6\times10^{23}f/A$ (cm⁻³)	Macroscopic cross-section $\Sigma=n\sigma$ (cm⁻¹)
0.1	H	1	332	n, γ	6×10^{23}	6×10^{22}	2×10^{-2}
0.18	C	12	3.4	absorption	5×10^{22}	9×10^{21}	3.06×10^{-5}
0.03	N	14	1850	absorption	4.3×10^{22}	1.29×10^{21}	2.39×10^{-3}
0.65	O	16	0.178	n, γ			
	O	17	235	n, α			
			0.265*		3.75×10^{22}	2.44×10^{22}	6.46×10^{-6}
0.015	Ca	40	2.5	n, α	1.5×10^{22}	2.25×10^{20}	5.62×10^{-7}
0.01	P	31	190	n, γ	1.94×10^{22}	1.94×10^{20}	3.69×10^{-5}
0.002	K	39	2100	absorption	1.54×10^{22}	3.08×10^{19}	6.47×10^{-5}
0.0025	S	32	520	absorption	1.88×10^{22}	4.7×10^{19}	2.44×10^{-5}
0.001	Na	23	536	n, γ	2.62×10^{22}	2.62×10^{19}	1.4×10^{-5}
0.0015	Cl	35.5	33.200	absorption	1.7×10^{22}	2.55×10^{19}	8.46×10^{-4}
						sum total	$\Sigma_T=2.34137\times10^{-2}$

* Weighted average.

This is the fraction of the neutrons thermalized in the body which end up in a sodium nucleus.

We still need to define or calculate or measure the fraction of the fast incident neutrons which are thermalized and absorbed in the body as a function of incident neutron energy. The amount of this fraction, which has been called the neutron capture probability ξ, has been calculated theoretically by several authors (Snyder, Neufeld, Auxier, Kogan, Smith). According to the latest calculations of Jones the capture probability in a cylinder 30 cm in diameter and 60 cm in height, which approximates the human body rather well, follows the dotted curve in fig. VII.5.

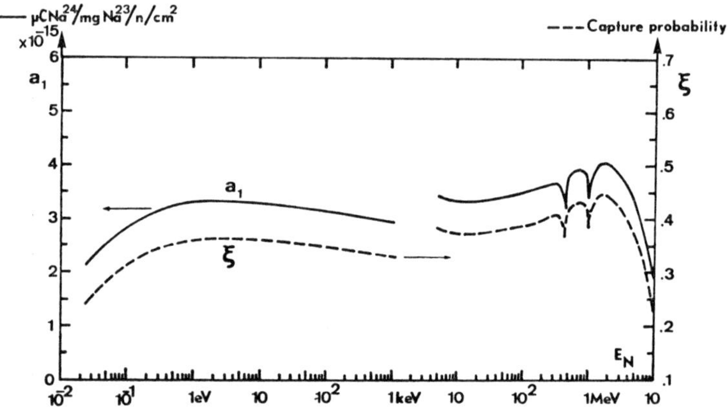

Fig. VII.5 Neutron capture probability ξ and specific activity a_1 per incident neutron per cm² for monoenergetic neutrons in a cylinder 30 cm in diameter and 60 cm in height.

The way to proceed in a practical case is now shown in the following theory, which is outlined in the paper by Hurst, Ritchie and Emerson. Let $\varphi(E)\,\mathrm{d}E$ be the neutron flux expressed in neutrons per second per square centimetre in the energy interval $(E, E+\mathrm{d}E)$. The human body will be considered as equivalent to a cylinder 60 cm in height and 30 cm in diameter with an area S along a section through the axis of rotational symmetry and a volume V. Let t be the irradiation time, which should be short compared with the half-life of ^{24}Na (15 h). Let us assume that the direction of incidence of the neutrons is perpendicular to the axis of rotational symmetry and that the neutron flux is parallel. The number of neutrons in the energy interval $(E, E+\mathrm{d}E)$ striking the cylinder is then $St\varphi(E)\,\mathrm{d}E$. Let $\xi(E)$ be the capture probability which is of course a function of energy. Then, according to what has been said above, $\xi(E)\Sigma_{\mathrm{Na}}/\Sigma_{\mathrm{T}}$ is the fraction of the neutrons which are thermalized and end up inside a sodium nucleus. Then the total number of ^{24}Na nuclei formed in the cylinder will be

$$N(^{24}\mathrm{Na}) = St\,\frac{\Sigma_{\mathrm{Na}}}{\Sigma_{\mathrm{T}}}\int_0^\infty \xi(E)\,\varphi(E)\,\mathrm{d}E. \tag{5.5}$$

To obtain the activity in disintegrations per second, one has to multiply this by the disintegration constant λ of ^{24}Na which is

$$\lambda = 0.693/T_{\frac{1}{2}} = 1.28\times 10^{-5}\ \mathrm{sec}^{-1}. \tag{5.6}$$

By a simple mathematical artifice it is possible to express the disintegration constant in microcuries (1 microcurie $= 3.7\times 10^4$ dis/sec). We get thus also

$$\lambda = 3.5\times 10^{-10}\ \mu\mathrm{Ci}. \tag{5.7}$$

The total activity per unit volume of the body will be, after multiplication by λ and division by the total volume V, for the number of ^{24}Na atoms produced

$$a = \frac{\lambda St\Sigma N_{\mathrm{a}}}{V\Sigma_{\mathrm{T}}}\int_0^\infty \xi(E)\,\varphi(E)\,\mathrm{d}E. \tag{5.8}$$

Here the value S/V for a cylinder of 60 cm height and 30 cm diameter is to be inserted

$$S/V = 2/\pi R = 0.0425\ \mathrm{cm}^{-1}. \tag{5.9}$$

As a cubic centimetre of the body contains 1 mg of Na on the average, a is also the activity per mg ^{23}Na.

It has been ascertained by various investigators, as mentioned by Jones, that:

(a) the ratio of ^{24}Na to ^{23}Na in circulated blood is equivalent to the ratio in the whole body;

(b) the inhomogeneity of distribution of ^{23}Na in living bodies has no significant effect on the efficiency of the production of ^{24}Na;

(c) a phantom filled with a dilute solution of NaCl or $NaNO_3$ can be

used to approximate a living body in experiments involving the production of ^{24}Na by neutron irradiation;
(d) the size of the body and its orientation with respect to the source of neutrons has an effect on the quantity of ^{24}Na produced;
(e) the exact time of sampling, within a range of 2 to 24 hours after irradiation, is not critical.

Thus the ratio S/V calculated for the cylinder mentioned will be taken as an intermediate value between the extreme values found whether the human body is facing the source or gets the flux from the side. The values S/V for these two cases are found to be approximately 0.078 and 0.038 for the 70 kg average human adult of density equal to 1 with apparent cross-sections of 5000 and 2660 cm^2.

When we put in all the numerical values we arrive at

$$a = 0.89 \times 10^{-14} t \int_0^\infty \xi(E)\varphi(E)\,\mathrm{d}E \quad \mu\text{Ci per mg } ^{23}\text{Na}. \qquad (5.10)$$

Incidentally, for a monoenergetic flux of neutrons, the integral per unit flux (1 neutron/sec cm^2) reduces to $\xi(E)$ and one has per neutron per cm^2 the simplified expression

$$a_1 = \frac{S\lambda\Sigma_{\mathrm{Na}}\xi(E)}{V\Sigma_{\mathrm{T}}} \quad \mu\text{Ci per mg } ^{23}\text{Na per neutron/cm}^2. \qquad (5.11)$$

It is this quantity which has been plotted in fig. VII.5 together with $\xi(E)$. However a_1 cannot be used directly in the case of a broad neutron spectrum to calculate the dose received, because the integrated neutron dose must always be calculated separately under consideration of both dose equivalent and spectral density for each energy, as will be outlined now.

We will now relate the quantity a found with (5.10) to the dose received. The dose deposition in soft tissue by neutrons is a function of energy. We gave thus to multiply $\varphi(E)\,\mathrm{d}E$ in every energy interval by an appropriate factor $d_0(E)$ expressed in rads or rems per incident neutron per cm^2 which relates the flux to the dose. This factor $d_0(E)$ is evidently the inverse of the quantities represented in fig. VII.1 of this chapter and valid either in rads or in rems for surface (first collision) dose and maximum dose inside the body. For convenience's sake we have drawn the quantities $d_0(E)$ in fig. VII.6.

The integrated neutron dose d_{n} the subject will have received is then calculated from

$$d_{\mathrm{n}} = t \int_0^\infty d_0(E)\varphi(E)\,\mathrm{d}E \tag{5.12}$$

and this can be expressed in rads or rems according to the concept chosen.

It is now useful to relate the dose to the activity per unit volume of the body, which is equivalent to the activity per mg of ^{23}Na. Thus one will get a proportionality factor which links dose to specific blood activity and can be checked in mock-up experiments with saline solution-filled phantoms imitating the form of the human body. So we write down the integrated neutron dose to specific activity ratio

$$\frac{d_{\mathrm{n}}}{a} = \frac{V\Sigma_{\mathrm{T}} \int_0^\infty d_0(E)\varphi(E)\,\mathrm{d}E}{S\Sigma_{\mathrm{Na}}\lambda \int_0^\infty \xi(E)\varphi(E)\,\mathrm{d}E}$$

$$= 1.12\times 10^{14} \frac{\int_0^\infty d_0(E)\varphi(E)\,\mathrm{d}E}{\int_0^\infty \xi(E)\varphi(E)\,\mathrm{d}E} \frac{\text{rads or rems}}{\mu\text{Ci per mg } ^{23}\text{Na}} \tag{5.13}$$

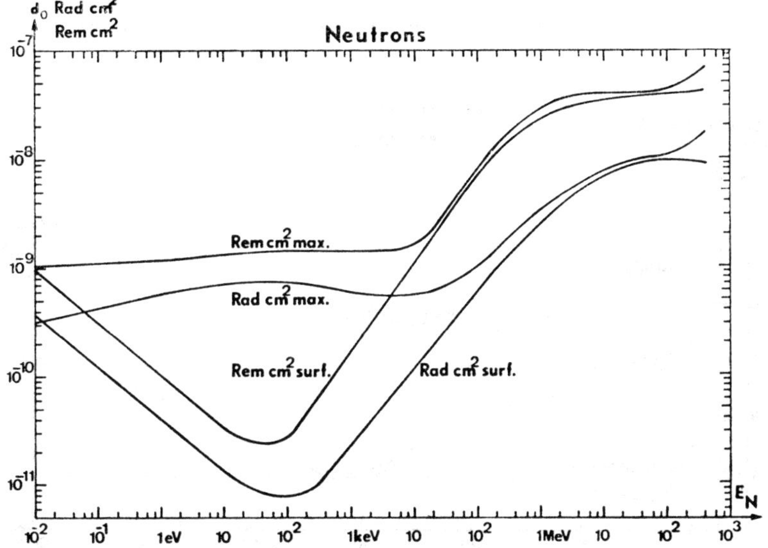

Fig. VII.6 Rad and rem doses per normally incident neutron per cm² as a function of neutron energy at the surface and at maximum inside the body.

After all this theory on blood sodium activation, it is worth-while to mention that there exists another way of finding the dose received by an exposed individual from the properties of the blood: it is to make a biological examination of it. In fact haematological changes represent not only the most serious clinical effects of the irradiation but also the best biological index of the exposure dose (including both gamma and neutron doses).

The various sorts of blood cells are different in life-span, production, ageing and radiation sensitivity. The changes after exposure are the consequence of anatomically visible damage in the bone marrow and the lymphoid tissue, the so-called haematopoetic organs. These tissues are in constant growth and deliver to the blood exactly the amount of cells that age away. Irradiation stops cell division and young cells are no longer produced to replace the older ones. By counting the red blood cells (haemoglobin) and the white blood cells, in particular the lymphocytes produced in the lymphoid tissue glands and the granulocytes produced in the bone marrow, it is possible to make a reasonable estimate of the dose in rads received since these changes are dose dependent. Usually the haemoglobin content, the numbers of lymphocytes and neutrophils (a major sub-group of the granulocytes) and the number of platelets are determined as a function of time after the irradiation. Fig. VII.7, 8, 9 present the curves for absorbed doses of 200, 300 and 450 rads according to Andrews, who has averaged the

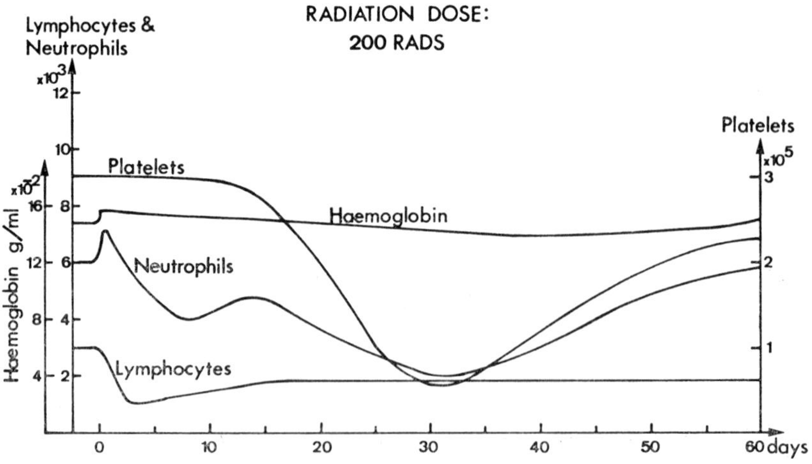

Fig. VII.7 Blood characteristics as a function of time after exposure to doses of 200 rads.

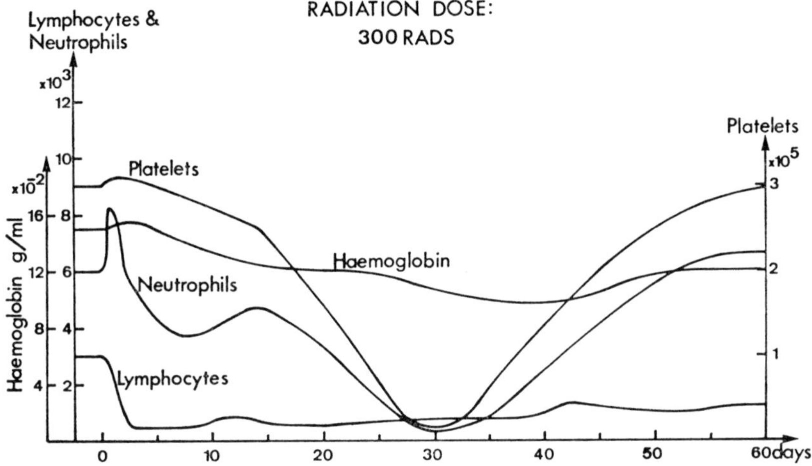

Fig. VII.8 Blood characteristics as a function of time after exposure to doses of 300 rads.

data obtained from human beings who were exposed in various radiation accidents. One notices that the height of the peak in the neutrophils counts appearing a few days after exposure is a sensitive indication of the magnitude of the irradiation. So is the significant drop in the lymphocytes counts, which appears even after as little as 25 rads of total body irradiation. In addition a rapid decrease of the

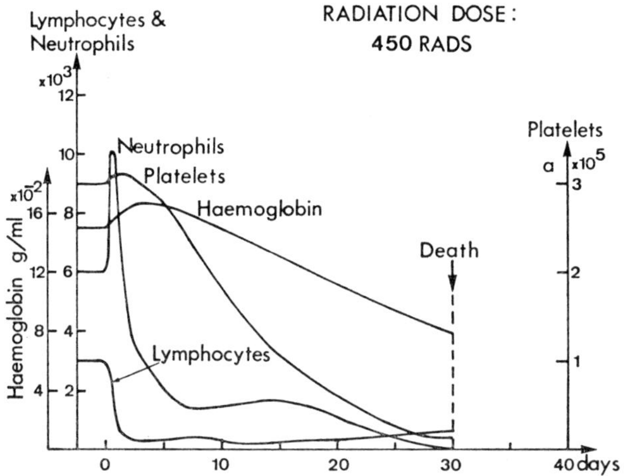

Fig. VII.9 Blood characteristics as a function of time after exposure to doses of 450 rads.

haemoglobin content and platelet numbers indicates a high dose, whereas the lymphocytes tend to increase again after a few days if the irradiation has not been much in excess of 200 rads.

As a supplementary indication, the next fig. VII.10 gives the depression (percentage decrease with respect to the values before irradiation) of the white blood cells (including lymphocytes, neutrophils, eosinophils and monocytes) averaged over the 13th to 18th days after

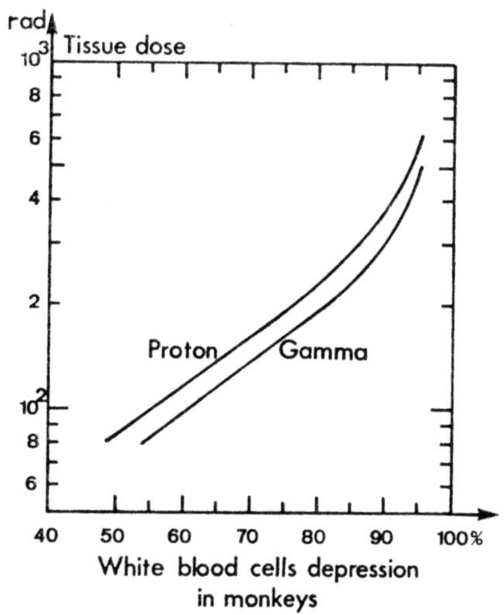

Fig. VII.10 Absorbed dose of 200 MeV protons and ^{60}Co gamma rays versus white blood cell depression in monkeys 15 days after exposure.

irradiation as a function of tissue dose for monkeys after Taketa et al. The incident fluxes were 200 MeV protons and ^{60}Co gamma rays. The strong dependence of the white blood cells counts on the absorbed dose is evident.

We would like now to consider the assessment of fast neutron doses in man by body hair activation. The reaction used is $^{32}S(n,p)^{32}P$ and the beta activity of ^{32}P is measured with a half-life of 14.3 days. Hair, nails and cartilage occurring near the surface of the body contain sufficient sulphur to be useful for fast neutron dose estimates. At the same time the phosphorus content of these objects is low enough that one need not fear formation of ^{32}P from ^{31}P by thermal neutrons (the P/S ratio is smaller than 1% for hair and 3% for cartilage). The sulphur concentration in hair, as measured by Petersen et al. on a number of samples, is remarkably constant regardless of its colour and body distribution and is equal to 47.7±5.5 mg per gram of hair. In cartilage the sulphur fraction measured in one case by Harris was 0.00926 by weight. The reaction cross-section for a fission spectrum is taken to be 0.225 barn. One can thus find the integrated neutron flux $\int \Phi \mathrm{d}t$ in neutrons per cm^2 from the measured activity a per gram

TABLE VII.8

Recorded d_γ/d_n ratios for some radiation accidents.

Reactor	Identification number	d_γ/d_n	Source
Y–12 Oak Ridge accident	Case I	2.87	Hurst, Ritchie and Emerson
	Case II	3.37	
	Case III	3.09	
Boris Kidrič–Vinča experimental run	Station 1	3.5	Hurst to Morgan
	Station 9	4	
	Station 10	4.2	
accident	Station 1	2.88	Hurst to Morgan
	Station 9	2.46	
	Station 10	2.78	

of hair in disintegrations per second at zero cooling time if one knows the beta counter efficiency using the formula

$$\int \Phi \, \mathrm{d}t = aA/\sigma s N_0 \lambda = 8.04 \times 10^9 a \text{ cm}^{-2} \tag{5.14}$$

where $A=32$ is the atomic weight of sulphur, N_0 Avogadro's number, s the sulphur content per gram of hair (0.0477) and $\lambda=5.6\times10^{-7}\,\mathrm{sec}^{-1}$ the decay constant of ^{32}P. By employing the flux-to-dose conversion factor for neutrons above the sulphur threshold (2.5 MeV) which is 3.83×10^{-9} rad per neutron per cm^2, as indicated by Hurst and Ritchie, one arrives at the following relationship between sulphur activity per gram of hair and received dose d

$$d = 34a \text{ rad} \tag{5.15}$$

which is valid for a fission spectrum.

As a general comment one can say that this method is much more straightforward than that using ^{24}Na activation in blood, as it is the incident flux which activates without having to be thermalized in the body before activation takes place. One is thus independent of the latter process and of the interference of thermal neutrons which may be present. Another feature of sulphur activation is that it can be used to assess the flux of evaporation neutrons in the neighbourhood of accelerators, since we have seen that evaporation and fission neutrons have practically the same spectral distributions.

One might add here that some work has also been done on the production of ^{32}P and ^{31}S in bone by thermal and fast neutrons (Chanteur et al.).

To the neutron dose assessed by blood, hair or bone activity measurements, one has still to add the dose caused by the gamma rays which usually accompany the neutron leakage spectrum. The gamma-to-neutron dose ratio varies from case to case, depending on the type of reactor involved, its shielding, the circumstances of the accident, the location of the exposed person etc. As an example table VII.8 gives a few values of this ratio for past nuclear accidents around reactors.

6 Activation of various body tissues by high energy particles inducing spallation

It is convenient to define a few parameters describing induced activity in tissues in order to take care of the facts that tissue is a mixture of elements and that the same radionuclide can be produced by several parent elements. Referring to the contents of ch. I we can write that the activity per gram at cooling time t_c for a target element i with atomic number A_i irradiated for a time t_i in a flux Φ (expressed in particles per sec per cm^2) is

$$\frac{dn_s}{dt} = \frac{\Phi N_0 \sigma_{s,i}}{A_i}\left[1 - \exp\left(-\frac{0.693t_i}{t_{\frac{1}{2},s}}\right)\right]\exp\left(-\frac{0.693t_c}{t_{\frac{1}{2},s}}\right) \qquad (6.1)$$

where N_0 is Avogadro's number, $\sigma_{s,i}$ the production cross-section of the radioisotope considered and $t_{\frac{1}{2},s}$ its half-life. At zero cooling time and with infinite irradiation time (saturation) the exponential functions on the right vanish. We will always refer to zero cooling time and saturation values when defining material parameters. So, when summing up over all target constituents i, each present with a weight fraction a_i, we find for a particular isotope s

$$(\mathrm{d}n_s/\mathrm{d}t)_{\substack{t_c=0\\ t_i=\infty}} = \Phi N \sum_i (a_i\sigma_{s,i}/A_i) = \Phi C_s \tag{6.2}$$

where

$$C_s = N \sum_i (a_i\sigma_{s,i}/A_i). \tag{6.3}$$

C_s is thus a first characteristic parameter of the material constituting the target for the production of the particular isotope s. It is expressed in disintegrations per gram per sec per unit flux. In other words C_s multiplied by the flux in particles per second and cm^2 gives the disintegration rate of the particular radionuclide s per gram of tissue.

Once an object has a given activity, it is useful to determine what is the radiation received from this object, which will have a given size and thickness, at a given distance, taking into account the kind of radiation emitted and its absorption in the object itself. Then this is what one measures experimentally.

For simplicity we remain at saturation and zero cooling time. Let $\varepsilon_{s,k}$ be the probability of emission per decay of a particular radiation of the type k by isotope s, and μ_k/ϱ the $1/e$ attenuation length in g cm^{-2} of this particular radiation in the target material.

It has been proved that if the radioactive body is large compared to the attenuation length of the radiation emitted in the decay process, the number of quanta of this radiation per unit surface of a counter per unit time, i.e. the counting rate, at a location from which the radioactive body is seen under a solid angle Ω, is given by

$$r = \frac{\Phi\Omega N}{4\pi} \sum_i \frac{a_i}{A_i} \sum_s \left(\sum_k \frac{\varepsilon_{s,k}}{\mu_k} \right) \sigma_{s,i} \ \mathrm{sec}^{-1}\,\mathrm{cm}^{-2} \tag{6.4}$$

where summation over i extends to all target constituents, over s to all isotopes produced and over k to all types of radiation emitted by these isotopes.

Here we recognize our previously defined parameter C_s, so that r becomes

$$r = \frac{\Phi\Omega}{4\pi} \sum_s \sum_k \frac{\varepsilon_{s,k}}{\mu_k} C_s. \tag{6.5}$$

It is convenient to define as the relevant material constant

$$K_s = \sum_k (\varepsilon_{s,k}\varrho/\mu_k) C_s \tag{6.6}$$

from which we get the counting rate of a 1 cm² counter per second

$$r = (\Phi\Omega/4\pi) \sum_s K_s \ \mathrm{sec}^{-1}\mathrm{cm}^{-2}. \tag{6.7}$$

Thus K_s is the flux of radiation quanta from the decay of isotope s received by a unit surface viewing the radioactive body under a solid angle 4π for unit irradiation flux, saturation conditions and zero cooling time, taking into account the self-absorption in the tissue. Clearly the formula is valid only for small $\Omega/4\pi$ values in this form, which gives a flux of radiation quanta per unit surface. However it leads to a formula valid for solid angles as large as desired if one considers a counting volume presenting the same area in any direction or if one goes over to absorbed doses instead of fluxes of quanta across a surface, because the doses add algebraically.

Let f_k be the factor indicating the number of radiation quanta of type k through the unit surface that will give an absorbed dose of 1 rad. Then we find the absorbed dose rate in rad/h from the radioactive sample to be at zero cooling time

$$R = 3600\Phi \frac{\Omega}{4\pi} \sum_s \sum_k \frac{\varepsilon_{s,k}\varrho}{f_k \mu_k} C_s \ \mathrm{rad/h} \tag{6.8}$$

which suggests the definition of a new material constant

$$L_s = 3600 C_s \sum_k \frac{\varepsilon_{s,k}\varrho}{f_k \mu_k} \tag{6.9}$$

so that

$$R = (\Phi\Omega/4\pi) \sum_s L_s. \tag{6.10}$$

The constant L_s, multiplied by the flux of particles and the relative solid angle $\Omega/4\pi$, gives the dose rate in rad/h from radionuclide s from a large body of active tissue, taking into account the self-absorption in the tissue. Clearly these three coefficients C_s, K_s and L_s can be applied to any particular kind of tissue or to the body as a whole.

In many applications the activating flux is delivered in a time which is very short compared to the half-lives of the isotopes produced. The characteristic material parameters defined above for infinite irradiation time have then to be multiplied by the time factor

$$F = 1 - \exp(-0.693 t_i/t_{\frac{1}{2},s}) \approx 0.693 t_i/t_{\frac{1}{2},s}. \tag{6.11}$$

Usually the time integral of the burst $\int \Phi(t)\,\mathrm{d}t$ is known or to be found, and may be introduced into the formula instead of the product Φt_{i} in the cases where the burst is short.

Hutton, Pasinetti and the author have made various experiments with 600 MeV protons to determine the values of these coefficients C_s, K_s and L_s for high energy particles. Thin slices of various tissues were prepared in wet condition. As these included bone, fat, muscle, liver, kidney, they should represent rather well the three families of animal tissues: bone, fat, parenchymatous. For comparison plexiglas and carbon were irradiated at the same time. Figs. VII.11 and 12 show the gamma and beta decay curves of the samples, which were irradiated for 20 minutes. Fig. VII.13 shows the gamma decay curves of bone samples which were irradiated for different times in a supplementary experiment.

As one sees from fig. VII.11 for instance, the decay curves for some

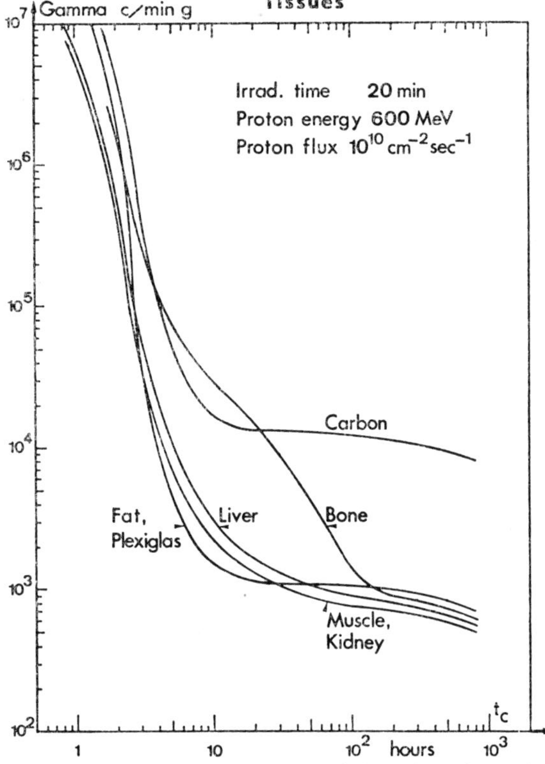

Fig. VII.11 Decay of registered gamma activity of various tissues irradiated by 600 MeV protons.

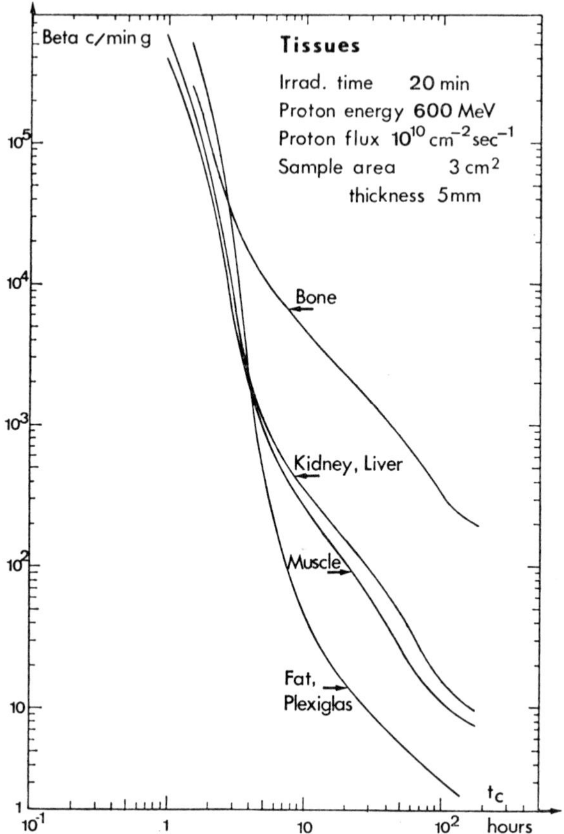

Fig. VII.12 Decay of registered beta activity of various tissues irradiated by 600 MeV protons.

tissues appear to be grouped together. In the first group we find fat, which decays in the same way as carbon and plexiglas, with no other component than ^{11}C and ^{7}Be. In the second group we find parenchymatous tissues (muscle, liver, kidney) which have more activity than does fat in the 1 to 10 hours region. From this data one can find the C_s values for the different kinds of tissue. This was done using the known gamma counter and beta counter efficiencies corrected for sample thickness. Table VII.9 gives the C_s values found for the various tissues by the gamma counts.

The C_s values that were found from the beta counts compare favourably with the C_s values from the gamma counts given in the previous table.

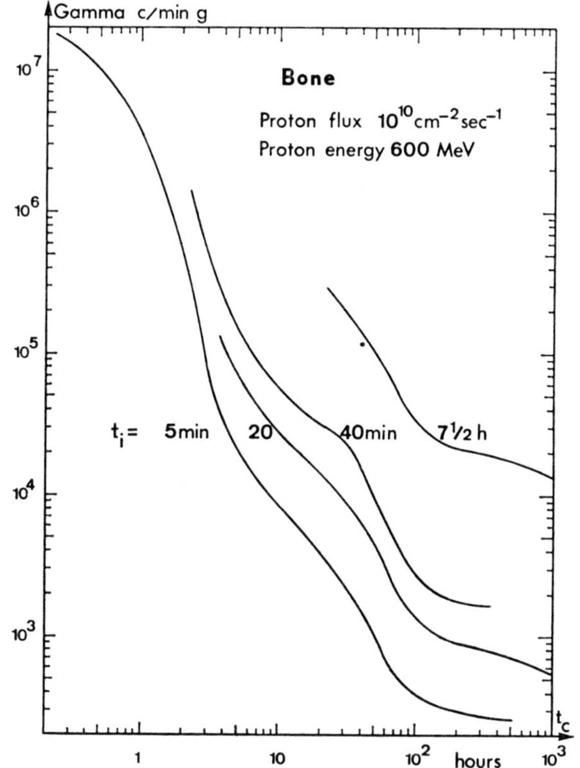

Fig. VII.13 Decay of registered gamma activity of bone samples irradiated by 600 MeV protons.

TABLE VII.9

C_s values for various animal tissues and other materials found from the gamma counts.

Substance	$^{11}C + ^{13}N$	$^{18}F + ^{31}S$	$^{24}Na + ^{28}Mg + ^{43}K$	^{7}Be
Bone	4.5×10^{-4}	1.2×10^{-5}	1.4×10^{-5}	3.8×10^{-4}
Liver	5×10^{-4}		0.62×10^{-6}	2.9×10^{-4}
Kidney	3.6×10^{-4}		0.33×10^{-6}	2.4×10^{-4}
Muscle	5×10^{-4}		0.21×10^{-6}	2.5×10^{-4}
Fat	1.4×10^{-3}			3.8×10^{-4}
Plexiglas	1.4×10^{-3}			3.4×10^{-4}
Carbon	1.1×10^{-3}			4×10^{-4}

From these results we can calculate a weighted value of the constants C_s and subsequently K_s and L_s for the whole body. For this calculation we have taken the percentages for bone, muscle and fat in the average human adult body and lumped the rest of the weight as parenchymatous tissue, as indicated in table VII.10.

TABLE VII.10

Tissue percentages taken for calculation of whole body parameters.

Muscle	43%
Fat	14%
Bone	10%
Rest (parenchyme)	33%

The next table VII.11 will show the whole-body values of C_s calculated in this manner for the various radionuclides produced. From these quasi-experimental values of C_s found by the superposition of the data obtained directly from the various tissues exposed to the radiation we can now calculate the other relevant parameters K_s and L_s for the whole body. The values obtained are also listed in table VII.11. The L_s values have been computed on the basis of the gamma counts only (which include the annihilation radiation from the positrons produced far from the surface), as one can check that the contribution to the dose from the electrons and positrons escaping from the surface amounts to a few per cent only.

TABLE VII.11

The constants C_s, K_s and L_s expected for the whole body.

Isotopes	C_s (gamma)	K_s (gamma)	L_s	K_s (beta)
$^{11}C+^{13}N$	6.0×10^{-4}	1.57×10^{-2}	1.49×10^{-8}	2.9×10^{-5}
$^{18}F+^{31}Si$	1.2×10^{-6}	3.05×10^{-5}	2.86×10^{-11}	8.75×10^{-8}
$^{24}Na+^{28}Mg+^{43}K$	3.9×10^{-6}	1.7×10^{-4}	5.92×10^{-10}	7.2×10^{-8}
^{7}Be	2.8×10^{-4}	4.23×10^{-4}	4.03×10^{-10}	
$^{32}P+^{33}P+^{35}S$				9.0×10^{-8}

In order to check the values obtained for the whole body by calculation from the results of the tissue irradiations, we have irradiated five mice, denominated A, B, C, D and E with doses equal to 60, 900, 1200, 15000 and 240000 rads (the flux-to-rad conversion factor utilized was 2.7×10^{-7} protons per cm^2 per rad).

For the gamma measurements, we simply laid each mouse on the cap of the 3″×3″ NaI(Tl) crystal. For the beta measurements we put the surface of the body of the mouse at the same distance from the counter as we did the surfaces of the various samples and tissues. We also registered the dose rate in mrad/h at the surface of mice D and E. Fig. VII.14 shows the gamma decay curves measured under these conditions and fig. VII.15 the beta decay curves.

It is now interesting to apply the values just obtained for C_s, K_s and L_s to personnel dosimetry by induced activity measurements in

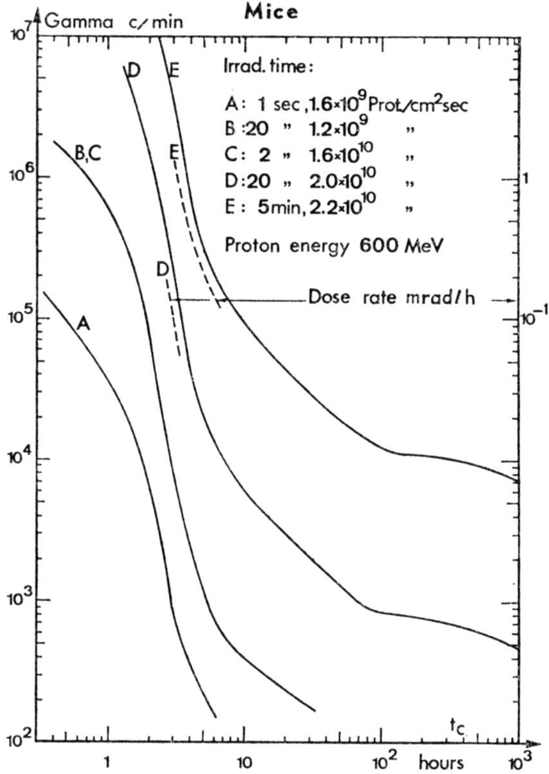

Fig. VII.14 Decay of registered gamma activity and dose rate of mice irradiated by 600 MeV protons.

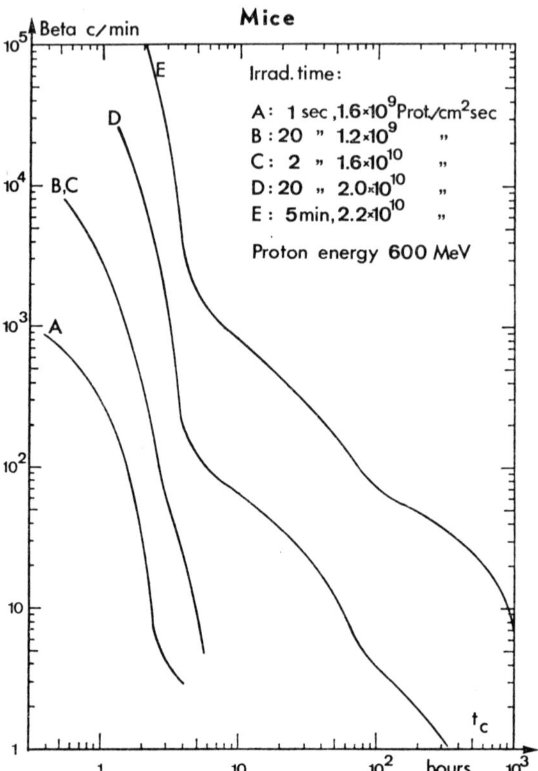

Fig. VII.15 Decay of beta activity of mice irradiated by 600 MeV protons.

case of radiation accidents. Two sorts of accidental exposure to very high energy particles could be considered as limiting cases. First, accidental exposure of short duration at a high radiation level; second, unnoticed exposure of long duration at a low radiation level. The first occurs when a person is exposed to an accelerator beam or happens to enter a radiation area without noticing the danger, or if a sudden burst of radiation is released accidentally. The exposure is sometimes limited to one part of the body, as in the case of a beam. The other type of exposure can happen in the case of insufficient shielding.

The data derived in the earlier sections cover both kinds of accidents. In the case of a short burst, one must apply the time factor given in (6.11). In the case of a long irradiation all isotopes, and not only the ^{7}Be, can contribute to the knowledge, provided the measurements are taken sufficiently soon after the exposure.

Experimental procedures for investigating the radioactivity in the human body are numerous. We can quote here several of these, three of which are of particular interest because of their simplicity, and which apply to gamma counts from the whole body or a part of it.

The first is the 1 metre radius arc method, in which the patient sits on a couch, the general line of which is approximately a circle of 1 m radius, at the centre of which the counter is.

The second is by placing the crystal on top of the thighs at the crotch and having the subject bent over the crystal so that the body almost completely surrounds the crystal. This also gives less background because of the shielding of the counter by the body itself.

The third is by placing the counter in contact with a part of the body. It is thus possible to make a plot of the distribution of the activity over each part of the body, as may be required in cases of local irradiation.

A fourth method is, of course, to place the person in a whole-body counter which has been suitably calibrated.

One can either note the total counts and plot them versus time or take gamma ray spectra at various intervals. Measurements of the induced activity in the human body, properly treated according to the general formulae given above and with correct calibration of the measuring apparatus should normally yield the flux of particles Φ to which the person has been subjected provided the irradiation time is known, or the flux integral $\int \Phi(t)\,dt$ if the irradiation was of short duration.

With the results obtained in this section for high energy particles one can construct as an exercise a typical gamma decay curve of body activity for short-burst irradiation and unit flux–time integral for a particular detector, for instance a 3″×3″ NaI(Tl) crystal in 2π geometry. We start from the expression for the photon flux per cm^2 of the detector, which is to be written, in the case of a short burst,

$$r = \left(\frac{\Phi t_{\mathrm{i}} \Omega}{4\pi}\right) \sum_s K_s \exp\left(-\frac{0.693 t_{\mathrm{c}}}{t_{\frac{1}{2},s}}\right) \frac{0.693}{t_{\frac{1}{2},s}} \,\mathrm{sec}^{-1}\,\mathrm{cm}^{-2}. \tag{6.12}$$

We take here $\Phi t_{\mathrm{i}}=1$ particle per cm^2 and $\Omega/4\pi=\frac{1}{2}$. This means that the crystal is placed in contact with the body. The number of counts per cm^2 effectively registered will be this quantity r times the efficiency of the counter. This efficiency η is different for each isotope and is defined as the number of counts per emission of a photon. The measured counting rate is thus $r\eta S$, where S is the area of the counter (here 45 cm²). Table VII.12 contains the whole-body K_s times $0.693/t_{\frac{1}{2},s}$ values and the η values that have been used.

TABLE VII.12

Parameters for short-burst irradiation.

Parameter	^{11}C	^{18}F	^{24}Na	^{7}Be
$0.693K_s/t_{\frac{1}{2},s}$	8.9×10^{-6}	3.2×10^{-9}	2.2×10^{-9}	6.4×10^{-11}
η	0.27	0.27	0.18	0.27

We computed the theoretical curve for unit integrated flux, i.e. 1 proton per cm², which we present in fig. VII.16. Then for any case of

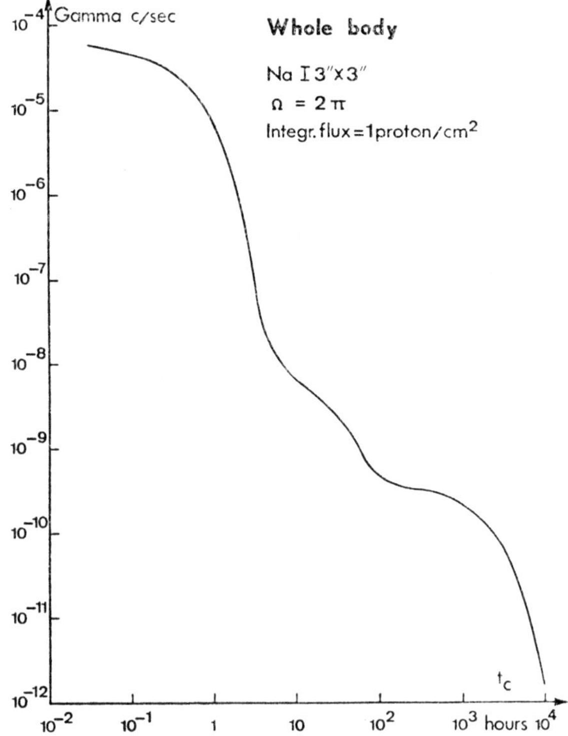

Fig. VII.16 Typical gamma decay curve of registered body activity $r\eta S$ on a 3″×3″ NaI(Tl) crystal, computed for an integrated flux of 1 proton per cm² of 600 MeV energy and a solid angle of 2π.

irradiation we can determine the dose by comparison. The sensitivity of the method depends on the isotope which can be measured, in other words on the time one waits after the irradiation to make the measurements. It also depends whether one uses total counts or takes a spectrum and measures the counts under a peak. For typical background conditions in a laboratory, the sensitivity would be 10 rads in the first case and 1 rad in the second two hours after the burst.

The experiments described in this section were made with a pure beam of 600 MeV protons. It is known that the spallation cross-sections for light weight elements do not vary appreciably with energy above this value, so that our results should be also valid for protons of kinetic energies above 600 MeV.

1. Calculate the various doses in rads and rems at the surface of the body or at the maximum inside per incident neutron for three typical spectral distributions of neutrons:
(a) thermal distribution,
(b) $\varphi(E)=E^{-1}$ between $E_1=0.4$ eV and $E_2=18$ MeV,
(c) fission spectrum $\varphi(E)=0.77E^{\frac{1}{2}}\exp(-0.776E)$ between the same limits.

For thermal neutrons simply read the values of the curves at the neutron energy 0.025 eV. For cases (b) and (c) calculate the expression

$$\int_{E_1}^{E_2} d_0(E)\varphi(E)\,\mathrm{d}E$$

and divide it by

$$\int_{E_1}^{E_2} \varphi(E)\,\mathrm{d}E.$$

The value of the latter integral is 17.6 in case b and 0.523 in case c.

2. Consider a nuclear burst flooding a person with 10^{11} thermal neutrons per cm^2. What is the surface dose in rads received? Calculate the blood sodium activation in microcuries ^{24}Na per milligram ^{23}Na at zero cooling time.

3. Answer the same questions for a burst giving 10^{11} neutrons per cm^2 with the spectral distribution E^{-1} between 0.4 eV and 18 MeV.

4. Answer the same questions for a burst giving 10^{11} fission neutrons per cm^2 with the spectral distribution given in (c), problem 1, within the same limits.

APPENDIX A

TABLES OF k_γ-FACTORS

The k_γ-factor is the gamma ray dose rate in rad/hour in soft tissue (0.93 rad equivalent to 1 röntgen) at a distance of 1 metre from a point source of strength 1 curie. In the case of positron emitters it has been assumed that all the positrons are absorbed in the source with the emission of 2 quanta of energy 0.51 MeV. All k_γ-factors have been calculated on the basis of decay schemes and branching ratios available to date in the literature, mainly from Strominger, Hollander, Seaborg and from Seelmann–Eggebert, Pfennig. In view of the difficulty of interpreting some decay schemes in terms of emission probabilities and owing to the discrepancies existing between various authors, the values given here should be used with caution.

The mean number ε of photons emitted per disintegration has been added to the table. It is equal to the sum of the emission probabilities for each gamma quantum in the decay scheme.

For more details, cf. subsection 5.6.1 of ch. I.

TABLE A.1

Isotopes up to ^{209}Bi with a half-life longer than 5 minutes.

Isotope		$t_{\frac{1}{2}}$	ε	k_γ	Isotope		$t_{\frac{1}{2}}$	ε	k_γ
Be	7	53.4 d	0.12	0.029	Na	22	2.6 y	2.80	1.104
C	11	0.3 h	2.00	0.515	Na	24	15.0 h	2.00	2.080
N	13	0.2 h	2.00	0.515	Mg	27	0.2 h	1.00	0.446
F	18	1.8 h	1.90	0.489	Mg	28	21.2 h	2.00	0.621

Isotope		$t_{\frac{1}{2}}$	ε	k_γ	Isotope		$t_{\frac{1}{2}}$	ε	k_γ
Al	26	7×10^5 y	2.70	1.366	Ni	57	1.5 d	2.14	0.995
Al	29	0.1 h	1.00	0.686	Ni	65	2.5 h	0.46	0.296
S	37	0.1 h	0.10	0.156	Cu	60	0.4 h	2.70	1.117
S	38	2.9 h	0.95	0.902	Cu	61	3.3 h	1.55	0.425
Cl	34	0.6 h	1.03	0.403	Cu	62	0.2 h	1.97	0.508
Cl	38	0.6 h	0.78	0.760	Cu	64	12.8 h	0.39	0.105
Cl	39	0.9 h	1.35	0.490	Cu	66	0.1 h	0.09	0.047
Ar	41	1.8 h	0.99	0.645	Cu	67	2.5 d	1.00	0.081
K	38	0.1 h	3.00	1.575	Zn	62	9.3 h	0.73	0.147
K	42	12.4 h	0.18	0.138	Zn	63	0.6 h	1.56	0.425
K	43	22.0 h	1.85	0.481	Zn	65	245.0 d	0.52	0.282
K	44	0.4 h	1.00	0.707	Zn	69	13.9 h	0.50	0.111
Ca	47	4.7 d	0.81	0.503	Zn	71	4.0 h	3.00	0.747
Ca	49	0.1 h	1.00	1.627	Zn	72	1.9 d	1.00	0.071
Sc	43	3.9 h	2.02	0.504	Ga	65	0.2 h	3.10	0.705
Sc	44	2.4 d	0.50	0.068	Ga	66	9.5 h	1.62	0.793
Sc	46	84.0 d	2.00	1.015	Ga	67	3.3 d	1.47	0.136
Sc	47	3.4 d	0.70	0.057	Ga	68	1.1 h	1.76	0.464
Sc	48	1.8 d	3.00	1.686	Ga	70	0.4 h	0.01	0.005
Ti	44	10^3 y	1.79	0.067	Ga	72	14.1 h	0.89	0.668
Ti	45	3.1 h	1.68	0.430	Ga	73	4.8 h	1.00	0.151
Ti	51	0.1 h	1.01	0.177	Ga	74	0.1 h	1.20	0.918
V	47	0.5 h	2.00	0.515	Ge	66	2.4 h	3.70	0.647
V	48	16.2 d	3.18	1.473	Ge	67	0.3 h	2.34	0.517
Cr	48	23.0 h	2.00	0.215	Ge	69	1.7 d	2.00	0.799
Cr	49	0.7 h	1.95	0.404	Ge	75	1.4 h	0.14	0.019
Cr	51	27.8 d	0.10	0.016	Ge	77	11.3 h	1.50	0.189
Mn	51	0.7 h	2.00	0.515	As	68	0.1 h	2.00	0.515
Mn	52	5.7 d	1.85	0.880	As	79	0.2 h	3.00	0.631
Mn	52	0.4 h	2.88	1.206	As	70	0.8 h	3.40	1.673
Mn	54	314.0 d	1.00	0.424	As	71	2.6 d	1.66	0.254
Mn	56	2.6 h	1.50	0.917	As	72	1.1 d	2.56	0.877
Fe	52	8.3 h	2.10	0.366	As	73	76.0 d	2.00	0.411
Fe	53	0.1 h	2.00	0.515	As	74	17.7 d	1.29	0.367
Fe	59	45.0 d	1.03	0.599	As	76	1.1 d	0.58	0.214
Fe	60	10^5 y	1.00	0.103	As	77	1.6 d	0.04	0.006
Co	55	18.0 h	2.65	0.911	As	78	1.5 h	0.50	0.156
Co	56	77.0 d	2.55	1.290	As	79	0.1 h	1.00	0.048
Co	57	267.0 d	1.00	0.061	Se	70	0.7 h	2.00	0.515
Co	58	71.0 d	1.30	0.486	Se	71	0.1 h	2.50	0.555
Co	58	9.0 h	0.50	0.060	Se	72	8.4 d	0.60	0.021
Co	60	0.2 h	1.00	0.032	Se	73	7.1 h	1.20	0.271
Co	60	5.3 y	2.00	1.264	Se	75	120.0 d	2.85	0.330
Co	61	1.7 h	1.00	0.036	Se	83	0.4 h	1.60	0.355
Co	62	0.2 h	2.25	1.437	Br	74	0.6 h	2.80	0.786
Ni	56	6.1 d	2.90	0.772	Br	75	1.7 h	2.80	0.693

Isotope	$t_{\frac{1}{2}}$	ε	k_γ	Isotope	$t_{\frac{1}{2}}$	ε	k_γ
Br 76	17.0 h	1.62	0.442	Zr 95	65.0 d	0.98	0.366
Br 77	2.4 d	1.47	0.395	Zr 97	17.0 h	0.10	0.058
Br 78	0.1 h	2.00	0.515	Nb 89	1.0 h	0.50	0.149
Br 80	4.4 h	0.20	0.011	Nb 89	1.9 h	0.90	0.232
Br 80	0.3 h	0.04	0.013	Nb 90	14.6 h	3.60	2.046
Br 82	1.5 d	3.22	1.323	Nb 91	62.0 d	0.42	0.257
Br 83	2.4 h	0.01	0.003	Nb 92	10.1 d	1.01	0.483
Br 84	0.5 h	0.32	0.225	Nb 93	3.7 y	0.10	0.008
Br 84	0.1 h	2.10	1.077	Nb 94	2×10^4 y	1.92	0.792
Kr 76	14.8 h	1.00	0.136	Nb 94	0.1 h	0.50	0.021
Kr 77	1.2 h	2.10	0.510	Nb 95	3.8 d	0.10	0.012
Kr 79	1.4 d	0.43	0.074	Nb 95	35.0 d	1.00	0.389
Kr 83	1.9 h	0.50	0.042	Nb 96	23.0 h	3.00	1.134
Kr 85	10.7 y	0.00	0.001	Nb 97	1.2 h	1.00	0.338
Kr 85	4.4 h	0.50	0.046	Mo 90	6.0 h	2.66	0.554
Kr 87	1.3 h	0.63	0.281	Mo 91	0.3 h	2.00	0.515
Kr 88	2.8 h	0.88	0.955	Mo 93	6.9 h	1.15	0.567
Rb 79	0.4 h	2.20	0.530	Mo 99	2.8 d	1.47	0.124
Rb 81	4.7 h	0.98	0.198	Mo 101	0.3 h	0.62	0.359
Rb 82	6.3 h	1.70	0.718	Tc 93	2.7 h	0.75	0.462
Rb 83	83.0 d	1.00	0.262	Tc 94	0.9 h	2.64	1.056
Rb 84	33.0 d	1.11	0.423	Tc 95	60.0 d	1.75	0.098
Rb 86	18.7 d	0.09	0.049	Tc 95	20.0 h	0.50	0.196
Rb 88	0.3 h	0.90	0.726	Tc 96	4.3 d	2.99	1.244
Rb 89	0.2 h	1.73	1.149	Tc 96	0.9 h	0.50	0.033
Sr 83	1.4 d	2.90	0.785	Tc 97	91.0 d	0.10	0.005
Sr 85	64.0 d	1.00	0.257	Tc 98	15×10^5 y	2.00	0.707
Sr 85	1.2 h	0.55	0.060	Tc 99	6.0 h	0.45	0.032
Sr 87	2.9 h	0.40	0.079	Tc 101	0.2 h	0.96	0.156
Sr 91	9.7 h	1.40	0.539	Tc 104	0.3 h	0.50	0.078
Sr 92	2.7 h	1.00	0.639	Ru 95	1.7 h	0.90	0.230
Y 84	3.7 h	2.00	0.515	Ru 97	2.9 d	0.15	0.019
Y 85	2.7 h	2.25	0.640	Ru 103	40.0 d	1.00	0.250
Y 86	14.6 h	2.56	1.338	Ru 105	4.4 h	1.80	0.391
Y 87	3.3 d	1.00	0.242	Rh 98	0.1 h	3.00	0.843
Y 87	14.0 h	0.40	0.077	Rh 99	16.0 d	2.00	0.368
Y 88	108.0 d	1.90	1.338	Rh 99	4.7 h	0.50	0.086
Y 90	3.1 h	0.50	0.050	Rh 100	21.0 h	1.65	0.825
Y 91	0.8 h	0.50	0.139	Rh 101	4.5 d	0.50	0.078
Y 92	3.5 h	0.14	0.066	Rh 101	5.0 y	1.00	0.081
Y 93	10.1 h	0.15	0.061	Rh 102	206.0 d	1.48	0.475
Y 94	0.3 h	0.54	0.245	Rh 103	1.0 h	0.50	0.023
Zr 86	17.0 h	5.1	2.671	Rh 105	1.5 d	0.10	0.016
Zr 87	1.7 h	1.76	0.489	Rh 106	2.2 h	1.34	0.503
Zr 88	85.0 d	1.00	0.197	Rh 107	0.4 h	0.90	0.141
Zr 89	3.8 d	0.51	0.137	Pd 99	0.4 h	3.50	1.328

Isotope	$t_{\frac{1}{2}}$	ε	k_γ	Isotope	$t_{\frac{1}{2}}$	ε	k_γ
Pd 100	4.1 d	1.00	0.045	Sn 123	125.0 d	0.01	0.005
Pd 101	8.5 h	0.60	0.133	Sn 125	9.4 d	0.05	0.029
Pd 103	17.0 d	1.10	0.065	Sn 125	0.2 h	1.01	0.177
Pd 109	13.5 h	0.09	0.004	Sn 126	10^5 y	2.00	0.550
Pd 111	0.4 h	0.10	0.020	Sb 116	1.0 h	0.80	0.467
Pd 111	5.5 h	1.10	0.168	Sb 116	0.3 h	1.30	0.576
Pd 112	21.0 h	0.20	0.042	Sb 117	2.8 h	1.05	0.094
Ag 103	1.1 h	1.50	0.247	Sb 118	5.1 h	1.00	0.568
Ag 104	1.2 h	1.50	0.415	Sb 119	1.6 d	0.14	0.018
Ag 105	40.0 d	2.50	0.431	Sb 120	5.8 d	1.67	0.610
Ag 106	8.3 d	1.05	0.430	Sb 122	2.8 d	0.76	0.224
Ag 110	253.0 d	1.40	0.592	Sb 124	60.2 d	1.51	0.741
Ag 111	7.5 d	0.08	0.014	Sb 126	12.5 d	3.00	0.898
Ag 112	3.2 h	0.50	0.195	Sb 126	0.3 h	0.80	0.220
Ag 113	5.3 h	0.50	0.076	Sb 127	3.9 d	1.82	0.457
Ag 115	0.4 h	0.10	0.012	Sb 128	0.2 h	0.60	0.184
Cd 104	1.0 h	1.50	0.059	Sb 128	9.0 h	1.00	0.268
Cd 105	0.9 h	1.20	0.256	Sb 129	4.3 h	0.60	0.165
Cd 107	6.5 h	0.05	0.002	Te 116	2.5 h	3.38	1.093
Cd 109	1.3 y	0.09	0.004	Te 117	1.1 h	1.10	0.336
Cd 111	0.8 h	0.60	0.068	Te 119	4.7 d	0.97	0.343
Cd 115	2.3 d	1.25	0.276	Te 119	16.0 h	0.57	0.182
Cd 115	43.0 d	0.02	0.010	Te 121	17.0 d	0.51	0.147
Cd 117	2.9 h	0.45	0.224	Te 121	154.0 d	1.90	0.334
In 107	0.6 h	3.00	0.626	Te 123	104.0 d	0.42	0.034
In 108	0.9 h	2.48	0.898	Te 127	105.0 d	0.61	0.022
In 108	0.7 h	1.40	0.391	Te 127	9.4 h	0.01	0.002
In 109	4.3 h	1.30	0.277	Te 129	33.0 d	1.20	0.255
In 110	4.9 h	3.50	1.078	Te 129	1.1 h	0.95	0.110
In 111	2.8 d	1.90	0.203	Te 131	1.2 d	0.51	0.136
In 112	0.2 h	0.27	0.075	Te 131	0.4 h	0.65	0.072
In 112	0.4 h	0.50	0.039	Te 132	3.3 d	1.03	0.117
In 113	1.7 h	0.33	0.065	Te 133	1.0 h	0.13	0.022
In 114	50.0 d	0.13	0.019	J 120	1.4 h	1.80	0.463
In 115	4.5 h	0.47	0.078	J 121	1.4 h	1.30	0.183
In 116	0.9 h	1.00	0.625	J 123	13.0 h	1.00	0.081
In 117	1.9 h	0.36	0.057	J 124	4.2 d	1.20	0.403
In 117	0.7 h	0.95	0.178	J 125	60.0 d	0.08	0.005
Sn 109	0.3 h	0.80	0.148	J 126	13.2 d	0.77	0.208
Sn 110	4.0 h	4.20	1.064	I 128	0.4 h	0.20	0.049
Sn 111	0.6 h	0.58	0.149	J 130	12.5 h	3.50	1.078
Sn 113	118.0 d	2.00	0.328	J 131	8.1 d	0.97	0.192
Sn 117	14.0 d	0.50	0.040	J 132	2.3 h	2.60	0.904
Sn 119	250.0 d	0.54	0.022	J 133	21.0 h	1.00	0.268
Sn 121	25.0 y	0.50	0.027	I 134	0.9 h	1.70	0.823
Sn 123	0.7 h	1.00	0.081	J 135	6.7 h	1.70	0.987

Isotope	$t_{\frac{1}{2}}$	ε	k_γ	Isotope	$t_{\frac{1}{2}}$	ε	k_γ
Xe 121	0.7 h	2.40	0.534	Ce 137	1.4 d	1.40	0.301
Xe 122	19.0 h	0.60	0.064	Ce 137	8.7 h	0.03	0.007
Xe 123	1.8 h	2.20	0.421	Ce 139	140.0 d	0.80	0.069
Xe 125	18.0 h	1.00	0.048	Ce 141	32.5 d	0.50	0.038
Xe 127	36.4 d	1.04	0.119	Ce 143	1.4 d	0.72	0.149
Xe 129	8.0 d	0.06	0.003	Ce 144	285.0 d	0.23	0.014
Xe 131	12.0 d	0.02	0.002	Ce 146	0.2 h	1.00	0.149
Xe 133	2.3 d	0.10	0.012	Pr 135	0.4 h	2.00	0.333
Xe 133	5.3 d	0.50	0.020	Pr 136	1.1 h	0.96	0.213
Xe 135	8.4 h	1.00	0.126	Pr 137	1.5 h	0.54	0.139
Xe 135	0.3 h	0.40	0.107	Pr 138	2.0 h	0.94	0.305
Xe 138	0.3 h	0.20	0.042	Pr 139	4.5 h	0.12	0.031
Cs 123	0.1 h	2.00	0.515	Pr 142	19.2 h	0.04	0.032
Cs 125	0.7 h	1.00	0.257	Pr 144	0.3 h	0.03	0.018
Cs 127	6.2 h	1.02	0.193	Pr 146	0.4 h	1.66	0.562
Cs 129	1.3 d	1.10	0.215	Nd 138	0.4 h	1.00	0.257
Cs 130	0.5 h	0.92	0.237	Nd 139	5.2 h	1.80	0.463
Cs 132	6.5 d	1.00	0.338	Nd 141	2.5 h	0.07	0.027
Cs 134	2.1 y	1.24	0.431	Nd 147	11.1 d	0.55	0.069
Cs 134	2.9 h	0.28	0.018	Nd 149	1.8 h	1.00	0.056
Cs 136	12.9 d	1.05	0.410	Nd 151	0.2 h	1.00	0.059
Cs 137	30.0 y	1.00	0.333	Pm 141	0.4 h	2.00	0.515
Cs 138	0.5 h	1.39	0.909	Pm 143	265.0 d	0.43	0.161
Ba 126	1.7 h	3.00	0.660	Pm 144	1.0 y	2.40	0.758
Ba 128	2.4 d	1.00	0.136	Pm 145	18.0 y	0.05	0.002
Ba 129	2.4 h	0.62	0.076	Pm 148	41.0 d	0.50	0.227
Ba 131	11.6 d	2.20	0.388	Pm 148	5.4 d	0.50	0.202
Ba 133	8.0 y	1.11	0.177	Pm 149	2.2 d	0.03	0.004
Ba 133	1.6 d	0.13	0.018	Pm 150	2.7 h	1.90	0.674
Ba 135	1.2 d	0.10	0.014	Pm 151	1.2 d	1.75	0.294
Ba 139	1.4 h	0.20	0.017	Sm 141	20.0 d	2.00	0.515
Ba 140	12.8 d	0.95	0.209	Sm 143	0.1 h	2.00	0.515
Ba 141	0.3 h	0.10	0.010	Sm 145	340.0 d	0.15	0.005
Ba 142	0.2 h	0.10	0.013	Sm 153	2.0 d	0.25	0.012
La 131	1.0 h	0.56	0.144	Sm 155	0.4 h	1.01	0.056
La 132	4.5 h	1.50	0.510	Sm 156	9.4 h	0.50	0.028
La 133	4.0 h	1.50	0.459	Eu 144	0.3 h	2.00	0.515
La 134	0.1 h	1.30	0.338	Eu 145	5.6 d	2.00	0.775
La 135	19.7 h	0.04	0.010	Eu 146	4.6 d	2.08	0.722
La 136	0.2 h	0.66	0.170	Eu 147	24.0 d	1.50	0.132
La 140	1.7 d	1.80	1.023	Eu 148	54.0 d	1.70	0.515
La 141	3.9 h	0.02	0.014	Eu 149	106.0 d	1.00	0.154
La 142	1.5 h	0.90	0.300	Eu 150	12.8 h	0.02	0.003
Ce 132	4.2 h	2.00	0.515	Eu 152	12.5 y	0.68	0.168
Ce 133	6.3 h	1.50	0.712	Eu 152	9.3 h	0.17	0.049
Ce 135	17.0 h	0.02	0.005	Eu 154	16.0 y	1.06	0.335

Isotope	$t_{\frac{1}{2}}$	ε	k_γ	Isotope	$t_{\frac{1}{2}}$	ε	k_γ
Eu 155	1.8 y	0.60	0.028	Tm 168	85.0 d	0.90	0.235
Eu 156	15.0 d	1.80	0.895	Tm 170	125.0 d	0.10	0.004
Eu 157	15.2 h	0.80	0.166	Tm 171	1.9 y	0.02	0.001
Gd 145	0.4 h	1.97	1.106	Tm 172	2.7 d	1.90	1.263
Gd 146	48.0 d	1.40	0.086	Tm 173	8.2 h	0.98	0.198
Gd 147	1.0 d	0.43	0.119	Yb 166	2.3 d	0 20	0.008
Gd 149	9.3 d	0.24	0.030	Yb 167	0.3 h	0.60	0.033
Gd 151	120.0 d	0.20	0.020	Yb 169	32.0 d	1.40	0.124
Gd 153	240.0 d	0.85	0.042	Yb 175	4.2 d	0.21	0.031
Gd 159	18.0 h	0.52	0.072	Yb 177	1.9 h	0.17	0.032
Tb 148	1.1 h	1.50	0.454	Lu 169	2.0 d	0.40	0.035
Tb 149	4.1 h	0.40	0.103	Lu 170	2.0 d	2.00	0.140
Tb 151	18.0 h	1.30	0.102	Lu 171	8.3 d	0.35	0.124
Tb 152	17.4 h	1.50	0.343	Lu 172	6.7 d	0.50	0.203
Tb 153	2.6 d	0.80	0.070	Lu 173	1.4 y	0.08	0.008
Tb 154	21.0 h	0.60	0.073	Lu 174	165.0 d	1.80	0.098
Tb 155	5.4 d	0.88	0.054	Lu 174	3.6 y	0.70	0.435
Tb 156	5.4 d	0.38	0.102	Lu 176	3.7 h	0.30	0.013
Tb 156	5.5 h	0.20	0.009	Lu 177	155.0 d	0.95	0.093
Tb 160	73.0 d	1.34	0.535	Lu 179	4.6 h	0.13	0.014
Tb 161	6.9 d	0.40	0.022	Hf 169	1.5 h	1.04	0.154
Tb 163	6.5 h	1.00	0.167	Hf 171	16.0 h	0.40	0.046
Dy 152	2.4 h	1.50	0.323	Hf 172	5.0 y	1.00	0.273
Dy 155	10.0 h	0.95	0.160	Hf 173	1.0 d	1.00	0.061
Dy 157	8.2 h	1.00	0.167	Hf 175	70.0 d	1.09	0.167
Dy 159	144.0 d	0.03	0.001	Hf 180	5.5 h	0.40	0.061
Dy 165	2.3 h	0.05	0.002	Hf 181	44.6 d	1.60	0.292
Dy 166	3.4 d	0.29	0.030	Hf 183	1.1 h	0.50	0.202
Ho 160	4.8 h	2.00	0.131	Ta 173	3.7 h	0.50	0.126
Ho 160	0.5 h	0.50	0.050	Ta 174	1.1 h	2.50	0.565
Ho 161	2.5 h	1.10	0.057	Ta 175	11.0 h	1.50	0.181
Ho 162	1.1 h	1.20	0.635	Ta 176	8.0 h	0.50	0.505
Ho 162	0.2 h	0.06	0.002	Ta 177	2.3 d	0.12	0.008
Ho 164	0.6 h	0.33	0.013	Ta 178	2.2 h	2.00	0.316
Ho 166	9×10^4 y	1.35	0.339	Ta 178	0.1 h	0.10	0.005
Ho 166	1.1 d	0.14	0.033	Ta 180	8.1 h	0.25	0.012
Ho 167	3.0 h	0.15	0.027	Ta 182	115.0 d	0.93	0.304
Er 158	2.5 h	1.50	0.356	Ta 183	5.0 d	2.20	0.230
Er 161	3.1 h	1.08	0.407	Ta 184	8.7 h	2.00	0.642
Er 163	1.2 h	1.00	0.386	Ta 185	0.8 h	1.22	0.083
Er 171	7.5 h	1.25	0.167	Ta 186	0.2 h	1.00	0.061
Er 172	2.0 d	1.00	0.207	W 176	2.3 h	1.00	0.050
Tm 163	1.8 h	0.80	0.097	W 178	22.0 d	0.41	0.021
Tm 165	1.2 d	1.00	0.075	W 179	0.1 h	0.50	0.056
Tm 166	7.7 h	0.94	0.688	W 179	0.7 h	0.70	0.058
Tm 167	9.6 d	0.81	0.077	W 181	125.0 d	0.15	0.011

Isotope	$t_{\frac{1}{2}}$	ε	k_γ	Isotope	$t_{\frac{1}{2}}$	ε	k_γ
W 187	1.0 d	0.94	0.270	Au 193	17.5 h	1.50	0.119
Re 177	0.3 h	2.00	0.515	Au 194	1.6 d	1.03	0.297
Re 178	0.2 h	2.00	0.515	Au 195	185.0 d	0.39	0.026
Re 180	19.9 h	0.22	0.057	Au 196	6.2 d	0.65	0.115
Re 181	19.0 h	1.00	0.187	Au 196	9.7 h	0.70	0.057
Re 182	2.7 d	0.65	0.218	Au 198	2.7 d	1.01	0.212
Re 182	13.0 h	1.00	0.505	Au 199	3.2 d	0.70	0.059
Re 183	70.0 d	0.85	0.089	Au 200	0.8 h	0.36	0.137
Re 184	169.0 d	1.80	0.617	Au 201	0.4 h	0.05	0.014
Re 184	38.0 d	1.14	0.439	Hg 191	1.0 h	1.00	0.126
Re 186	3.7 d	0.25	0.017	Hg 192	5.0 h	1.00	0.120
Re 188	17.0 h	0.17	0.018	Hg 193	6.0 h	2.00	0.138
Re 189	23.0 h	0.25	0.028	Hg 193	11.0 h	1.50	0.249
Os 182	21.0 h	1.10	0.233	Hg 195	1.7 d	1.00	0.096
Os 183	9.8 h	0.50	0.170	Hg 195	9.6 h	0.75	0.220
Os 183	13.4 h	0.75	0.144	Hg 197	1.0 d	0.15	0.010
Os 185	94.2 d	0.95	0.325	Hg 197	2.7 d	0.23	0.009
Os 190	0.2 h	1.77	0.387	Hg 203	47.0 d	0.86	0.122
Os 191	15.0 d	1.00	0.066	Tl 195	1.2 h	1.20	0.106
Os 193	1.3 d	0.59	0.052	Tl 196	1.8 h	1.00	0.217
Ir 184	3.2 h	0.50	0.068	Tl 197	2.8 h	0.30	0.023
Ir 185	13.9 h	1.00	0.091	Tl 198	1.9 h	0.50	0.103
Ir 186	15.8 h	1.10	0.111	Tl 198	5.3 h	0.75	0.155
Ir 187	10.6 h	1.50	0.220	Tl 199	7.4 h	0.42	0.073
Ir 188	1.7 d	1.20	0.223	Tl 200	1.1 d	1.18	0.409
Ir 189	13.3 d	0.80	0.101	Tl 201	3.1 d	0.19	0.016
Ir 190	12.3 d	0.50	0.121	Tl 202	12.0 d	1.00	0.222
Ir 190	3.2 h	1.70	0.381	Pb 196	0.6 h	0.60	0.065
Ir 192	74.0 d	0.91	0.173	Pb 197	0.7 h	2.40	0.401
Ir 192	650.0 y	2.30	0.380	Pb 198	2.4 h	1.00	0.192
Ir 193	11.9 d	0.50	0.020	Pb 199	1.5 h	1.24	0.286
Ir 194	19.0 h	0.34	0.057	Pb 200	21.6 h	1.70	0.141
Ir 195	4.2 h	0.80	0.087	Pb 201	9.4 h	1.20	0.199
Ir 197	0.1 h	0.50	0.454	Pb 202	3.6 h	1.30	0.477
Pt 188	10.0 d	1.20	0.126	Pb 203	2.2 d	0.90	0.130
Pt 189	10.8 h	1.00	0.310	Pb 204	1.1 h	1.50	0.548
Pt 191	3.0 d	1.15	0.235	Bi 201	1.8 h	0.60	0.191
Pt 195	4.1 d	0.50	0.025	Bi 203	11.8 h	0.20	0.084
Pt 197	19.9 h	0.24	0.012	Bi 204	11.3 h	1.50	0.548
Pt 197	1.4 h	0.50	0.088	Bi 205	15.0 d	1.40	0.862
Pt 200	11.5 h	0.50	0.202	Bi 206	6.3 d	1.30	0.602
Au 189	0.7 h	1.00	0.107	Bi 207	30.0 y	1.72	0.698
Au 191	3.0 h	0.80	0.128	Bi 208	37×10^4 y	1.00	1.317
Au 192	4.8 h	0.60	0.095				

TABLE A.2

The naturally radioactive families.

1 *The uranium series* $(4n+2)$

Isotope	$t_{\frac{1}{2}}$	ε	k_γ-factor (rad/hCi at 1 m)
92 U 238	4.5×10^9 y	0.23	0.008
90 Th 234	24.1 d	0.46	0.023
91 Pa 234 m	1.18 m	0.98	0.294
91 Pa 234	6.66 h	1	0.480
92 U 234	2.48×10^5 y	0.28	0.007
90 Th 230	7.52×10^4 y	0.24	0.008
88 Ra 226	1622 y	0.04	0.004
86 Rn 222	3.825 d	–	
84 Po 218	3.05 m	–	
82 Pb 214	26.8 m	1	0.175
85 At 218	1.3 s	–	
83 Bi 214	19.7 m	0.99	0.643
81 Tl 210	1.32 m	2.97	1.538
84 Po 214	1.6×10^{-4} s	–	
82 Pb 210	22 y	0.15	0.006
80 Hg 206	8.5 m	–	
83 Bi 210	5.01 d	–	
81 Tl 206	4.3 m	–	
84 Po 210	138.4 d	–	
82 Pb 206	stable		
In equilibrium with decay products:			
U 238 (uranium I)		4.37	1.168
Th 230 (ionium)		2.42	0.836
Ra 226 (radium)		2.18	0.828
Ra 226 (radium experimental value*)		–	0.75
Rn 222 (radon)		2.14	0.824

* Attix, F., et al., Journal of Research National Bureau of Standards Washington **59** (1957) 293.

2 *The thorium series (4n)*

Isotope	$t_{\frac{1}{2}}$	ε	k_γ-factor (rad/hCi at 1 m)
90 Th 232	1.39×10^{10} y	0.24	0.008
88 Ra 228	6.7 y	–	
89 Ac 228	6.13 h	2.4	0.471
90 Th 228	1.9 y	0.28	0.012
88 Ra 224	3.64 d	0.05	0.006
86 Rn 220	54.5 s	0.003	0.001
84 Po 216	0.158 s	–	
82 Pb 212	10.6 h	0.84	0.105
83 Bi 212	60.6 m	0.31	0.034
81 Tl 208	3.1 m	2.07	1.473
84 Po 212	3×10^{-7} s	–	
82 Pb 208	stable		
In equilibrium with decay products:			
Th 232 (thorium)		4.82	1.134
Ra 228 (mesothorium I)		4.58	1.126
Th 228 (radiothorium)		2.18	0.655

3 *The actinium series ($4n+3$)*

Isotope	$t_{\frac{1}{2}}$	ε	k_γ-factor (rad/hCi at 1 m)
92 U 235	7.13×10^8 y	0.81	0.067
90 Th 231	25.6 h	2.5	0.147
91 Pa 231	3.48×10^4 y	0.93	0.067
89 Ac 227	22 y	–	
87 Fr 223	22 m	0.68	0.028
85 At 219	0.9 m	–	
83 Bi 215	8 m	–	
90 Th 227	18.17 d	0.99	0.082
88 Ra 223	11.7 d	1.17	0.118
86 Rn 219	3.92 s	0.18	0.030
84 Po 215	1.8×10^{-3} s	–	
82 Pb 211	36.1 m	0.25	0.093
85 At 215	10^{-4} s	–	
83 Bi 211	2.15 m	0.17	0.033
81 Tl 207	4.79 m	0.002	0.001
84 Po 211	0.52 s	0.01	0.004
82 Pb 207	stable		
In equilibrium with decay products:			
U 235 (actino-uranium)		7.00	0.637
Pa 231 (protactinium)		3.69	0.423
Ac 227 (actinium)		2.76	0.356

4 *The neptunium series (4n+1)*

Isotope	$t_{\frac{1}{2}}$	ε	k_γ-factor (rad/hCi at 1 m)
93 Np 237	2.2×10^6 y	0.28	0.015
91 Pa 233	27 d	1.17	0.228
92 U 233	1.6×10^5 y	0.1	0.004
90 Th 229	7.30×10^3 y	0.9	0.065
88 Ra 225	14.8 d	0.6	0.029
89 Ac 225	10 d	0.44	0.024
87 Fr 221	4.8 m	0.16	0.018
85 At 217	0.018 s	–	
83 Bi 213	47 m	0.31	0.074
81 Tl 209	2.2 m	3	1.098
84 Po 213	4.2×10^{-6} s	–	
82 Pb 209	3.3 h	–	
83 Bi 209	stable		
In equilibrium with decay products:			
U 233		4.02	0.479
Th 229		2.47	0.232

TABLE A.3

Various other elements of high atomic number.

Isotope	$t_{\frac{1}{2}}$	ε	k_γ-factor (rad/hCi at 1m)
84 Po 206	8.8 d	1.90	0.201
84 Po 209	103 y	0.01	0.003
85 At 210	8.3 h	2.78	1.308
86 Rn 211	16 h	1.58	0.587
91 Pa 230	17 d	1.02	0.297
93 Np 239	2.4 d	0.94	0.096
94 Pu 238	89.6 y	0.28	0.011
94 Pu 239	2.4×10^4 y	0.28	0.079
94 Pu 240	6.6×10^3 y	0.24	0.009
94 Pu 242	3.7×10^5 y	0.24	0.009
95 Am 241	470 y	0.43	0.016
97 Bk 249	310 d	0.01	0.002

APPENDIX B

INDUCED ACTIVITY TABLES AND DANGER PARAMETER GRAPHS FOR THE SPALLATION CASE

Induced activity data for various materials and 5000 days irradiation time are given in tables B.1–B.8. The cooling times are 0, 1, 6 hour, 1, 7, 30, 180 and 360 days resp. All data are computed for a flux of 10^6 particles/sec cm^2. Isotopes having a half-life shorter than 5 minutes have not been considered. Under the third heading of the tables, the values of the product of the gamma danger parameter by the activating flux are given in mrad/h. To find the values of the gamma danger parameter proper in mrad/h per particle/sec cm^2 one has thus to divide these figures by 10^6. For further details, see ch. II § 3.

In view of application to practical cases, figs. B.1–B.30 show the danger parameter for other irradiation times than infinite. The selected target elements were Al, Fe, Ni, Cu, Ag, W, Pb and the bombarding energies chosen 50, 500 and 2900 MeV. One has added to those target elements the main compounds found in shielding materials, such as C, O_2, SiO_2, $CaCO_3$ and $BaSO_4$. For the targets C, Al, Ni, Fe, O_2, SiO_2, the danger parameter curves at 2900 MeV bombarding energy approach closely the curves calculated for 500 MeV and have been omitted.

Cooling time 0h

TABLE B.1

Target	Gamma dose rate (mrad/h g at 1 cm)				Gamma danger parameter times activating flux (mrad/h)				Gamma emission rate per sec and g				Total beta emission rate per sec and g				Beta minus emission rate per sec and g			
Irradiation (MeV):	50	100	600	2900	50	100	600	2900	50	100	600	2900	50	100	600	2900	50	100	600	2900
12MG	.508	.512	.164	.099	108.1	108.6	34.6	20.8	4127	4181	1354	822	1396	1412	455	276	195	190	57	33
13AL	.431	.628	.319	.214	124.2	177.2	86.4	57.0	1721	2732	1607	1128	806	1223	668	458	708	965	423	268
14SI	.172	.349	.264	.195	47.5	94.8	68.9	50.2	708	1540	1336	1033	348	708	562	423	311	567	361	250
15P	.173	.240	.218	.188	41.6	59.2	56.4	48.5	973	1316	1175	1030	887	1073	675	530	885	1046	555	394
16S	.068	.133	.181	.172	15.9	31.8	45.3	43.1	383	730	973	944	484	734	604	510	483	718	504	385
17CL	.012	.032	.106	.135	2.6	7.6	27.5	35.2	107	234	598	757	2218	2768	1439	1035	2186	2722	1364	928
18A	.117	.122	.091	.121	34.9	35.9	25.4	33.4	460	524	499	679	715	1050	1069	988	706	1016	998	884
19K	.016	.029	.073	.114	3.9	6.9	18.0	28.7	110	230	461	671	402	888	1075	977	373	821	988	863
20CA	.242	.245	.127	.139	60.6	60.9	31.3	34.3	1348	1391	752	806	1016	1338	1149	1029	709	1014	993	882
22TI	.483	.591	.294	.226	99.3	119.8	60.8	50.7	4341	5571	2803	1947	2004	2546	1526	1309	1336	1515	911	888
23V	.297	.481	.346	.276	64.4	103.2	73.4	61.6	2484	4171	3156	2390	697	1345	1374	1328	260	578	722	827
24CR	.159	.316	.306	.266	32.8	64.8	63.0	57.6	1495	2950	2886	2355	429	944	1227	1278	123	346	611	775
25MN	.235	.369	.322	.291	50.3	78.3	67.7	63.4	2095	3338	3003	2602	543	904	1081	1235	12	73	381	648
26FE	.350	.462	.347	.308	69.8	92.7	70.3	64.7	3134	4188	3230	2761	359	680	981	1192	6	46	330	613
27CO	.189	.305	.303	.303	38.9	64.2	63.6	64.9	2594	3515	2964	2789	320	536	775	1082	238	330	335	577
28NI	.390	.481	.347	.321	80.5	98.3	69.7	65.9	3359	4230	3179	2878	687	848	862	1121	457	484	369	594
29CU	.251	.374	.328	.332	50.8	76.8	67.7	70.2	2768	3957	3284	3165	1294	1649	1062	1185	585	767	485	610
30ZN	.256	.362	.319	.334	46.5	67.9	63.2	68.8	3094	4183	3358	3278	1571	1985	1246	1284	451	633	469	598
32GE	.203	.332	.340	.372	37.2	62.5	67.1	76.7	2701	4336	4127	4092	1103	1603	1401	1466	242	337	336	507
33AS	.615	.824	.565	.500	53.4	95.4	92.5	93.2	9505	12108	7477	5976	955	1729	1705	1695	57	144	262	461
34SE	.147	.300	.430	.478	15.4	34.5	65.6	85.4	3152	5806	6684	6384	283	632	1209	1542	103	160	220	413
35BR	.348	.562	.572	.564	54.3	83.1	87.1	96.9	5520	9328	8981	7834	704	1282	1576	1766	52	105	194	396
38SR	.398	.765	.753	.727	39.8	73.9	83.4	100.0	4871	8983	9438	9350	1012	1619	1607	1884	62	84	127	308
40ZR	1.623	2.074	1.330	1.076	397.6	481.2	255.4	190.8	12166	16914	13228	11967	648	1278	1645	1986	62	82	101	268
41NB	.875	1.504	1.355	1.156	220.8	366.7	279.7	216.1	6172	11365	12535	12210	767	1367	1666	2025	149	188	128	267
42MO	.504	.839	1.026	1.061	125.0	202.5	215.4	202.1	3880	6654	9455	11096	495	821	1265	1829	120	153	117	245
46PD	.127	.231	.462	.823	20.6	40.1	91.6	157.6	2675	4312	5947	9513	271	480	787	1471	185	311	346	380
47AG	.281	.408	.497	.829	42.1	63.1	91.1	154.2	5395	7743	7305	10131	489	770	884	1497	167	253	312	370
48CD	.031	.074	.242	.584	5.4	12.8	45.2	110.6	549	1242	3518	7179	98	191	459	1071	71	101	163	267
49IN	.135	.273	.389	.713	17.6	38.2	65.2	130.7	2799	5280	6246	9310	555	819	767	1314	383	527	335	371
50SN	.049	.124	.296	.639	6.3	16.3	47.1	115.3	1026	2489	5030	8665	299	494	640	1200	277	418	360	399
51SB	.102	.210	.329	.634	22.1	44.2	61.3	118.1	1159	2514	4693	8360	43	161	487	1087	13	67	224	343
53I	.072	.197	.425	.697	11.7	31.7	73.5	126.0	1207	3324	6527	9638	488	699	914	1348	361	357	203	309
56BA	.035	.102	.310	.619	4.0	11.8	40.9	99.3	762	2187	6280	10357	44	119	574	1244	34	57	82	208
66DY	.020	.060	.218	.571	1.9	5.2	24.5	80.0	511	1673	5156	10363	34	84	533	1699	28	41	116	258
73TA	.129	.299	.454	.642	20.0	51.6	80.8	108.1	2381	5191	8200	11492	50	191	401	996	22	44	68	159
74W	.051	.137	.345	.588	5.8	20.0	60.8	99.4	1418	3177	6647	10743	119	181	313	889	115	140	94	161
78PT	.130	.280	.471	.693	13.2	29.7	63.6	106.1	2524	5341	8400	12110	411	602	541	884	350	462	339	308
79AU	.175	.309	.467	.695	18.0	32.6	58.8	102.2	4893	7604	9164	12582	278	443	486	843	270	398	336	318
80HG	.103	.205	.390	.654	8.3	17.8	44.0	91.6	2955	5543	8528	12501	821	1030	666	891	820	1021	577	450
82PB	.246	.418	.478	.688	41.5	66.7	65.1	96.1	2994	5541	8102	12405	512	748	692	908	511	740	635	567
83BI	.491	.795	.696	.804	108.3	170.9	123.8	125.5	4124	7139	8960	12889	91	274	532	837	91	271	487	522

Cooling time 1 h

TABLE B.2

Target	Gamma dose rate (mrad/h g at 1 cm)				Gamma danger parameter times activating flux (mrad/h)				Gamma emission rate per sec and g				Total beta emission rate per sec and g				Beta minus emission rate per sec and g			
Irradiation (MeV):	50	100	600	2900	50	100	600	2900	50	100	600	2900	50	100	600	2900	50	100	600	2900
12MG	.503	.507	.163	.098	106.7	107.2	34.2	20.6	4109	4163	1349	819	1387	1403	453	274	186	182	54	31
13AL	.413	.603	.308	.207	118.8	170.0	83.2	55.0	1657	2645	1569	1104	774	1179	648	446	676	921	404	255
14SI	.161	.332	.254	.188	44.7	90.2	66.1	48.2	650	1457	1293	1004	303	649	534	406	265	508	333	233
15P	.024	.082	.160	.152	6.3	22.1	42.6	40.1	137	406	829	821	54	171	340	330	51	143	219	194
16S	.010	.045	.132	.140	2.4	11.9	34.3	35.7	54	225	687	752	151	227	323	325	150	211	223	200
17CL	.003	.010	.076	.109	.6	2.7	20.4	28.8	27	69	401	589	2143	2592	1225	857	2134	2577	1164	758
18A	.038	.040	.048	.085	11.4	11.7	13.8	24.1	150	168	254	464	312	620	825	780	310	610	786	701
19K	.003	.007	.044	.085	.7	1.5	11.6	22.3	24	56	251	474	307	746	902	805	301	730	857	721
20CA	.039	.040	.048	.085	10.3	10.6	12.4	21.6	154	172	253	463	326	627	821	779	322	617	783	700
22TI	.472	.570	.261	.188	97.3	115.9	53.1	41.2	4204	5342	2539	1665	1936	2428	1364	1115	1325	1485	828	761
23V	.287	.462	.316	.239	62.8	99.8	67.1	52.7	2354	3901	2854	2076	626	1204	1205	1127	251	564	669	722
24CR	.146	.292	.276	.228	30.7	61.0	57.0	49.1	1277	2572	2536	2011	334	776	1046	1070	117	335	566	676
25MN	.128	.218	.240	.227	27.3	46.3	50.5	49.2	1118	1930	2200	2007	212	414	765	954	7	63	348	562
26FE	.284	.352	.271	.244	56.5	70.7	54.9	51.2	2484	3101	2455	2158	117	274	668	906	3	40	302	532
27CO	.171	.266	.249	.245	34.6	55.4	52.0	52.5	2468	3199	2441	2261	249	385	551	827	182	254	286	494
28NI	.358	.428	.289	.263	73.2	87.0	58.0	53.7	3116	3786	2619	2337	561	645	625	861	357	380	316	509
29CU	.182	.280	.263	.269	37.6	58.5	54.5	56.8	1855	2764	2545	2514	841	1073	727	876	480	650	424	528
30ZN	.098	.177	.220	.252	19.8	36.1	45.1	52.8	990	1739	2133	2365	485	739	652	839	329	490	394	511
32GE	.164	.270	.263	.288	30.7	51.9	53.4	60.7	2149	3445	3086	3057	678	990	836	938	65	109	201	388
33AS	.593	.763	.471	.400	48.4	82.4	73.8	73.3	9283	11488	6376	4821	860	1447	1195	1154	23	52	148	346
34SE	.144	.286	.374	.395	14.8	31.8	54.5	68.5	3115	5653	6012	5401	267	566	904	1087	98	142	156	320
35BR	.291	.470	.477	.459	45.0	68.0	70.2	76.9	4703	8030	7742	6534	340	719	1045	1189	50	95	140	307
38SR	.385	.737	.689	.629	38.3	70.2	72.9	82.5	4497	8348	8444	8030	1000	1554	1280	1355	61	75	92	235
40ZR	1.592	2.016	1.255	.973	392.3	471.3	242.7	172.5	11721	16059	12112	10552	542	1086	1326	1471	60	79	80	207
41NB	.844	1.447	1.280	1.053	213.6	354.5	266.1	197.5	5884	10754	11528	10832	714	1244	1394	1539	148	186	112	211
42MO	.472	.787	.957	.961	118.1	191.8	202.8	184.1	3506	6014	8533	9765	416	666	999	1359	118	151	105	197
46PD	.122	.216	.417	.735	19.9	37.7	83.3	141.6	2524	4027	5317	8348	255	432	636	1112	176	296	333	346
47AG	.261	.372	.445	.738	39.2	58.1	82.3	138.3	4917	6926	6416	8841	404	627	698	1128	157	239	299	337
48CD	.027	.067	.217	.521	5.0	12.0	41.2	99.4	437	1055	3081	6263	91	171	373	814	69	97	156	244
49IN	.085	.193	.316	.617	10.8	26.8	53.7	114.2	1828	3795	5036	7881	94	171	383	884	78	115	158	265
50SN	.041	.101	.246	.554	4.8	12.7	39.0	100.6	892	2120	4227	7402	238	340	393	826	235	324	254	309
51SB	.073	.135	.226	.516	15.6	27.3	39.9	95.1	869	1767	3539	6870	12	58	248	712	10	50	157	266
53I	.068	.176	.337	.569	11.2	28.6	57.0	102.1	1124	2964	5367	7972	472	619	645	929	361	356	173	249
56BA	.035	.098	.267	.512	3.9	11.3	34.2	80.7	750	2113	5597	8807	40	100	400	836	33	56	75	172
66DY	.019	.057	.190	.461	1.8	4.9	20.7	63.3	484	1595	4723	8754	34	80	378	1089	27	41	112	243
73TA	.105	.253	.402	.553	17.4	45.5	72.9	94.5	1842	4287	7288	10039	35	119	248	637	19	40	63	147
74W	.036	.105	.297	.503	5.1	17.7	54.9	87.1	915	2190	5522	9172	114	158	211	576	112	135	89	150
78PT	.122	.259	.436	.619	12.2	27.3	59.4	96.0	2378	4993	7770	10815	385	549	449	614	344	453	331	297
79AU	.170	.296	.435	.624	17.5	31.3	55.2	92.8	4787	7322	8536	11301	253	404	411	594	247	374	324	306
80HG	.098	.195	.365	.591	7.5	16.6	41.3	83.5	2849	5331	8012	11327	746	944	595	670	746	938	544	427
82PB	.162	.302	.402	.603	25.7	45.4	52.8	84.0	2129	4320	7098	11032	482	695	617	715	482	690	583	526
83BI	.475	.748	.632	.722	105.2	162.3	113.6	114.0	3955	6649	8133	11586	81	247	471	658	81	245	444	483

Cooling time 6h

TABLE B.3

Target / Irradiation (MeV):	Gamma dose rate (mrad/h g at 1 cm) 50	100	600	2900	Gamma danger parameter times activating flux (mrad/h) 50	100	600	2900	Gamma emission rate per sec and g 50	100	600	2900	Total beta emission rate per sec and g 50	100	600	2900	Beta minus emission rate per sec and g 50	100	600	2900
12MG	.481	.486	.156	.095	100.4	101.1	32.4	19.5	4032	4088	1326	806	1348	1365	441	268	147	144	43	25
13AL	.334	.496	.261	.177	95.6	138.4	69.3	46.2	1377	2265	1402	998	634	989	565	393	537	731	321	203
14SI	.131	.273	.215	.161	36.0	73.4	55.1	40.5	540	1247	1155	908	248	544	465	358	210	402	264	184
15P	.020	.067	.135	.130	5.1	18.0	35.5	33.7	112	342	736	739	42	139	293	289	39	111	173	153
16S	.008	.037	.112	.119	1.9	9.7	28.5	29.9	44	190	610	677	134	193	278	284	133	178	178	159
17CL	.000	.006	.063	.092	.1	1.7	16.7	24.0	2	31	343	523	2048	2415	1090	753	2048	2413	1034	657
18A	.000	.001	.029	.066	.1	.2	8.3	18.5	1	3	159	376	126	421	700	672	126	421	674	602
19K	.000	.001	.034	.071	.0	.2	8.8	18.2	0	4	184	400	290	699	813	715	290	699	783	641
20CA	.001	.001	.029	.066	.2	.4	7.4	16.5	4	6	158	374	138	426	696	670	138	426	670	601
22TI	.441	.522	.227	.159	92.0	107.5	46.4	34.7	3804	4718	2146	1391	1751	2127	1135	928	1305	1434	742	662
23V	.279	.442	.289	.211	61.2	96.3	61.9	46.7	2257	3663	2542	1807	577	1086	1029	951	237	538	610	636
24CR	.137	.273	.250	.201	29.2	57.9	52.2	43.3	1138	2300	2214	1728	277	658	877	894	111	320	515	596
25MN	.105	.183	.210	.196	22.5	39.2	44.6	42.8	900	1576	1860	1697	138	287	609	780	7	60	317	496
26FE	.271	.329	.244	.215	53.9	66.0	49.8	45.2	2361	2861	2154	1859	76	189	531	740	3	38	275	469
27CO	.136	.215	.212	.211	26.0	43.0	43.9	44.9	2248	2864	2132	1949	109	182	388	649	48	71	193	403
28NI	.294	.357	.250	.227	58.7	71.0	49.8	46.3	2649	3254	2266	2007	315	372	452	680	122	136	218	416
29CU	.133	.223	.225	.233	29.2	48.3	47.3	49.3	1087	1893	2075	2111	457	651	511	674	325	475	327	433
30ZN	.074	.139	.186	.217	15.6	29.5	38.8	45.7	628	1190	1707	1966	284	451	447	636	236	363	303	417
32GE	.120	.204	.208	.237	22.8	39.8	42.7	50.3	1568	2598	2397	2452	391	591	527	656	27	56	143	308
33AS	.552	.684	.390	.331	40.4	66.1	57.2	58.9	8846	10640	5496	4057	732	1184	878	846	13	26	102	273
34SE	.138	.267	.324	.335	13.7	28.0	44.2	56.0	3029	5408	5435	4722	237	486	701	820	91	129	128	260
35BR	.258	.408	.399	.381	39.5	57.6	55.9	61.9	4233	7155	6745	5596	236	494	733	853	45	86	115	249
38SR	.360	.684	.613	.542	35.0	62.2	59.9	66.6	3947	7352	7273	6825	971	1475	1074	1045	58	66	67	183
40ZR	1.516	1.870	1.108	.838	378.5	444.6	216.2	148.0	10716	14120	10101	8745	334	700	919	1037	57	74	64	164
41NB	.739	1.270	1.112	.901	187.2	312.3	231.7	168.2	5047	9142	9557	8949	566	941	998	1096	143	179	98	171
42MO	.385	.659	.818	.816	96.3	160.9	173.1	155.1	2817	4916	7000	8027	326	502	714	966	112	143	94	161
46PD	.113	.199	.361	.625	18.4	34.2	71.6	119.3	2349	3727	4657	7036	243	413	549	854	167	283	318	315
47AG	.236	.330	.381	.626	35.6	51.6	70.5	116.5	4353	5982	5440	7381	300	466	548	846	140	219	283	307
48CD	.024	.059	.188	.443	4.6	11.0	35.9	84.2	338	875	2613	5234	80	146	301	616	64	89	146	222
49IN	.049	.117	.228	.498	5.5	14.9	38.6	91.9	1204	2608	3755	6285	58	103	269	645	55	83	138	237
50SN	.033	.077	.182	.444	3.5	8.7	28.0	80.2	783	1773	3301	5932	223	310	313	619	223	305	238	284
51SB	.038	.066	.139	.395	7.9	12.8	22.9	71.7	448	901	2369	5263	8	44	191	530	8	43	145	243
53I	.061	.146	.243	.430	10.3	24.3	40.9	76.8	958	2372	3884	5989	447	512	404	622	360	355	167	231
56BA	.034	.091	.213	.389	3.8	10.5	26.5	60.5	725	1970	4539	6752	35	82	247	515	30	51	71	160
66DY	.014	.044	.157	.356	1.5	3.8	16.8	47.3	380	1293	4089	7127	31	67	266	688	27	39	98	217
73TA	.051	.152	.295	.416	9.9	28.7	52.5	69.7	835	2597	5670	7949	13	31	73	335	12	27	51	128
74W	.024	.067	.216	.376	3.6	11.8	39.2	63.8	605	1440	4204	7189	101	122	91	311	101	121	77	132
78PT	.106	.221	.370	.503	10.4	22.7	49.7	76.7	2120	4353	6662	8980	343	464	354	412	328	431	312	275
79AU	.154	.258	.367	.507	15.7	27.4	46.5	74.5	4416	6504	7257	9362	185	313	321	402	183	302	290	277
80HG	.085	.168	.307	.482	6.3	14.1	34.8	67.2	2520	4652	6767	9372	659	836	508	492	659	834	490	388
82PB	.054	.136	.262	.450	5.8	15.6	31.4	61.2	991	2479	5122	8596	479	676	553	561	478	675	545	485
83BI	.445	.665	.505	.569	99.6	147.1	93.5	90.9	3662	5760	6395	9201	80	240	421	516	80	239	415	445

Target	Gamma dose rate (mrad/h g at 1 cm)				Gamma danger parameter times activating flux (mrad/h)				Gamma emission rate per sec and g				Total beta emission rate per sec and g				Beta minus emission rate per sec and g			
Irradiation (MeV):	50	100	600	2900	50	100	600	2900	50	100	600	2900	50	100	600	2900	50	100	600	2900
12MG	.434	.440	.143	.087	86.8	87.9	28.4	17.2	3862	3922	1276	778	1264	1283	417	253	64	62	18	10
13AL	.164	.264	.159	.113	45.2	69.7	39.2	27.1	770	1438	1039	768	331	576	383	278	233	318	139	88
14SI	.064	.145	.131	.103	17.0	37.0	31.1	23.8	302	791	856	699	129	316	316	253	91	175	115	80
15P	.010	.036	.082	.083	2.6	9.2	20.0	19.7	62	209	537	563	22	80	197	203	20	52	77	67
16S	.004	.020	.068	.076	1.0	5.0	16.1	17.5	24	116	445	516	118	151	195	203	117	136	95	78
17CL	.000	.003	.038	.058	.1	.8	9.4	14.1	1	19	250	398	1974	2300	984	654	1974	2297	928	558
18A	.000	.000	.018	.042	.0	.1	4.6	10.8	0	1	115	287	123	409	650	599	123	409	624	530
19K	.000	.000	.021	.045	.0	.1	5.0	10.6	0	2	134	305	286	682	755	638	286	682	725	564
20CA	.000	.000	.018	.042	.1	.1	4.1	9.7	1	2	114	285	125	405	643	596	125	404	617	528
22TI	.401	.459	.186	.122	84.6	95.8	38.5	26.1	3332	3977	1702	1081	1565	1810	906	750	1265	1347	642	560
23V	.265	.415	.258	.177	58.2	90.6	55.5	38.7	2135	3400	2214	1519	534	980	865	790	213	488	531	542
24CR	.130	.256	.223	.169	27.8	54.5	46.9	36.0	1070	2126	1926	1453	257	594	737	743	100	290	449	508
25MN	.096	.169	.189	.167	20.6	36.1	40.2	36.1	826	1448	1643	1447	126	261	517	651	6	54	277	422
26FE	.265	.318	.225	.187	52.7	63.7	45.9	38.8	2308	2759	1954	1617	67	168	449	616	3	35	240	400
27CO	.121	.192	.191	.183	22.5	37.7	39.4	38.4	2146	2700	1942	1711	54	102	304	528	0	6	145	331
28NI	.263	.321	.226	.198	52.5	64.0	45.0	39.9	2335	2897	2021	1744	218	262	362	556	39	49	166	343
29CU	.110	.195	.204	.205	25.5	43.9	43.6	43.6	748	1479	1795	1824	322	504	423	558	302	446	296	375
30ZN	.060	.119	.167	.191	13.4	26.4	35.6	40.4	435	907	1458	1689	157	292	346	515	145	259	247	349
32GE	.093	.154	.162	.194	17.2	29.1	32.8	40.9	1277	2068	1902	2004	323	463	387	505	13	29	101	246
33AS	.476	.566	.309	.268	26.1	43.8	41.1	45.8	7924	9207	4524	3335	485	808	622	631	4	11	70	217
34SE	.127	.237	.271	.278	11.7	22.7	33.9	44.4	2873	5011	4763	4049	185	367	515	618	76	109	102	210
35BR	.202	.324	.322	.311	30.1	43.4	41.9	48.2	3460	6006	5723	4720	135	301	500	620	33	68	91	201
38SR	.337	.626	.531	.457	30.9	51.9	45.4	50.8	3639	6588	6167	5708	902	1324	868	803	55	59	46	141
40ZR	1.426	1.699	.944	.698	359.6	408.5	182.9	120.2	9767	12342	8291	7126	237	494	649	743	49	64	48	128
41NB	.541	.961	.880	.721	134.3	232.4	177.9	128.9	3668	6862	7461	7110	360	608	679	771	128	158	80	136
42MO	.244	.461	.630	.643	59.6	110.2	128.7	116.6	1795	3426	5339	6306	207	327	484	677	97	126	79	129
46PD	.100	.168	.285	.494	15.5	27.2	53.8	89.9	2187	3382	3925	5729	230	386	471	660	156	259	283	272
47AG	.227	.309	.318	.502	33.9	47.5	56.2	89.3	4214	5712	4804	6140	252	404	469	658	105	177	246	263
48CD	.023	.056	.161	.359	4.4	10.4	29.5	65.2	325	834	2330	4381	72	132	262	486	57	79	127	190
49IN	.027	.064	.160	.385	2.3	6.6	26.3	68.2	819	1741	2833	5007	53	92	232	508	51	76	122	204
50SN	.029	.062	.133	.343	2.9	6.3	19.3	59.2	714	1516	2618	4754	219	303	289	505	219	299	226	255
51SB	.013	.025	.088	.296	2.4	4.0	13.1	51.4	195	442	1693	4114	7	43	176	432	7	42	138	218
53I	.053	.126	.194	.335	9.2	21.4	32.3	57.8	783	1977	3149	4789	432	482	352	511	357	351	162	211
56BA	.032	.082	.171	.302	3.6	9.4	21.2	45.8	684	1772	3641	5278	25	62	193	407	22	38	61	144
66DY	.007	.023	.116	.270	.9	2.1	12.5	34.9	219	807	3171	5635	27	46	149	418	26	36	71	165
73TA	.020	.078	.197	.293	2.9	10.5	28.9	43.3	437	1715	4367	6161	7	18	46	204	7	18	41	102
74W	.016	.040	.140	.261	2.0	5.0	20.4	38.5	482	1094	3252	5570	86	102	67	194	86	102	63	105
78PT	.087	.172	.276	.365	8.1	16.9	35.3	51.5	1772	3494	5108	6873	296	388	280	295	295	386	273	235
79AU	.119	.193	.268	.365	11.4	19.8	32.8	50.0	3675	5093	5418	7070	104	203	238	283	104	203	232	228
80HG	.063	.122	.220	.344	4.6	10.1	24.6	45.2	1888	3418	4872	6928	526	663	397	362	526	663	394	316
82PB	.031	.077	.161	.302	2.4	7.1	17.8	38.7	658	1594	3327	6042	476	663	494	454	476	663	493	420
83BI	.412	.587	.385	.408	92.8	132.5	75.2	65.9	3354	4922	4585	6587	80	235	376	417	80	235	375	386

Cooling time 7d

TABLE B.5

Target / Irradiation (MeV):	Gamma dose rate (mrad/h g at 1 cm) 50	100	600	2900	Gamma danger parameter times activating flux (mrad/h) 50	100	600	2900	Gamma emission rate per sec and g 50	100	600	2900	Total beta emission rate per sec and g 50	100	600	2900	Beta minus emission rate per sec and g 50	100	600	2900
12MG	.396	.403	.131	.080	76.1	77.4	25.2	15.4	3717	3780	1233	752	1195	1215	396	242	0	0	0	0
13AL	.032	.085	.081	.063	6.4	16.8	15.9	12.4	302	798	756	589	97	256	243	189	0	0	0	0
14SI	.013	.047	.066	.057	2.4	8.9	12.6	10.9	117	438	622	535	37	140	200	172	0	0	0	0
15P	.001	.009	.040	.045	.2	1.8	7.8	8.8	8	86	374	421	2	27	120	135	0	0	0	0
16S	.000	.005	.033	.041	.1	1.0	6.3	7.8	3	47	310	386	83	95	123	136	82	80	23	12
17CL	.000	.001	.019	.032	.0	.2	3.7	6.3	0	7	174	298	1637	1877	773	503	1637	1875	717	408
18A	.000	.000	.009	.023	.0	.0	1.8	4.8	0	0	80	214	113	366	543	485	113	366	517	416
19K	.000	.000	.010	.024	.0	.0	1.9	4.7	0	0	93	228	264	611	631	516	264	611	601	443
20CA	.000	.000	.008	.023	.0	.0	1.6	4.3	0	0	79	213	111	357	536	482	111	357	511	413
22TI	.297	.305	.102	.063	64.1	65.7	21.7	13.4	2237	2338	805	523	1215	1288	582	507	1162	1207	530	441
23V	.197	.303	.171	.110	43.2	66.6	37.3	23.9	1566	2411	1363	895	368	666	559	529	128	331	392	404
24CR	.097	.187	.148	.105	20.7	40.1	31.6	22.3	781	1501	1186	857	177	404	477	499	60	197	331	378
25MN	.049	.094	.117	.101	10.5	20.2	25.1	21.6	430	815	967	837	64	148	324	431	3	37	204	314
26FE	.237	.267	.161	.123	46.9	53.1	32.9	25.4	2072	2325	1359	1035	33	94	281	407	2	23	176	298
27CO	.113	.174	.148	.128	20.8	33.8	30.4	26.6	2058	2521	1534	1202	43	72	193	350	0	4	106	247
28NI	.241	.288	.177	.140	48.6	57.7	35.4	28.1	2113	2571	1557	1211	190	214	241	373	37	45	126	257
29CU	.108	.188	.178	.159	25.0	42.6	38.2	33.9	717	1394	1529	1390	314	488	359	416	299	440	273	308
30ZN	.055	.111	.146	.149	12.6	25.0	31.2	31.5	371	809	1234	1288	104	229	278	377	93	200	210	279
32GE	.044	.070	.099	.132	7.5	12.6	20.3	28.3	628	947	1040	1257	237	315	262	351	1	6	70	187
33AS	.326	.335	.163	.160	1.7	4.7	15.0	25.5	5882	6006	2457	1903	43	117	210	326	0	1	46	163
34SE	.105	.182	.177	.182	8.3	13.4	17.0	26.2	2551	4203	3407	2750	98	172	231	335	44	66	67	156
35BR	.070	.150	.182	.189	8.0	14.2	17.9	26.3	1636	3569	3742	3072	25	83	197	318	11	31	57	148
38SR	.293	.536	.417	.332	24.8	39.3	28.1	29.9	3099	5470	4661	4031	729	998	539	461	44	46	31	103
40ZR	1.247	1.425	.734	.515	322.5	353.1	143.0	86.5	7862	9400	5903	4908	116	278	367	410	21	31	28	90
41NB	.338	.622	.610	.498	83.4	150.9	120.2	84.7	2150	4114	4840	4676	112	209	323	397	70	81	42	91
42MO	.129	.281	.431	.442	31.2	67.2	85.8	76.2	916	1968	3440	4135	72	126	234	350	53	70	45	87
46PD	.079	.119	.175	.319	11.6	17.5	30.5	54.2	1766	2621	2699	3812	197	310	301	373	127	194	169	159
47AG	.184	.242	.212	.331	26.2	35.1	34.6	54.8	3571	4730	3561	4227	123	233	285	369	30	81	136	150
48CD	.020	.044	.109	.237	3.8	7.9	18.7	40.3	287	690	1737	3034	56	94	165	277	46	61	78	112
49IN	.008	.025	.095	.248	.9	3.4	15.8	41.7	214	554	1638	3241	47	74	150	292	46	64	79	122
50SN	.025	.050	.091	.226	2.6	5.3	12.8	36.9	597	1155	1776	3187	201	276	229	320	201	274	188	180
51SB	.005	.013	.058	.195	1.0	1.9	8.3	31.9	93	264	1125	2747	6	38	139	274	6	38	115	154
53I	.028	.079	.127	.218	4.7	12.7	20.0	35.5	457	1333	2191	3243	360	378	234	314	332	324	143	158
56BA	.024	.055	.096	.181	2.7	6.2	11.6	26.6	532	1219	2089	3165	4	12	72	208	3	7	37	100
66DY	.004	.011	.071	.170	.6	1.1	7.8	21.6	125	427	2031	3680	22	30	58	170	22	30	45	93
73TA	.012	.052	.122	.170	1.1	5.1	14.0	20.5	295	1218	2903	3919	6	15	34	96	6	15	33	66
74W	.013	.029	.087	.152	1.5	2.9	9.7	18.0	410	873	2227	3584	75	87	53	98	75	87	52	71
78PT	.043	.085	.158	.212	3.7	8.1	20.1	27.3	919	1779	2905	4143	237	301	184	167	237	301	184	152
79AU	.048	.071	.133	.201	1.5	3.9	15.5	25.1	2272	2712	2842	4093	38	102	139	154	38	102	139	139
80HG	.025	.051	.110	.188	1.7	4.0	12.4	23.2	759	1479	2407	3877	126	171	148	162	126	171	148	150
82PB	.011	.026	.061	.149	.9	2.1	6.2	17.7	226	528	1294	3069	465	615	341	263	465	615	341	254
83BI	.299	.398	.217	.220	67.1	91.4	46.1	36.4	2418	3175	2238	3422	78	218	259	241	78	218	259	234

Cooling time 30d

TABLE B.6

Target	Gamma dose rate (mrad/h g at 1 cm)				Gamma danger parameter times activating flux (mrad/h)				Gamma emission rate per sec and g				Total beta emission rate per sec and g				Beta minus emission rate per sec and g			
Irradiation (MeV):	50	100	600	2900	50	100	600	2900	50	100	600	2900	50	100	600	2900	50	100	600	2900
12MG	.390	.396	.129	.079	74.8	76.1	24.8	15.1	3655	3717	1213	740	1175	1194	389	237	0	0	0	0
13AL	.032	.084	.079	.062	6.2	16.5	15.6	12.2	296	784	743	579	95	252	238	186	0	0	0	0
14SI	.012	.046	.065	.056	2.3	8.7	12.4	10.7	115	430	611	526	37	138	196	169	0	0	0	0
15P	.001	.009	.039	.044	.2	1.8	7.7	8.6	8	84	367	414	2	27	118	133	0	0	0	0
16S	.000	.005	.032	.040	.1	.9	6.2	7.7	3	46	304	379	28	41	105	126	27	26	7	4
17CL	.000	.001	.018	.031	.0	.2	3.6	6.2	0	7	171	293	857	938	389	280	857	935	334	186
18A	.000	.000	.008	.022	.0	.0	1.8	4.7	0	0	79	210	87	257	335	304	87	257	309	236
19K	.000	.000	.010	.024	.0	.0	1.9	4.7	0	0	91	224	203	431	389	324	203	431	360	251
20CA	.000	.000	.008	.022	.0	.0	1.6	4.2	0	0	78	209	85	251	331	302	85	251	305	234
22TI	.230	.229	.071	.047	50.0	49.7	15.5	10.0	1678	1669	525	368	964	1003	418	344	964	1003	411	304
23V	.084	.141	.089	.063	18.5	31.1	19.5	13.7	657	1090	675	494	175	360	342	332	87	240	287	273
24CR	.041	.087	.076	.060	8.8	18.6	16.4	12.7	329	678	584	470	84	217	291	312	40	143	243	255
25MN	.006	.021	.049	.051	1.2	4.5	10.7	11.0	61	191	393	410	8	45	181	260	2	26	149	212
26FE	.204	.210	.100	.075	40.0	41.3	20.2	15.3	1787	1836	837	618	4	29	157	246	1	17	129	201
27CO	.096	.144	.103	.083	17.3	27.5	20.8	17.2	1862	2220	1138	817	34	50	114	214	0	3	78	166
28NI	.199	.235	.125	.093	40.0	46.8	24.9	18.6	1752	2112	1110	804	156	166	152	233	35	41	95	175
29CU	.103	.174	.146	.119	24.0	39.7	31.5	25.4	674	1270	1245	1036	301	459	299	299	289	421	247	240
30ZN	.052	.103	.121	.112	12.0	23.4	26.0	23.8	348	742	1014	967	99	215	232	272	90	191	190	217
32GE	.036	.057	.081	.102	6.2	10.4	17.1	22.1	472	680	773	922	218	287	232	272	0	5	62	145
33AS	.258	.259	.125	.121	.4	2.0	11.4	19.3	4632	4603	1794	1375	10	53	155	238	0	0	41	127
34SE	.081	.137	.130	.133	5.6	8.6	11.3	18.7	2070	3347	2582	2036	36	63	122	213	17	26	39	112
35BR	.039	.098	.130	.138	3.2	7.2	11.3	18.4	1114	2660	2822	2282	8	29	103	202	4	12	33	106
38SR	.231	.424	.326	.254	18.6	29.2	20.0	21.2	2426	4285	3603	3059	534	719	360	302	20	21	16	72
40ZR	1.039	1.175	.590	.405	272.2	295.9	117.1	68.8	6279	7355	4510	3708	61	174	237	265	9	14	14	62
41NB	.255	.477	.474	.384	64.2	118.1	94.9	65.8	1546	2959	3571	3470	49	94	188	246	42	46	22	62
42MO	.081	.193	.323	.334	20.4	47.6	64.9	57.6	531	1251	2448	3018	31	59	136	216	27	37	24	58
46PD	.063	.092	.118	.227	10.0	14.2	20.8	38.4	1172	1764	1780	2632	160	232	177	216	101	144	96	89
47AG	.105	.139	.132	.229	13.7	18.7	21.1	37.5	2176	2867	2227	2836	41	104	147	205	20	56	77	83
48CD	.015	.027	.064	.160	2.8	4.6	10.6	26.8	220	453	1069	2012	38	54	82	149	36	47	49	65
49IN	.003	.010	.053	.165	.5	1.8	8.8	27.6	36	163	896	2084	40	54	78	159	40	53	52	72
50SN	.021	.040	.061	.155	2.2	4.4	8.3	24.9	479	900	1204	2141	147	200	149	189	146	200	134	117
51SB	.001	.006	.037	.132	.1	.7	5.1	21.4	38	155	744	1837	4	28	90	161	4	28	81	100
53I	.012	.042	.072	.138	1.7	6.0	10.1	21.5	255	859	1420	2119	255	251	116	159	254	248	106	107
56BA	.015	.030	.046	.104	1.6	3.1	5.1	14.9	346	678	1069	1840	1	4	26	96	1	4	24	65
66DY	.003	.006	.047	.109	.4	.7	5.0	13.5	82	247	1368	2446	17	24	41	90	17	24	39	74
73TA	.010	.041	.085	.110	.9	4.1	9.3	12.3	230	965	2094	2632	5	13	27	59	5	13	27	53
74W	.011	.023	.061	.098	1.2	2.4	6.5	10.8	319	659	1587	2401	61	71	43	63	61	71	42	57
78PT	.019	.036	.093	.135	1.7	3.8	13.2	18.0	403	695	1535	2546	175	213	119	104	175	213	119	101
79AU	.034	.042	.078	.127	.3	1.3	9.7	16.2	1819	1935	1631	2533	28	72	89	96	28	72	89	93
80HG	.012	.028	.067	.120	.8	2.3	8.2	15.3	409	888	1436	2425	6	21	59	84	6	21	59	82
82PB	.003	.007	.029	.090	.2	.6	3.4	11.2	65	155	626	1824	443	571	271	183	443	571	271	181
83BI	.169	.192	.093	.115	36.0	42.0	18.9	17.9	1459	1620	1022	1925	74	202	206	168	74	202	206	167

Cooling time 180d

TABLE B.7

Target	Gamma dose rate (mrad/h g at 1 cm)				Gamma danger parameter times activating flux (mrad/h)				Gamma emission rate per sec and g				Total beta emission rate per sec and g				Beta minus emission rate per sec and g			
Irradiation (MeV):	50	100	600	2900	50	100	600	2900	50	100	600	2900	50	100	600	2900	50	100	600	2900
12MG	.349	.355	.116	.071	67.0	68.2	22.2	13.6	3276	3331	1087	663	1053	1070	349	213	0	0	0	0
13AL	.028	.075	.071	.055	5.6	14.8	14.0	10.9	265	703	666	519	85	226	214	166	0	0	0	0
14SI	.011	.041	.058	.050	2.1	7.8	11.1	9.6	103	385	548	472	33	123	176	151	0	0	0	0
15P	.001	.008	.035	.040	.2	1.6	6.9	7.7	7	75	329	371	2	24	105	119	0	0	0	0
16S	.000	.004	.029	.036	.1	.8	5.5	6.9	2	41	272	340	1	13	87	109	0	0	0	0
17CL	.000	.001	.016	.028	.0	.1	3.2	5.5	0	6	153	262	90	91	76	98	90	89	26	13
18A	.000	.000	.008	.020	.0	.0	1.6	4.2	0	0	71	188	23	63	82	101	23	63	59	40
19K	.000	.000	.009	.021	.0	.0	1.7	4.2	0	0	82	201	59	109	96	108	59	109	70	44
20CA	.000	.000	.007	.020	.0	.0	1.4	3.8	0	0	70	187	23	62	81	101	23	62	58	40
22TI	.067	.066	.022	.022	14.5	14.4	4.8	4.6	486	484	165	182	310	332	138	121	310	332	132	85
23V	.006	.016	.016	.019	1.4	3.6	3.6	4.2	45	119	120	161	27	78	96	106	27	78	92	77
24CR	.003	.010	.014	.018	.6	2.1	2.9	3.8	21	71	102	151	12	46	81	100	12	46	78	72
25MN	.000	.002	.008	.015	.0	.4	1.8	3.2	1	14	63	126	0	8	50	83	0	8	48	60
26FE	.144	.142	.048	.033	28.3	27.9	9.6	6.7	1261	1239	411	285	0	5	43	78	0	5	41	57
27CO	.041	.064	.044	.035	6.3	11.2	8.4	6.9	1085	1261	575	390	8	12	31	67	0	0	25	47
28NI	.071	.093	.051	.038	14.2	18.3	9.9	7.4	655	880	474	346	59	62	47	75	30	32	36	52
29CU	.086	.132	.087	.062	20.5	31.1	19.4	13.7	533	886	694	530	261	376	199	150	258	368	188	129
30ZN	.040	.076	.072	.059	9.3	17.6	16.1	12.9	256	510	571	499	85	176	155	137	81	168	145	118
32GE	.025	.039	.052	.058	4.2	7.1	11.0	12.7	324	465	487	521	150	199	159	155	0	4	48	80
33AS	.066	.070	.053	.059	.2	1.3	7.4	11.1	1188	1225	650	601	6	36	106	135	0	0	32	70
34SE	.027	.044	.049	.059	1.9	3.0	5.6	10.0	752	1185	940	817	0	6	57	108	0	0	17	56
35BR	.015	.035	.050	.060	1.2	2.7	5.5	9.7	441	1014	1063	922	0	2	48	103	0	0	14	53
38SR	.057	.107	.089	.082	3.7	5.9	5.0	8.3	570	1047	1005	996	95	123	64	92	0	0	4	35
40ZR	.389	.433	.202	.141	103.0	111.2	42.9	26.2	2279	2595	1463	1252	11	30	42	78	1	1	3	29
41NB	.071	.153	.157	.131	18.3	38.6	33.2	24.3	423	917	1128	1157	9	17	34	72	8	8	4	27
42MO	.025	.063	.105	.113	6.4	15.8	22.4	21.1	170	404	771	1003	3	7	23	62	2	3	3	24
46PD	.036	.053	.051	.083	5.8	8.2	8.7	14.7	598	962	817	991	80	110	63	68	47	64	35	32
47AG	.013	.026	.043	.078	1.8	3.8	7.1	13.6	240	504	699	947	13	40	50	63	8	24	28	30
48CD	.008	.012	.022	.055	1.6	2.2	3.7	9.7	129	195	359	677	24	31	32	48	23	29	22	26
49IN	.002	.005	.018	.057	.3	.9	3.2	10.1	23	79	310	707	36	44	36	54	36	44	29	31
50SN	.009	.016	.023	.054	.9	1.8	3.1	9.2	191	364	459	756	21	32	34	52	21	32	29	33
51SB	.000	.002	.014	.046	.0	.3	1.9	7.9	16	61	283	648	0	4	20	44	0	4	18	28
53I	.005	.018	.029	.051	.7	2.4	4.0	8.1	107	379	612	813	44	43	21	39	44	43	20	27
56BA	.012	.020	.019	.037	1.2	2.0	2.1	5.5	271	453	430	657	1	2	5	23	1	2	5	16
66DY	.001	.002	.021	.049	.1	.2	2.2	6.2	37	130	680	1136	6	11	29	48	6	11	28	46
73TA	.003	.017	.042	.052	.4	2.4	6.0	7.0	74	362	907	1151	3	7	15	32	3	7	15	31
74W	.004	.009	.029	.046	.5	1.2	4.1	6.2	115	230	651	1027	23	27	19	32	23	27	19	31
78PT	.005	.008	.030	.051	.4	1.0	4.7	7.6	89	139	436	913	46	54	31	35	46	54	31	35
79AU	.019	.020	.027	.049	.1	.4	3.5	6.8	1013	1014	597	948	9	21	24	33	9	21	24	32
80HG	.006	.013	.026	.047	.3	1.0	3.2	6.5	218	461	594	947	1	6	16	29	1	6	16	29
82PB	.000	.001	.009	.034	.0	.0	1.1	4.6	3	16	221	694	380	475	205	117	380	475	205	117
83BI	.135	.131	.044	.047	27.2	26.4	8.3	7.4	1228	1196	504	780	64	169	156	107	64	169	156	107

TABLE B.8

Target	Gamma dose rate (mrad/h g at 1 cm)				Gamma danger parameter times activating flux (mrad/h)				Gamma emission rate per sec and g				Total beta emission rate per sec and g				Beta minus emission rate per sec and g			
Irradiation (MeV):	50	100	600	2900	50	100	600	2900	50	100	600	2900	50	100	600	2900	50	100	600	2900
12MG	.306	.311	.102	.062	58.8	59.8	19.5	11.9	2872	2921	953	581	923	938	306	187	0	0	0	0
13AL	.025	.066	.062	.049	4.9	12.9	12.3	9.6	233	616	584	455	74	198	187	146	0	0	0	0
14SI	.010	.036	.051	.044	1.8	6.8	9.7	8.4	90	338	480	413	29	108	154	133	0	0	0	0
15P	.001	.007	.031	.035	.1	1.4	6.0	6.8	6	66	288	325	2	21	92	104	0	0	0	0
16S	.000	.004	.026	.032	.1	.7	4.8	6.0	2	36	239	298	1	12	77	95	0	0	0	0
17CL	.000	.001	.014	.025	.0	.1	2.8	4.8	0	6	134	230	21	23	50	77	21	21	6	3
18A	.000	.000	.007	.018	.0	.0	1.4	3.7	0	0	62	165	5	15	34	63	5	15	14	9
19K	.000	.000	.008	.019	.0	.0	1.5	3.7	0	0	72	176	17	30	41	67	17	29	17	11
20CA	.000	.000	.007	.018	.0	.0	1.2	3.3	0	0	61	164	5	15	34	63	5	15	14	10
22TI	.015	.015	.006	.013	3.3	3.3	1.4	2.6	110	110	50	113	86	97	49	60	86	97	44	29
23V	.001	.004	.004	.011	.3	.8	1.0	2.3	10	26	34	95	7	22	33	51	7	22	30	26
24CR	.001	.002	.004	.010	.1	.5	.8	2.1	4	16	29	89	3	13	28	48	3	13	25	24
25MN	.000	.000	.002	.008	.0	.1	.5	1.7	0	3	18	74	0	2	17	40	0	2	15	20
26FE	.097	.095	.029	.021	19.0	18.6	5.8	4.1	847	829	255	182	0	1	15	38	0	1	13	19
27CO	.020	.033	.023	.020	2.6	5.3	4.4	3.9	637	737	332	234	1	2	10	32	0	0	8	16
28NI	.031	.045	.027	.021	6.3	8.9	5.2	4.1	296	442	256	200	33	34	20	37	28	28	17	20
29CU	.078	.116	.067	.045	18.7	27.6	15.4	10.0	471	734	493	358	242	344	169	112	241	342	166	98
30ZN	.032	.062	.055	.042	7.5	14.5	12.5	9.4	199	397	401	336	78	160	131	102	75	156	129	90
32GE	.016	.025	.035	.040	2.7	4.5	7.6	8.8	207	296	319	342	96	128	112	109	0	4	43	61
33AS	.013	.017	.027	.036	.1	.8	5.1	7.7	236	275	276	327	4	23	75	95	0	0	28	54
34SE	.008	.014	.021	.033	.7	1.1	3.4	6.6	245	379	345	372	0	3	40	76	0	0	15	43
35BR	.005	.011	.021	.033	.4	1.0	3.1	6.3	151	340	386	405	0	1	34	72	0	0	13	40
38SR	.011	.022	.022	.031	.6	1.1	1.6	4.5	110	213	258	348	13	17	17	51	0	0	4	26
40ZR	.121	.133	.060	.048	32.2	34.7	13.4	9.8	687	768	415	417	1	4	10	42	0	0	2	22
41NB	.018	.043	.045	.044	4.7	11.0	10.1	8.9	103	248	314	382	1	3	8	39	1	2	2	20
42MO	.008	.019	.031	.038	1.9	4.7	6.8	7.7	63	134	224	335	0	1	5	33	0	0	1	17
46PD	.022	.033	.026	.033	3.3	4.6	3.8	5.9	419	700	531	484	43	59	32	36	25	34	18	20
47AG	.004	.012	.020	.031	.6	1.7	3.0	5.4	76	262	421	452	7	21	25	34	4	12	14	18
48CD	.005	.007	.011	.022	1.0	1.3	1.7	3.9	87	128	221	330	15	19	17	26	14	18	13	16
49IN	.001	.003	.009	.023	.2	.6	1.5	4.1	15	52	192	347	34	38	24	31	34	38	20	20
50SN	.003	.006	.010	.022	.3	.7	1.3	3.7	66	131	208	346	5	9	14	26	5	9	12	17
51SB	.000	.001	.006	.018	.0	.1	.8	3.1	8	24	129	297	0	1	9	22	0	1	7	15
53I	.002	.008	.013	.021	.3	1.1	1.7	3.3	46	166	272	368	5	5	5	18	5	5	5	12
56BA	.011	.019	.015	.019	1.1	1.9	1.6	2.6	258	430	332	370	1	1	2	11	1	1	2	7
66DY	.000	.001	.012	.028	.0	.1	1.3	3.7	16	69	379	657	4	8	24	37	4	8	24	36
73TA	.002	.011	.031	.036	.3	1.9	5.0	5.3	31	193	611	756	1	3	9	22	1	3	9	22
74W	.002	.005	.021	.031	.2	.7	3.3	4.7	42	100	422	667	7	9	9	21	7	9	9	20
78PT	.001	.003	.012	.027	.1	.3	2.0	4.1	28	45	179	489	14	17	9	16	14	17	9	16
79AU	.009	.010	.011	.025	.1	.1	1.5	3.7	514	509	272	505	4	8	8	15	4	8	8	15
80HG	.003	.006	.011	.024	.1	.4	1.3	3.4	108	225	274	494	0	2	5	13	0	2	5	13
82PB	.000	.000	.004	.017	.0	.0	.5	2.4	0	5	100	361	344	428	180	95	344	428	180	95
83BI	.133	.129	.039	.031	26.9	26.1	7.7	5.3	1214	1178	408	472	57	152	137	88	57	152	137	88

Fig. B.1

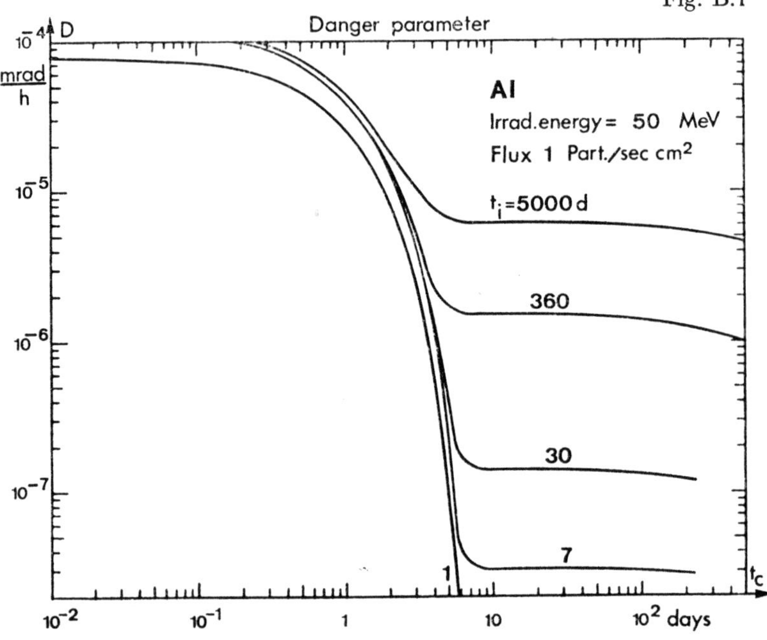

Danger parameter
D
mrad/h
Al
Irrad.energy = 50 MeV
Flux 1 Part./sec cm²
tᵢ=5000 d
360
30
7
1
tc
10⁻⁴
10⁻⁵
10⁻⁶
10⁻⁷
10⁻²
10⁻¹
1
10
10² days

Fig. B.2

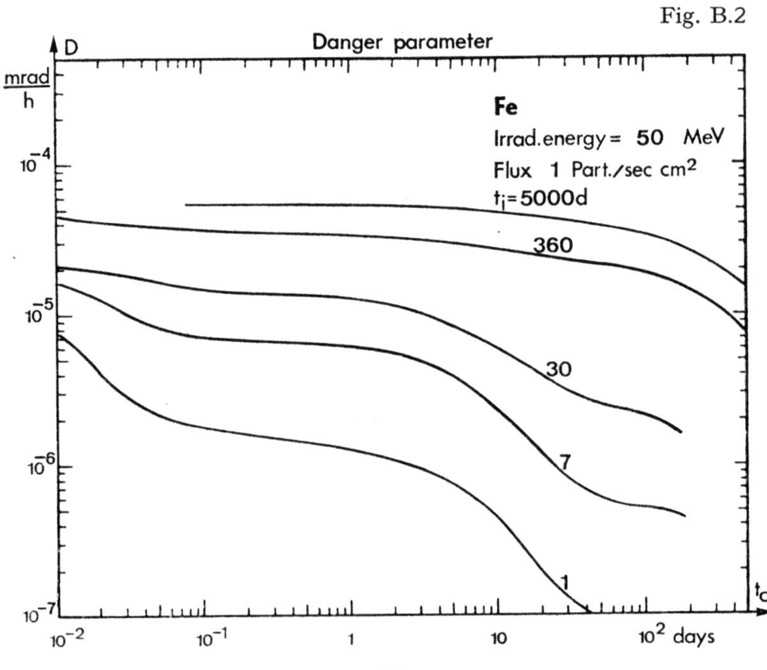

Danger parameter
D
mrad/h
Fe
Irrad.energy = 50 MeV
Flux 1 Part./sec cm²
tᵢ=5000d
360
30
7
1
tc
10⁻⁴
10⁻⁵
10⁻⁶
10⁻⁷
10⁻²
10⁻¹
1
10
10² days

Fig. B.3

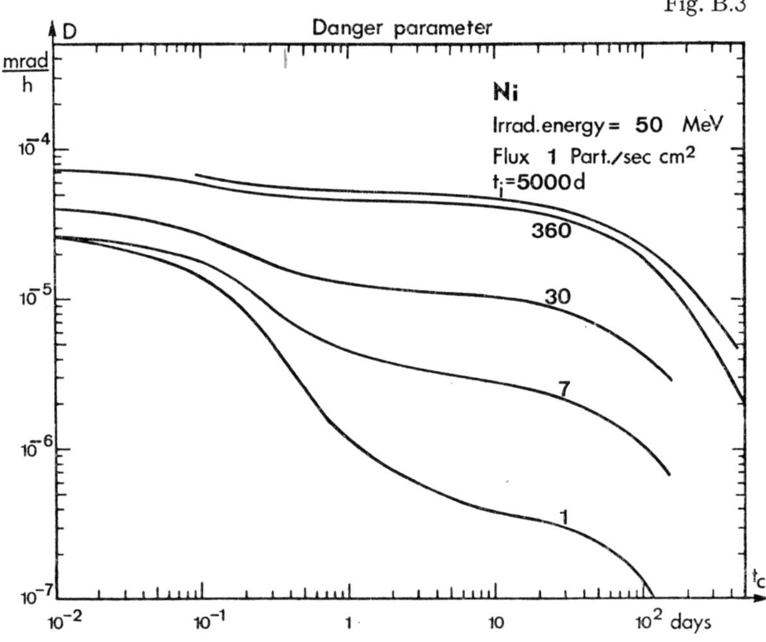

Danger parameter
D
mrad/h
Ni
Irrad.energy = 50 MeV
Flux 1 Part./sec cm²
tᵢ=5000d
360
30
7
1
10⁻⁴
10⁻⁵
10⁻⁶
10⁻⁷
10⁻²
10⁻¹
1
10
10² days
t_c

Fig. B.4

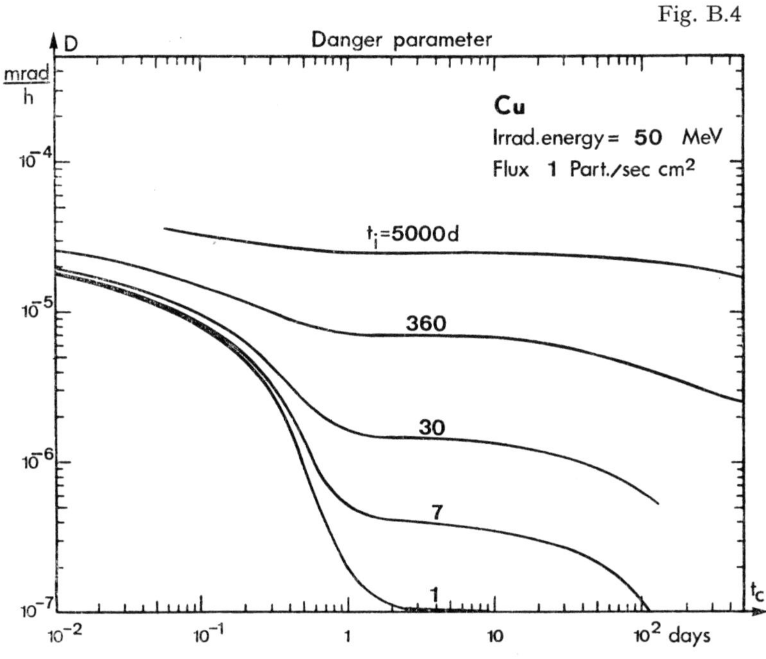

Danger parameter
D
mrad/h
Cu
Irrad.energy = 50 MeV
Flux 1 Part./sec cm²
tᵢ=5000d
360
30
7
1
10⁻⁴
10⁻⁵
10⁻⁶
10⁻⁷
10⁻²
10⁻¹
1
10
10² days
t_c

Fig. B.5

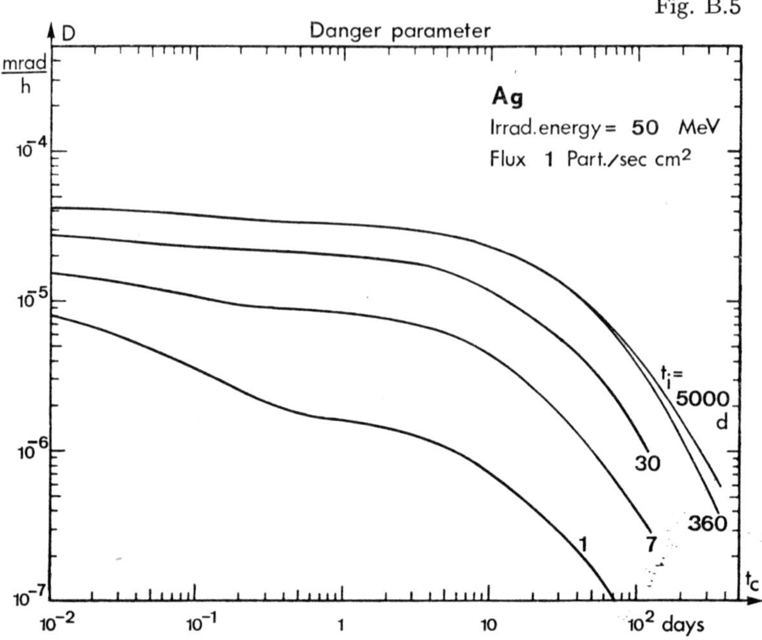

D
Danger parameter
mrad/h
Ag
Irrad.energy = 50 MeV
Flux 1 Part./sec cm2
10^-4
10^-5
10^-6
10^-7
t_i = 5000 d
30
360
7
1
t_c
10^-2
10^-1
1
10
10^2 days

Fig. B.6

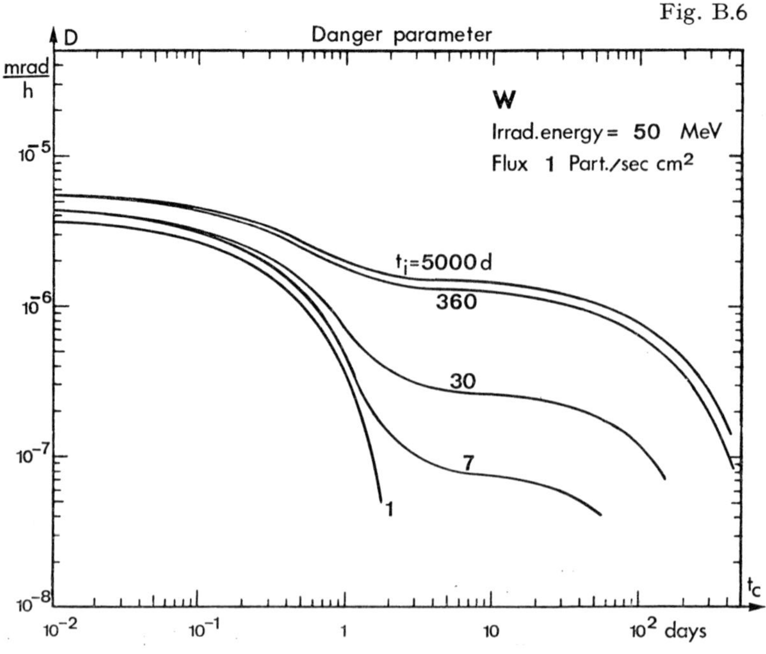

D
Danger parameter
mrad/h
W
Irrad.energy = 50 MeV
Flux 1 Part./sec cm2
10^-5
10^-6
10^-7
10^-8
t_i=5000 d
360
30
7
1
t_c
10^-2
10^-1
1
10
10^2 days

Fig. B.7

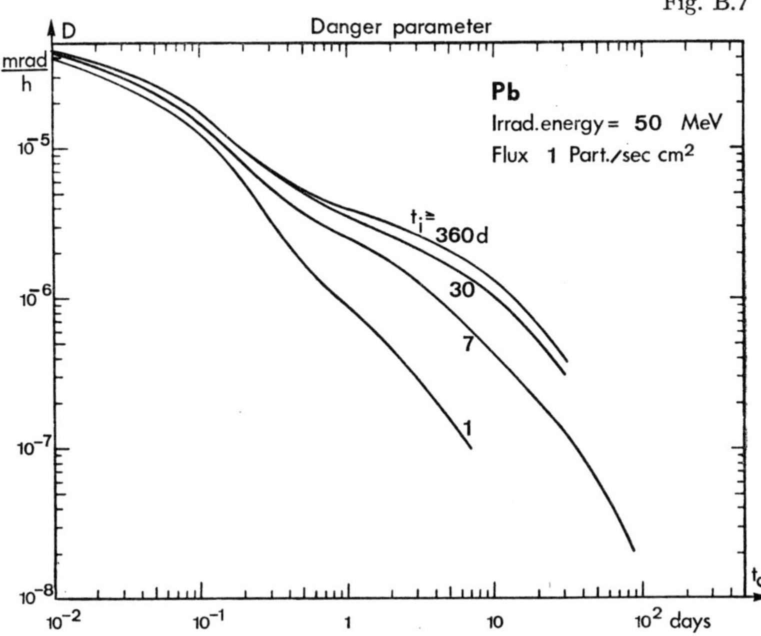
D
Danger parameter
mrad/h
Pb
Irrad.energy = 50 MeV
Flux 1 Part./sec cm²
t_i ≥ 360 d
30
7
1
10⁻⁵
10⁻⁶
10⁻⁷
10⁻⁸
10⁻²
10⁻¹
1
10
10² days
t_c

Fig. B.8

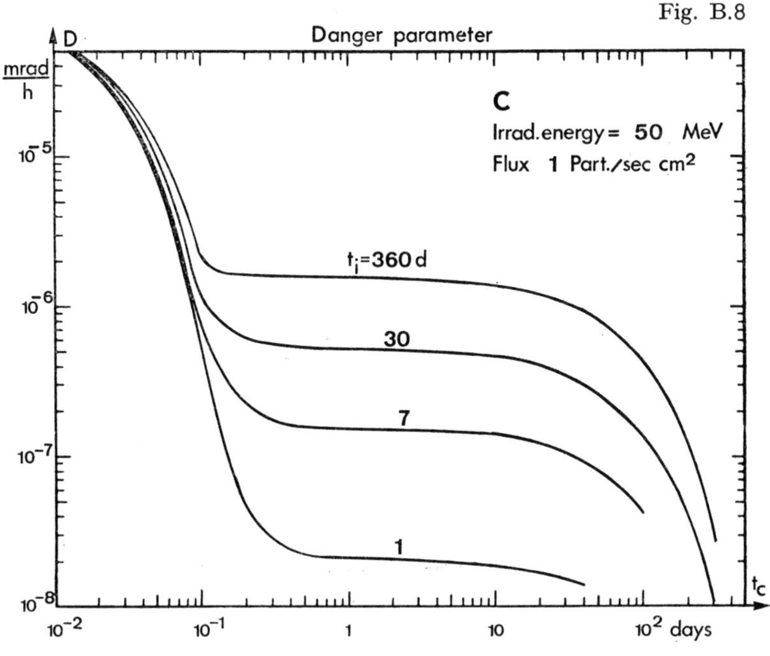
D
Danger parameter
mrad/h
C
Irrad.energy = 50 MeV
Flux 1 Part./sec cm²
t_i = 360 d
30
7
1
10⁻⁵
10⁻⁶
10⁻⁷
10⁻⁸
10⁻²
10⁻¹
1
10
10² days
t_c

Fig. B.9

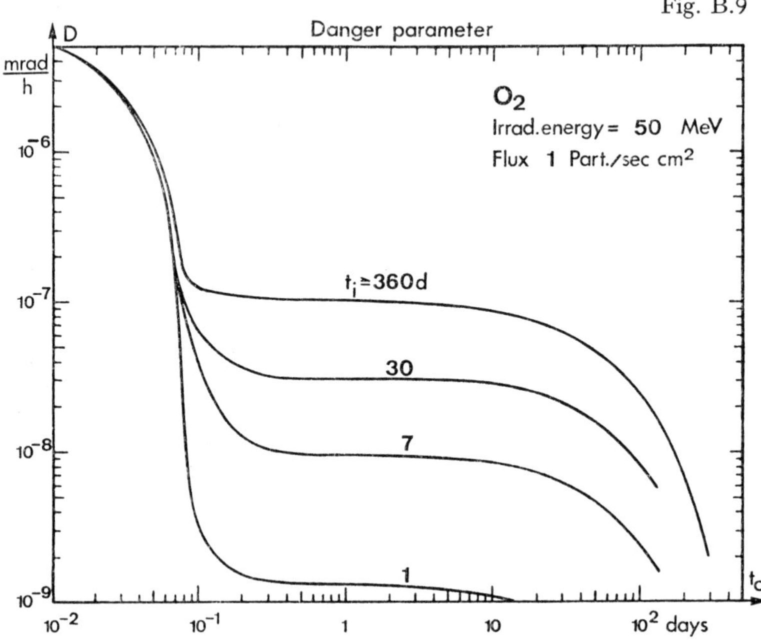

Fig. B.10

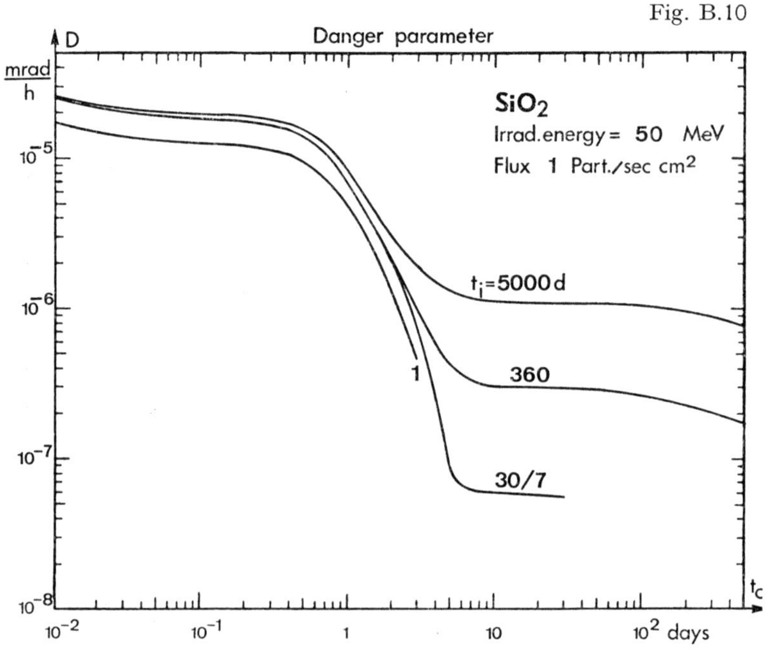

Fig. B.11

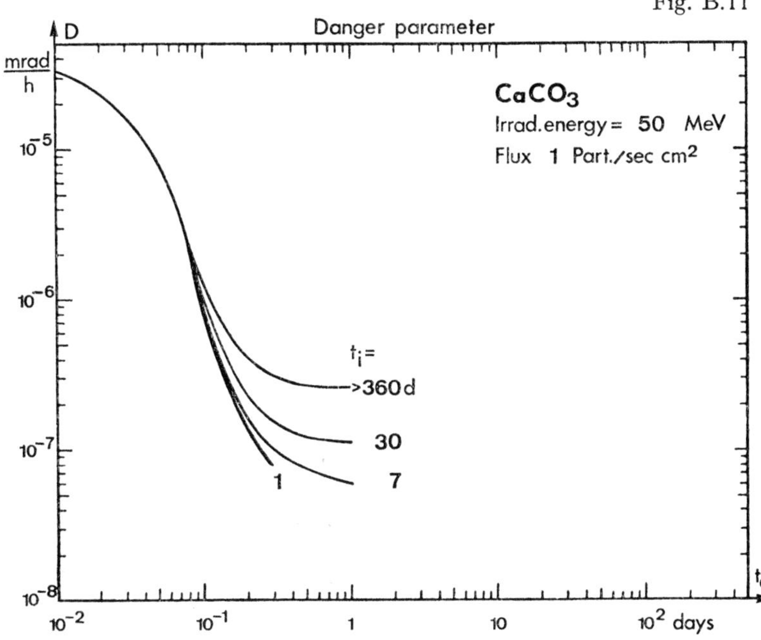

Fig. B.12

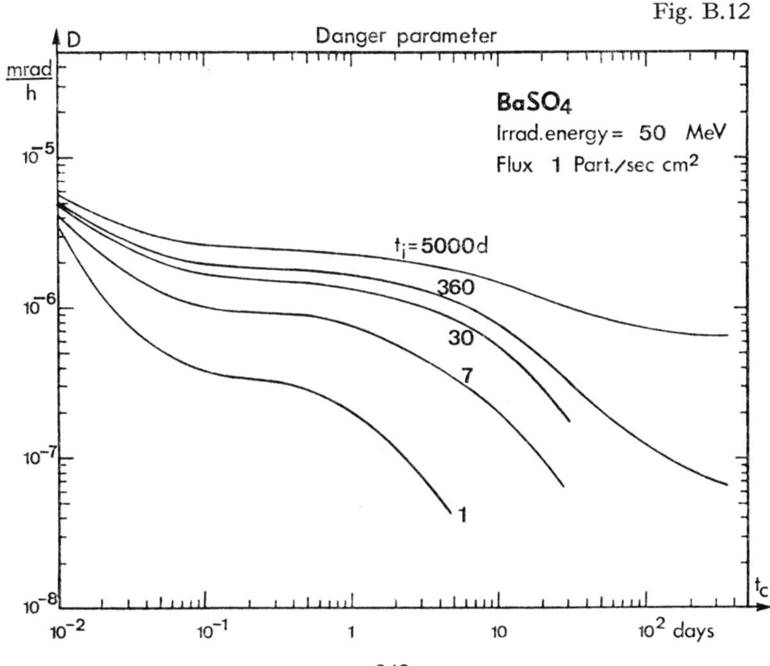

Fig. B.13

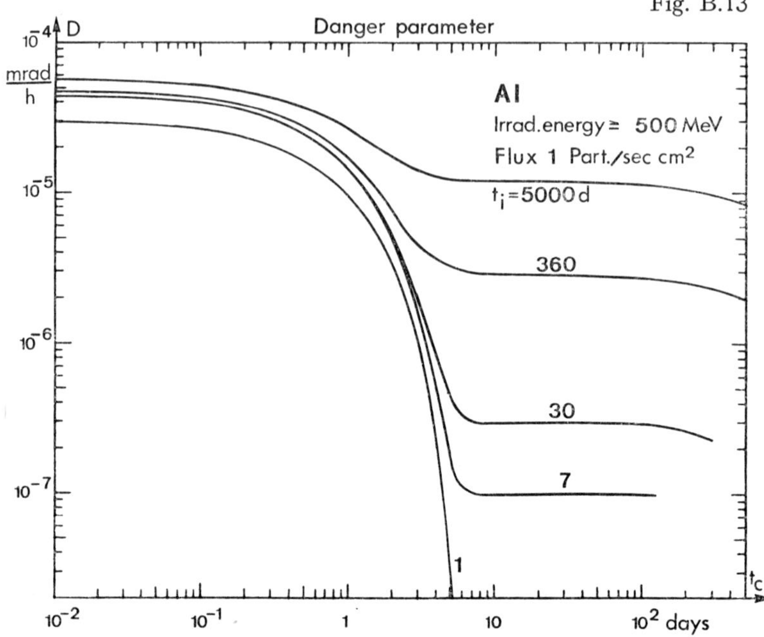

Fig. B.14

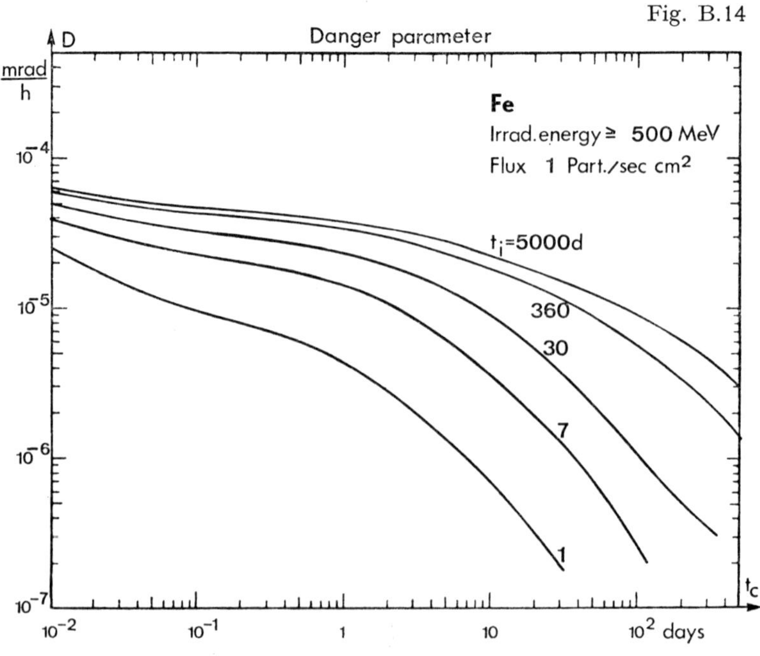

Fig. B.15

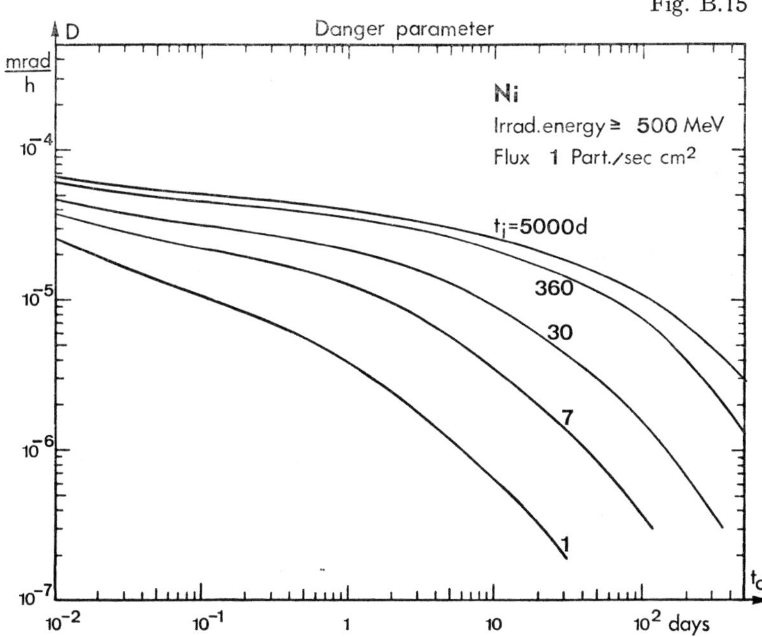

Fig. B.16

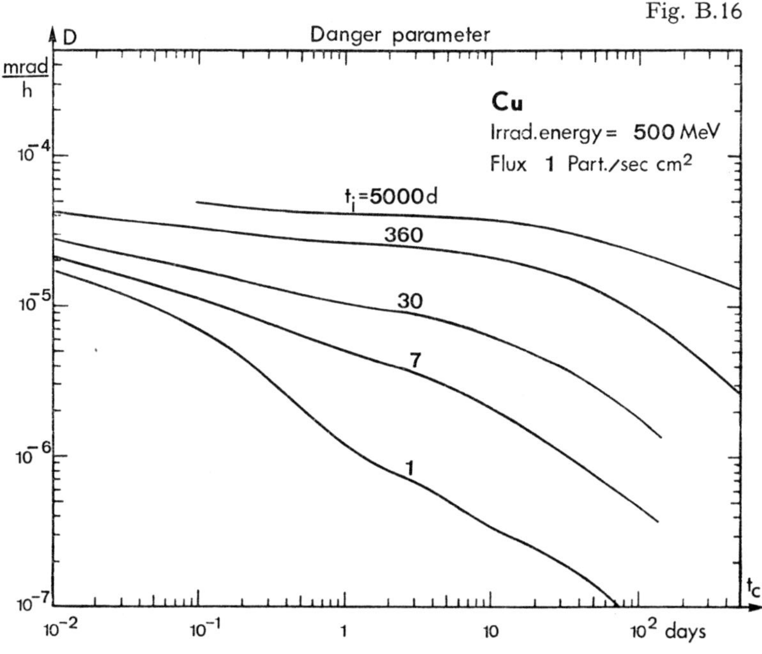

Fig. B.17

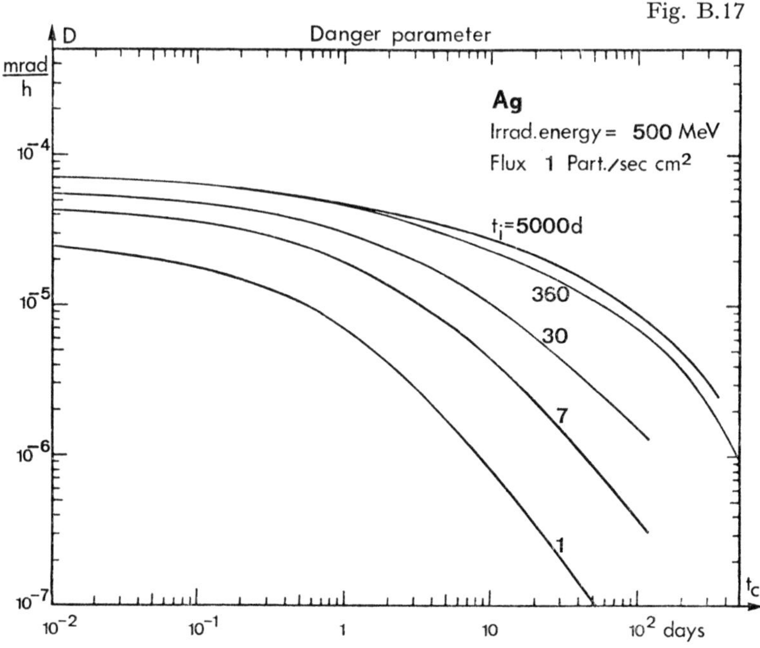

Fig. B.18

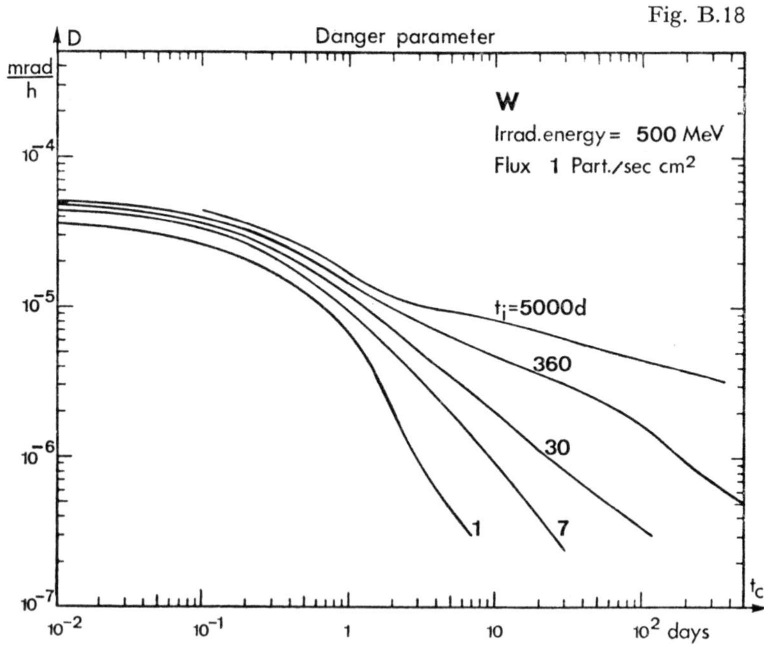

Fig. B.19

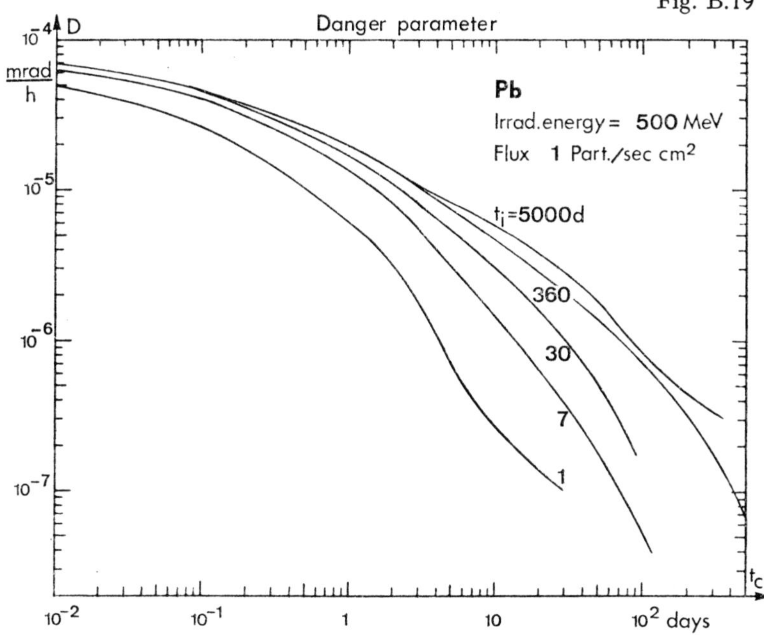

Fig. B.20

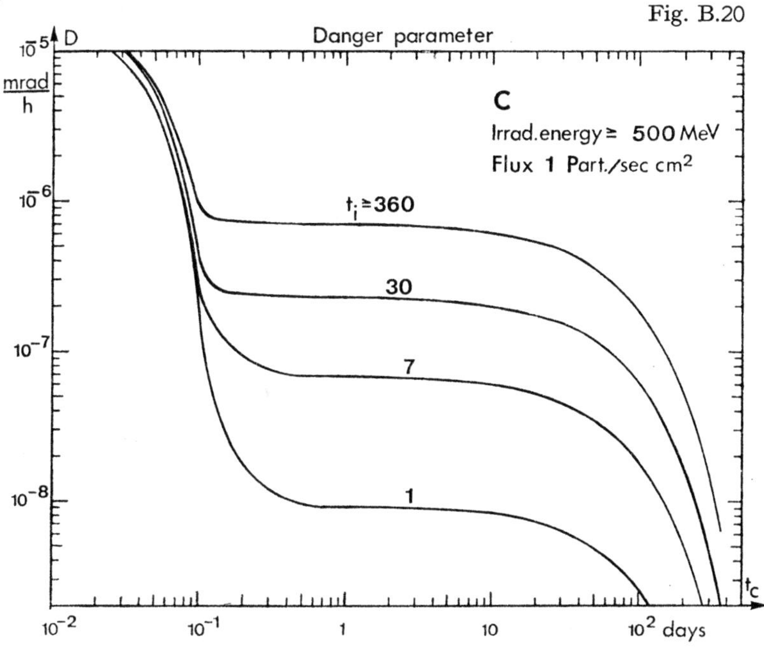

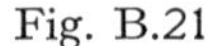
Fig. B.21

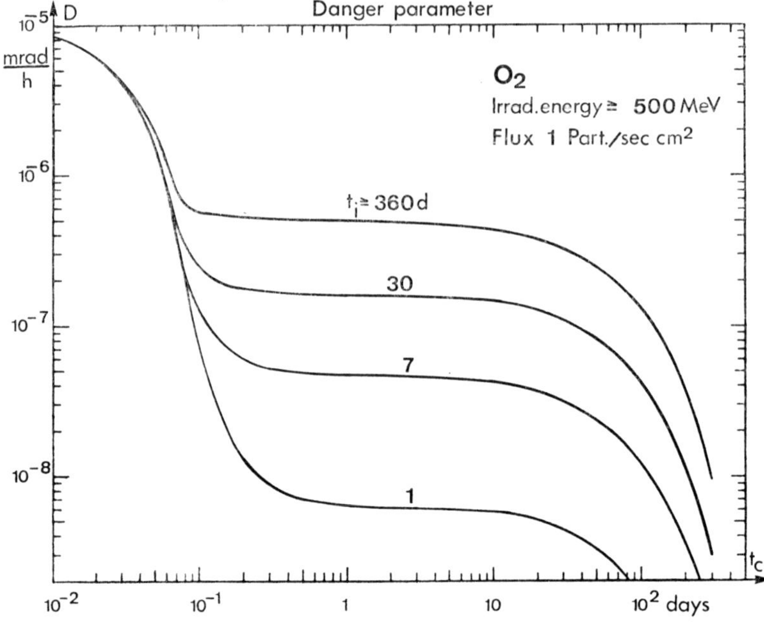

Fig. B.22

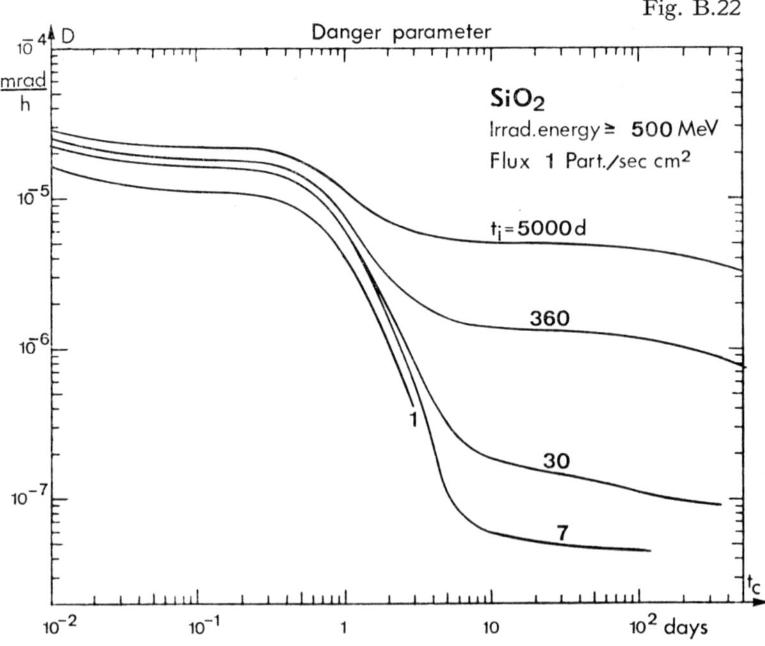

Fig. B.23

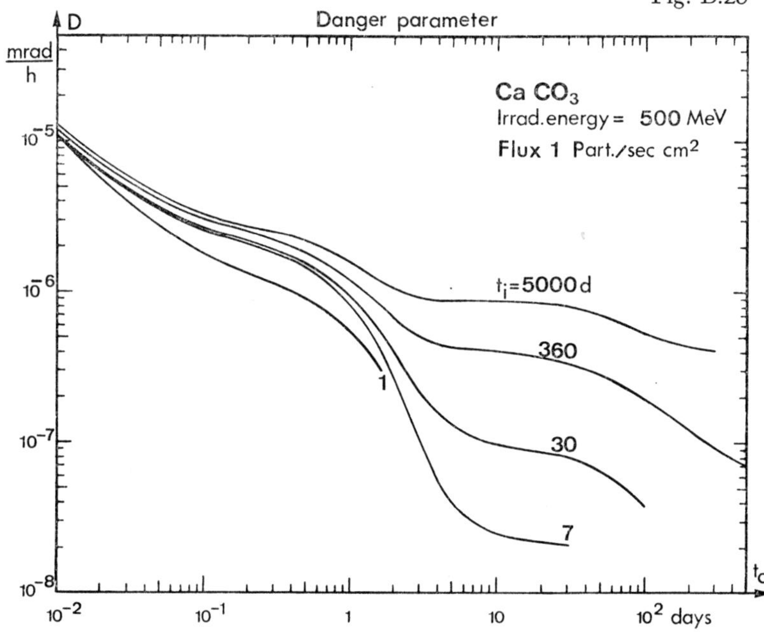

Fig. B.24

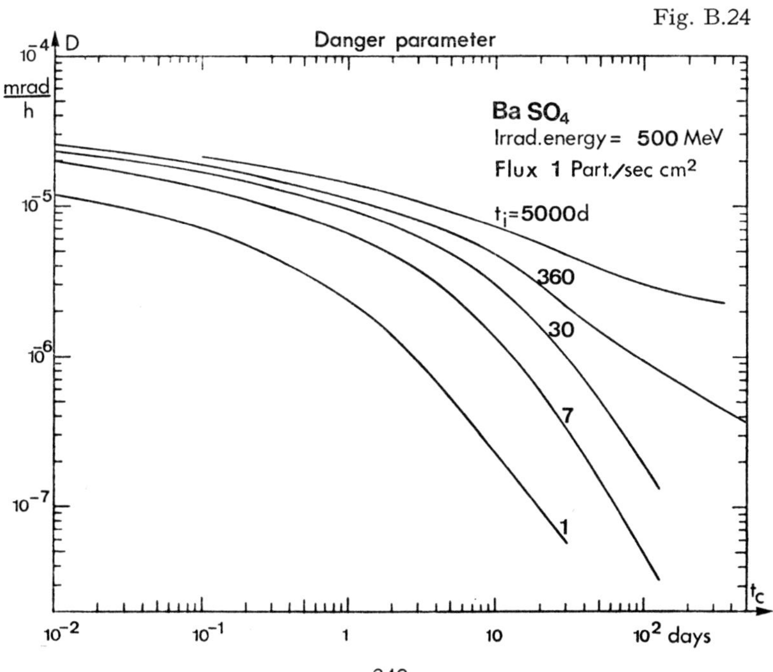

Fig. B.25

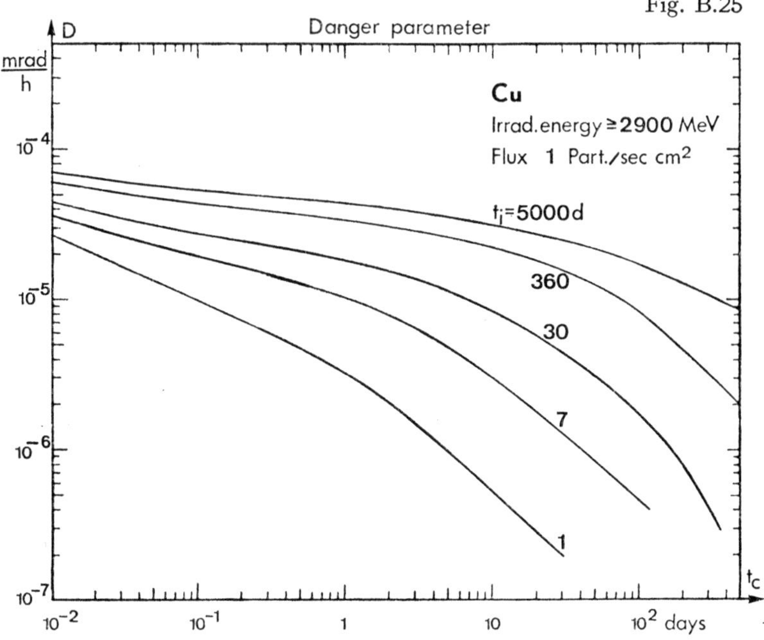

Fig. B.26

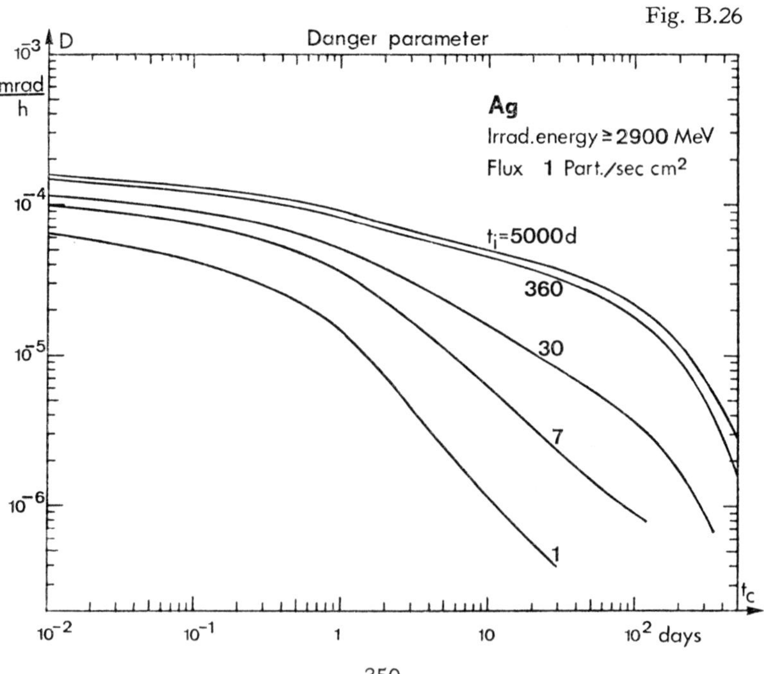

Fig. B.27

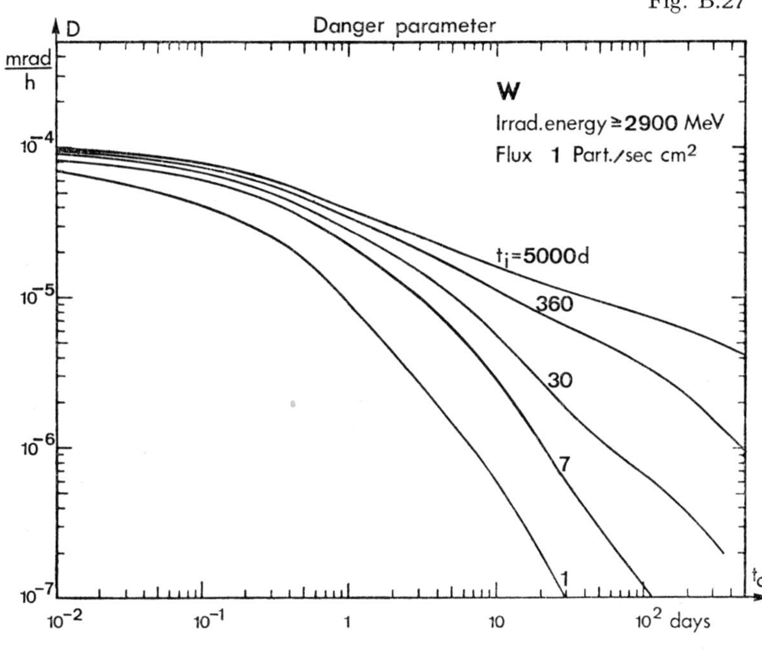

Fig. B.28

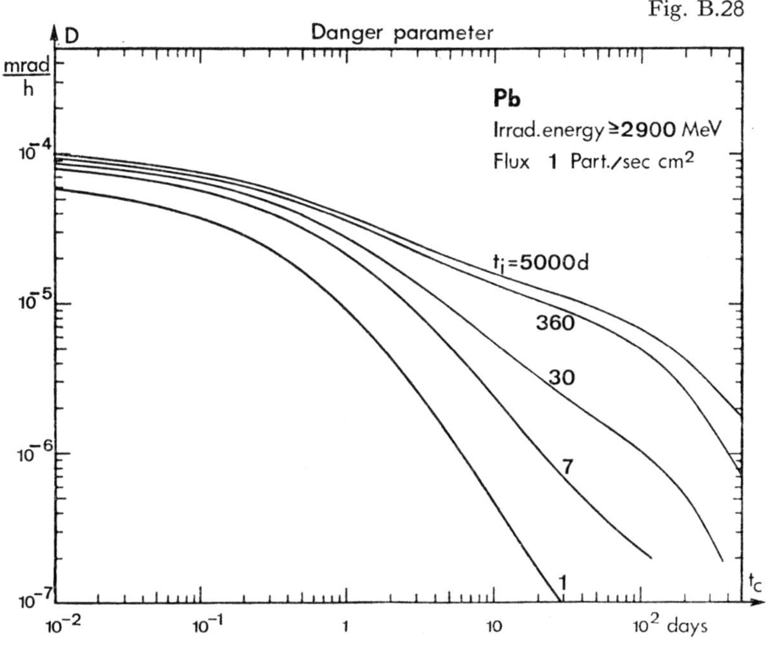

Fig. B.29

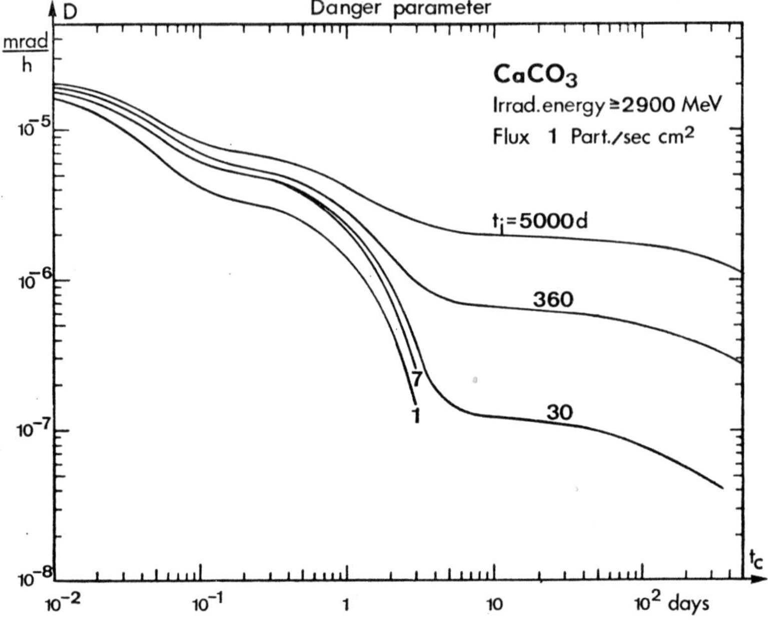

Fig. B.30

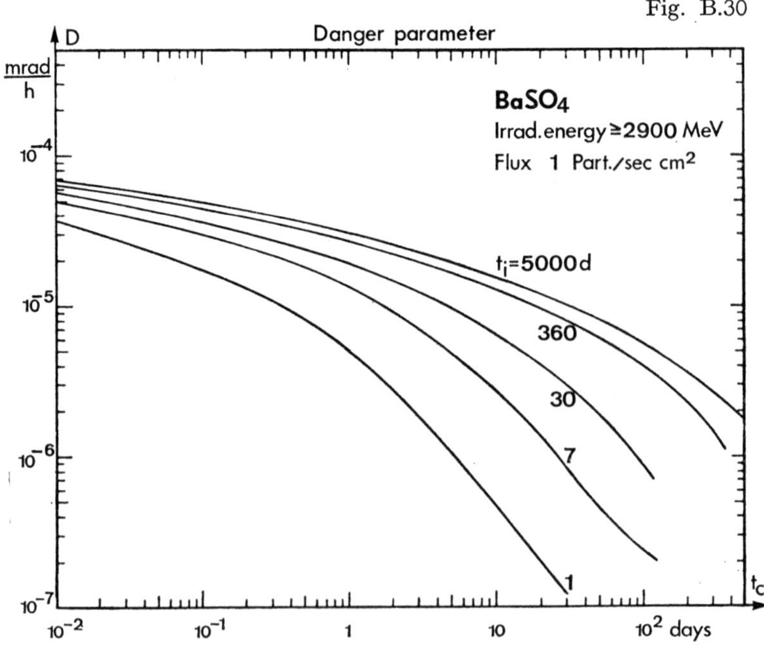

APPENDIX C

SATURATION VALUES OF GAMMA DOSE RATES FOR NEUTRON ACTIVATION

The saturation values of the gamma dose rates of selected target elements activated by neutrons of spectral energy distributions as described in ch. III, eqs. (3.1) to (3.6) and fig. III.21 (full line) are given below for unit flux (1 neutron/sec cm²) for each isotope produced together with its half-life. Only isotopes of half-life longer than 10 minutes are included.

Target	Product	Half-life	Rad/h g at 1 m, $t_i = \infty$, $t_c = 0$		
			Thermal neutrons	Epithermal neutrons	Fast neutrons
Mg	^{24}Na	15 h			1.1×10^{-15}
Al	^{24}Na	15 h			7.5×10^{-16}
Si	^{31}Si	2.62 h	8.3×10^{-19}	3×10^{-19}	
P	^{31}Si	2.62 h			4.5×10^{-18}
S	^{31}Si	2.62 h			2.7×10^{-20}
Ti	^{46}Sc	84 d			2.6×10^{-16}
	^{47}Sc	3.4 d			6×10^{-19}
	^{48}Sc	1.8 d			7.2×10^{-17}
Cr	^{51}Cr	27.8 d	3.8×10^{-15}	1.3×10^{-16}	
Mn	^{56}Mn	2.58 h	3.1×10^{-12}	2.9×10^{-13}	
Fe	^{59}Fe	45 d	5.5×10^{-16}	5.6×10^{-17}	
	^{54}Mn	280 d			3.8×10^{-16}
	^{56}Mn	2.58 h			2×10^{-16}
Co	^{60}Co	5.27 y	1.2×10^{-11}	2.1×10^{-12}	
	^{59}Fe	45 d			2.1×10^{-15}
	^{56}Mn	2.58 h			3×10^{-16}

Target	Product	Half-life	Rad/h g at 1 m, $t_i=\infty$, $t_c=0$ Thermal neutrons	Epithermal neutrons	Fast neutrons
Ni	^{65}Ni	2.56 h	1×10^{-15}	1.8×10^{-16}	
	^{58m}Co	9.1 h			6.5×10^{-16}
	^{58}Co	71 d			9.2×10^{-15}
	^{60m}Co	10.5 m			2.5×10^{-17}
	^{60}Co	5.27 y			9.7×10^{-16}
	^{61}Co	1.65 h			4.8×10^{-19}
	^{62}Co	13.9 m			2.2×10^{-18}
	^{59}Fe	45 d			3.2×10^{-19}
Cu	^{64}Cu	12.9 h	8.4×10^{-14}	5.6×10^{-15}	
	^{60}Co	5.27 y			1.6×10^{-16}
	^{65}Ni	2.56 h			4×10^{-17}
Zn	^{65}Zn	245 d	1.7×10^{-14}	5.9×10^{-15}	
	^{69m}Zn	13.9 h	9.8×10^{-15}	2×10^{-15}	
	^{64}Cu	12.9 h			6.2×10^{-16}
	^{67}Cu	61 h			3.4×10^{-19}
	^{65}Ni	2.56 h			1.9×10^{-19}
Zr	^{95}Zr	65 d	2.3×10^{-15}	5.1×10^{-15}	
	^{97}Zr	17 h	7.4×10^{-16}	1.7×10^{-15}	
	^{87m}Sr	2.8 h			6.6×10^{-19}
	^{89}Sr	51 d			8.1×10^{-23}
	^{91}Sr	9.7 h			1.5×10^{-18}
	^{90}Y	64.2 h			1.6×10^{-20}
	^{91m}Y	50 m			6.1×10^{-18}
	^{91}Y	58 d			5×10^{-20}
	^{92}Y	3.5 h			2.1×10^{-18}
Mo	^{99}Mo	67 h	7.4×10^{-16}	2.2×10^{-15}	
	^{101}Mo	14.6 m	1.2×10^{-15}	4.8×10^{-15}	
	^{92}Nb	10.1 d			3.8×10^{-17}
	^{97}Nb	72 m			5.8×10^{-18}
	^{89}Zr	79 h			6.1×10^{-19}
	^{95}Zr	65 d			5.6×10^{-18}
	^{97}Zr	17 h			1.3×10^{-18}
W	^{181}W	145 d	2.8×10^{-19}	1.2×10^{-19}	
	^{185}W	74 d	7.4×10^{-19}	2.1×10^{-19}	
	^{187}W	24 h	6.8×10^{-14}	5.8×10^{-14}	
	^{182}Ta	115 d			2.5×10^{-18}
	^{183}Ta	5 d			2.3×10^{-19}
	^{184}Ta	8.7 h			2.9×10^{-19}
Th	^{233}Th	22.1 m	5.1×10^{-16}	5.3×10^{-16}	
	^{233}Pa	27.4 d	9.5×10^{-14}	9.8×10^{-14}	

APPENDIX D

GAMMA ACTIVITY INDUCED BY GIANT RESONANCE REACTIONS

Tables and graphs for the total gamma activity induced by giant resonance reactions in various thick targets absorbing the electromagnetic cascade completely for an electron current of 1 electron per second, computed per MeV primary electron energy are presented. The activity figures in the table are given for infinite irradiation time and zero cooling time.

Target (natural)	End-product	Half-life (days)	Gamma activity ($t_i=\infty$, $t_c=0$)		
			Disint/sec	Photons/sec	Rad/h at 1 m
Al	^{24}Na	0.63	2.5×10^{-6}	5×10^{-6}	1.4×10^{-16}
Ca	^{38}K	5.4×10^{-3}	6.8×10^{-6}	2×10^{-5}	2.7×10^{-16}
	^{42}K	0.52	1.3×10^{-7}	2.3×10^{-8}	4.9×10^{-19}
	^{43}K	0.92	2×10^{-6}	3.7×10^{-6}	2.6×10^{-17}
Fe	^{52}Fe	0.345	8×10^{-6}	1.6×10^{-5}	7.4×10^{-17}
	^{53}Fe	6×10^{-3}	5.8×10^{-6}	1.2×10^{-5}	8.1×10^{-17}
	^{52m}Mn	1.45×10^{-2}	1×10^{-5}	2.9×10^{-5}	3.2×10^{-16}
	^{52}Mn	5.7	1×10^{-5}	1.85×10^{-5}	2.4×10^{-16}
	^{54}Mn	314	3.7×10^{-6}	3.7×10^{-6}	4.3×10^{-17}
	^{56}Mn	0.107	1.2×10^{-6}	1.8×10^{-6}	3×10^{-17}
Co	^{57}Co	267	1.3×10^{-5}	1.3×10^{-5}	2×10^{-17}
	^{58}Co	17	1.1×10^{-4}	1.4×10^{-4}	1.5×10^{-15}

Target (natural)	End-product	Half-life (days)	Gamma activity ($t_i = \infty$, $t_c = 0$)		
			Disint/sec	Photons/sec	Rad/h at 1 m
Ni	^{56}Co	77.3	2.5×10^{-6}	6.4×10^{-6}	2.9×10^{-16}
	^{57}Co	267	3.4×10^{-5}	3.5×10^{-5}	5.7×10^{-17}
	^{58}Co	71	9.2×10^{-7}	1.2×10^{-6}	1.2×10^{-17}
	^{60}Co	1920	6.6×10^{-7}	1.3×10^{-6}	2.2×10^{-17}
	^{61}Co	0.115	1.5×10^{-6}	1.5×10^{-6}	1.5×10^{-18}
	^{62}Co	9.7×10^{-3}	3.3×10^{-8}	7.4×10^{-8}	1.2×10^{-18}
	^{56}Ni	6.1	8.1×10^{-6}	2.3×10^{-5}	1.7×10^{-16}
	^{57}Ni	1.54	7.5×10^{-5}	1.6×10^{-4}	2×10^{-15}
Cu	^{63}Ni	0.69	1.6×10^{-7}	1.6×10^{-7}	1.6×10^{-19}
	^{61}Cu	0.14	8.6×10^{-6}	1.3×10^{-5}	1×10^{-16}
	^{62}Cu	6.9×10^{-3}	8.3×10^{-5}	1.6×10^{-4}	1.1×10^{-15}
	^{64}Cu	0.54	3.4×10^{-5}	1.3×10^{-5}	9×10^{-17}
Zn	^{62}Zn	0.39	8×10^{-6}	5.8×10^{-6}	3.2×10^{-17}
	^{65}Zn	245	4.5×10^{-5}	2.3×10^{-5}	3.5×10^{-16}
	^{63}Zn	1.6×10^{-2}	7.5×10^{-5}	1.2×10^{-4}	8.6×10^{-16}
	^{67}Cu	2.54	6×10^{-6}	6×10^{-6}	1.3×10^{-17}
	^{64}Cu	0.54	8×10^{-7}	3.1×10^{-7}	2.3×10^{-18}
	^{62}Cu	4.1×10^{-3}	1.5×10^{-6}	3×10^{-6}	2×10^{-17}
	^{66}Cu	3.5×10^{-3}	1×10^{-6}	3.3×10^{-7}	2.6×10^{-18}
W	^{181}W	130	4.9×10^{-6}	6.4×10^{-6}	3.3×10^{-17}
	^{182}Ta	115	2.1×10^{-7}	2×10^{-7}	1.7×10^{-18}
	^{183}Ta	5	5×10^{-7}	1.1×10^{-6}	3.3×10^{-18}
	^{185}Ta	0.03	4.3×10^{-7}	5.2×10^{-7}	1×10^{-18}
	^{180}Hf	0.23	1.4×10^{-9}	5.6×10^{-10}	2.7×10^{-21}
	^{178}W	22	4.7×10^{-8}	1.9×10^{-7}	2.6×10^{-20}
	^{179}W	0.021	1.6×10^{-7}	1×10^{-7}	2.5×10^{-19}
Au	^{195}Au	183	5×10^{-5}	2×10^{-5}	3.5×10^{-17}
	^{196}Au	6.2	1.5×10^{-4}	1×10^{-4}	2.3×10^{-16}
	^{194}Ir	12	4×10^{-9}	2×10^{-9}	2.1×10^{-20}
Pb	^{203}Pb	1.24	2.1×10^{-6}	1.9×10^{-6}	7.4×10^{-18}
	^{204}Pb	0.047	1.1×10^{-5}	1.7×10^{-5}	1.6×10^{-16}
	^{202}Pb	0.15	7.5×10^{-7}	1×10^{-6}	1×10^{-16}
	^{203}Hg	47	8×10^{-10}	7×10^{-10}	2.6×10^{-20}

Fig. D.1

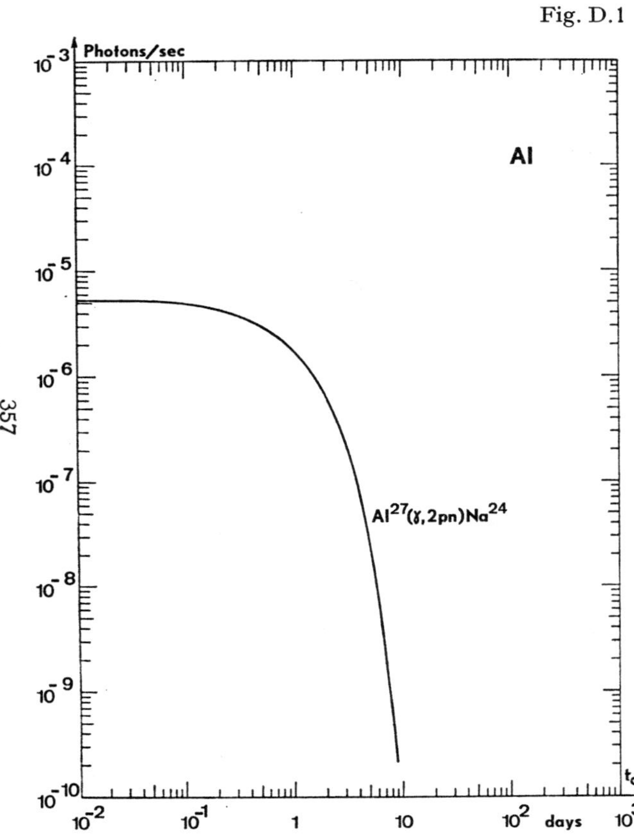
Photons/sec
Al
Al27(γ,2pn)Na24
10-3
10-4
10-5
10-6
10-7
10-8
10-9
10-10
10-2
10-1
1
10
102
103
days
tc

Fig. D.2

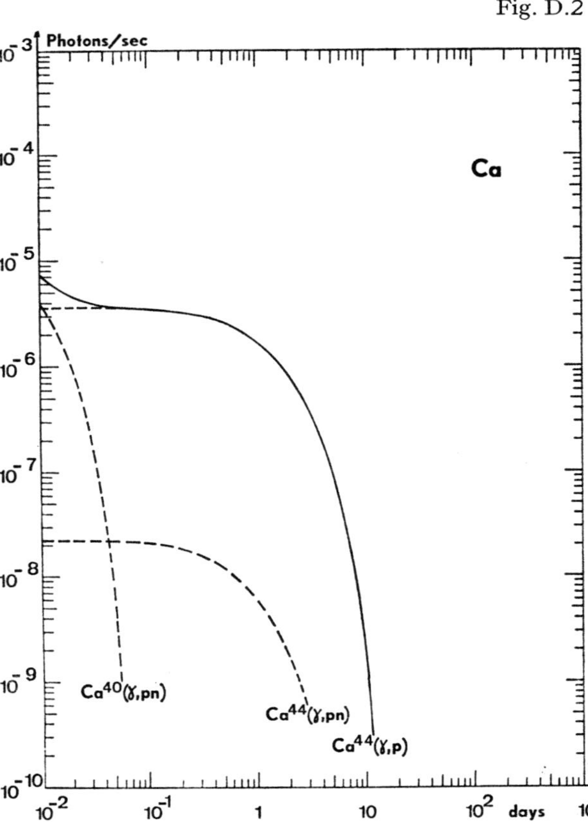
Photons/sec
Ca
Ca40(γ,pn)
Ca44(γ,pn)
Ca44(γ,p)
10-3
10-4
10-5
10-6
10-7
10-8
10-9
10-10
10-2
10-1
1
10
102
103
days
tc

Fig. D.3

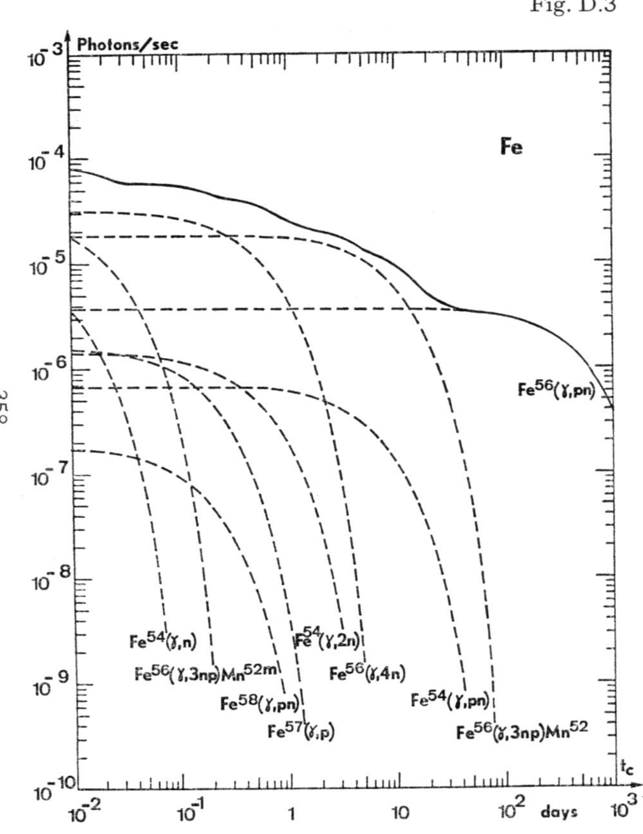
Photons/sec
Fe
10^{-3}
10^{-4}
10^{-5}
10^{-6}
10^{-7}
10^{-8}
10^{-9}
10^{-10}
$Fe^{56}(\gamma,pn)$
$Fe^{54}(\gamma,n)$
$Fe^{56}(\gamma,3np)Mn^{52m}$
$Fe^{58}(\gamma,pn)$
$Fe^{57}(\gamma,p)$
$Fe^{54}(\gamma,2n)$
$Fe^{56}(\gamma,4n)$
$Fe^{54}(\gamma,pn)$
$Fe^{56}(\gamma,3np)Mn^{52}$
t_c
10^{-2}
10^{-1}
1
10
10^{2} days
10^{3}

Fig. D.4

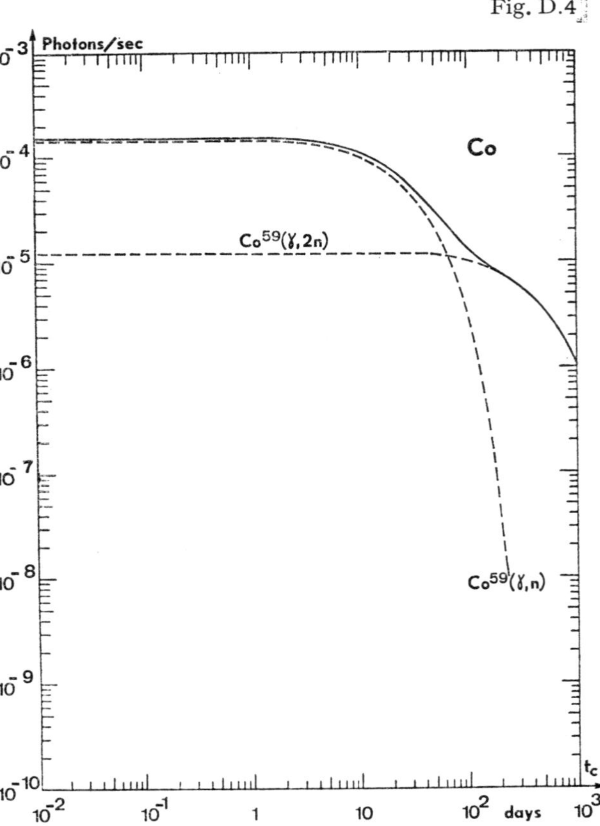
Photons/sec
Co
10^{-3}
10^{-4}
10^{-5}
10^{-6}
10^{-7}
10^{-8}
10^{-9}
10^{-10}
$Co^{59}(\gamma,2n)$
$Co^{59}(\gamma,n)$
t_c
10^{-2}
10^{-1}
1
10
10^{2} days
10^{3}

Fig. D.5

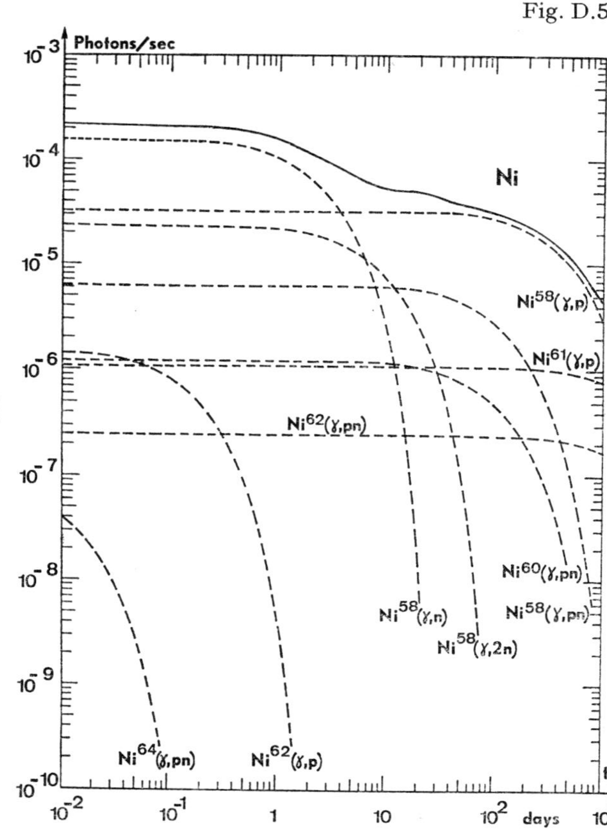
Photons/sec
Ni
Ni^{58}(γ,p)
Ni^{61}(γ,p)
Ni^{62}(γ,pn)
Ni^{60}(γ,pn)
Ni^{58}(γ,n)
Ni^{58}(γ,pn)
Ni^{58}(γ,2n)
Ni^{64}(γ,pn)
Ni^{62}(γ,p)
10^{-3}
10^{-4}
10^{-5}
10^{-6}
10^{-7}
10^{-8}
10^{-9}
10^{-10}
10^{-2}
10^{-1}
1
10
10^{2} days
10^{3}
t_c

Fig. D.6

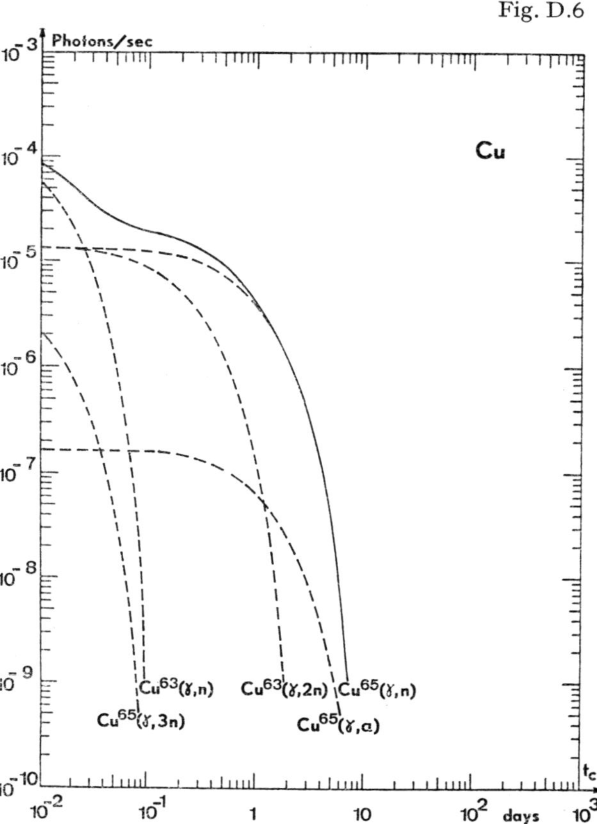
Photons/sec
Cu
Cu^{63}(γ,n)
Cu^{63}(γ,2n)
Cu^{65}(γ,n)
Cu^{65}(γ,3n)
Cu^{65}(γ,α)
10^{-3}
10^{-4}
10^{-5}
10^{-6}
10^{-7}
10^{-8}
10^{-9}
10^{-10}
10^{-2}
10^{-1}
1
10
10^{2} days
10^{3}
t_c

Fig. D.7

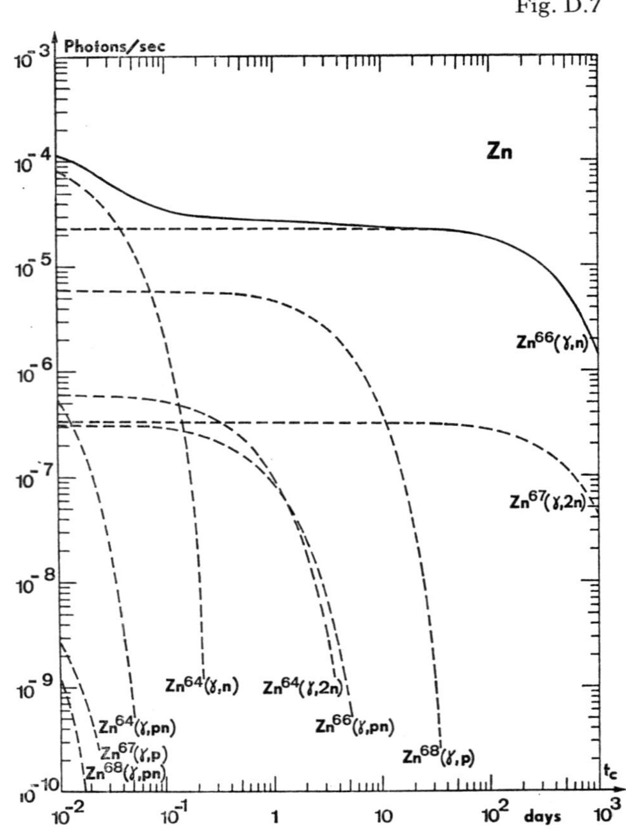

Photons/sec
Zn
10^{-3}
10^{-4}
10^{-5}
10^{-6}
10^{-7}
10^{-8}
10^{-9}
10^{-10}
$Zn^{66}(\gamma,n)$
$Zn^{67}(\gamma,2n)$
$Zn^{64}(\gamma,n)$
$Zn^{64}(\gamma,2n)$
$Zn^{64}(\gamma,pn)$
$Zn^{66}(\gamma,pn)$
$Zn^{67}(\gamma,p)$
$Zn^{68}(\gamma,p)$
$Zn^{68}(\gamma,pn)$
t_c
10^{-2}
10^{-1}
1
10
10^{2} days
10^{3}

Fig. D.8

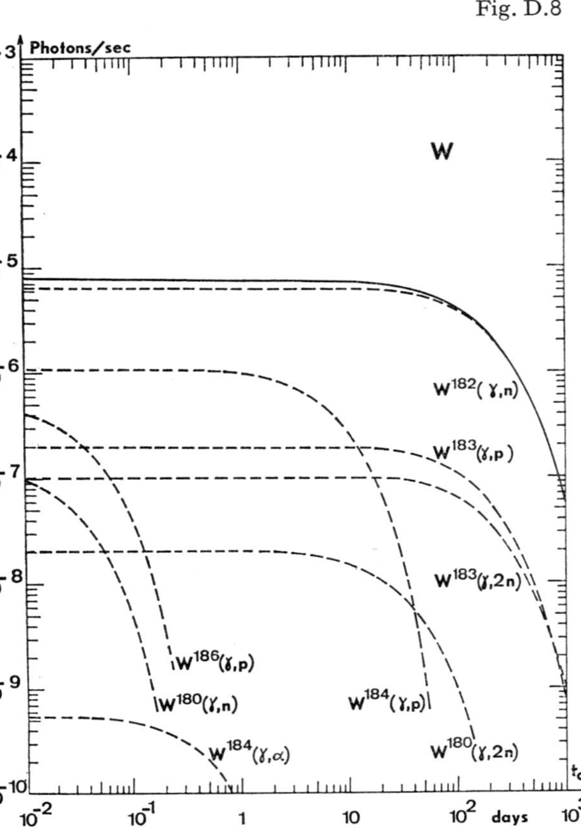

Photons/sec
W
10^{-3}
10^{-4}
10^{-5}
10^{-6}
10^{-7}
10^{-8}
10^{-9}
10^{-10}
$W^{182}(\gamma,n)$
$W^{183}(\gamma,p)$
$W^{183}(\gamma,2n)$
$W^{186}(\gamma,p)$
$W^{180}(\gamma,n)$
$W^{184}(\gamma,p)$
$W^{184}(\gamma,\alpha)$
$W^{180}(\gamma,2n)$
t_c
10^{-2}
10^{-1}
1
10
10^{2} days
10^{3}

Fig. D.9

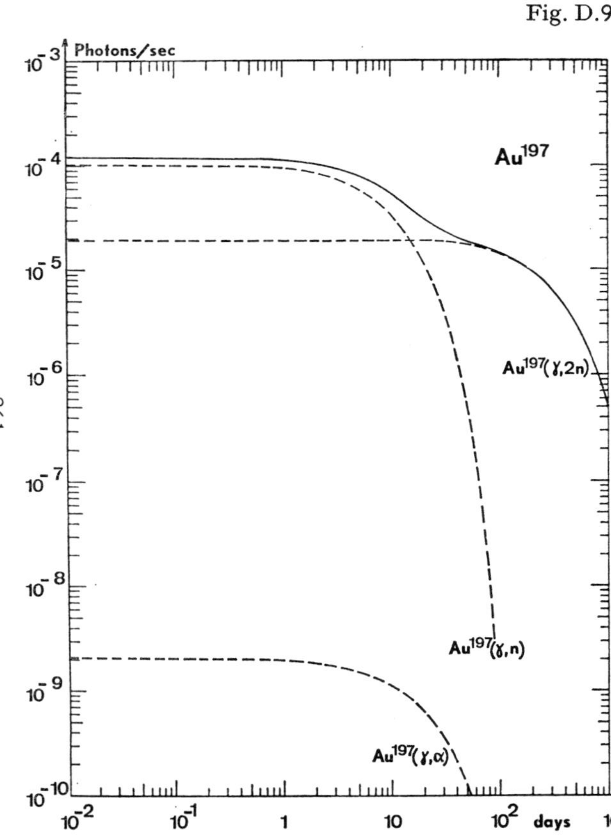

Fig. D.10

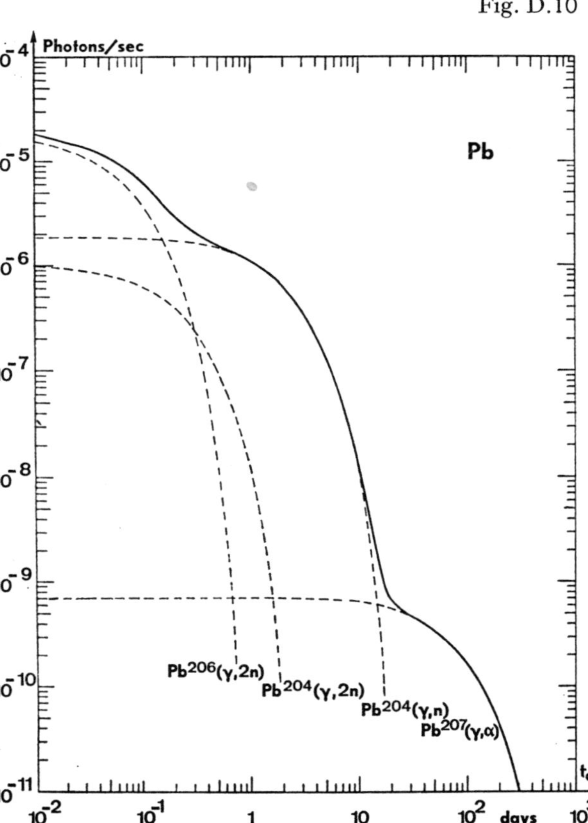

Fig. D.11

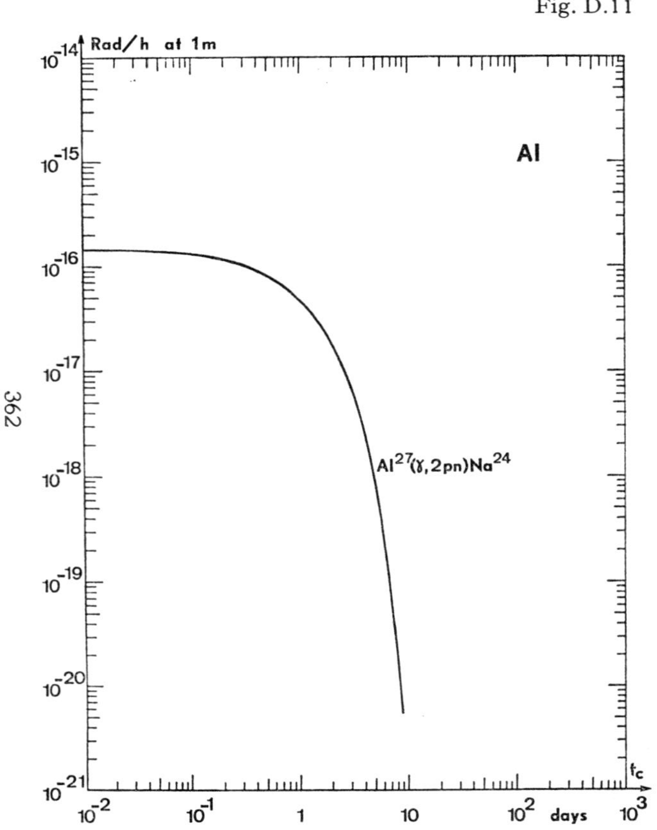

Fig. D.12

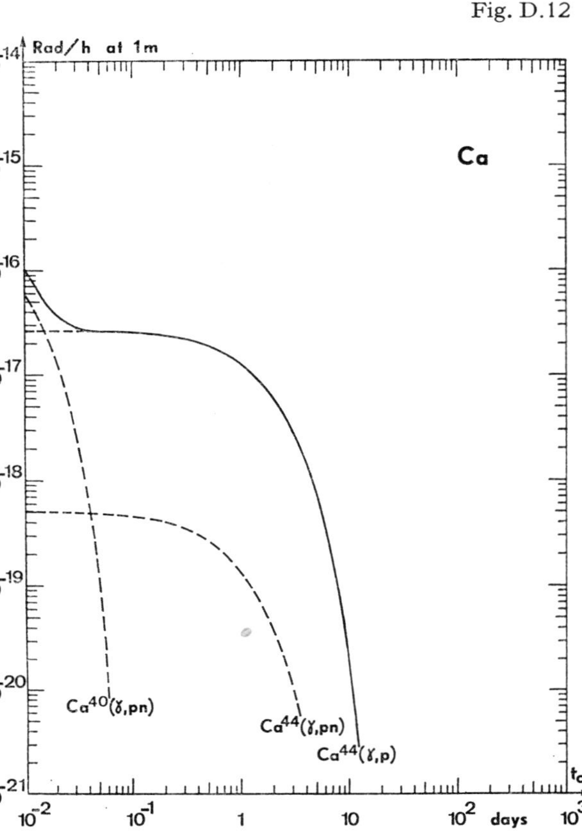

Fig. D.13

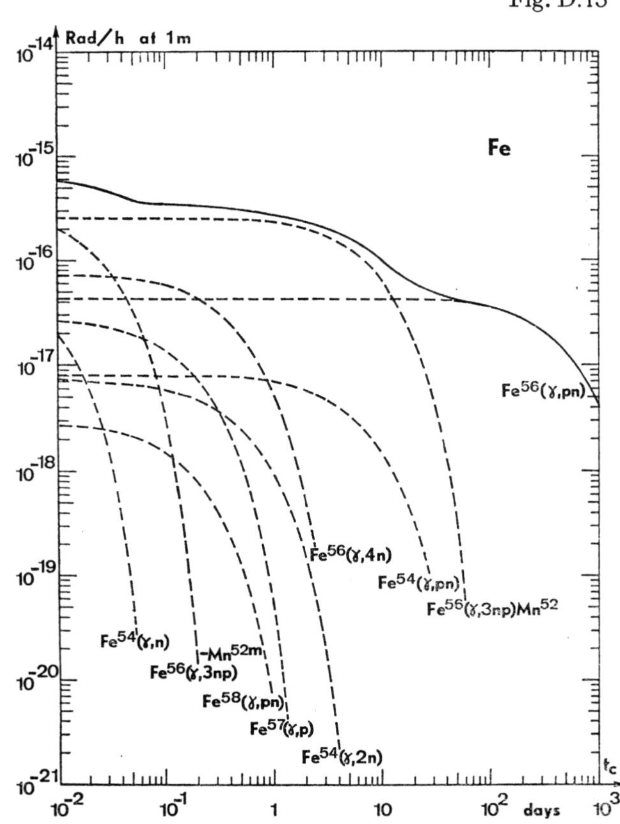

Fig. D.14

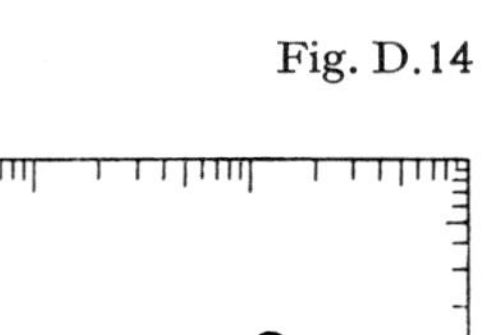

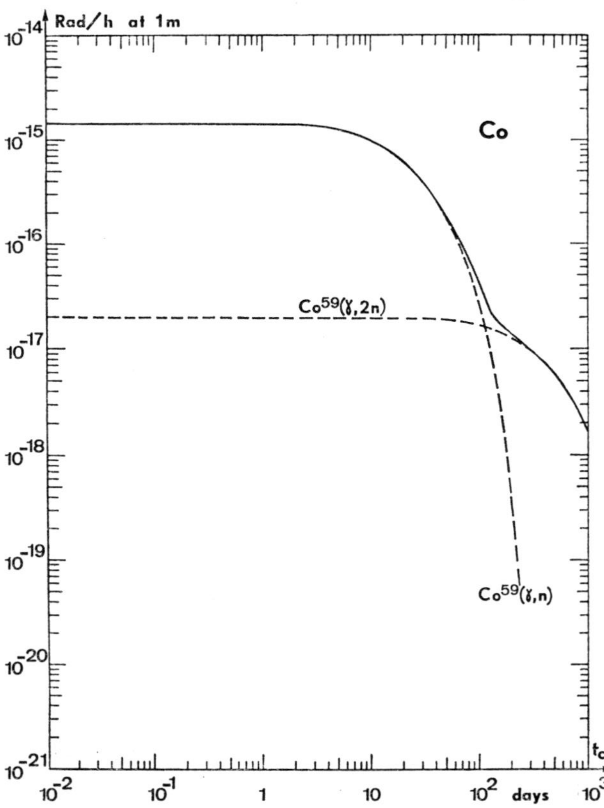

Fig. D.15

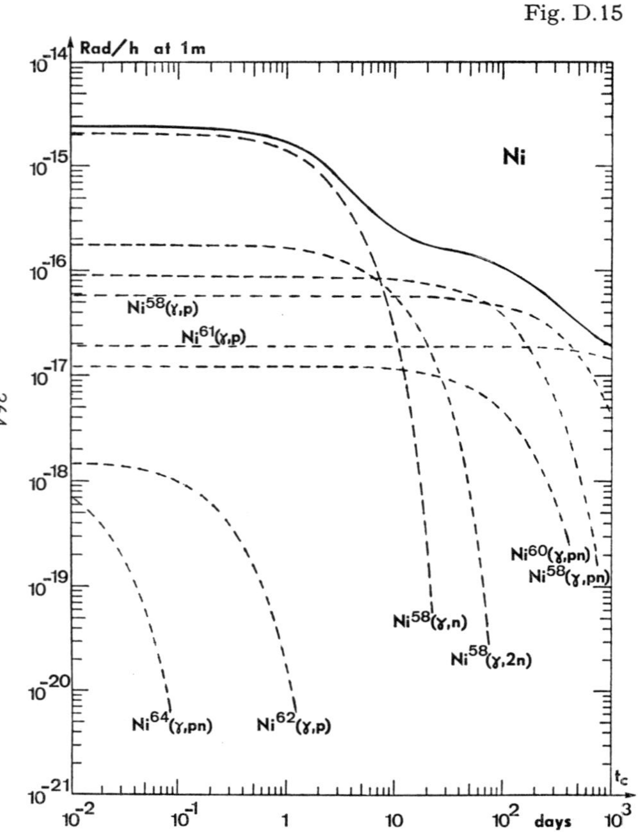

Fig. D.16

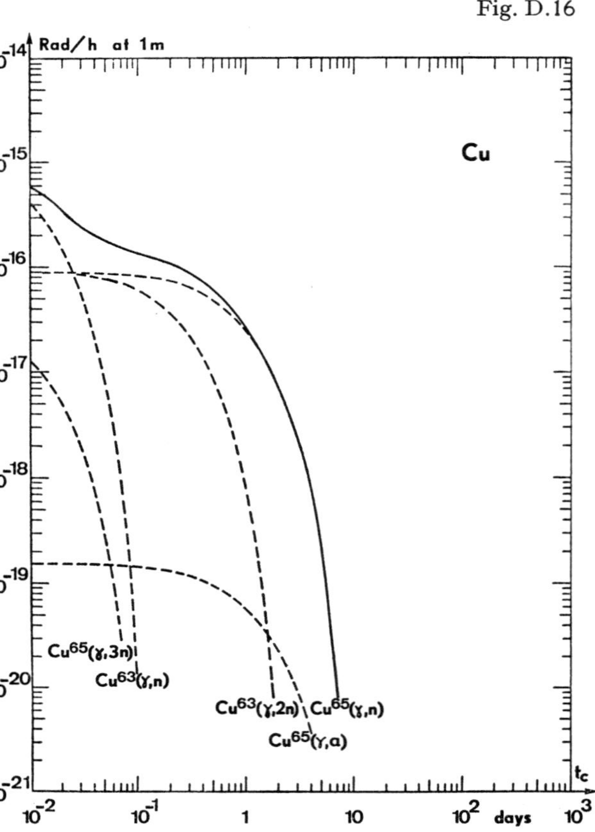

Fig. D.17

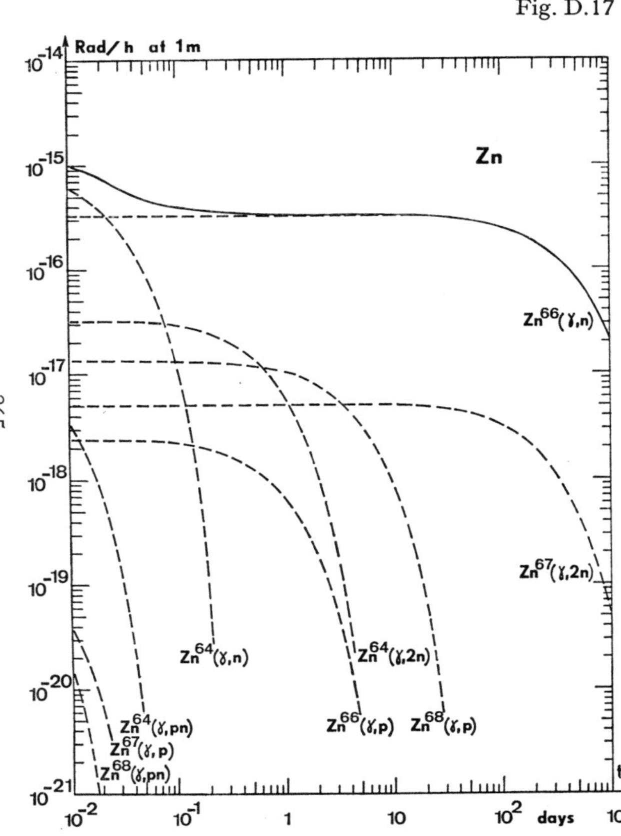

Fig. D.18

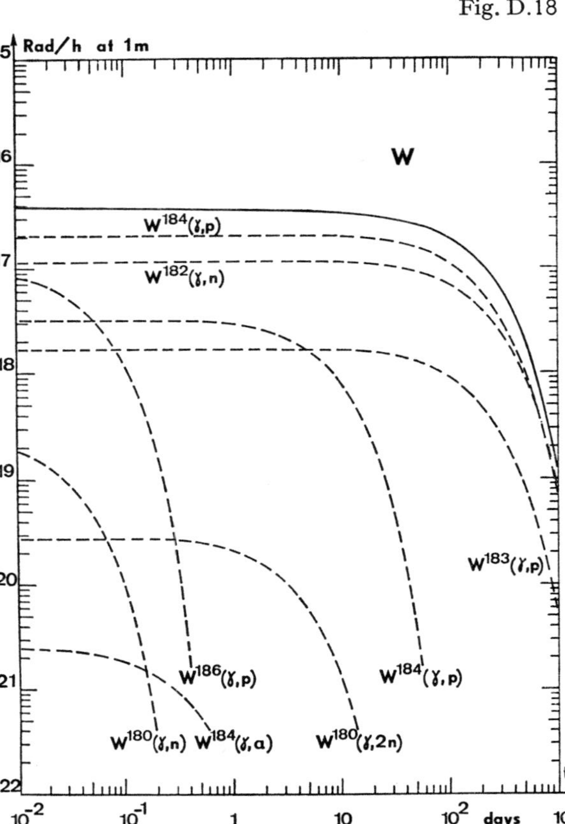

Fig. D.19

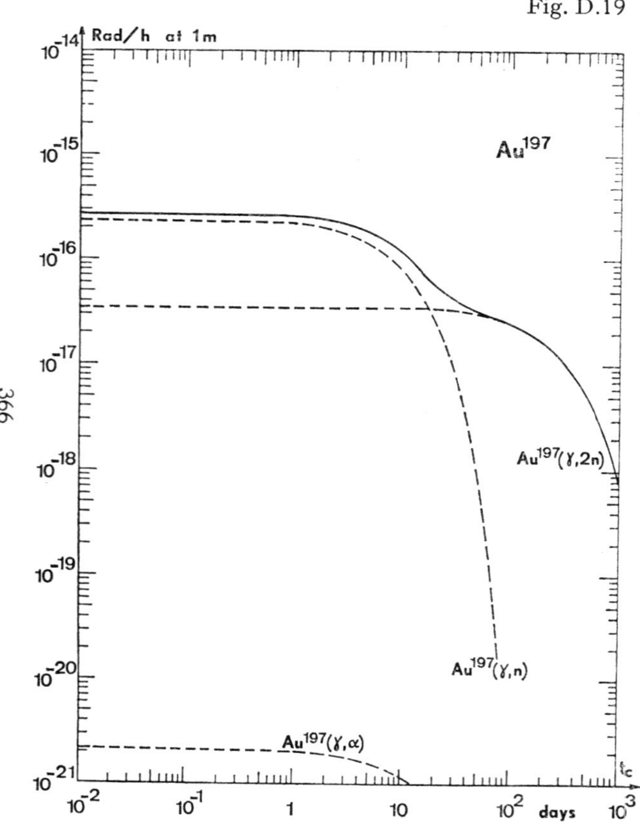

Fig. D.20

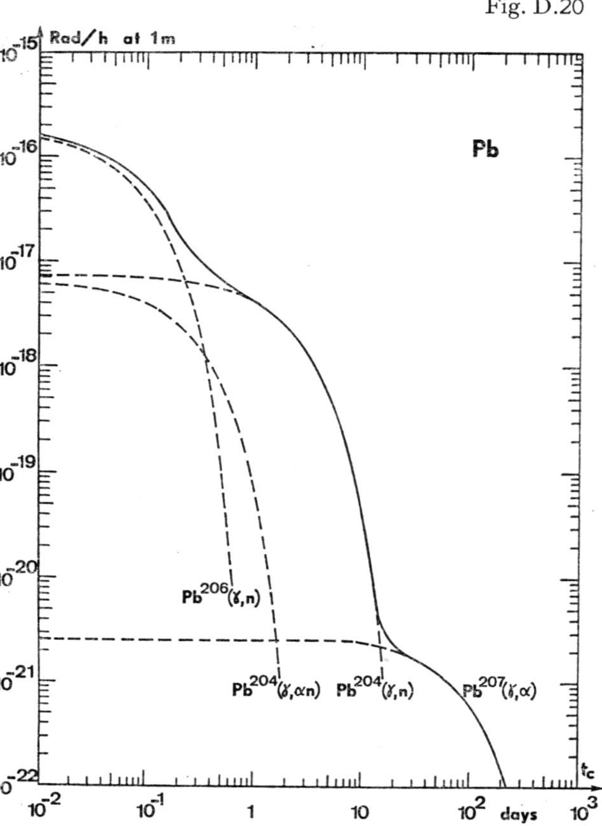

APPENDIX E

CHART OF THE NUCLIDES

Acknowledgement

On the following pages are reproduced the data as presented on the Chart of the Nuclides, 9th edition, revised to July 1966, prepared by D. T. Goldman and J. R. Roesser and distributed by Educational Relations, General Electric Company, Schenectady, N.Y. We are grateful to

Knolls Atomic Power Laboratory
Schenectady, New York
Operated by the General Electric Company
for the United States Atomic Energy Commission

for their permission to reproduce the Chart.

On the next four pages are reproduced the list of atomic elements, the key to the chart and a survey map. On this map, the heavy numerals refer to the subsequent 31 sections; proton numbers increase from bottom to top, neutron numbers from left to right, parallel to the spine of the book.

LIST OF ATOMIC ELEMENTS

Element	Symbol	Number
Actinium	Ac	89
Aluminum	Al	13
Americium	Am	95
Antimony	Sb	51
Argon	Ar	18
Arsenic	As	33
Astatine	At	85
Barium	Ba	56
Berkelium	Bk	97
Beryllium	Be	4
Bismuth	Bi	83
Boron	B	5
Bromine	Br	35
Cadmium	Cd	48
Calcium	Ca	20
Californium	Cf	98
Carbon	C	6
Cerium	Ce	58
Cesium	Cs	55
Chlorine	Cl	17
Chromium	Cr	24
Cobalt	Co	27
Copper	Cu	29
Curium	Cm	96
Dysprosium	Dy	66
Einsteinium	Es	99
Erbium	Er	68
Europium	Eu	63
Fermium	Fm	100
Fluorine	F	9
Francium	Fr	87
Gadolinium	Gd	64
Gallium	Ga	31
Germanium	Ge	32
Gold	Au	79
Hafnium	Hf	72
Helium	He	2
Holmium	Ho	67
Hydrogen	H	1
Indium	In	49
Iodine	I	53
Iridium	Ir	77
Iron	Fe	26
Krypton	Kr	36
Lanthanum	La	57
Lawrencium	Lw	103
Lead	Pb	82
Lithium	Li	3
Lutetium	Lu	71
Magnesium	Mg	12
Manganese	Mn	25
Mendelevium	Md	101
Mercury	Hg	80
Molybdenum	Mo	42
Neodymium	Nd	60
Neon	Ne	10
Neptunium	Np	93
Nickel	Ni	28
Niobium	Nb	41
Nitrogen	N	7
Nobelium	No	102
Osmium	Os	76
Oxygen	O	8
Palladium	Pd	46
Phosphorus	P	15
Platinum	Pt	78
Plutonium	Pu	94
Polonium	Po	84
Potassium	K	19
Praseodymium	Pr	59
Promethium	Pm	61
Protactinium	Pa	91
Radium	Ra	88
Radon	Rn	86
Rhenium	Re	75
Rhodium	Rh	45
Rubidium	Rb	37
Ruthenium	Ru	44
Samarium	Sm	62
Scandium	Sc	21
Selenium	Se	34
Silicon	Si	14
Silver	Ag	47
Sodium	Na	11
Strontium	Sr	38
Sulfur	S	16
Tantalum	Ta	73
Technetium	Tc	43
Tellurium	Te	52
Terbium	Tb	65
Thallium	Tl	81
Thorium	Th	90
Thulium	Tm	69
Tin	Sn	50
Titanium	Ti	22
Tungsten	W	74
Uranium	U	92
Vanadium	V	23
Xenon	Xe	54
Ytterbium	Yb	70
Yttrium	Y	39
Zinc	Zn	30
Zirconium	Zr	40

Relative Locations of the Products of Various Nuclear Processes

			He^3 in	α in
	β^- out	p in	d in	t in
	n out	**Original Nucleus**	n in	
t out	d out	p out	β^+ out ϵ	
α out	He^3 out			

n = neutron

p = proton

d = deuteron

t = triton (H^3)

α = alpha particle

β^- = negative electron

β^+ = positron

ϵ = electron capture

Displacements Caused by Nuclear Bombardment Reactions

α, 3n	α, 2n He^3, n	α, n	
p, n	p, γ d, n He^3, np	α, np t, n He^3, p	
γ, n n, 2n	**Original Nucleus**	d, p n, γ t, np	t, p
γ, np	γ, p	n, p	
n, α	n, He^3		

SYMBOLS

RADIATIONS AND DECAY

α	alpha particle
β^-	negative electron
β^+	positron
γ	gamma ray
n	neutron
p	proton
ε	electron capture
IT	isomeric transition
D	radiation delayed
SF	spontaneous fission
E	disintegration energy

TIME

μs	microseconds (10^{-6}s)
s	seconds
m	minutes
h	hours
d	days
y	years

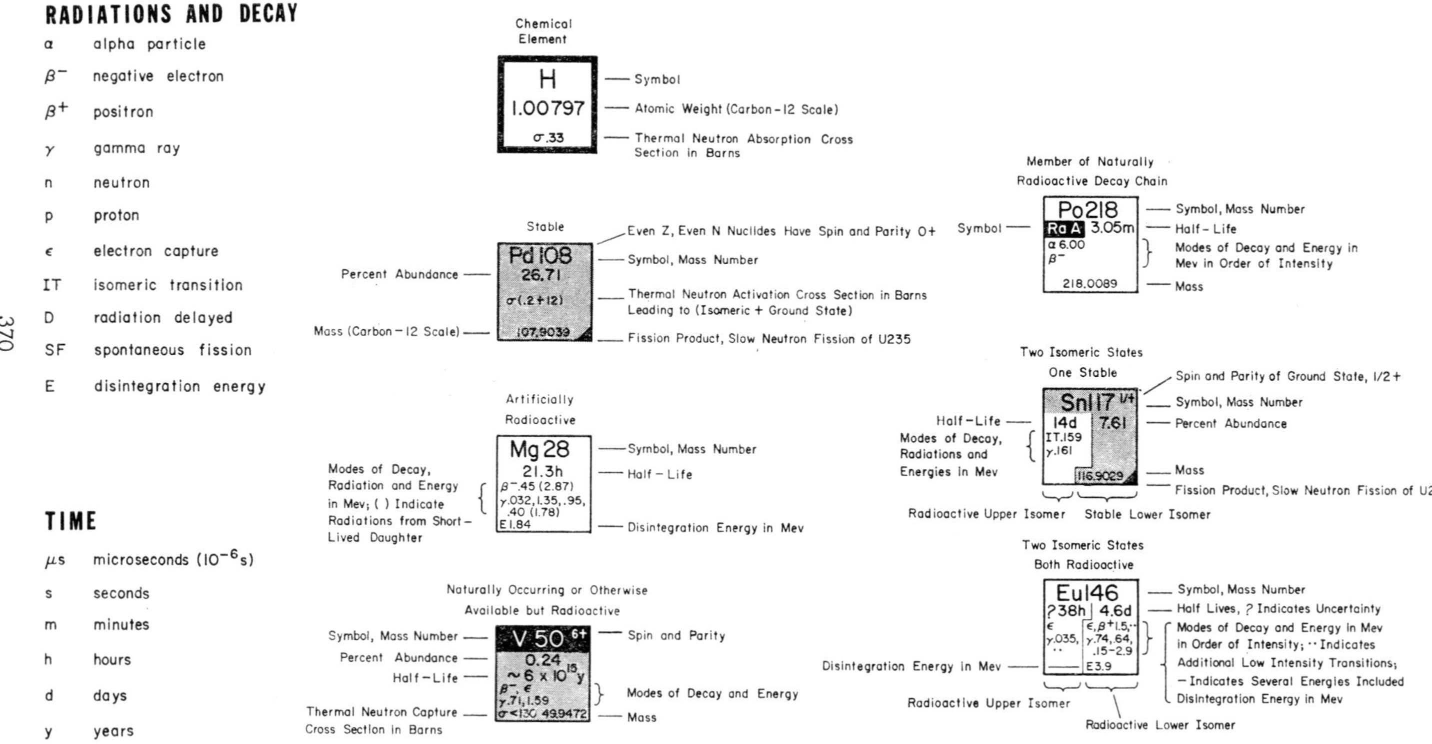

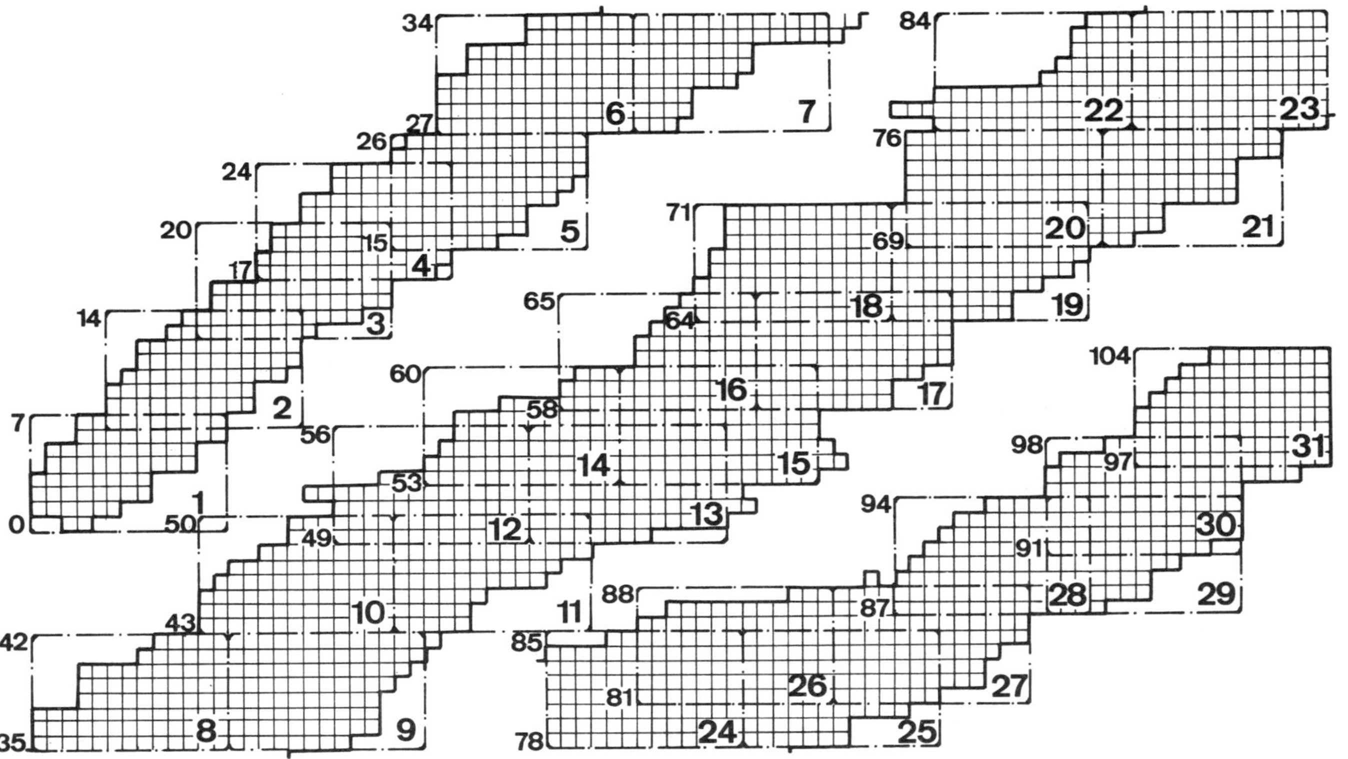
34
84
6
7
22
23
27
26
76
24
5
71
20
15
69
20
21
4
17
65
18
19
64
3
14
104
60
16
17
2
7
58
56
98
31
97
14
15
53
1
94
13
0
50
12
30
49
91
88
28
29
11
87
10
43
42
85
27
81
26
8
9
35
78
24
25

				N 14.0067 σ 1.9			N 12 1+ 0.011s β^+ 16.3 (3α ~4) E17.3	N 13 1/− 9.96m β^+ 1.19 E2.22	N 14 1+ 99.63 σ_{np} 1.8 σ .08 14.00307	N 15 1/− 0.37 σ .00002 15.00011	N 16 2− 7.1s β^- 4.3, 10.4, ·· γ 6.13, 7.12 (α 1.7) E10.4	N 17 4.14s β^- 3.3, 4.1, 2.7−8.7 (n 1.2, .43) γ .87, ·· E8.7
				C 12.01115 σ .0034	C 9 0.13s β^+ 3.5 (p 9.3, 12.3)	C 10 19s β^+ 1.9, ·· γ .72, 1.04 E 3.6	C 11 3/− 20.5m β^+ .96 no γ E1.98	C 12 98.89 σ .0034 12.00000	C 13 1/− 1.11 σ .0009 13.00335	C 14 5730y β^- .156 no γ $\sigma < 10^{-6}$ E.156	C 15 1/+ 2.25s β^- 4.5, 9.77 γ 5.30 E9.77	C 16 0.74s β^- (n) E8.0
		B 10.811 σ 759			B 8 0.78s β^+ 1.4 (2α 3) E18	B 9 $\geq 3 \times 10^{-19}$ s p, (2α) 9.01333	B 10 3+ 19.78 $\sigma_{n\alpha}$ 3840 σ .5 10.01294	B 11 3/− 80.22 σ .005 11.00931	B 12 1+ 0.020s β^- 13.4, 9.0 γ 4.4 (α .2) E13.4	B 13 3/− 0.019s β^- 13.4, 9.7 γ 3.68 E13.4		
		Be 9.0122 σ .009		Be 6 $\gtrsim 4 \times 10^{-21}$ s p, α, Li[5] 6.0197	Be 7 3/− 53d ε γ .48 σ_{np} 5400 E.86	Be 8 $\sim 3 \times 10^{-16}$ s 2α .09 8.00531	Be 9 3/− 100 σ .009 9.01219	Be 10 2.7×10^6 y β^- .56 no γ E.56	Be 11 13.6s β^- 11.5, 9.3, ·· γ 2.12, 6.8, 4.6−8.0 E11.5	Be 12 0.011s β^- (n)		
Li 6.939 σ 71				Li 5 3/− $\sim 10^{-21}$ s p, α 5.0125	Li 6 1+ 7.42 $\sigma_{n\alpha}$ 950 σ .045 6.01512	Li 7 3/− 92.58 σ .036 7.01600	Li 8 2+ 0.85s β^- 13 (2α 3) E16.0	Li 9 0.17s β^- 11, 13 (n, 2α) E13.6				
He 4.0026 σ .007			He 3 1/+ 0.00013 σ_{np} 5327 3.01603	He 4 ~100 σ 0 4.00260	He 5 3/− 2×10^{-21} s n, α 5.0123	He 6 0.81s β^- 3.51 no γ E3.51		He 8 0.122s β^- 10.5 γ .98 (n) E11.5				
H 1.00797 σ .33		H 1 1/+ 99.985 σ .33 1.007825	H 2 1+ 0.015 σ .0005 2.01410	H 3 1/+ 12.26y β^- .0186 no γ σ < .000007 E.0186								
			n 1 1/+ 12m β^- .782 E.782 1.008665									

1

Z: 0, 2, 4, 6

N: 0, 2, 4, 6, 8, 10

Z													
14						**Si** 28.086 σ .16		Si 25 0.23s β^+ (p 4.28, 3.46, 5.62,···)	Si 26 2s β^+ 3.8, 2.9 γ .82,·· E 5.1	Si 27 5/+ 4.2s β^+ 3.8, 1.5 γ .84, 1.01 E 4.81	Si 28 92.21 σ .08 27.97693	Si 29 1/+ 4.70 σ .28 28.97649	Si 30 3.09 σ .11 29.97376
					Al 26.9815 σ .23			Al 24 0.13s | 2.1s IT.44 | β^+ 8.8, 4.5,·· β^+ 13.3,·· | γ 1.37–7.1 E 13.9 (α ~2)	Al 25 5/+ 7.2s β^+ 3.24,·· γ .58 – 1.6 E 4.26	0+ Al 26 5+ 6.5s | 7.4x10^{5}y β^+ 3.21 | β^+ 1.16 no γ | ε γ 1.83, 1.12,·· E 4.00	Al 27 5/+ 100 σ .23 26.98154	Al 28 2.30m β^- 2.87 γ 1.78 E 4.64	Al 29 6.6m β^- 2.5, 1.4 γ 1.28, 2.43 E 3.7
12			**Mg** 24.312 σ .063			Mg 21 0.12s β^+ (p 3.44, 4.03, 4.81, 6.45··)	Mg 22 3.9s γ .074, .59 E 5.04	Mg 23 3/+ 12s β^+ 3.0,·· γ .44 E 4.06	Mg 24 78.70 σ .03 23.98504	Mg 25 5/+ 10.13 σ .27 24.98584	Mg 26 11.17 σ .03 25.98259	Mg 27 1/+ 9.5m β^- 1.75, 1.59 γ .84, 1.01,·· σ < .03 E 2.61	Mg 28 21.3h β^- .45, (2.87) γ .032, 1.35, .95, .40, (1.78) E 1.84
			Na 22.9898 σ .53			Na 20 0.4s β^+ (α 2.14, 2.49, 3.80, 4.44) E 15.3	Na 21 3/+ 23s β^+ 2.50,·· γ .35 E 3.52	Na 22 3+ 2.60y β^+ .54,··, ε γ 1.28 σ 9000 E 2.83	Na 23 3/+ 100 σ (.40 + .13) 22.98977	Na 24 4+ 0.02s | 15.0h IT.47 | β^- 1.39,·· γ 2.75, 1.37,·· E 5.51	Na 25 5/+ 60s β^- 3.8, 2.8,·· γ .98, .58, .40, 1.61,·· E 3.8	Na 26 1.0s β^- 6.7,·· γ 1.83,·· E 8.5	
10		**Ne** 20.183 σ .032		Ne 17 0.10s β^+ > 5 (p 2.3, 4.0, 4.8, 5.4, 7.4)	Ne 18 1.46s β^+ 3.42, 2.37 γ 1.04 E 4.4	Ne 19 1/+ 18s β^+ 2.23 E 3.24	Ne 20 90.92 19.99244	Ne 21 3/+ 0.257 20.99395	Ne 22 8.82 σ .036 21.99138	Ne 23 38s β^- 4.4, 3.9,·· γ .44, 1.65 E 4.4	Ne 24 3.38m β^- 1.98, 1.10 γ .47, .88 E 2.46		**16**
	F 18.9984 σ .010			F 16 ~10^{-19}s p .81 16.01171	F 17 66s β^+ 1.74 E 2.76	F 18 1+ 110m β^+ .65, ε E 1.66	F 19 1/+ 100 σ .010 18.99840	F 20 11s β^- 5.40 γ 1.63 E 7.03	F 21 4.4s β^- 4.0, 5.4,·· γ .35, 1.38 E 5.7	F 22 4s β^- 9.1, 11.2 γ 2.1, 1.3 E 12.5	**14**		
8	**O** 15.9994 σ < .0002	O 13 .0087s β^+ (p 6.9, 7.6)	O 14 71s β^+ 1.81, 4.14 γ 2.31 E 5.14	O 15 1/– 124s β^+ 1.74 no γ E 2.76	O 16 99.759 σ .0002 15.99492	O 17 5/+ 0.037 $\sigma_{n\alpha}$ .24 16.99913	O 18 0.204 σ .0002 17.99916	O 19 5/+ 29s β^- 3.25, 4.60 γ .20, 1.36,·· E 4.81	O 20 14s β^- 2.69 γ 1.06 E 3.8				
	N 14.0067 σ 1.9	N 12 1+ 0.011s β^+ 16.3 (3α ~4) E 17.3	N 13 1/– 9.96m β^+ 1.19 E 2.22	N 14 1+ 99.63 σ_{np} 1.8 σ .08 14.00307	N 15 1/– 0.37 σ .00002 15.00011	N 16 2– 7.1s β^- 4.3, 10.4,·· γ 6.13, 7.12 (α 1.7) E 10.4	N 17 4.14s β^- 3.3, 4.1, 2.7–8.7 (n 1.2, .43) γ .87,·· E 8.7	N 18 0.63s β^- 9.4 γ .82, 1.65, 1.98, 2.47 E 13.9	**12**				
			6		**8**		**10**						

20						Ca 40.08 σ.43		Ca 37 0.17s β^+ 5.1,·· γ (p 3.2,··)	Ca 38 0.7s γ 3.5 E7	Ca 39 $^{3/+}$ 0.9s β^+ 5.5 no γ E6.5	Ca 40 96.97 σ.2 39.96259	Ca 41 $^{7/-}$ 7.7×10^4 y ϵ E.41	Ca 42 0.64 σ 50 41.95863
						K 39.102 σ 2.1			K 37 $^{3/+}$ 1.2s β^+ 5.1, 2.3 γ 2.8 E6.1	$^{1+}$ K 38 $^{3+}$ 0.95s | 7.7m β^+ 5.1 | β^+ 2.7 γ 2.2 E6.1 | E5.9	K 39 $^{3/+}$ 93.10 σ 2.2 38.96371	K 40 $^{4-}$ 0.0118 1.3×10^9 y β^- 1.32, ϵ σ 70 γ 1.46 σ_{np} 4 E^- 1.32 E^+ 1.51	K 41 $^{3/+}$ 6.88 σ 1.1 40.96183
18				Ar 39.948 σ.63		Ar 33 0.18s β^+ (p 3.26, 3.9)		Ar 35 $^{3/+}$ 1.8s β^+ 4.96,·· γ 1.19, 1.73 E5.98	Ar 36 0.337 σ 6 $\sigma_{n\alpha}$ .006 35.96755	Ar 37 $^{3/+}$ 35.1 d ϵ E.81	Ar 38 0.063 σ.8 37.96273	Ar 39 $^{7/-}$ 270y β^-.57 no γ $\sigma \sim 1000$ E.57	Ar 40 99.60 $\sigma_{n\alpha}$ .63 39.96238
				Cl 35.453 σ 33		Cl 32 0.29s β^+ 9.5, 4.7, ··· γ 2.24, 4.78, 2.45, ··· ($\alpha \sim 3$) E13	Cl 33 $^{3/+}$ 2.5s β^+ 4.5 γ 2.8 E5.57	$^{3+}$ Cl 34 $^{0+}$ 32.0m | 1.5s β^+ 2.5, 1.3 | β^+ 4.5 IT .14 γ 2.04,·· | E5.5	Cl 35 $^{3/+}$ 75.53 σ 44 σ_{np} .4 34.96885	Cl 36 $^{2+}$ 3×10^5 y β^-.71, ϵ no γ σ 100 E^-.71 E^+ 1.14	Cl 37 $^{3/+}$ 24.47 σ(.003 + .43) 36.96590	Cl 38 $^{2-}$ 0.74s | 37.3m IT .66 | β^- 4.8, 1.1, 2.8 γ 2.2, 1.6 E4.9	Cl 39 $^{3/+}$ 55m β^- 1.91, 2.18, 3.44 γ 1.27, .25, 1.52 E3.44
16		S 32.064 σ.51		S 29 0.20s β^+ γ (p)	S 30 1.4s β^+ 4.42, 5.08 γ.67 E6.1	S 31 $^{1/+}$ 2.6s β^+ 4.4,·· γ 1.27 E5.4	S 32 95.0 $\sigma_{n\alpha}$ .004 31.97207	S 33 $^{3/+}$ 0.76 σ_{np} .002 $\sigma_{n\alpha}$ .2 32.97146	S 34 4.22 σ.27 33.96786	S 35 $^{3/+}$ 86.7d β^-.167 no γ E.167	S 36 0.014 σ.14 35.96709	S 37 5.1m β^- 1.6, 4.8 γ 3.1 E4.8	S 38 2.9h β^- 1.1,·· (4.8,··) γ 1.88 (2.2, 1.6) E3.0
		P 30.9738 σ.19		P 28 0.28s β^+ 11, ~8,·· γ 1.78, 2.6–7.6 E14	P 29 $^{1/+}$ 4.4s β^+ 3.95,·· γ 1.28, 2.43 E4.95	P 30 $^{1+}$ 2.5m β^+ 3.3,·· γ 2.24 E4.3	P 31 $^{1/+}$ 100 σ.19 30.97376	P 32 $^{1+}$ 14.3d β^- 1.71 no γ E1.71	P 33 25d β^-.25 no γ E.25	P 34 $^{1+}$ 12.4s β^- 5.1, 3.0 γ 2.1, 4.0 E5.1			
14	Si 28.086 σ.16	Si 25 0.23s β^+ (p 4.28, 3.46, 5.62,···)	Si 26 2s β^+ 3.8, 2.9 γ.82,·· E5.1	Si 27 $^{5/+}$ 4.2s β^+ 3.8, 1.5 γ.84, 1.01 E4.81	Si 28 92.21 σ.08 27.97693	Si 29 $^{1/+}$ 4.70 σ.28 28.97649	Si 30 3.09 σ.11 29.97376	Si 31 $^{3/+}$ 2.62h β^- 1.48,·· γ 1.27 E1.48	Si 32 ~700y β^-.1 (1.71) no γ E.1				
	Al 26.9815 σ.23	Al 24 0.13s | 2.1s IT .44 | β^+ 8.8, 4.5,·· β^+ 13.3,·· | γ 1.37–7.1 E13.9 | ($\alpha \sim 2$)	Al 25 $^{5/+}$ 7.2s β^+ 3.24,·· γ.58–1.6 E4.26	$^{0+}$ Al 26 $^{5+}$ 6.5s | 7.4×10^5 y β^+ 3.21 | β^+ 1.16 no γ | ϵ γ 1.83, 1.12,·· E4.00	Al 27 $^{5/+}$ 100 σ.23 26.98154	Al 28 2.30m β^- 2.87 γ 1.78 E4.64	Al 29 6.6m β^- 2.5, 1.4 γ 1.28, 2.43 E3.7	Al 30 3.3s β^- 5.05,·· γ 2.26, 3.52 E7.3					
			12		14		16		18		20		22

4

Z \ N	14	15	16	17	18	19	20	21	22	23	24	25	26
24						**Cr** 51.996 σ 3.1			Cr 46 1.1s β^+	?Cr 47 0.4s β^+	Cr 48 23h ϵ γ.30, .12 E1.4	Cr 49 42m β^+1.54, 1.39, .73 γ.09, .06, .15 E2.56	Cr 50 4.31 σ 17.0 49.94605
						V 50.942 σ 4.9			V 45 ~1s β^+	V 46 0.42s β^+6.05 E7.06	V 47 32m β^+1.94, ϵ γ1.54, 1.87 E2.92	V 48 16.1d β^+.70, ϵ γ.99, 1.31, 2.25 E4.01	V 49 7/− 330d ϵ noγ E.61
22				**Ti** 47.90 σ 6.1		Ti 41 0.091s β^+ (p 4.75, 2.37−5.43)		Ti 43 0.4s β^+5.6 E6.6	Ti 44 47y ϵ(β^+1.47) γ.079, .070, (1.16, ··) E.155	Ti 45 7/− 3.08h β^+1.02 ϵ γ E2.06	Ti 46 7.93 σ.6 45.95263	Ti 47 5/− 7.28 σ1.7 46.95177	Ti 48 73.94 σ 8.0 47.94795
				Sc 44.956 σ 23		Sc 40 4− 0.18s β^+9.2 γ3.73, .73, ··· E14	Sc 41 7/− 0.59s β^+5.5 E6.5	Sc 42 62s: β^+2.82; γ1.52, 1.23, .44; E7.13 0.66s: β^+5.38; E6.40	Sc 43 7/− 3.9h β^+1.19, .82 γ.37 E2.22	6+ Sc 44 2+ 2.4d: IT.271 4.0h: β^+1.47, ϵ; γ1.16, .68−2.7; E3.65	Sc 45 7/− 100 σ(10 + 13) 44.95592	Sc 46 4+ 20s: IT.14; ϵ 83.8d: β^-.36, ··; γ1.12, .89, ··; σ.25; E2.37	Sc 47 3.4d β^-.44, .60 γ.16 E.60
20		**Ca** 40.08 σ.43		Ca 37 0.17s β^+5.1, ·· γ (p 3.2, ··)	Ca 38 0.7s γ3.5 E7	Ca 39 3/+ 0.9s β^+5.5 noγ E6.5	Ca 40 96.97 σ.2 39.96259	Ca 41 7/− 7.7 x 10^4y ϵ E.41	Ca 42 0.64 σ 50 41.95863	Ca 43 7/− 0.145 42.95878	Ca 44 2.06 σ.7 43.95549	Ca 45 163d β^-.25 noγ E.25	Ca 46 0.0033 σ.3 45.9537
		K 39.102 σ 2.1			K 37 3/+ 1.2s β^+5.1, 2.3 γ 2.8 E6.1	1+ K 38 3+ 0.95s: β^+5.1; E6.1 7.7m: β^+2.7; γ2.2; E5.9	K 39 3/+ 93.10 σ 2.2 38.96371	K 40 4− 0.0118 1.3 x 10^9y β^-1.32, ϵ σ 70 γ1.46 σ_{np}4 E^-1.32 E^+1.51	K 41 3/+ 6.88 σ 1.1 40.96183	K 42 2− 12.4h β^-3.53, 2.01 γ1.52, ·· E3.53	K 43 3/+ 22h β^-.83, .46−1.82 γ.37, .61, .22−1.00 E1.82	K 44 22m β^-4.91, ·· γ1.16, 2.1, .48−5.0 E6.1	K 45 20m β^-2.1, 1.1, 4.0 γ.18, 1.7, ·· E4.2
18	**Ar** 39.948 σ.63	Ar 33 0.18s β^+ (p 3.26, 3.9)		Ar 35 3/+ 1.8s β^+4.96, ·· γ.19, 1.73 E5.98	Ar 36 0.337 σ 6 $\sigma_{n\alpha}$.006 35.96755	Ar 37 3/+ 35.1d ϵ E.81	Ar 38 0.063 σ.8 37.96273	Ar 39 7/− 270y β^-.57 noγ σ ~ 1000 E.57	Ar 40 99.60 $\sigma_{n\alpha}$.63 39.96238	Ar 41 7/− 1.83h β^-1.20, 2.49 γ1.29 σ.5 E2.49	Ar 42 33y β^-(3.5, 2.0) (γ1.52) E.6		
	Cl 35.453 σ 33	Cl 32 0.29s β^+9.5, 4.7, ··· γ2.24, 4.78, 2.45, ··· (α ~ 3) E13	Cl 33 3/+ 2.5s β^+4.5 γ2.8 E5.57	3+ Cl 34 0+ 32.0m: β^+2.5, 1.3; IT.14; γ2.04, ·· 1.5s: β^+4.5; E5.5	Cl 35 3/+ 75.53 σ 44 σ_{np}.4 34.96885	Cl 36 2+ 3 x 10^5y β^-.71, ϵ noγ σ 100 E^-.71 E^+1.14	Cl 37 3/+ 24.47 σ(.003 + .43) 36.96590	Cl 38 2− 0.74s: IT.66 37.3m: β^-4.8, 1.1, 2.8; γ2.2, 1.6; E4.9	Cl 39 3/+ 55m β^-1.91, 2.18, 3.44 γ1.27, .25, 1.52 E3.44	Cl 40 1.4m β^-3.2, 7.5 γ2.75, 1.46, 6.0 E7.5			
			16		18		20		22		24		26

5

26		**Fe** 55.847 σ 2.53		Fe 52 8.3h β+.80,(2.63), ε γ.17,.38D E2.37	Fe 53 8.5m β+~2.8,2.4,1.6 γ.38,·· E4.0	Fe 54 5.82 σ 2.9 53.93962	Fe 55 3/− 2.4y ε no γ E.231	Fe 56 91.66 σ 2.5 55.93494	Fe 57 1/− 2.19 σ 2.5 56.93540	Fe 58 0.33 σ 1.2 57.93328	Fe 59 3/− 45d β−.46,.27,1.57 γ1.10,1.29,.19,·· E1.57	Fe 60 ~10^5y β−.14(.3,··) γ.027(1.33,1.17,.059) E.14	Fe 61 6.0m β− 2.8 γ.29,·· E4
	Mn 54.9380 σ 13.3		Mn 50 0+ 2m β+ γ.66−1.45 | 0.29s β+6.6 E7.6	Mn 51 45m β+2.2 γ.74,1.17 E3.2	2+ Mn 52 6+ 21m β+2.63 IT.38 γ1.43 | 5.7d ε,β+.58 γ1.43,.94,.74,·· E4.71	Mn 53 7/− ~2 x 10^6y ε no γ σ~170 E.60	Mn 54 3+ 312 d ε γ.84 E1.38	Mn 55 5/− 100 σ 13.3 54.93805	Mn 56 3+ 2.58h β−2.85,1.04,.73,·· γ.845,1.81,2.11,·· E3.70	Mn 57 1.7m β−2.6,1.1 γ1.22,.014,.137,.22−.71 E2.7	Mn 58 1.1m β− E6.5		
24	?Cr 47 0.4s β+	Cr 48 23h ε γ.30,.12 E1.4	Cr 49 42m β+1.54,1.39,.73 γ.09,.06,.15 E2.56	Cr 50 4.31 σ 17.0 49.94605	Cr 51 7/− 27.8d ε γ.32 E.75	Cr 52 83.76 σ.8 51.94051	Cr 53 3/− 9.55 σ 18 52.94065	Cr 54 2.38 σ.38 53.93888	Cr 55 3/− 3.5m β−2.59 no γ E2.59	Cr 56 5.9m β−1.5 γ.083,.026 E1.6			
	V 46 0.42s β+6.05 E7.06	V 47 32m β+1.94, ε γ1.54,1.87 E2.92	V 48 16.1d β+.70, ε γ.99,1.31,2.25 E4.01	V 49 7/− 330d ε no γ E.61	V 50 6+ 0.24 ~6 x 10^{15}y β−, ε γ.71,1.59 σ<130 E−1.03 E+2.21 49.94716	V 51 7/− 99.76 σ 4.9 50.94396	V 52 3.77m β−2.47,·· γ1.43,·· E3.97	V 53 2.0m β−2.53 γ1.01,·· E3.54	V 54 55s β−3.3 γ.99,.84,2.21 E7				
22	Ti 45 7/− 3.08h β+1.02 ε γ E2.06	Ti 46 7.93 σ.6 45.95263	Ti 47 5/− 7.28 σ1.7 46.95177	Ti 48 73.94 σ 8.0 47.94795	Ti 49 7/− 5.51 σ 1.9 48.94787	Ti 50 5.34 σ.14 49.94479	Ti 51 7/− 5.80m β−2.13,1.52 γ.32,.93,.61 E2.46	Ti 52 β− E1.97				**34**	
	6+ Sc 44 2+ 2.4d IT.271 | 4.0h β+1.47, ε γ1.16,.68−2.7 E3.65	Sc 45 7/− 100 σ(10+13) 44.95592	Sc 46 4+ 20s IT.14 ε | 83.8d β−.36,·· γ1.12,.89,·· σ.25 E2.37	Sc 47 3.4d β−.44,.60 γ.16 E.60	Sc 48 6+ 44h β−.65,.48 γ1.31,1.04,.99,.17 E3.98	Sc 49 57.5m β−2.01,·· γ1.76 E2.01	Sc 50 0.35s IT.26 | 1.8m β−3.6 γ.52,1.56,1.12 E6.5	Sc 51 β− γ		**32**			
20	Ca 43 7/− 0.145 42.95878	Ca 44 2.06 σ.7 43.95549	Ca 45 163d β−.25 no γ E.25	Ca 46 0.0033 σ.3 45.9537	Ca 47 4.5d β−.67,1.98 γ1.31,·· E1.98	Ca 48 0.18 σ1.1 47.95253	Ca 49 8.8m β−2.0,1.0 γ3.10,4.05,4.7 E5.26	Ca 50 9s β− γ.07,(.26) E4.9					
	K 42 2− 12.4h β−3.53,2.01 γ1.52,·· E3.53	K 43 3/+ 22h β−.83,.46−1.82 γ.37,.61,.22−1.00 E1.82	K 44 22m β−4.91,·· γ1.16,2.1,.48−5.0 E6.1	K 45 20m β−2.1,1.1,4.0 γ.18,1.7,·· E4.2	K 46	K 47 18s β−4.1,3.5,·· γ2.0,2.6 E6.64		**30**					
		24		**26**		**28**							

Z			28		30		32		34		36		38
34							Se 78.96 σ 12				Se 70 44m β+	Se 71 5m β+3.4 γ.16 E4.4	Se 72 8.4d ε (β+2.50,3.34,··) γ.046 (.84,.63,··) E~.6
							As 74.9216 σ 4.5			As 68 ~7m β+	As 69 15m β+2.9 γ.23 E3.9	As 70 50m β+1.4,2.5 ε γ1.04,2.0,.18–1.7 E6.24	As 71 5/– 62h ε β+.81 γ.175,.023 E2.00
32			Ge 72.59 σ 2.4					Ge 65 1.5m β+3.7,·· γ.67,1.72 E6.4	Ge 66 2.4h β+1.3,2.0 ε γ.38,.046,.068–.71 E3.0	Ge 67 19m β+3.2,2.3,·· ε γ.17,.92,.34–3.4 E4.40	Ge 68 280d ε noγ E.7	Ge 69 37h ε,β+1.21,.61,·· γ1.12,.58,.88,.09–2.00 E2.23	Ge 70 20.52 σ(.3 + 3.2) 69.92425
			Ga 69.72 σ 3.0				Ga 63 35s E5.5	Ga 64 2.6m β+6.0,2.8,·· γ.98,3.3,.8–3.0 E7.07	Ga 65 15m β+2.11,1.39,2.24,.82,·· ε γ.12,.054–1.86 E3.26	Ga 66 0+ 9.5h β+4.15,·· ε γ1.04,2.75,.83–4.8 E5.17	Ga 67 3/– 78h ε γ.092D,.182,.30,.090–.87 E1.00	Ga 68 1+ 68m β+1.90,·· ε γ1.08,.81–1.88 E2.92	Ga 69 3/– 60.4 σ 1.9 68.92556
30	Zn 65.37 σ 1.10				Zn 60 2.1m	Zn 61 89s β+4.38,·· γ.48,1.64,.98,.69 E5.4	Zn 62 9.3h ε,β+.67,·· γ.042,.59,·· E1.69	Zn 63 3/– 38m β+2.35,·· ε γ.67,.97,.81–2.9 E3.37	Zn 64 48.89 σ.46 63.92914	Zn 65 5/– 243d ε,β+.33 γ1.12 E1.35	Zn 66 27.81 65.92605	Zn 67 5/– 4.11 66.92714	Zn 68 18.57 σ(.1 + 1.0) 67.92486
	Cu 63.54 σ 3.8			Cu 58 3.2s β+7.5,4.6 γ1.45,2.9 E8.57	Cu 59 81s β+3.78,·· γ1.30,.87,.46,.34 E4.80	Cu 60 2+ 24m β+2.00,3.00,3.92,·· ε γ1.33,1.76,.82,2.15–4.0 E6.13	Cu 61 3/– 3.3h β+1.22,.94,·· ε γ.28,.66,.068,·· E2.24	Cu 62 1+ 9.9m β+2.91,·· ε γ E3.93	Cu 63 3/– 69.09 σ 4.5 62.92959	Cu 64 1+ 12.9h ε γ1.34 β–.57 β+.66 E–.57 E+1.68	Cu 65 3/– 30.91 σ 2.3 64.92779	Cu 66 1+ 5.1m β–2.63,1.59,·· γ1.04,.83 σ130 E2.63	Cu 67 61h β–.40,.48,.57 γ.182,.090,.092D,·· E.57
28	Ni 58.71 σ 4.6		Ni 56 6.1d ε γ.16,.82,.76,.48,.27,1.57,.98 E2.1	Ni 57 36h ε,β+.84,.71,·· γ1.38,.127,1.92,1.76,·· E3.23	Ni 58 67.88 σ 4.4 57.93534	Ni 59 3/– 8 x 10^{4}y ε noγ E1.07	Ni 60 26.23 σ 2.6 59.93079	Ni 61 3/– 1.19 σ 2 60.93106	Ni 62 3.66 σ 15 61.92834	Ni 63 92y β–.067 noγ E.067	Ni 64 1.08 σ 1.5 63.92796	Ni 65 2.56h β–2.13,.6,1.0,·· γ1.49,1.12,.37,·· σ 20 E2.13	Ni 66 55h β–.20 noγ E.20
	Co 58.9332 σ 37	Co 54 0+ 0.19s β+7.23 E8.25	Co 55 18h β+1.51,1.04,·· ε γ.93,1.41,.48 E3.46	Co 56 4+ 77.3d ε,β+1.46,·· γ.85,1.24,.73–3.45 E4.57	Co 57 7/– 272d ε γ.122,.014,.137,·· E.84	Co 58 5+ 9h IT.025 σ 140000 2+ 71d ε,β+.48,·· γ.81,1.65 σ ~ 2500 E2.31	Co 59 7/– 100 σ(18 + 19) 58.93319	Co 60 2+ 10.5m IT.059 β–1.54,·· γ1.33,·· σ~100 5+ 5.24y β–.31,·· γ1.33,1.17,·· σ 6 E2.82	Co 61 1.65h β–1.22 γ.068 E1.29	Co 62 1.9m β– γ 13.9m β–2.88,.88 γ1.17,1.47,1.74,2.03,·· E5.22	Co 63 1.4h 52s β–3.6 noγ	Co 64 2m IT γ 7.8m β– γ	

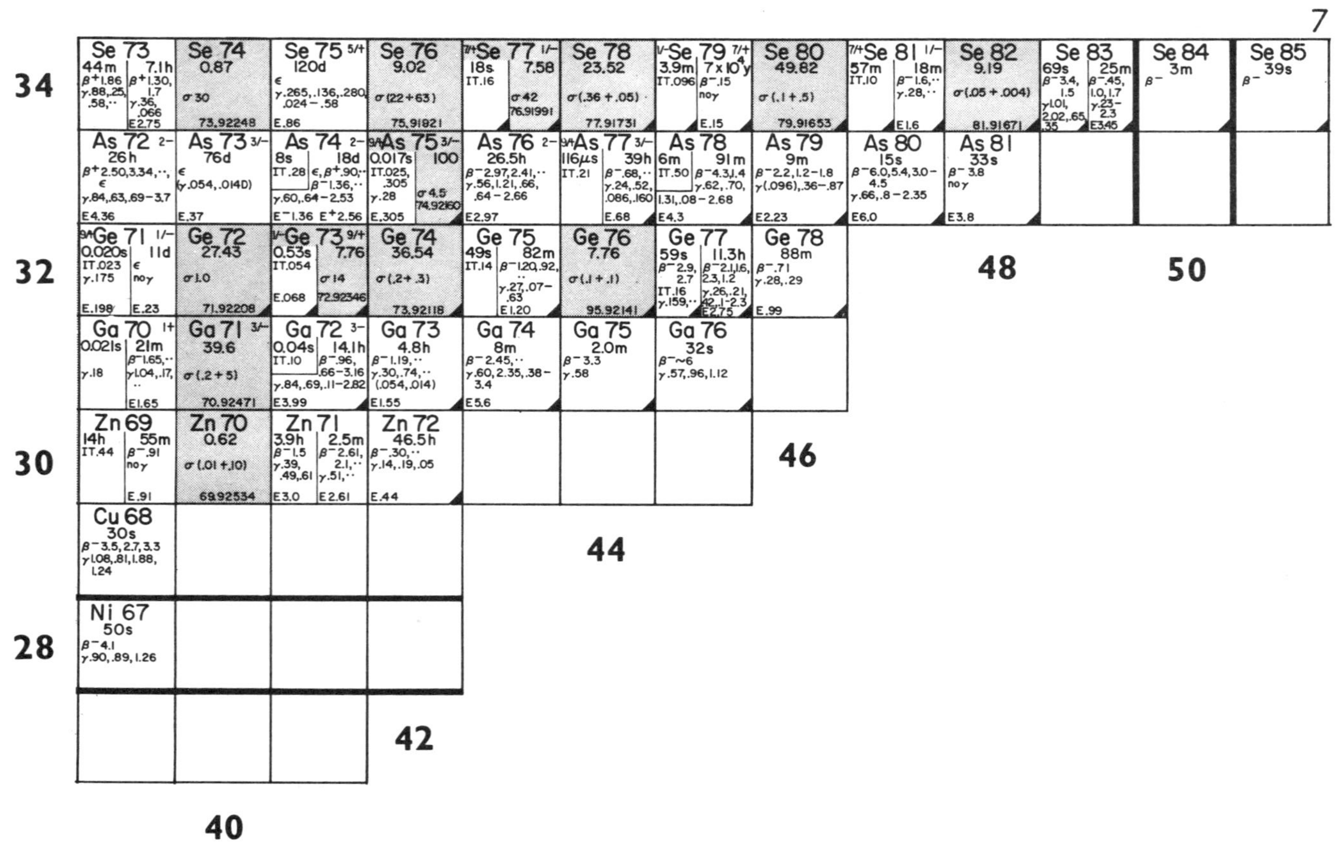

7

Z													
34	Se 73 44m: β^+1.86, γ.88,.25, .58,·· 7.1h: β^+1.30, 1.7, γ.36, .066 E2.75	Se 74 0.87 σ 30 73.92248	Se 75 5/+ 120d ε γ.265,.136,.280, .024 – .58 E.86	Se 76 9.02 σ (22+63) 75.91821	7/+ Se 77 1/– 18s: IT.16 7.58: σ 42 76.91991	Se 78 23.52 σ (.36 + .05) 77.91731	1/– Se 79 7/+ 3.9m: IT.096 7 x 10⁴y: β^-.15, no γ E.15	Se 80 49.82 σ (.1 + .5) 79.91653	7/+ Se 81 1/– 57m: IT.10 18m: β^-1.6,··, γ.28,·· E1.6	Se 82 9.19 σ (.05 + .004) 81.91671	Se 83 69s: β^-3.4, 1.5, γ1.01, 2.02,.65, .35 25m: β^-.45, 1.0,1.7, γ.23–2.3 E3.45	Se 84 3m β^-	Se 85 39s β^-
	As 72 2– 26h β^+2.50,3.34,··, ε γ.84,.63,.69–3.7 E4.36	As 73 3/– 76d ε (γ.054,.0140) E.37	As 74 2– 8s: IT.28 18d: ε, β^+.90,··, β^-1.36,·· γ.60,.64–2.53 E^-1.36 E^+2.56	9/+ As 75 3/– 0.017s: IT.025, .305, γ.28 100: σ 4.5 74.92160 E.305	As 76 2– 26.5h β^-2.97,2.41,·· γ.56,1.21,.66, .64–2.66 E2.97	9/+ As 77 3/– 116μs: IT.21 39h: β^-.68,··, γ.24,.52, .086,.160 E.68	As 78 6m: IT.50 91m: β^-4.3,1.4, γ.62,.70, 1.31,.08–2.68 E4.3	As 79 9m β^-2.2,1.2–1.8 γ(.096),.36–.87 E2.23	As 80 15s β^-6.0,5.4,3.0–4.5 γ.66,.8–2.35 E6.0	As 81 33s β^-3.8 no γ E3.8			
32	9/+ Ge 71 1/– 0.020s: IT.023, γ.175 E.198 11d: ε, no γ E.23	Ge 72 27.43 σ 1.0 71.92208	1/– Ge 73 9/+ 0.53s: IT.054 E.068 7.76: σ 14 72.92346	Ge 74 36.54 σ (.2+.3) 73.92118	Ge 75 49s: IT.14 82m: β^-1.20,.92,··, γ.27,.07–.63 E1.20	Ge 76 7.76 σ (.1 + .1) 95.92141	Ge 77 59s: β^-2.9, 2.7, IT.16, γ.159,·· 11.3h: β^-2.1,1.6, 2.3,1.2, γ.26,.21, .42,.1–2.3 E2.75	Ge 78 88m β^-.71 γ.28,.29 E.99					
	Ga 70 1+ 0.021s: γ.18 21m: β^-1.65,··, γ1.04,.17,·· E1.65	Ga 71 3/– 39.6 σ (.2 + 5) 70.92471	Ga 72 3– 0.04s: IT.10 14.1h: β^-.96, .66–3.16 γ.84,.69,.11–2.82 E3.99	Ga 73 4.8h β^-1.19,·· γ.30,.74,·· (.054,.014) E1.55	Ga 74 8m β^-2.45,·· γ.60, 2.35,.38–3.4 E5.6	Ga 75 2.0m β^-3.3 γ.58	Ga 76 32s β^-~6 γ.57,.96,1.12						
30	Zn 69 14h: IT.44 55m: β^-.91, no γ E.91	Zn 70 0.62 σ (.01 +.10) 69.92534	Zn 71 3.9h: β^-1.5, γ.39, .49,.61 E3.0 2.5m: β^-2.61, 2.1,··, γ.51,·· E2.61	Zn 72 46.5h β^-.30,·· γ.14,.19,.05 E.44									
	Cu 68 30s β^-3.5,2.7,3.3 γ1.08,.81,1.88, 1.24												
28	Ni 67 50s β^-4.1 γ.90,.89,1.26												
N		40		42		44		46		48		50	

8

Z													
42									Mo 95.94 σ 2.6	Mo 88 27m β+ 2.5 γ 2.69	Mo 89 7m β+ 4.05, 4.95	Mo 90 5.7h ε, β+ 1.2 γ (.25, .12) E 2.5	Mo 91 65s: IT.65, β+ 2.45, 2.78, 3.99, ε, γ 1.54, 1.21 15.5m: β+ 3.44, ε, no γ, E 4.46
								Nb 92.906 σ 1.1			Nb 88 21m β+ 3.2 γ .20, .72, .97, 1.42 E7	Nb 89 42m: β+ (γ .59) 1.9h: β+ 2.9, E 3.9	Nb 90 24s: IT.12 0.010s: IT.25 14.6h: β+ 1.50, ‥ γ 1.14, 2.32D, E 6.12
40				Zr 91.22 σ .18	Zr 81 11m	Zr 82 7.5m	Zr 83 7m	Zr 84 16m	Zr 85 6m	Zr 86 16.5h ε (β+ 1.19, ‥) γ .24, .03, ‥ (1.08, 1.16) E~'	Zr 87 1.6h β+ 2.10, ‥ γ 1.2, ‥ (.4) E 3.50	Zr 88 85d ε γ .39D	Zr 89 4.2m: IT.59, ε, β+ .9, 2.4, γ 1.5 78.4h: ε, β+ .90, γ (.91), 1.7, ‥ E 2.84
				Y 88.906 σ 1.3			Y 82 9m β+ 2	Y 83 7.4m	Y 84 42m β+ 2.5, 3.5, ‥ γ .80, .98, 1.04, ‥ E 6.3	Y 85 2.7h: β+ 1.5, ‥, ε, γ .50, (.23), ‥ 5.0h: β+ 2.2, 2.0, ‥ ε, γ .23–2.5, E 3.26	8+ Y 86 4− 49m: IT.01, γ .21 15h: ε, β+ 1.19, .6–3.1, γ 1.08, 1.16, .18–4.9, E 5.27	Y 87 14h: IT.38 80h: ε, β+ .7, γ .48, (.39), E 1.7	1+ Y 88 4− 300μs: IT.39 0.014s: IT.24, γ .45 107d: ε, β+ .78, γ 1.84, .90, ‥ E 3.62
38				Sr 87.62 σ 1.3		Sr 80 1.7h ε γ .58	Sr 81 29m β+	Sr 82 25d ε (β+ 3.2) (γ .78, 1.4)	Sr 83 33h ε, β+ 1.15, .8, ‥ γ .755, .385, .040	Sr 84 0.56 σ (.65 + .8) 83.91343	Sr 85 70m: IT.008, .23, γ .23, ε, γ .15 65d: ε, γ .514D, ‥ E 1.1	Sr 86 9.86 σ (1.3 + ?) 85.90928	1/− Sr 87 9/+ 2.8h: IT.39, ε 7.02 86.90889
	Rb 85.47 σ .73					Rb 79 21m β+ γ .15, .19	Rb 80 1+ 34s β+ 4.1, ε γ .62 E 5.1	9/+ Rb 81 3/− 32m: β+ 1.4, IT.085 4.7h: ε, β+ 1.03, ‥ γ .19, .45, .25, 1.1, E 2.24	5− Rb 82 1+ 6.3h: ε, β+ .8, γ .78, .62, .55, .70–1.47 75s: β+ 3.15, ‥ γ .78, 1.4, E 4.17	Rb 83 5/− 83d ε γ .52, .53, .55, .009, (.032, .009)	6+ Rb 84 2− 20m: IT.22, .46, γ .24, ε, γ .88 33d: ε, β+ .8, 1.6, γ .88, 1.01, 1.90, β− .90, E− .90, E+ 2.68	Rb 85 5/− 72.15 σ (.1 + .9) 84.91180	Rb 86 2− 1.0m: IT.56 18.7d: β− 1.77, .7, ‥ γ 1.08, ‥ E 1.77
36	Kr 83.80 σ 24	Kr 74 15m β+ 3.1 no γ E 4.1	Kr 75 5m E5	Kr 76 14.8h ε γ .040–.65 E~1	Kr 77 5/− 1.2h β+ 1.86, 1.67, ‥, ε γ .108, .131, .024, .149–.87 E 2.88	Kr 78 0.35 σ (? + 2) 77.92040	7/+ Kr 79 1/− 55s: IT.127 34h: ε, β+ .60, .34, γ .261, .606, .398, .14–1.3, E 1.62	Kr 80 2.27 σ 15 79.91638	1/− Kr 81 7/+ 13s: IT.19 2.1x10^5 y: ε, no γ, E .25	Kr 82 11.56 σ (3 + 42) 81.91348	1/− Kr 83 9/+ 1.86h: IT.032, γ .009, E .041 11.55: σ 180 82.91413	Kr 84 56.90 σ (.10 + .04) 83.91150	1/− Kr 85 9/+ 4.4h: β− .83, γ .15, IT.31 10.76y: β− .67, ‥ γ .52, E .67, σ < 15
	Br 79.909 σ 6.7	Br < 74 4m	Br 74 42m β+ γ .64 E 6.8	Br 75 1.6h β+ 1.70, 1.10, .65 ε, γ .285, .62 E 2.72	Br 76 1+ 16.5h β+ 3.58, .92, ‥, ε γ .56, .65, 1.21, .75–4.4 E 4.6	Br 77 3/− 4.2m: IT.107 58h: ε, β+ .34, γ .24, .52, .09–1.02, E 1.36	3− Br 78 1+ 120μs: IT.149 6.5m: β+ 2.55, ‥ γ .62, E 3.57	9/+ Br 79 3/− 4.8s: IT.21 50.54: σ (2.9 + 8.5) 78.91833	5− Br 80 1+ 4.5h: IT.049, γ .037, E .087 18m: β− 2.00, 1.38, γ .62, ‥ ε, β+ .87, γ .65 E+ 1.87 E− 2.01	9/+ Br 81 3/− 37μs: IT.27, γ .28, E .55 49.46: σ (3 + .2) 80.91629	Br 82 5− 6.1m: IT.046, β−, γ .78, 1.48 35.3h: β− .44, γ .78, .55, .62, .70–1.48, E 3.09	Br 83 2.4h β− .94, ‥ γ .051 (γ .032, .009) E 1.0	Br 84 6m: β− 1.9, .8, 3.2, γ .88, 1.46, ‥ 32m: β− 4.7, 2.8, ‥ γ .88, 2.1, 1.9, .27–3.9, E 4.70
		38		40		42		44		46		48	

Z \ N	50	51	52	53	54	55	56	57	58	59	60	61	62
42	Mo 92 15.84 σ(<.006+<.3) 91.90681	Mo 93 6.9h | ~10^4y IT.26 | ϵ γ.69, 1.48 E.42	Mo 94 9.04 93.90509	Mo 95 5/+ 15.72 σ 14 94.90584	Mo 96 16.53 σ 1 95.90467	Mo 97 5/+ 9.46 σ 2 96.90602	Mo 98 23.78 σ.51 97.90541	Mo 99 67h β^-1.23,.45,·· γ(.14,··),.74,.041–.95 E1.37	Mo 100 9.63 σ.2 99.90747	Mo 101 14.6m β^-2.23,·· γ1.02,.59,2.08,.08–1.66 E2.82	Mo 102 11m β^-1.2 (4.1) E1.2		Mo 104 1.6m β^-2.2,·· γ.89,.36
41	Nb 91 62d | long IT.105 | ϵ ϵ,γ1.21 E1.1	Nb 92 10.2d | ~10^7y ϵ,β^+ | ϵ γ.90,.93,1.84 | γ.56,.93 E2.03 | E2.01	Nb 93 9/+ 3.7y | 100 IT.029 | σ(1+.1) 92.90638	Nb 94 6+ 6.3m | 2.0x10^4y IT .042 β^-1.2, γ.87,·· | β^-.5,·· γ.87,.70 E2.1 σ~15	Nb 95 90h | 35d IT.23 | β^-.16,·· γ.77 σ~7 E.92	Nb 96 23h β^-.75,.57 γ.77,.56,1.08,.22–1.19 E3.1	Nb 97 1m | 72m IT.75 | β^-1.27,·· γ.66,·· E1.93	Nb 98 51.5m β^-3.1,·· γ.78,.72,.33–2.7 E4.6	Nb 99 10s | 2.5m β^- IT | β^-3.2,·· γ.10,.26 E3.2	Nb 100 ?11.5m | 3m β^-4.2 γ.53,·· | β^- γ.53,.36,.45,.14–2.9 E6	Nb 101 1m β^-		
40	5- Zr 90 0.8s | 51.46 IT2.32,.14 γ2.18 | σ.1 89.90470	Zr 91 5/+ 11.23 σ 1 90.90564	Zr 92 17.11 σ.2 91.90503	Zr 93 5/+ 9.5 x 10^5 y β^-.063,.034 (γ.029) σ < 4 E.063	Zr 94 17.40 σ.08 93.90631	Zr 95 65d β^-.40,.36,.89,(.16) γ.72,.76,(.23,.77) E1.12	Zr 96 2.80 σ.05 95.90829	Zr 97 1/+ 17h β^-1.91,··,(1.27,··) γ.5–2.6(.75) E2.66		Zr 99 < 1.6s β^-			
39	Y 89 1/- 16s | 100 IT.91 | σ(.001+1.3) 88.90587	Y 90 2- 3.2h | 64.2h IT.48 γ.20 β^- 62 γ2.3 | β^-2.28 γ1.75 E2.28	9/+ Y 91 1/- 50m | 59d IT.55 | β^-1.54,·· γ1.21 σ1.4 E1.54	Y 92 2- 3.53h β^-3.63,1.32,1.59–2.71 γ.932,1.39,.56,.448–2.4 E3.63	Y 93 10.1h β^-2.89,·· γ.27,.94,.38–2.4 E2.89	Y 94 20m β^-5.0,·· γ.92,.56,1.13,1.65–3.5 E5.0	Y 95 11m β^- γ.95,2.18,1.32,.43–3.58 E4	Y 96 2.3m β^-3.5 γ.7,1.0,·· E7	Y97 6s				
38	Sr 88 82.56 σ.006 87.90564	Sr 89 5/+ 50.6d β^-1.46 no γ σ.4 E1.46	Sr 90 28.8y β^-.54,(2.28) no γ σ 1 E.54	Sr 91 5/+ 9.7h β^-1.09,1.36,2.67,·· γ(.55),.65–1.65 E2.67	Sr 92 2.7h β^-.545,1.5 γ1.37,.44,.23 E1.92	Sr 93 8.3m β^-2.9,2.6,2.2 γ.6,.8,.31–2.1 E4.8	Sr 94 1.2m β^-2.1 γ1.42 E3.5	Sr 95 0.8m β^-					
37	Rb 87 3/- 27.85 5.2x10^{10}y β^-.27 no γ σ.12 E.27	Rb 88 2- 18m β^-5.2,3.3,2 γ1.85,.91,2.7,1.39–4.9 E5.2 σ1.0	Rb 89 15m β^-3.9,2.8,·· γ1.05,1.26,.66,2.20,1.55–3.5 E3.9	Rb 90 2.9m β^-6.6,5.8,2.2,·· γ.84,.53,–5.23 E6.6	Rb 91 72s β^-4.6 γ.095,.35 E5.5	Rb 92 5s β^- E8	Rb 93 6s β^-	Rb 94 3s β^-	Rb 95 < 2.5s β^-				
36	Kr 86 17.37 σ.06 85.91062	Kr 87 76m β^-3.8,1.3,3.3 γ.40,2.57,.85,·· E3.9 σ<600	Kr 88 2.8h β^-.52,2.7,·· γ2.4,.19,.85,.028–2.2 E2.9	Kr 89 3.2m β^-4.0,~2 γ.60,.22,.38–1.52 E4.6	Kr 90 33s β^-2.8,·· γ.120,.54,1.11,1.54,.11–3.6 E4.56	Kr 91 10s β^-3.6,·· γ	Kr 92 3s β^-	Kr 93 2s β^-	Kr 94 1s β^-	Kr 95 Short β^-			
35	Br 85 3.0m β^-2.5 γ.78,.90 E2.8	Br 86 54s β^-3,5,7.1 γ1.6, 1.4.2.8 E7.1	Br 87 55s β^-2.6,8.0 γ1.4,2.6,4.2,··· (n.3) E8.0	Br 88 16s β^- γ.76 (n)	Br 89 4.5s β^- (n.5)	Br 90 1.6s β^- (n)							

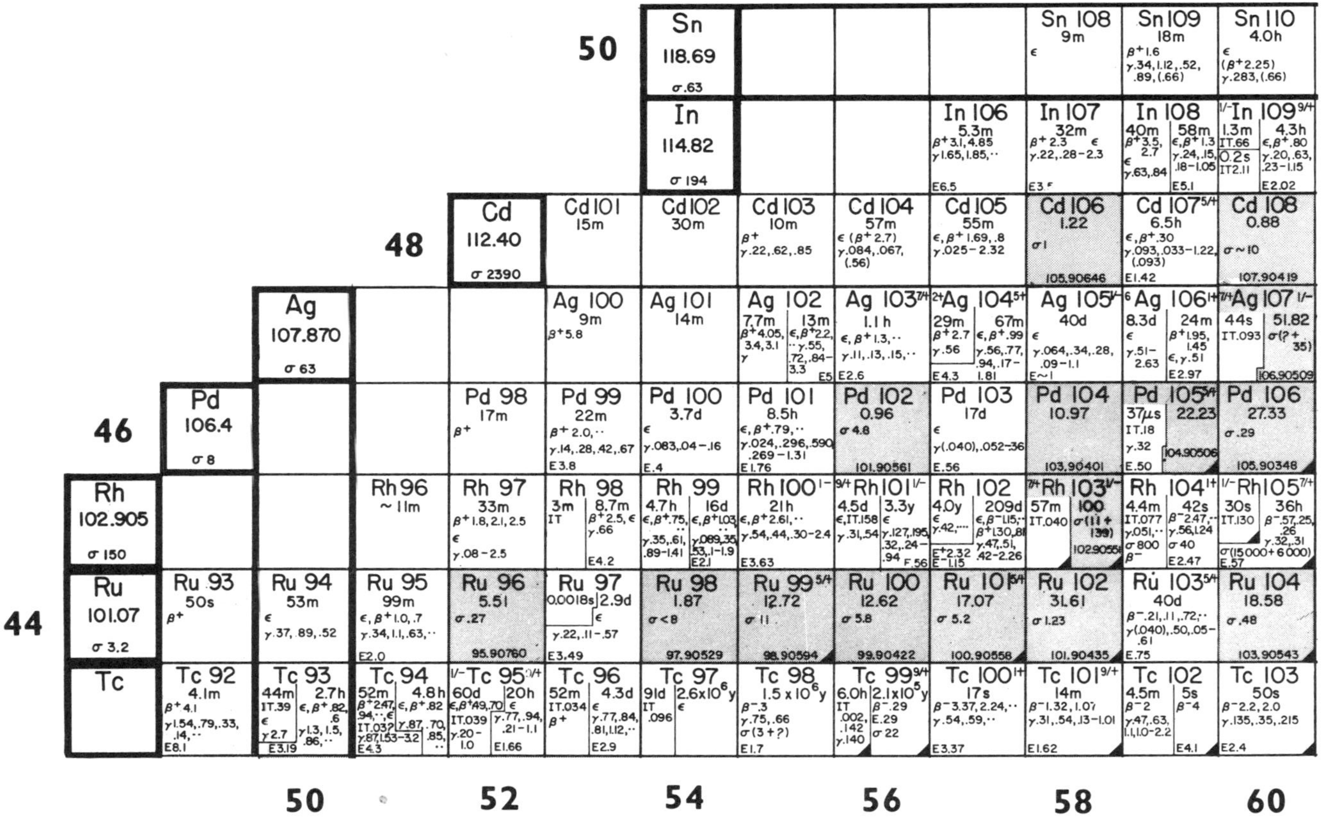

10

						50	Sn 118.69 σ.63				Sn 108 9m ε	Sn 109 18m β^+1.6 γ.34,1.12,.52, .89,(.66)	Sn 110 4.0h ε (β^+2.25) γ.283,(.66)
							In 114.82 σ 194			In 106 5.3m β^+3.1,4.85 γ1.65,1.85,·· E6.5	In 107 32m β^+2.3 ε γ.22,.28−2.3 E3.F	In 108 40m | 58m β^+3.5, 2.7 \| ε,β^+1.3 ε | γ.24,.15, γ.63,.84 | .18−1.05 E5.1	1/− In 109 9/+ 1.3m | 4.3h IT.66 | ε,β^+.80 0.2s | γ.20,.63, IT2.11 | .23−1.15 E2.02
				48	Cd 112.40 σ 2390	Cd 101 15m	Cd 102 30m	Cd 103 10m β^+ γ.22,.62,.85	Cd 104 57m ε (β^+2.7) γ.084,.067, (.56)	Cd 105 55m ε,β^+1.69,.8 γ.025−2.32	Cd 106 1.22 σ1 105.90646	Cd 107 5/+ 6.5h ε,β^+.30 γ.093,.033−1.22, (.093) E1.42	Cd 108 0.88 σ~10 107.90419
			Ag 107.870 σ 63			Ag 100 9m β^+5.8	Ag 101 14m	Ag 102 7.7m | 13m β^+4.05, \| ε,β^+2.2, 3.4,3.1 | ·· γ.55, γ | .72,.84− 3.3 E5	Ag 103 7/+ 1.1h ε,β^+1.3,·· γ.11,.13,.15,·· E2.6	2+ Ag 104 5+ 29m | 67m β^+2.7 \| ε,β^+.99 γ.56 | γ.56,.77, .94,.17− E4.3 | 1.81	Ag 105 1/− 40d ε γ.064,.34,.28, .09−1.1 E~1	6 Ag 106 1+ 8.3d | 24m ε | β^+1.95, 1.45 γ.51− | ε,γ.51 2.63 | E2.97	7/+ Ag 107 1/− 44s | 51.82 IT.093 | σ(?+ 35) 106.90509
	46	Pd 106.4 σ 8			Pd 98 17m β^+	Pd 99 22m β^+2.0,·· γ.14,.28,.42,.67 E3.8	Pd 100 3.7d ε γ.083,.04−.16 E.4	Pd 101 8.5h ε,β^+.79,·· γ.024,.296,.590, .269−1.31 E1.76	Pd 102 0.96 σ4.8 101.90561	Pd 103 17d ε γ(.040),.052−.36 E.56	Pd 104 10.97 103.90401	Pd 105 5/+ 37μs | 22.23 IT.18 γ.32 | 104.90506 E.50	Pd 106 27.33 σ.29 105.90348
	Rh 102.905 σ 150			Rh 96 ~11m	Rh 97 33m β^+1.8,2.1,2.5 ε γ.08−2.5	Rh 98 3m | 8.7m IT | β^+2.5,ε γ.66 E4.2	Rh 99 4.7h | 16d ε,β^+.75, | ε,β^+1.03, ·· | ·· γ.35,.61, | γ.089,.35, .89−1.41 | .53,.1−1.9 E2.1	Rh 100 1− 21h ε,β^+2.61,·· γ.54,.44,.30−2.4 E3.63	9/+ Rh 101 1/− 4.5d | 3.3y ε,IT.158 | ε γ.31,.54 | γ.127,.195, .32,.24− .94 F.56	Rh 102 4.0y | 209d ε | ε,β^-1.15,·· γ.42,··· | β^+1.30,.81 γ.47,.51, E+2.32 | .42−2.26 E−1.15	7/+ Rh 103 1/− 57m | 100 IT.040 | σ(11+ 139) 102.9055	Rh 104 1+ 4.4m | 42s IT.077 | β^-2.47,·· γ.051,·· | γ.56,1.24 σ800 | σ40 β^- | E2.47	1/− Rh 105 7/+ 30s | 36h IT.130 | β^-.57,.25, .26 γ.32,.31 σ(15000+6000) E.57
44	Ru 101.07 σ 3.2	Ru 93 50s β^+	Ru 94 53m ε γ.37,.89,.52	Ru 95 99m ε,β^+1.0,.7 γ.34,1.1,.63,·· E2.0	Ru 96 5.51 σ.27 95.90760	Ru 97 0.0018s | 2.9d ε γ.22,.11−.57 E3.49	Ru 98 1.87 σ<8 97.90529	Ru 99 5/+ 12.72 σ11 98.90594	Ru 100 12.62 σ5.8 99.90422	Ru 101 5/+ 17.07 σ5.2 100.90558	Ru 102 31.61 σ1.23 101.90435	Ru 103 5/+ 40d β^-.21,.11,.72;·· γ(.040),.50,.05− .61 E.75	Ru 104 18.58 σ.48 103.90543
	Tc	Tc 92 4.1m β^+4.1 γ1.54,.79,.33, .14,·· E8.1	Tc 93 44m | 2.7h IT.39 | ε,β^+.82, ε | .6 γ2.7 | γ1.3,1.5, .86,·· E3.19	Tc 94 52m | 4.8h β^+2.47, \| ε,β^+.82 .94,··,ε IT.03? | γ.87,.70, γ.87,1.53−3.2 | .85, E4.3 | ··	1/− Tc 95 9/+ 60d | 20h ε,β^+.49,.70 | ε IT.039 | γ.77,.94, γ.20− | .21−1.1 1.0 | E1.66	Tc 96 52m | 4.3d IT.034 | ε β^+ | γ.77,.84, .81,1.12,·· E2.9	Tc 97 91d | 2.6×10^6y IT | ε .096	Tc 98 1.5×10^6y β^-.3 γ.75,.66 σ(3+?) E1.7	Tc 99 9/+ 6.0h | 2.1×10^5y IT | β^-.29 .002, | E.29 .142 | σ22 γ.140	Tc 100 1+ 17s β^-3.37,2.24,·· γ.54,.59,·· E3.37	Tc 101 9/+ 14m β^-1.32,1.07 γ.31,.54,.13−1.01 E1.62	Tc 102 4.5m | 5s β^-2 | β^-4 γ.47,.63, 1.1,1.0−2.2 E4.1	Tc 103 50s β^-2.2,2.0 γ.135,.35,.215 E2.4
			50		52		54		56		58		60

Z	N=61	N=62	N=63	N=64	N=65	N=66	N=67	N=68	N=69	N=70	N=71	N=72	N=73
50	Sn 111 35m ε,β+1.51 E2.52	Sn 112 0.96 σ(.4+.9) 111.90483	Sn 113 20m IT.079 ε | 118d ε γ.255,(.39) E1.02	Sn 114 0.66 113.90277	Sn 115 1/+ 159μs IT.11 γ.12,.50 | 0.35 114.90335	Sn 116 14.30 σ(.006+?) 115.90174	11/− Sn 117 1/+ 14d IT.159 γ.161 | 7.61 116.90296	Sn 118 24.03 σ(.01+?) 117.90161	11/− Sn 119 1/+ 250d IT.065 γ.024 | 8.58 118.90331	Sn 120 32.85 σ(~.001+.14) 119.90220	Sn 121 ~25y β−.35 γ.037 | 26.8h β−.38 no γ E.38	Sn 122 4.72 σ(.001+.2) 121.90344	11/− Sn 123 129d β−1.42,·· γ1.08 | 40m β−1.26 γ.16 E1.42
	7 In 110 2+ 4.9h ε γ.12,.66,.11−.94 | 66m β+2.25 ε,γ.66,·· E3.93	1/− In 111 9/+ ~10m IT.53 | 2.81d ε γ.247,.173,·· E1	In 112 1+ 0.042s IT.31; 21m IT.155 | 14m β−.66 ε,β+1.6 γ.62,.71 E−.66 E+2.6	1/− In 113 9/+ 1.73h IT.39 | 4.28 σ(?+8+4) 112.90409	In 5+ 114 1+ 2.5s IT.150, .56 | 50d IT.191 ε,γ.72 ε,β+,4 | 72s β−1.98,·· γ1.3,.2 E−1.98 E+1.44	1/− In 115 9/+ 4.4h IT.34 β−.84 | 95.72 5×10^{14}y β−.5 σ(4+154+45) 114.90387	In 116 5+ 2.2s IT.16 | 54m β−1.00,.85,.60 γ1.29,1.10,·· | 14s β−3.29,·· γ1.3,·· E3.29	1/− In 117 9/+ 1.9h β−1.77,1.61 IT.31 γ.16,·· | 45m β−.74 γ.56,.16 E1.47	In 118 1+ 4.4m β−1.3,2.0,·· γ1.22,1.05,.2−2.0 | 5.1s β−4.2,·· γ1.22 E4.2	In 119 18m β−2.7,1.8 IT.3 γ.9 | 2.0m β−1.6 γ.82,·· E2.4	In 120 3.2s β−5.6,·· γ1.18 | 44s β−2.2 γ.73−1.18 E5	In 121 3.1m β−3.1 | 30s β− γ.94 E4	In 122 7.5s β−~5 γ1.14,1.0 E6
48	1/+ Cd 109 5/+ 12μs IT.058 | 453d ε (γ.088) E.16	Cd 110 12.39 σ(.1+3) 109.90301	11/− Cd 111 1/+ 49m IT.150 γ.247D | 12.75 110.90419	Cd 112 24.07 σ(.03+?) 111.90276	11/− Cd 113 1/+ 14y β−.58 IT.27 | 12.26 σ20,000 112.90441	Cd 114 28.86 σ(.14+1.1) 113.90336	Cd 115 43d β−1.63,·· γ.94,.17−1.56 | 2.3d β−1.11,·· γ.52,··,(.34) E1.63 E1.45	Cd 116 7.58 σ(?+1.4) 115.90476	11/− Cd 117 1/+ 3.2h β−1.63,1.84,4−.99 γ.3−2.5 E2.7 | 2.5h β−.67,1.82,2.2,·· γ.3−1.6 E2.52	Cd 118 50m β−1.8,2.3,(4.2) γ.42	Cd 119 2.7m β− | 9.5m β− (γ.82,··) E3.5	?Cd 120 <1m β−	?Cd 121 3.5m β− γ.85
	Ag 108 1+ ~100y ε γ.72,.61,.43,·· IT.030,.079 | 2.4m β−1.64,·· ε,β+.7 γ.63,·· E−1.64 E+1.92	7/+ Ag 109 1/− 40s IT.088 | 48.18 σ(3+89) 108.90476	6− Ag 110 1+ 260d β−.086,.53,·· IT.12 γ.66,.88,·· σ80 | 24s β−2.87,·· γ.66,·· E2.87	Ag 111 1/− 74s IT.065 | 7.5d β−1.05,.71,·· γ.34,.25D E1.05	Ag 112 2− 3.2h β−4.01,·· γ.62,1.4,.69−3.28 E4.01	Ag 113 1/− 1.2m IT β− γ.31−.70 | 5.3h β−2.0,·· γ.31,.12−1.18 E2.0	Ag 114 5s β−4.6,·· γ.56 E4.6	Ag 115 20s β− no γ | 21m β−1.1,3.2,.7−3.0 γ.22,.10−2.5 E3.2	Ag 116 2.5m β−5.1,·· γ1.8,.5−2.6	Ag 117 1.1m β−			
46	Pd 107 22s IT.22 | 7×10^{6}y β−.035 no γ E.035	Pd 108 26.71 σ(.2+12) 107.90389	11/− Pd 109 5/+ 4.8m IT.18 | 13.5h β−1.03,·· γ(.088),·· E1.12	Pd 110 11.81 σ(.04+.2) 109.90516	11/− Pd 111 5/+ 5.5h IT.17,.28 β−.37−1.12 γ.16,.38,.28−1.9 (.07) | 22m β−2.11,.33−1.62 γ.16,.38,.63,.28−1.44 (.07) E2.2	Pd 112 21.0h β−.30 γ.018 E.30	Pd 113 1.5m β−	Pd 114 2.4m β− no γ	Pd 115 45s β−				
	Rh 106 1+ 2.2h β−.79−1.62 γ.51,.22−1.22 | 30s β−3.53,·· γ.51,.62,.7−3.4 E3.53	Rh 107 21.7m β−1.2,.83−1.5 γ.31,.11−1.1 E1.5	Rh 108 1+ 17s β−4.5,3.5,4.1,·· γ.43,.62,.51,1.52,·· E4.5	Rh 109 50s IT.11 | 30s β− γ.49,.31 E2.5	Rh 110 5s β−5.5,·· γ.38 E5.5								
44	Ru 105 5/+ 4.43h β−1.15,1.08,1.87,1.74,·· γ.72,.21−1.7,(.13) E1.87 σ2	Ru 106 1.0y β−.04,(3.53,··) (γ.51,.62−2.66) σ.15 E.04	Ru 107 4.2m β−3.2,1.9−3.0 γ.19,.37−1.3 E3.2	Ru 108 4.5m β−1.3,1.2 γ.17 E1.3									
	Tc 104 18m β−2.4,·· γ.36,.53,.88,1.63,·· 6−4.7 E5.9	Tc 105 8m β−3.4,1.8 γ.11,·· E3.4	Tc 106 38s γ.28	Tc 107 29s									
		62		64		66		68		70		72	

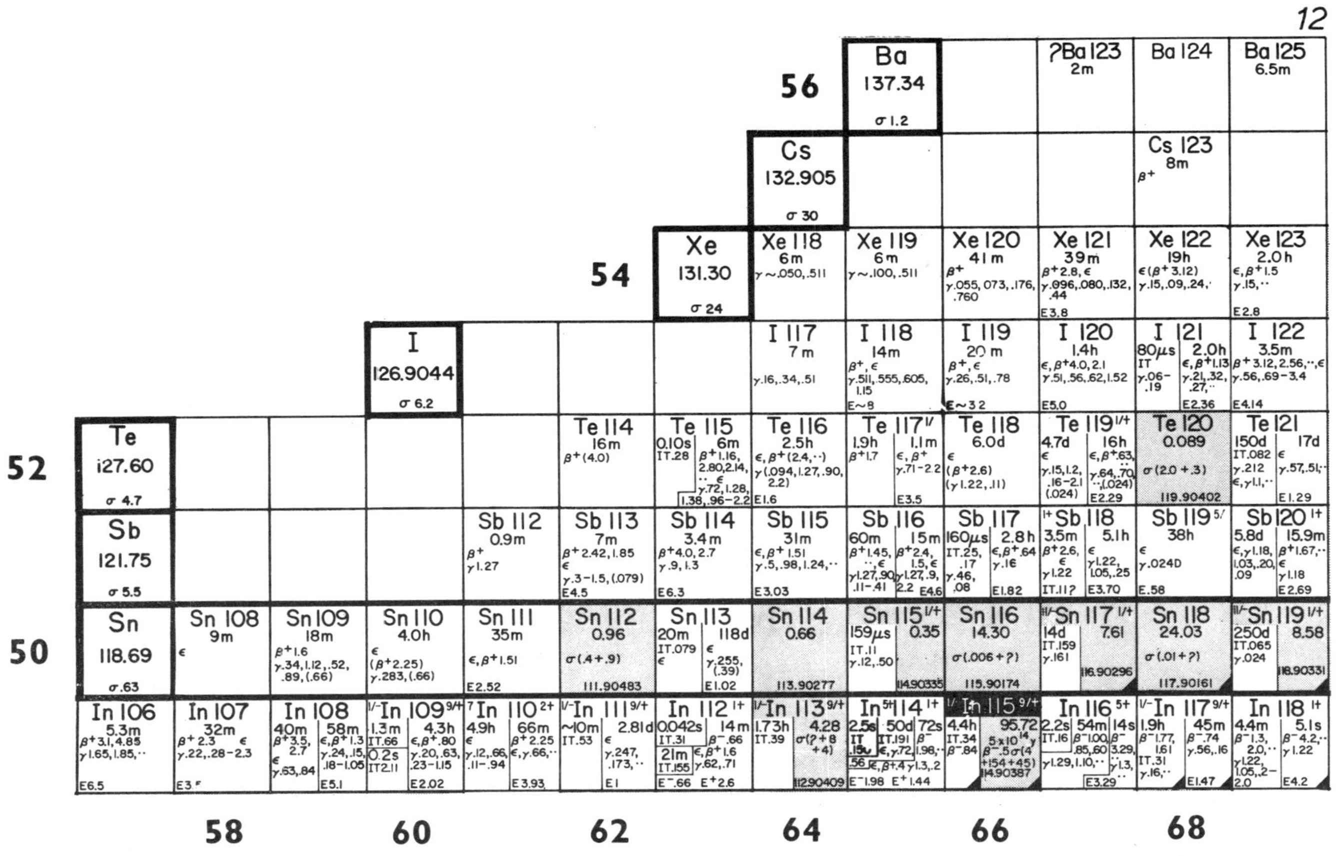

Z	N=57	58	59	60	61	62	63	64	65	66	67	68	69
56									**Ba** 137.34 σ 1.2		?Ba 123 2m	Ba 124	Ba 125 6.5m
								Cs 132.905 σ 30				Cs 123 8m β+	
54							**Xe** 131.30 σ 24	Xe 118 6m γ∼.050,.511	Xe 119 6m γ∼.100,.511	Xe 120 41 m β+ γ.055, 073,.176, .760	Xe 121 39m β+2.8, ε γ.096,.080,.132, .44 E3.8	Xe 122 19h ε(β+3.12) γ.15,.09,.24,·	Xe 123 2.0h ε, β+1.5 γ.15,·· E2.8
				I 126.9044 σ 6.2				I 117 7 m γ.16,.34,.51	I 118 14m β+, ε γ.511,.555,.605, 1.15 E∼8	I 119 20 m β+, ε γ.26,.51,.78 E∼3.2	I 120 1.4h ε, β+4.0, 2.1 γ.51,.56,.62,1.52 E5.0	I 121 80μs IT γ.06–.19 | 2.0h ε, β+1.13 γ.21,.32, .27,·· E2.36	I 122 3.5m β+3.12, 2.56,··, ε γ.56,.69–3.4 E4.14
52	**Te** 127.60 σ 4.7					Te 114 16m β+(4.0)	Te 115 0.10s IT.28 | 6m β+1.16, 2.80, 2.14, ·· ε γ.72, 1.28, 1.38,.96–2.2	Te 116 2.5h ε, β+(2.4,··) γ(.094, 1.27,.90, 2.2) E1.6	Te 117 1/ 1.9h β+1.7 | 1.1m ε, β+ γ.71–2.2 E3.5	Te 118 6.0d ε (β+2.6) (γ1.22,.11)	Te 119 1/+ 4.7d ε γ.15, 1.2, .16–2.1 (.024) | 16h ε, β+.63, ·· γ.64,.70, ··,(.024) E2.29	Te 120 0.089 σ(2.0+.3) 119.90402	Te 121 150d IT.082 γ.212 ε, γ1.1,·· | 17d ε γ.57,.51,·· E1.29
	Sb 121.75 σ 5.5				Sb 112 0.9m β+ γ1.27	Sb 113 7m β+2.42, 1.85 ε γ.3–1.5, (.079) E4.5	Sb 114 3.4m β+4.0, 2.7 γ.9, 1.3 E6.3	Sb 115 31m ε, β+1.51 γ.5,.98, 1.24,·· E3.03	Sb 116 60m β+1.45, ··, ε γ1.27,.90 .11–.41 | 15m β+2.4, 1.5, ε γ1.27,.9, 2.2 E4.6	Sb 117 160μs IT.25, .17 γ.46, .08 | 2.8h ε, β+.64 γ.16 E1.82	1+ Sb 118 3.5m β+2.6, ε γ1.22 IT.11? | 5.1h ε γ1.22, 1.05,.25 E3.70	Sb 119 5/ 38h ε γ.024D E.58	Sb 120 1+ 5.8d ε, γ1.18, 1.03,.20, .09 | 15.9m β+1.67,·· ε γ1.18 E2.69
50	**Sn** 118.69 σ.63	Sn 108 9m ε	Sn 109 18m β+1.6 γ.34, 1.12,.52, .89,(.66)	Sn 110 4.0h ε (β+2.25) γ.283,(.66)	Sn 111 35m ε, β+1.51 E2.52	Sn 112 0.96 σ(.4+.9) 111.90483	Sn 113 20m IT.079 ε | 118d ε γ.255, (.39) E1.02	Sn 114 0.66 113.90277	Sn 115 1/+ 159μs IT.11 γ.12,.50 | 0.35 114.90335	Sn 116 14.30 σ(.006+?) 115.90174	11/– Sn 117 1/+ 14d IT.159 γ.161 | 7.61 116.90296	Sn 118 24.03 σ(.01+?) 117.90161	11/– Sn 119 1/+ 250d IT.065 γ.024 | 8.58 118.90331
	In 106 5.3m β+3.1, 4.85 γ1.65, 1.85,·· E6.5	In 107 32m β+2.3 ε γ.22,.28–2.3 E3.5	In 108 40m β+3.5, 2.7 ε γ.63,.84 | 58m ε, β+1.3 γ.24,.15, .18–1.05 E5.1	1/– In 109 9/+ 1.3m IT.66 | 0.2s IT2.11 | 4.3h ε, β+.80 γ.20,.63, .23–1.15 E2.02	7 In 110 2+ 4.9h ε γ.12,.66, .11–.94 | 66m β+2.25 ε, γ.66,·· E3.93	1/– In 111 9/+ ∼10m IT.53 | 2.81d ε γ.247, .173,·· E1	In 112 1+ 0.042s IT.31 | 21m IT.155 | 14m β−.66 ε, β+1.6 γ.62,.71 E−.66 E+2.6	1/– In 113 9/+ 1.73h IT.39 | 4.28 σ(?+8 +4) 112.90409	In 5+ 114 1+ 2.5s IT .15u .56 | 50d IT.191 ε, γ.72 ε, β+.4 γ1.3,.2 | 72s β− 1.98,·· E−1.98 E+1.44	1/ In 115 9/+ 4.4h IT.34 β−.84 | 95.72 5×10[14]y β−.5 σ(4 +154+45) 114.90387	In 116 5+ 2.2s IT.16 | 54m β−1.00, .85,.60 γ1.29, 1.10,·· | 14s β− 3.29, ·· γ1.3, ·· E3.29	1/– In 117 9/+ 1.9h β−1.77, 1.61 IT.31 γ.16,·· | 45m β−.74 γ.56,.16 E1.47	In 118 1+ 4.4m β−1.3, 2.0,·· γ1.22, 1.05,.2–2.0 | 5.1s β−4.2,·· γ1.22 E4.2
		58		60		62		64		66		68	

Z \ N	70	71	72	73	74	75	76	77	78	79	80	81	82
56	Ba 126 97m ε,(β^+ 3.8) γ.27,.70,(.39)	Ba 127 10m β^+	Ba 128 2.4d ε,(β^+2.9,··) γ.27,(1.3,.45,··)	Ba 129 2.13h ε γ.182,·· 2.2h β^+1.43, 1.25,.98 γ.21,.22,·· E2.45	Ba 130 0.101 σ 8.8 129.9062	Ba 131 14.6m IT.078 γ.107 11.6d ε γ.50, .122D,.216,.055–1.03	Ba 132 0.097 σ(<.2+7) 131.9051	Ba 133 39h IT.276 γ.012 7.2y ε γ.081D,.36,.30,.08–.38 E.288 E.489	Ba 134 2.42 σ(2+.16+<4) 133.9043	Ba 135 3/+ 20.33s γ~.7~.8 29h IT.268 6.59 σ5 134.9056	Ba 136 0.37s IT γ.82, 1.06 7.81 σ(.010+<1) 135.9044	11/- Ba 137 3/+ 2.6m IT.662 11.32 σ4 136.9056	Ba 138 71.66 σ.4 137.9050
55	Cs 125 45m ε,β^+2.05 γ.112 E3.07	Cs 126 1+ 1.6m β^+3.8,··,ε γ.39 E4.8	Cs 127 1/+ 6.2h ε,β^+.7,1.06,·· γ.41,.125,.17–.44 E2.1	Cs 128 1+ 3m β^+2.9,2.5,1.9, 1.5, ε γ.13,.45,.98 E3.9	Cs 129 1/+ 32h ε γ.37,.41,.55,.32, .28	Cs 130 1+ 30m ε β^+1.97 β^-.44 E^-.44 E^+2.99	Cs 131 5/+ 9.7d ε no γ E.35	Cs 132 2− 6.58d ε,β^-,β^+.4 γ.67,.46,.51–2.0 E^+2.08 E^-1.2	Cs 133 7/+ 100 σ(2.6+28) 132.90534	8− Cs 134 4+ 2.9h IT.13, .14 β^-.55 γ.010 2.1y β^-.65,.09, .21–1.45 γ.60,.80, .20–1.37 σ136 E2.06	Cs 135 7/+ 53m IT.84 γ.78 2.0x10^6 y β^-.21 no γ σ 8.7 E.21	Cs 136 5− 13d β^-.34,.66 γ1.06,.82,.065–2.5 E2.83	Cs 137 7/+ 30y β^-.51,1.17 (γ.662) σ.11 E1.18
54	Xe 124 0.096 σ110 123.9061	Xe 125 60s IT.075 γ.111 17h ε γ.187,.056, .243, .096–.46 E2	Xe 126 0.090 σ~2.2 125.90429	Xe 127 75s IT.175 γ.125 36.4d ε γ.20,.17, .15,.06 E.7	Xe 128 1.92 σ<5 127.90354	11/- Xe 129 1/+ 8d IT.196 γ.040 26.44 σ25 128.90478	Xe 130 4.08 σ<5 129.90351	11/- Xe 131 3/+ 12d IT.164 21.18 σ85 130.90509	Xe 132 26.89 σ(<5+.2) 131.90416	11/- Xe 133 3/+ 2.3d IT.233 5.27d β^-.35,·· γ.081D,·· σ190 E.43	Xe 134 10.44 σ(<5+.2) 133.90540	Xe 135 16m IT.53 9.2h β^-.91,.55 γ.25D,.61,·· σ 2.7x10^6 E1.16	Xe 136 8.87 σ.28 135.90722
53	I 123 5/+ 13h ε γ.159,··	I 124 2− 4.2d ε,β^+1.55,2.15,·· γ.60,1.7,.65–2.9 E3.2	I 125 5/+ 60 d ε γ.035 σ 900 E.15	I 126 2− 12.8d ε,β^-.38–1.25 β^+1.13,·· γ.39,.67,.48–1.41 E^-1.25 E^+2.15	I 127 5/+ 100 σ 6.2 126.90447	I 128 1+ 25.0m β^-2.14,1.67,··,ε γ.45,.54–.99 E^-2.14 E^+1.27	I 129 7/+ 1.6 x 10^7 y β^-.15 γ.04 σ(19+9) E.19	I 130 5 9.2m IT β^- σ18 12h β^-.60,1.02, 1.78 γ.419,.538, .669,.743,1.15 E2	I 131 7/+ 8.05d β^-.61,.25–.81 γ.36,.080D–.72 σ~.7 E.97	I 132 4+ 2.3h β^-1.53,1.16,2.12, .80,1.61,2.14,·· γ.67,.78,.53,.96, .62–2.5 E3.56	I 133 7/+ 21h β^-1.23,1.54,.5,.89 γ.53,.87,1.3,·· (n) E1.8	I 134 53m β^-2.4,1.25,.5, 2.2 γ.85,.61,.121–1.8 E4.1	I 135 7/+ 6.7h β^-1.0,.5,1.4 γ1.14,1.28,1.80, .42–2.4 (.53) E3
52	Te 122 2.46 σ(1+2) 121.90307	11/- Te 123 1/+ 117d IT.089 γ.159D 0.87 1.2x10^13 y ε σ400 E.03 122.90428	Te 124 4.61 σ(5+2) 123.90284	11/- Te 125 1/+ 58d IT.11 γ.035D 6.99 σ1.5 124.90442	Te 126 18.71 σ(.1+.9) 125.90332	Te 127 105d IT.089 β^-,γ 9.3h β^-.69,·· γ.06–.42 E.69	Te 128 31.79 σ(.017+.14) 127.90448	Te 129 34d IT.106 β^-1.6, γ.46, .027–1.255 67m β^-1.45,1.0, ·· γ.027,.46, .21–1.2 E1.48	Te 130 34.48 σ(.03+.2) 129.90624	Te 131 1.2d β^-.42, .57–2.5 IT.18 γ.77,.1–2.3 25m β^-2.1,1.7, γ.15,.45, .3–1.15 E2.28	Te 132 78h β^-.22(1.16–2.12) γ.228,.050,··(.78 .53–2.5) E.51	Te 133 50m β^-1.3, 2.4 IT.334 γ.31–.97 ~2m β^-1.3, 2.4 γ.6,1.0,··	Te 134 42m β^- γ.030–.262
51	Sb 121 5/+ 57.25 σ(.06+?+6) 120.90382	Sb 122 2− 4.1m IT.026 530μs IT.08 γ.06 2.8d β^-1.4,1.97,·· ε,β^+.58 γ.56,.7–1.1 E^-1.97 E^+1.60	Sb 123 7/+ 42.75 σ(.015+.03+3.3) 122.90421	Sb 124 3− 21m IT.025 1.2m IT.010 β^-1.2,·· 60.2d β^-.61,·· γ.60–2.1 σ6.5 E2.92 γ.50–.65	Sb 125 7/+ 2.7y β^-.30,.12,.61,·· γ.43,.60,.035–.64 (.110,.035) σ<20 E.76	Sb 126 19m β^-2.0 γ.42,.67 ·· IT≤.03 12.5d β^-2.0,·· γ.42,.67, .3–1.0 E3.7	Sb 127 3.9d β^-1.60,1.1,··,(.69,··) γ.47,.69,.25,··, (.42,··) E1.60	Sb 128 10m β^-2.5, 2.8 γ.75,.31 9.6h β^-1.0–1.5 γ.31–.90 E4.3	Sb 129 4.5h β^-1.87,··,(1.45,··) γ.18–2.1, (.027–1.2) E2.5	Sb 130 7m β^- γ.20–1.2 39m β^- γ.19–.94 E5	Sb 131 25m β^- γ.95, .65	Sb 132 45s 2.1m β^-	Sb 133 4.1m β^-
50	Sn 120 32.85 σ(~.001+.14) 119.90220	Sn 121 ~25y β^-.35 γ.037 26.8h β^-.38 no γ E.38	Sn 122 4.72 σ(.001+.2) 121.90344	11/- Sn 123 129d β^-1.42,·· γ1.08 40m β^-1.26 γ.16 E1.42	Sn 124 5.94 σ(.1+.004) 123.90527	Sn 125 11/- 9.7m β^-2.04, γ.33,1.4, ·· 9.62d β^-2.34, γ1.07,.23 –1.97 E2.34	Sn 126 ~10^5 y β^-(2.0) γ.092,.07(.42, .67)	Sn 127 4.1m β^-2.7 E3.2 2.1h β^- γ1.10,.82, .12–2.8 E3	Sn 128 62m β^-.8,.7,(2.5,2.8) γ.04–.57,(.75,.31) E1.3	Sn 129 1h β^- 6.2m β^- γ1.1–3.1	Sn 130 2.6m β^-	Sn 131 ≲2m β^-	Sn 132 2.2m β^-
49	In 119 18m β^-2.7, 1.8 IT.3 γ.9 2.0m β^-1.6 γ.82,·· E2.4	In 120 3.2s β^-5.6,·· γ1.18 44s β^-2.2 γ.73–1.18 E5	In 121 3.1m β^-3.1 30s β^- γ.94 E4	In 122 7.5s β^-~5 γ1.14,1.0 E6	In 123 36s β^-4.6 10s β^- γ1.10 E7.4	In 124 3.6s β^-5.3 γ1.13,.99,3.21 E7.4							

56 54 52 50

70 72 74 76 78 80 82

60											Nd 144.24 σ 50		
										Pr 140.907 σ 19			Pr 134 17m γ.13,.22,.94
58						Ce 140.12 σ .7			?Ce 129 ~13m ϵ γ.08	Ce 130 [illegible] ϵ, β^+ γ.130	Ce 131 30m β^+4.2	Ce 132 4.2h ϵ, β^+ γ077,.097,.174	Ce 133 6.3h ϵ β^+1.3(1.2) γ1.8(.8)
			La 138.91 σ 8.9		La 124 7m	?La 125 <1m	La 126 1m β^+, ϵ γ.26,··	La 127 3.5m	La 128 4.6m β^+3.2, ϵ γ.28,.48,.64,··· E5.6	La 129 10m β^+, ϵ γ.28 E4	La 130^{3+} 9m β^+, ϵ γ.36,·· E6	La 131 61m ϵ, β^+1.42,1.94,·· γ.115,.36,.42, .28,·· E2.96	La 132^{3+} 4.8h β^+3.8,··,ϵ γ.47,.56,.66,.90. ·· E4.8
56			Ba 137.34 σ 1.2		?Ba 123 2m	Ba 124	Ba 125 6.5m	Ba 126 97m ϵ,(β^+3.8) γ.27,.70,(.39)	Ba 127 10m β^+	Ba 128 2.4d ϵ,(β^+2.9,··) γ.27,(1.3,.45,··)	Ba 129 2.13h | 2.2h ϵ γ.182,·· | β^+1.43, 1.25,.98 γ.21,.22,·· E2.45	Ba 130 0.101 σ 8.8 129.9062	Ba 131 14.6m | 11.6d IT.078 γ.107 | ϵ γ.50, .122D,.216,.055– 1.03
		Cs 132.905 σ 30				Cs 123 8m β^+		Cs 125 45m ϵ, β^+2.05 γ.112 E3.07	Cs 126^{1+} 1.6m β^+3.8,··,ϵ γ.39 E4.8	Cs 127$^{1/+}$ 6.2h ϵ, β^+.7,1.06,·· γ.41,.125,.17-44 E2.1	Cs 128^{1+} 3m β^+2.9,2.5,1.9, 1.5, ϵ γ.13,.45,.98 E3.9	Cs 129$^{1/+}$ 32h ϵ γ.37,.41,.55,.32, .28	Cs 130^{1+} 30m ϵ β^+1.97 β^-.44 E^-.44 E^+2.99
54	Xe 131.30 σ 24	Xe 118 6m γ~.050,.511	Xe 119 6m γ~.100,.511	Xe 120 41m β^+ γ.C55, 073,.176, .760	Xe 121 39m β^+2.8, ϵ γ.096,.080,.132, .44 E3.8	Xe 122 19h ϵ(β^+3.12) γ.15,.09,.24,··	Xe 123 2.0h ϵ, β^+1.5 γ.15,·· E2.8	Xe 124 0.096 σ 110 123.9061	Xe 125 60s | 17h IT.075 γ.111 | ϵ γ.187,056, .243, .096–.46 E2	Xe 126 0.090 σ~2.2 125.90429	Xe 127 75s | 36.4d IT.175 γ.125 | ϵ γ.20,.17, .15,.06 E.7	Xe 128 1.92 σ < 5 127.90354	$^{11/}$Xe 129$^{1/+}$ 8d | 26.44 IT.196 γ.040 | σ 25 128.90478
	I 126.9044 σ 6.2	I 117 7m γ.16,.34,.51	I 118 14m β^+, ϵ γ.511,.555,.605, 1.15 E~8	I 119 20m β^+, ϵ γ.26,.51,.78 E~3 2	I 120 1.4h ϵ, β^+4.0,2.1 γ.51,.56,.62,1.52 E5.0	I 121 80μs | 2.0h IT γ.06– .19 | ϵ, β^+1.13 γ.21,.32, .27,·· E2.36	I 122 3.5m β^+3.12,2.56,··,ϵ γ.56,.69–3.4 E4.14	I 123$^{5/+}$ 13h ϵ γ.159,··	I 124^{2-} 4.2d ϵ, β^+1.55,2.15,·· γ.60,1.7,.65–2.9 E3.2	I 125$^{5/+}$ 60d ϵ γ.035 σ 900 E.15	I 126^{2-} 12.8d ϵ, β^-.38–1.25 β^+1.13,·· γ.39,.67,.48–1.41 E^-1.25 E^+2.15	I 127$^{5/+}$ 100 σ 6.2 126.90447	I 128^{1+} 25.0m β^-2.14,1.67,··,ϵ γ.45,.54–.99 E^-2.14 E^+1.27
		64		66		68		70		72		74	

Z \ N	76	77	78	79	80	81	82	83	84	85	86	87	88
60		Nd 137 55m	Nd 138 5h ε	Nd 139 5.2h ε, β+ 3.3, ··(1.0) γ1.3, .11–2.5	Nd 140 600μs | 3.3d IT.43 | ε γ1.0, .77 | γ.11–.5	Nd 141$^{3/4+}$ 64s | 2.5h IT.76 | ε, β+.78 γ1.15, 1.30, ·· E1.8	Nd 142 27.11 σ17 141.90764	Nd 143$^{7/-}$ 12.17 σ330 $σ_{na}$.02 142.90978	Nd144 23.85 2.1 x 10^{15}y α1.9 σ5 143.91004	Nd 145$^{7/-}$ 8.30 σ50 144.91254	Nd 146 17.22 σ2 145.91319	Nd 147$^{5/-}$ 11.1d β−.81, .36, ·· γ.091, .53, .12–.69 E.90	Nd 148 5.73 σ4 147.91687
	Pr 135 22m β+2.5 γ.22, .08, .30	Pr 136 13.5m	Pr 137 70m ε β+1.7 E2.8	Pr 138^{1+} 2.0h ε, β+1.4 γ.16–1.7	Pr 139$^{5/4+}$ 4.5h ε, β+1.0 γ1.3, 1.6 E4.3	Pr 140^{1+} 3.4m β+2.32, ·· ε γ1.6, 1.9, .3 E3.3	Pr 141$^{5/4+}$ 100 σ19 140.90760	Pr 142^{2-} 19.2h β−2.15, .58 γ1.57 σ20 E2.15	Pr 143$^{7/+}$ 13.7d β−.93 no γ σ89 E.93	Pr 144^{0-} 17.3m β−2.99, .8, 2.3, ·· γ.69, 2.18, 1.49 E2.99	Pr 145 5.9h β−1.80 γ.072–1.15 E1.80	Pr 146 24m β−2.3, 3.8 γ.46, 1.49, .75, ·· E4.2	Pr 147 12m β−2.1, 1.5, 2.7 γ.08, .57, .11–1.9
58	Ce 134 72h ε no γ E.120	Ce 135 18h ε β+.81 γ.26, .60, .30, ···	Ce 136 0.193 σ(.6+6.0) 135.907	Ce 137 34h | 9h IT.26 | ε ε | γ.010, .46 γ.7, .8, .9	Ce 138 0.250 σ(.04+1.0) 137.9057	II/− Ce 139$^{3/4+}$ 55s | 138d IT.74 | ε γ.166 E.27	Ce 140 88.48 σ.6 139.90539	Ce 141$^{7/-}$ 32.5d β−.435, .58 γ.145 σ30 E.58	Ce142 11.07 5 x 10^{15}y α1.5 σ1 141.9091	Ce 143 33h β−1.09, 1.38, ·· γ.058, .29, .66, ·· σ6 E1.44	Ce 144 285d β−.3, .19, ··, (2.98, ··) γ.034–.134, (.69–2.18) E.32 σ1.0	Ce 145 3.0m β−2.0 γ.72, .44, ··· E2.0	Ce 146 14m β−.7 γ.32, .05–.27 E1.0
	La 133 4h ε, β+1.2 γ.8 E2.2	La 134^{1+} 6.5m β+2.58, 1.98, 1.41, 1.22, ε γ.56, .60, .76, 1.17, 1.47, ·· E3.77	La 135 19.4h ε γ.48, .108–.87 E1.1	La 136^{1+} 9.5m ε β+1.8, ·· γ.83 E2.9	La 137 6 x 10^{4}y ε	La138^{5-} 0.089 1.1 x 10^{11}y ε, β−.205 γ1.4, .8 E+1.78 E−1.02 137.90691	La 139$^{7/+}$ 99.911 σ8.9 138.90614	La 140^{3-} 40.2h β−1.34, 1.10, .42, ·· γ1.6, .49, .81, .33, .065–3.1 E3.77	La 141 3.9h β−2.4, ·· γ1.37 E2.43	La 142^{2-} 1.4h β−2.1, 2.3, .87, 4.5, 1.8, ·· γ.64, 2.41, 2.55, .90, .86–3.65 E4.5	La 143 14m β−3.3 γ.63, .20–2.85 E3.3	La 144 Short β− E6	
56	Ba 132 0.097 σ(<.2+7) 131.9051	Ba 133 39h | 7.2y IT.276 | ε γ.012 | γ.081D, .36, .30, .08–.38 E.288 | E.489	Ba 134 2.42 σ(2+.16+<4) 133.9043	Ba 135$^{3/4+}$?0.33s | 6.59 γ~.7~.8 29h IT.268 σ5 134.9056	Ba 136 0.37s | 7.81 IT | σ(.010+<1) γ.82, 1.06 135.9044	II/− Ba 137$^{3/4+}$ 2.6m | 11.32 IT.662 | σ4 136.9056	Ba 138 71.66 σ.4 137.9050	Ba 139 83m β−2.34, 2.17, .9, ·· γ.166D, 1.43, ·· σ4 E2.34	Ba 140 12.8d β−.48, 1.02, 1.03, .59 γ.03, .01, .16, .30, .44, .54, ·· E1.05 σ<20	Ba 141 18m β−2.8, ·· γ.19, .12–1.6 E3.0	Ba 142 11m β− γ.26, .89, 1.20, .08, .97–1.68	Ba 143 12s β−	Ba 144 Short β−
	Cs 131$^{5/+}$ 9.7d ε no γ E.35	Cs 132^{2-} 6.58d ε, β−, β+.4 γ.67, .46, .51–2.0 E+2.08 E−1.2	Cs 133$^{7/+}$ 100 σ(2.6+28) 132.90534	8− Cs 134^{4+} 2.9h | 2.1y IT.13, .14 | β−.65, .09, .21–1.45 β−.55 | γ.60, .80, .20–1.37 γ.010 | σ136 E2.06	Cs 135$^{7/+}$ 53m | 2.0x10^{6}y IT.84 | β−.21 γ.78 | no γ σ8.7 E.21	Cs 136^{5-} 13d β−.34, .66 γ(1.06, .82), .065–2.5 E2.83	Cs 137$^{7/+}$ 30y β−.51, 1.17 (γ.662) σ.11 E1.18	Cs 138 32.2m β−3.40, ·· γ1.43, 1.0, .46, 2.2, .11–3.3 E4.83	Cs 139 9.5m β−4, ~3.4, ~2.7 γ1.28, .63 E4	Cs 140 66s β−>4.4, ·· γ.59, .88–3.15 E6	Cs 141 24s β−	Cs 142 2.3s β−	Cs 143 2s β−
54	Xe 130 4.08 σ<5 129.90351	II/− Xe 131$^{3/+}$ 12d | 21.18 IT.164 | σ85 130.90509	Xe 132 26.89 σ(<5+.2) 131.90416	II/− Xe 133$^{3/+}$ 2.3d | 5.27d IT.233 | β−.35, ·· γ.081D, ·· σ190 E.43	Xe 134 10.44 σ(<5+.2) 133.90540	Xe 135 16m | 9.2h IT.53 | β−.91, .55 γ.250, .61, ·· σ 2.7x10^{6} E1.16	Xe 136 8.87 σ.28 135.90722	Xe 137 4.2m β−4.0, 3.7 γ.44 E4.0	Xe 138 17m β−.82, 2.40, 2.56, 2.82 γ.42 2.0, .16–1.74 E2.8	Xe 139 41s β−~4.6 γ.22, .30, .17, .40 E4.6	Xe 140 16s β−	Xe 141 2s β−	Xe 142 ~1.5s β−
	I 129$^{7/+}$ 1.6 x 10^{7}y β−.15 γ.04 σ(19+9) E.19	I 130^{5} 9.2m | 12[illegible] IT | β−.60, 1.02, 1.78 β− | γ.419, .538, .669, .743, 1.15 σ18 E2.[illegible]	I 131$^{7/+}$ 8.05d β−.61, .25–.81 γ.36, .080D–.72 σ~.7 E.97	I 132^{4+} 2.3h β−1.53, 1.16, 2.12, .80, 1.61, 2.14, ·· γ.67, .78, .53, .96, .62–2.5 E3.56	I 133$^{7/+}$ 21h β−1.23, 1.54, .5, .89 γ.53, .87, 1.3, ·· (n) E1.8	I 134 53m β−2.4, 1.25, .5, 2.2 γ.85, .61, .121–1.8 E4.1	I 135$^{7/+}$ 6.7h β−1.0, .5, 1.4 γ1.14, 1.28, 1.80, .42–2.4 (.53) E3	I 136 83s β−4.2, 4.4, 5.6, ·· γ1.32, .20–3.2 E7.0	I 137 24s β− (n.6)	I 138 6.3s β− (n)	I 139 2.0s β− (n)		

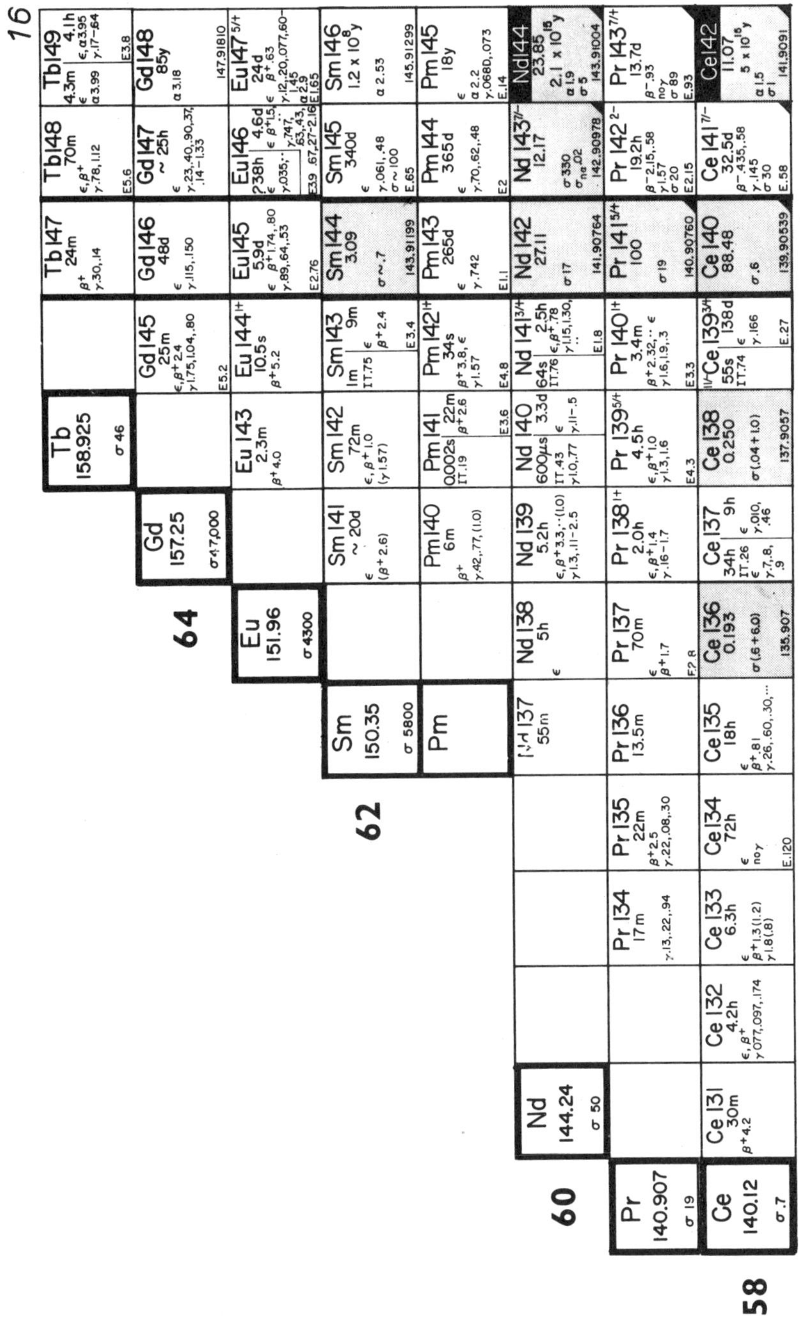

Z \ N	72	73	74	75	76	77	78	79	80	81	82	83	84
65									**Tb** 158.925 σ 46		Tb 147 24m β+ γ.30,.14	Tb148 70m ε,β+ γ.78,1.12 E5.6	Tb149 4.3m | 4.1h ε | ε,α3.95 α3.99 | γ.17–.64 E3.8
64								**Gd** 157.25 σ47,000		Gd 145 25m ε,β+2.4 γ1.75,1.04,.80 E5.2	Gd 146 48d ε γ.115,.150	Gd 147 ~ 25h ε γ.23,.40,.90,.37, .14–1.33	Gd 148 85y α 3.18 147.91810
63							**Eu** 151.96 σ 4300		Eu 143 2.3m β+4.0	Eu 144 1+ 10.5s β+5.2	Eu145 5.9d ε β+1.74,.80 γ.89,.64,.53 E2.76	Eu146 ?38h | 4.6d ε | ε β+1.5, γ.035,·· | γ.747, .63,.43, E3.9 .67,.27–2.16	Eu147 5/+ 24d ε β+.63 γ.12,.20,.077,.60– 1.45 α 2.9 E1.65
62						**Sm** 150.35 σ 5800		Sm 141 ~ 20d ε (β+2.6)	Sm 142 72m ε, β+1.0 (γ1.57)	Sm 143 1m | 9m IT.75 | ε β+2.4 E3.4	Sm 144 3.09 σ ~.7 143.91199	Sm 145 340d ε γ.061,.48 σ ~ 100 E.65	Sm 146 1.2 x 10^8y α 2.53 145.91299
61						**Pm**		Pm140 6m β+ γ.42,.77,(1.0)	Pm 141 0.002s | 22m IT.19 | β+2.6 E3.6	Pm 142 1+ 34s β+3.8, ε γ1.57 E4.8	Pm 143 265d ε γ.742 E1.1	Pm 144 365d ε γ.70,.62,.48 E2	Pm 145 18y ε α 2.2 γ.068D,.073 E.14
60		**Nd** 144.24 σ 50				Nd 137 55m	Nd 138 5h ε	Nd 139 5.2h ε,β+3.3,··(1.0) γ1.3,.11–2.5	Nd 140 600μs | 3.3d IT.43 | ε γ1.0,.77 | γ.11–.5	Nd 141 3/+ 64s | 2.5h IT.76 | ε,β+.78 γ1.15,1.30,·· E1.8	Nd 142 27.11 σ 17 141.90764	Nd 143 7/– 12.17 σ 330 $σ_{nα}$ .02 142.90978	Nd144 23.85 2.1 x 10^{15}y α 1.9 σ 5 143.91004
59	**Pr** 140.907 σ 19			Pr 134 17m γ.13,.22,.94	Pr 135 22m β+2.5 γ.22,.08,.30	Pr 136 13.5m	Pr 137 70m ε β+1.7 E2.8	Pr 138 1+ 2.0h ε,β+1.4 γ.16–1.7	Pr 139 5/+ 4.5h ε,β+1.0 γ1.3,1.6 E4.3	Pr 140 1+ 3.4m β+2.32,·· ε γ1.6,1.9,.3 E3.3	Pr 141 5/+ 100 σ 19 140.90760	Pr 142 2– 19.2h β–2.15,.58 γ1.57 σ 20 E2.15	Pr 143 7/+ 13.7d β–.93 no γ σ 89 E.93
58	**Ce** 140.12 σ .7	Ce 131 30m β+4.2	Ce 132 4.2h ε, β+ γ.077,.097,.174	Ce 133 6.3h ε β+1.3(1.2) γ1.8(.8)	Ce134 72h ε no γ E.120	Ce 135 18h ε β+.81 γ.26,.60,.30,···	Ce 136 0.193 σ(.6+6.0) 135.907	Ce 137 34h | 9h IT.26 | ε ε | γ.010, γ.7,.8, | .46 .9	Ce 138 0.250 σ(.04+1.0) 137.9057	11/–Ce 139 3/+ 55s | 138d IT.74 | ε γ.166 E.27	Ce 140 88.48 σ .6 139.90539	Ce 141 7/– 32.5d β–.435,.58 γ.145 σ 30 E.58	Ce142 11.07 5 x 10^{15}y α 1.5 σ 1 141.9091

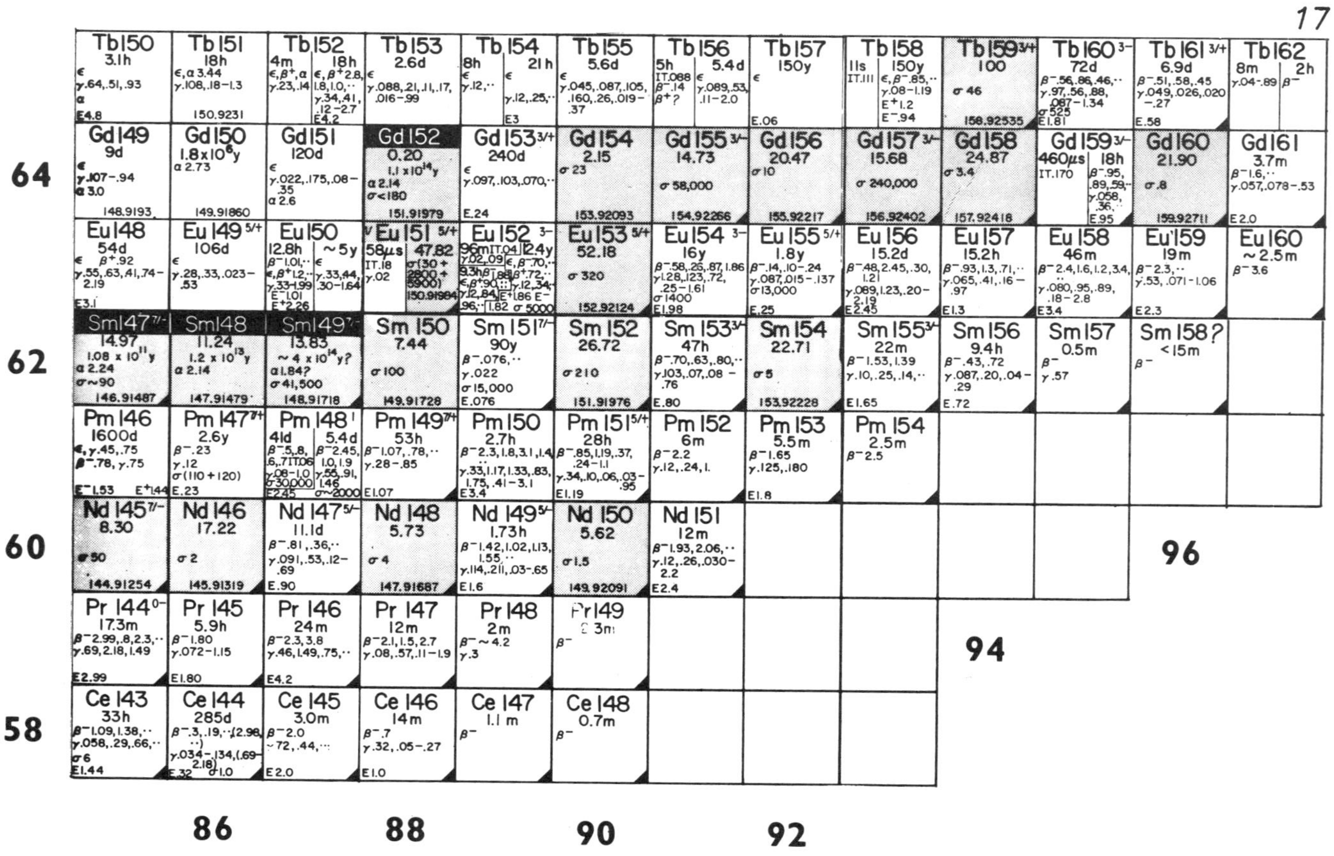

17

Z													
	Tb150 3.1h ε γ.64,.51,.93 α E4.8	Tb151 18h ε, α 3.44 γ.108,.18–1.3 150.9231	Tb152 4m: ε, β^+, α; γ.23,.14 18h: ε, β^+2.8, 1.8,1.0,··; γ.34,.41, .12–2.7; E4.2	Tb153 2.6d ε γ.088,.21,.11,.17, .016–.99	Tb154 8h: ε; γ.12,·· 21h: ε; γ.12,.25,··; E3	Tb155 5.6d ε γ.045,.087,.105, .160,.26,.019–.37	Tb156 5h: IT.088; β^-.14; β^+? 5.4d: ε; γ.089,.53, .11–2.0	Tb157 150y ε E.06	Tb158 11s: IT.111 150y: ε, β^-.85,··; γ.08–1.19; E^+1.2; E^-.94	Tb159 3/+ 100 σ 46 158.92535	Tb160 3– 72d β^-.56,.86,.46,·· γ.97,.56,.88, .087–1.34 σ 525 E1.81	Tb161 3/+ 6.9d β^-.51,.58,.45 γ.049,.026,.020 –.27 E.58	Tb162 8m: γ.04–.89 2h: β^-
64	Gd149 9d ε γ.107–.94 α 3.0 148.9193	Gd150 1.8x10^{6}y α 2.73 149.91860	Gd151 120d ε γ.022,.175,.08–.35 α 2.6	Gd152 0.20 1.1 x 10^{14}y α 2.14 σ<180 151.91979	Gd153 3/+ 240d ε γ.097,.103,.070,·· E.24	Gd154 2.15 σ 23 153.92093	Gd155 3/– 14.73 σ 58,000 154.92266	Gd156 20.47 σ 10 155.92217	Gd157 3/– 15.68 σ 240,000 156.92402	Gd158 24.87 σ 3.4 157.92418	Gd159 3/– 460μs: IT.170 18h: β^-.95, .89,.59,··; γ.058, .36,··; E.95	Gd160 21.90 σ .8 159.92711	Gd161 3.7m β^-1.6,·· γ.057,.078–.53 E2.0
	Eu148 54d ε β^+.92 γ.55,.63,.41,.74–2.19 E3.1	Eu149 5/+ 106d ε γ.28,.33,.023–.53	Eu150 12.8h: β^-1.01,··; ε, β^+1.2,··; γ.33–1.99; E^-1.01; E^+2.26 ~5y: ε; γ.33,.44, .30–1.64	Eu151 5/+ 58μs: IT.18; γ.02 47.82: σ(30 + 2800 + 5900) 150.91984	Eu152 3– 96m: IT.04; γ.02,.09 9.3h: β^-1.88; ε, β^+.90,··; γ.12,.84, .96,··; 1.82 12.4y: ε, β^-.70,··; β^+.72,··; γ.12,.34,··; E^+1.86 E^-; σ 5000	Eu153 5/+ 52.18 σ 320 152.92124	Eu154 3– 16y β^-.58,.26,.87,1.86 γ1.28,.123,.72, .25–1.61 σ 1400 E1.98	Eu155 5/+ 1.8y β^-.14,.10–.24 γ.087,.015–.137 σ 13,000 E.25	Eu156 15.2d β^-.48,2.45,.30, 1.21 γ.089,1.23,.20–2.19 E2.45	Eu157 15.2h β^-.93,1.3,.71,·· γ.065,.41,.16–.97 E1.3	Eu158 46m β^-2.4,1.6,1.2,3.4,·· γ.080,.95,.89, .18–2.8 E3.4	Eu159 19m β^-2.3,·· γ.53,.071–1.06 E2.3	Eu160 ~2.5m β^-3.6
62	Sm147 7/– 14.97 1.08 x 10^{11}y α 2.24 σ~90 146.91487	Sm148 11.24 1.2 x 10^{13}y α 2.14 147.91479	Sm149 7/– 13.83 ~4 x 10^{14}y? α 1.84? σ 41,500 148.91718	Sm150 7.44 σ 100 149.91728	Sm151 7/– 90y β^-.076,·· γ.022 σ 15,000 E.076	Sm152 26.72 σ 210 151.91976	Sm153 3/– 47h β^-.70,.63,.80,·· γ.103,.07,.08–.76 E.80	Sm154 22.71 σ 5 153.92228	Sm155 3/– 22m β^-1.53,1.39 γ.10,.25,.14,·· E1.65	Sm156 9.4h β^-.43,.72 γ.087,.20,.04–.29 E.72	Sm157 0.5m β^- γ.57	Sm158? <15m β^-	
	Pm146 1600d ε, γ.45,.75 β^-.78, γ.75 E^-1.53 E^+1.44	Pm147 7/+ 2.6y β^-.23 γ.12 σ(110 + 120) E.23	Pm148 1 41d: β^-.5,.8, .6,.7; IT.06; γ.08–1.0; σ 30,000; E2.45 5.4d: β^-2.45, 1.0,1.9; γ.55,.91, 1.46; σ~2000	Pm149 7/+ 53h β^-1.07,.78,·· γ.28–.85 E1.07	Pm150 2.7h β^-2.3,1.8,3.1,1.4,·· γ.33,1.17,1.33,.83, 1.75,.41–3.1 E3.4	Pm151 5/+ 28h β^-.85,1.19,.37, .24–1.1 γ.34,.10,.06,.03–.95 E1.19	Pm152 6m β^-2.2 γ.12,.24,1.	Pm153 5.5m β^-1.65 γ.125,.180 E1.8	Pm154 2.5m β^-2.5				
60	Nd145 7/– 8.30 σ 50 144.91254	Nd146 17.22 σ 2 145.91319	Nd147 5/– 11.1d β^-.81,.36,·· γ.091,.53,.12–.69 E.90	Nd148 5.73 σ 4 147.91687	Nd149 5/– 1.73h β^-1.42,1.02,1.13, 1.55,·· γ.114,.211,.03–.65 E1.6	Nd150 5.62 σ 1.5 149.92091	Nd151 12m β^-1.93,2.06,·· γ.12,.26,.030–2.2 E2.4						
	Pr144 0– 17.3m β^-2.99,.8,2.3,·· γ.69,2.18,1.49 E2.99	Pr145 5.9h β^-1.80 γ.072–1.15 E1.80	Pr146 24m β^-2.3,3.8 γ.46,1.49,.75,·· E4.2	Pr147 12m β^-2.1,1.5,2.7 γ.08,.57,.11–1.9	Pr148 2m β^-~4.2 γ.3	Pr149 [illegible] 3m β^-							
58	Ce143 33h β^-1.09,1.38,·· γ.058,.29,.66,·· σ 6 E1.44	Ce144 285d β^-.3,.19,··(2.98,··) γ.034–.134,(.69–2.18) σ 1.0 E.32	Ce145 3.0m β^-2.0 γ.72,.44,·· E2.0	Ce146 14m β^-.7 γ.32,.05–.27 E1.0	Ce147 1.1m β^-	Ce148 0.7m β^-							
		86		88		90		92		94		96	

Z													
			Lu 174.97 σ 80	Lu 155 0.07s α 5.63	Lu 156 0.23s | ~0.5s α 5.54 | α 5.43								
70			**Yb** 173.04 σ 37	Yb 154 0.39s α 5.33	Yb 155 1.65s α 5.21							Yb 162 ~ 24m ε γ.041	
			Tm 168.934 σ 125	Tm 153 1.58s α 5.11	Tm 154 2.98s | 5s α 5.04 | α 4.96							Tm 161 30m ε γ.084–.172 E3.5	Tm 162 80m | 22m ε γ.10,.24 | ε β^+3.8, 2.3,.90 γ.102,·· E4.9
68		**Er** 167.26 σ 160		Er 152 10.7s α 4.93 β^+	Er 153 36s α 4.80 β^+	Er 154 4.5m α 4.26		Er 156 ≤12m	Er 157 24m	Er 158 2.4h ε β^+.8 γ.067, 387	Er 159 ~ 1h ε γ.21,.05–.30	Er 160 29h ε (β^+.30–1.90) no γ (γ.73,.97,.107–2.76)	Er 161 3.1h ε,β^+1.21,.82 γ.067,.211,.83, 1.12,··
		Ho 164.930 σ 65	Ho 150? ~ 20s	Ho 151 42s | 36s α 4.60 | α 4.51	Ho 152 2.4m | 52s α 4.38 | α 4.45	Ho 153 6.5m | 9m α 4.00 | ε α 3.92	Ho 154 7m	Ho 155 ?16m | 46m β^+ or ε α 3.96 | β^+2.1 γ.14	Ho 156 57m β^+2.1 ε γ.14,.27,.69,··	Ho 157 18m	Ho 158 0.5h | 10.9m ε,β^+1.3 γ.07–1.07 IT.067 \| ε,β^+ γ.10–2.9 E4.04	Ho 159 33m ε γ.20 .30,··	Ho 160 4.8h | 25.6m IT.060 ε,β^+ γ.11–2.8 | ε,β^+.57, γ.73,.97, .65,.197 E3.3
66	**Dy** 162.50 σ 940		Dy 149 ~15m ε	Dy 150 7.4m β^+ α 4.2 γ.39	Dy 151 18m β^+ ε α 4.06 γ.15	Dy 152 2.4h ε γ.256 α 3.7 E.4 151.92473	Dy 153 6h ε α 3.5 γ.081,.100,.082–.54 152.9257	Dy 154 13h | ~10^6y α 3.35 | α 2.85 153.9243	Dy 155 10h ε β^+.85,·· γ.23,.090–1.67 E? 10	**Dy 156** 0.052 2 x 10^{14}y α~3 σ~3 155.9238	Dy 157 8.5h ε γ.327,.083,.061, .144,.182–.266	**Dy 158** 0.090 σ 100 157.9244	Dy 159 144d ε γ.058,.36,.29, .20,·· E.38
	Tb 158.925 σ 46	Tb 147 24m β^+ γ.30,.14	Tb 148 70m ε,β^+ γ.78,1.12 E5.6	Tb 149 4.3m | 4.1h ε α 3.99 | ε,α3.95 γ.17–.64 E3.8	Tb 150 3.1h ε γ.64,.51,.93 α E4.8	Tb 151 18h ε,α 3.44 γ.108,.18–1.3 150.9231	Tb 152 4m | 18h ε,β^+,α γ.23,.14 | ε,β^+2.8, 1.8,1.0,·· γ.34,.41, .12–2.7 E4.2	Tb 153 2.6d ε γ.088,.21,.11,.17, .016–.99	Tb 154 8h | 21h ε γ.12,·· | ε γ.12,.25,·· E3	Tb 155 5.6d ε γ.045,.087,.105, .160,.26,.019–.37	Tb 156 5h | 5.4d IT.088 β^-.14 β^+? | ε γ.089,.53, .11–2.0	Tb 157 150y ε E.06	Tb 158 11s | 150y IT.111 | ε,β^-.85,·· γ.08–1.19 E^+1.2 E^-.94
64	Gd 145 25m ε,β^+2.4 γ1.75,1.04,.80 E5.2	Gd 146 48d ε γ.115,.150	Gd 147 ~ 25h ε γ.23,.40,.90,.37, .14–1.33	Gd 148 85y α 3.18 147.91810	Gd 149 9d ε γ.107–.94 α 3.0 148.9193	Gd 150 1.8x10^6y α 2.73 149.91860	Gd 151 120d ε γ.022,.175,.08–.35 α 2.6	**Gd 152** 0.20 1.1 x 10^{14}y α 2.14 σ<180 151.91979	Gd 153 3/+ 240d ε γ.097,.103,.070,·· E.24	**Gd 154** 2.15 σ 23 153.92093	**Gd 155** 3/− 14.73 σ 58,000 154.92266	**Gd 156** 20.47 σ 10 155.92217	**Gd 157** 3/− 15.68 σ 240,000 156.92402
		82		84		86		88		90		92	

Z \ N	94	95	96	97	98	99	100	101	102	103	104	105	106
			Lu 167 55m ε β+.6,1.1,1.6, 2.06 γ.030,.24,.056−.40 E3.1	Lu 168 7m | 2h ε,β+1.2 | ε γ.99,.90,.087,·· | γ.087	Lu 169 1.5d ε γ.087,.024,.063,.071−2.2 E2.3	Lu 170 2.0d ε β+1.8 γ.084,.193,.245−3.09 E3.5	Lu 171 8.2d ε γ.74,.076,.067,.009−.85	1− Lu 172 4− 4m | 6.7d IT.042 | ε γ.079,1.09,.90,.091,.056−1.58	Lu 173 7/+ 1.4y ε γ.079,.101,.273,.171,·· E.69	Lu 174 75μs | 1300d IT.133 | ε 165d IT.059 ε γ.044−.99 | γ.077,1.2 E1.54	Lu 175 7/+ 97.41 σ (18+5) 174.9406	1− Lu 176 7− 3.7h | 2.59 β−1.23,1.32 | 2.2×10^10y γ.088 | β−.42 γ.31,.20,.09 σ(~1+2100) E1.02	23/− Lu 177 7/+ ~155d | 6.7d β−.153,.167 | β−.50,.18,.25,.38 IT.12 γ.055−.47 | γ.113,.208,.07−.32 E.50
70	Yb 164 75m ε β+2.9,··	Yb 165 10m β+ γ.080−1.44	Yb 166 58h ε γ.08 E.3	Yb 167 18m ε β+.65 γ.106,.113,.026−1.52 E1.97	Yb 168 0.135 σ 11,000 167.9342	1/− Yb 169 50s | 32d IT.024 | ε γ.008,.109,.021−.309	Yb 170 3.03 169.9350	Yb 171 1/− 14.31 170.9381	Yb 172 21.82 171.9366	Yb 173 5/− 16.13 172.9383	Yb 174 0.001s | 31.84 IT1.00,1.28 | σ(46+9) γ.27,.18,.076 173.9397	Yb 175 0.067s | 4.2d IT.51 | β−.47,.073,·· γ.40,.11,.28,.14−.25 E.47	Yb 176 12s | 12.73 IT.19 γ.29,.39 | σ 7 175.9427
	Tm 163 1.8h ε β+1.05,.40 γ.104,.242,.022−1.80 E2.3	Tm 164 1.9m β+2.9,1.3 γ.091,.21−2.38 E4.0	Tm 165 29h ε γ.054,.243,.015−1.38 E3.04	Tm 166 2+ 7.7h ε,β+1.9,1.2,·· γ.08,.18,.09−2.10	Tm 167 1/+ 9.6d ε γ.21,.057,.53	Tm 168 86d ε γ.83,.20,.45,.080−1.64 E1.72	Tm 169 1/+ 100 σ 125 168.93424	Tm 170 1− 125d β−.97,.88 γ.084 σ 150 E.97	Tm 171 1/+ 1.9y β−.098,.03 γ.067 E.098	Tm 172 2− 64h β−1.79,1.87,·· γ.079,.18−1.61 E1.87	Tm 173 8.2h β−.86,.90 γ.40,.47 E1.32	Tm 174 5.2m β−1.2,.7 γ.37,.50,.87,(.076−1.28) E3.0	Tm 175 20m β−2.0 γ(.51) E2.5
68	Er 162 0.136 σ 2 161.9287	Er 163 75m ε β+.19 γ.43,1.1,.3 E1.21	Er 164 1.56 σ 1.7 163.9293	Er 165 5/− 10.36h ε noγ E.37	Er 166 33.41 σ 12 165.93031	1/− Er 167 7/+ 2.5s | 22.94 IT.208 | σ 700 166.93216	Er 168 27.07 σ 2 167.9324	Er 169 1/− 9.4d β−.34,·· γ.008 E.34	Er 170 14.88 σ 9 169.9355	Er 171 5/− 7.5h β−1.06,.5−1.49 γ.112,.308,.124,.09−.96 E1.49	Er 172 49h β−.28,.36,.89 γ.05,.41,.61,·· E.89		
	Ho 161 7/− 2.5h ε γ.026,.078,.075,.044−.176	Ho 162 1+ 68m | 15m ε | ε IT.01 | β+1.08,1.13 γ.08,.04,1.34 | γ.08 E2.15	Ho 163 0.8s | <60y IT.3 | ε E.01	Ho 164 35m | ? β−.99,.90, ε IT(?).046 γ.073,.09,.037 E−1.02 E+1.12	Ho 165 7/− 100 σ(~1+64) 164.93042	Ho 166 0− 1.2×10^3y | 27.2h β−~.06 | β−1.85,1.76,·· γ.08,.184,.71,·· | γ.08,1.4,.7−1.8 E1.85	Ho 167 3.1h β−1.0,.36 γ.079,.083,.057,.074−.532 E1.0	Ho 168 3m β−2.0 γ.85	Ho 169 4.8m β−1.20,1.95,·· γ.76,.16−.92 E2.	Ho 170 44s β−3.3,·· γ.43 E3.3			106
66	Dy 160 2.29 σ 55 159.92520	Dy 161 5/+ 18.88 σ 600 160.92694	Dy 162 25.53 σ 150 161.92680	Dy 163 5/− 24.97 σ 125 162.92875	Dy 164 28.18 σ(2000+700) 163.93920	1/− Dy 165 7/+ 75.4s | 2.35h IT.108 | β−1.30,1.20,·· β−.88,·· γ.51,.36,.15,·· | γ.095,.028−1.1 E1.30 σ 4700	Dy 166 82h β−.40,.48,.06 γ.082,.03−.43 E.48	Dy 167 4.4m β− γ.19,.26,.31,.51,.57			104		
	Tb 159 3/+ 100 σ 46 158.92535	Tb 160 3− 72d β−.56,.86,.46,·· γ.97,.56,.88,.087−1.34 σ 525 E1.81	Tb 161 3/+ 6.9d β−.51,.58,.45 γ.049,.026,.020−.27 E.58	Tb 162 8m | 2h γ.04−.89 | β−	Tb 163 16m | 6.5h β− | β−1.4,1.7,1.5,~1.1 γ.08−.50 | γ.33,.03,.16−.51 E1.7	Tb 164 1d β− E4			102				
64	Gd 158 24.87 σ 3.4 157.92418	Gd 159 3/− 460μs | 18h IT.170 | β−.95,.89,.59,·· γ.058,.36,·· E.95	Gd 160 21.90 σ .8 159.92711	Gd 161 3.7m β−1.6,·· γ.057,.078−.53 E2.0	Gd 162 >1y β− γ.041−1.39								
	94		96		98		100						

		96		98		100		102		104		106	
76	**Os** 190.2 σ 15									Os180 22m ε	Os 181 2.9m ε γ.12,.14 / 105m ε γ.117,.142	Os 182 <1m / 21h ε γ.51,.18,.028–.37(.02–1.4)	1/- Os 183 9/+ 10h ε IT.171 γ1.1,1.03,.07 / 14h ε γ.38,.11,.15–1.44
	Re 186.2 σ 85							Re 177 17m β+7.4	Re 178 15m β+3.1 ε	Re 179 19.7m ε γ.29,.43,.42,···	Re 180 2.4m ε,β+1.1 γ.90,.104	Re 181 19h ε γ.020–1.54	Re 182 13h ε,β+1.7,.55 γ.10,1.2,··· / 64h ε γ.020–1.44 E2.86
74	**W** 183.85 σ 19					W174 29m ε	?W175 1.5h β+ γ	W 176 2.5h ε γ.10,.03–0.95	W 177 2.2h ε γ~.5,1.2	W178 22d ε (β+.88) γ.093,1.35	W179 5m IT.222 / 38m ε γ.031 E~1.2	W180 0.005s IT.15–.5 γ.10 / 0.14 σ<20 179.9470	5/- W 181 9/+ 14μs IT.37,.25 γ.113 / 130d ε γ.006,.15,.14 E.17
	Ta 180.948 σ 21				Ta172 44m β+	Ta173 2.5h ε γ.037–.181	Ta174 1.3h β+ γ.091–.35	Ta175 10.5h ε γ.082,.207,.267,.104,.186–1.15 E2.3	Ta176 8.0h ε γ.088,.100–2.08	Ta177 56.4h ε γ.113,.072–1.058 E1.16	1+ Ta178 9.4m ε,β+.88 γ.093,1.35,·· / 2.2h ε,β+ γ.332,(.09–.43) E1.9	Ta179 ~1.6y ε E.119	Ta180 8.1h ε β−.62,.71 γ.093D,.102 / 0.0123 179.9475
72	**Hf** 178.49 σ 105	Hf168 22m ε β+1.7 γ.13,.17	Hf169 1.5h ε β+2.4,1.3 γ.05,.12	Hf170 12h ε β+ γ.12,.16,.21,.045–.77	Hf171 14h ε γ.098,.63,.12,.17–1.1	Hf172 ~5y ε γ.123,.080	Hf173 1/- 24h ε γ.140,.124,.135,.163–1.2	Hf174 0.18 2.0×10^15 y α~2.5 σ 400 173.9403	Hf175 70d ε γ.343,.089,.113–.432 E.590	Hf176 5.20 σ<30 175.9417	Hf177 7/- 18.50 σ(1.4+370) 176.9435	8- Hf178 3.8s IT.089 γ.093–.43 / 27.14 σ(50+30) 177.9439	1/- Hf179 9/+ 18s IT.161,.375 γ.217 / 13.75 σ(.2+65) 178.9460
	Lu 174.97 σ 80	Lu167 55m ε β+.6,1.1,1.6,2.06 γ.030,.24,.056–.40 E3.1	Lu168 7m ε,β+1.2 γ.99,.90,.087,·· / 2h ε γ.087	Lu169 1.5d ε γ.087,.024,.063,.071–2.2 E2.3	Lu170 2.0d ε β+1.8 γ.084,.193,.245–3.09 E3.5	Lu171 8.2d ε γ.74,.076,.067,.009–.85	1- Lu172 4- 4m IT.042 / 6.7d ε γ.079,1.09,.90,.091,.056–1.58	Lu173 7/+ 1.4y ε γ.079,.101,.273,.171,·· E.69	Lu174 75μs IT.133 165d IT.059 ε γ.044–.99 / 1300d ε γ.077,1.2 E1.54	Lu175 7/+ 97.41 σ(18+5) 174.9406	1- Lu176 7- 3.7h β−1.23,1.32 γ.088 / 2.59 2.2×10^10 y β−.42 γ.31,.20,.09 σ(~1+2100) E1.02	23/- Lu177 7/+ ~155d β−.153,.67 IT.12 γ.055–.47 / 6.7d β−.50,.18,.25,.38 γ.113,.208,.07–.32 E.50	Lu178 30m? β−1.5 γ.089,.093–.43) / 5m? β−2.25,.77 γ.33,(.089,.093–.43) E2.25
70	Yb165 10m β+ γ.080–1.44	Yb166 58h ε γ.08 E.3	Yb167 18m ε β+.65 γ.106,.113,.026–1.52 E1.97	Yb168 0.135 σ 11,000 167.9342	1/- Yb169 50s IT.024 / 32d ε γ.008,.109,.021–.309	Yb170 3.03 169.9350	Yb171 1/- 14.31 170.9381	Yb172 21.82 171.9366	Yb173 5/- 16.13 172.9383	Yb174 0.001s IT1.00,1.28 γ.27,.18,.076 / 31.84 σ(46+9) 173.9397	Yb175 0.067s IT.51 / 4.2d β−.47,.073,·· γ.40,.11,.28,.14–.25 E.47	Yb176 12s IT.19 γ.29,.39 / 12.73 σ 7 175.9427	1/- Yb177 6.4s IT.23 γ.10 / 1.9h β−1.40,1.26,··· γ.15,.12,1.1,.14–1.24 E1.40
	Tm164 1.9m β+2.9,1.3 γ.091,.21–2.38 E4.0	Tm165 29h ε γ.054,.243,.015–1.38	Tm166 2+ 7.7h ε,β+1.9,1.2,·· γ.08,.18,.09–2.10 E3.04	Tm167 1/+ 9.6d ε γ.21,.057,.53	Tm168 86d ε γ.83,.20,.45,.080–1.64 E1.72	Tm169 1/+ 100 σ 125 168.93424	Tm170 1- 125d β−.97,.88 γ.084 σ 150 E.97	Tm171 1/+ 1.9y β−.098,.03 γ.067 E.098	Tm172 2- 64h β−1.79,1.87,··· γ.079,.18–1.61 E1.87	Tm173 8.2h β−.86,.90 γ.40,.47 E1.32	Tm174 5.2m β−1.2,.7 γ.37,.50,.87,(.076–1.28) E3.0	Tm175 20m β−2.0 γ(.51) E2.5	Tm176 1.5m β−4.2 E4.2

Z													
76	1/- Os 183 9/+ 10h ε IT.171 γ1.1,1.03,.07 | 14h ε γ.38,.11,.15–1.44	Os 184 0.018 σ<200 183.9527	Os 185 1/- 94d ε γ.65,.88,.87,.072–.75 E.98	Os 186 1.59 185.9539	Os 187 1/- 1.64 186.9558	Os 188 13.3 187.9560	9/- Os 189 3/- 5.7h IT.03 | 16.1 σ(.008+?) 188.9583	Os 190 10m IT.04 γ.50,.62,.36,.19 | 26.4 σ(8.6+3.9) 189.9586	Os 191 14h IT.014 | 15d β−.14 γ(.042,.129,··) E.31	Os 192 41.0 σ1.6 191.9614	Os 193 32h β−1.13,·· γ.139,.073,.107–.56 σ200 E1.13	Os 194 6.0y β−.097,.054,··· γ.043,·· E.097	Os 195 6m β−2
	Re 182 13h ε,β+1.7,.55 γ.10,1.2,··· | 64h ε γ.020–1.44 E2.86	Re 183 5/+ 0.001s IT.194 γ.11–.30 | 70d ε γ.046,.162,.053,.041–.41	Re 184 165d ε IT.08 γ.03–1.4 | 38d ε γ.90,.111,.79,.06–1.4 E~1.6	Re 185 5/+ 37.07 σ110 184.9531	Re 186 1/- 90h β−1.07,.93,··,ε γ.137D,.123,·· E+.54 E−1.07	Re 187 5/+ 62.93 5 x 10^10 y β−.0025 σ(1.3+70) 186.9558	Re 188 1/- 18.6m IT.002,.02 γ.06,.11,.09 | 16.8h β−2.12,2.0,·· γ.155D,.218–1.96 σ<2 E2.12	Re 189 ?140d β | 24h β−1.0,.8,·· γ.031,.070,.030 E1.0–2.5	Re 190 2.8h β−~1.6 γ | 3m β−1.7,·· γ.19,.57,.39,.83,·· E3.1	Re 191 10m β−1.8	Re 192 6.2s γ		
74	5/- W 181 9/+ 14μs IT.37,.25 γ.113 | 130d ε γ.006;.15,.14 E.17	W 182 26.41 σ(.5+20) 181.9483	W 183 1/- 5.1s IT.102 γ.16,.21 | 14.40 σ11 182.9503	W 184 30.64 σ(.01+2.1) 183.9510	W 185 3/- 1.7m IT γ.075–.175 | 74d β−.43,·· γ.125 E.43	W 186 28.41 σ40 185.9544	W 187 3/- 24.0h β−.63,1.31,.33 γ.69,.13,.072–.87 σ~90 E1.31	W 188 69d β−.35,.06,·· γ.29,·· E.43	W 189 11m β−2.5,··· γ.130,.178,.258,.417 E2.5				
	Ta 180 8.1h ε β−.62,.71 γ.093D,.102 | 0.0123 179.9475	1/+ Ta 181 7/+ 20μs IT γ | 7μs IT.006 | 99.988 σ(.07+21) 180.9480	Ta 182 16m IT.18,.36 γ.15–.32 E1.81 | 115d β−.52,.25,.43,·· γ.10,1.12,1.22,1.19,·· σ8000	Ta 183 7/+ 5.0d β−.615,·· γ.046,.108,.053,.246,.041–.41 E1.07	Ta 184 8.7h β−1.19,1.45–2.64 γ.111,.41,.25,.92,.16–1.44 E2.9	Ta 185 49m β−1.72·· γ.175,.10,.075,.11–.24 E1.96	Ta 186 10m β−2.2 γ.20,.73,.12–1.1 E3.7						
72	1/- Hf 179 9/+ 18s IT.161,.375 γ.217 | 13.75 σ(.2+65) 178.9460	8- Hf 180 5.5h IT.058,.50 γ.093–.44 E1.14 | 35.24 σ10 179.9468	Hf 181 1/- 43d β−.41,·· γ.48,.006–.70 σ~40 E1.02	Hf 182 9 x 10^6 y β−~.4 γ.27 E~.5	Hf 183 91d β−2.2,1.9 γ.25,.59 ··· | 64m β−1.4 γ.095,.405,.25 ··· E2 1								
	Lu 178 30m? β−1.5 (γ.089,.093–.43) E2.25 | 5m? β−2.25,.77 γ.33,(.089,.093–.43)	Lu 179 4.6h β−1.34,1.08 γ.22 E1.34	Lu 180 2.5m β−3.3 E3.3										
70	1/- Yb 177 6.4s IT.23 γ.10 E1.40 | 1.9h β−1.40,1.26,··· γ.15,.12,1.1,.14–1.24												
	Tm 176 1.5m β−4.2 E4.2												
N		108		110		112		114		116		118	

		98		100		102		104		106		108	
84											Po	Po 192 0.5s α 6.58	Po 193 4s α 7.0
										Bi 208.980 σ .034			
82										Pb 207.19 σ .17			
									Tl 204.37 σ 3.3				
80								Hg 200.59 σ 360	Hg 185 53s α 5.64		Hg 187 3m ε γ .18, .26, .40	Hg 188 or 187 3m α 5.14	Hg 189 9.3m ε γ .029, .14, .22
	Au 196.967 σ 98.8	Au 177 1.4s α 6.11	Au 178 2.7s α 5.91	Au 179 7.1s α 5.84		Au 181 10s α 5.60, 5.47		Au 183 44s α 5.34		Au 185 4.33m α 5.07	Au 186 12m ε	Au 187 8m ε no γ	Au 188 8m ε E5.2
78	Pt 195.09 σ 10	Pt 176 6s α 5.74	Pt 177 6.5s α 5.52	Pt 178 22s α 5.43	Pt 179 33s α 5.15	Pt 180 50s α 5.14	Pt 181 48s α 5.02	Pt 182 3m ε α 4.82	Pt 183 6m ε α 4.74, 4.70	Pt 184 42m: ε, γ .68, 1.72, 1.85 20m: ε, α 4.48, γ .155, .190	Pt 185 1h ε γ .035, .63, 1.56	Pt 186 2.0h ε α 4.23 γ .065, .14, .19, .68, 1.40	Pt 187 2.1h ε γ .11, .18, 2.01
	Ir 192.2 σ 460								Ir 182 15m ε, β+ γ .13, .28	Ir 183 58m ε γ .24, ..	Ir 184 3.2h ε, β+ γ .26, .12, .39–4.3	Ir 185 14h ε γ .037–.254	Ir 186 ?1.7h: γ .14 15h: ε, β+ 1.92, .., γ .07–1.7, E3.8

Z \ N	110	111	112	113	114	115	116	117	118	119	120	121	122
84	Po 194 0.5s α 6.85	Po 195 1s α 6.72, 6.63	Po 196 5.9s α 6.53	Po 197 17s α 6.39, 6.30	Po 198 1.75m α 6.18	Po 199 4.24m α 6.06	Po 200 11.4m α 5.86	Po 201 3/ 17.5m ε α 5.68, 5.57, 5.77	Po 202 45m ε α 5.58	Po 203 5/ 45m ε α 5.48	Po 204 3.5h ε α 5.37	Po 205 5/- 644μs IT.16 γ.71 | 1.8h ε α 5.2	Po 206 8.8d ε, α 5.22 γ.060, .083-1.32 E 2.03 205.9803
83					Bi <198 1.7m α 6.2	Bi 198 8m ε α 5.81 E7.4	Bi 199 9/ 24m α 5.52 | 27m ε	Bi 200 7 35m ε γ1.03, .46	Bi 201 9/- 52m ε α5.27 | 1.8h ε (γ.63) E4	Bi 202 5 1.6h ε γ .42, .96	Bi 203 9/- 11.8h ε, β+ 1.35, .73 α 4.85 γ.06-1.9 E3.19	Bi 204 6+ 11.2h ε γ.08, .38, .22, .91, .10-1.2 E4	Bi 205 9/- 15.3d ε, β+ .98 γ.70, 1.77, .12-2.6, (.026, .99) E2.70
82			Pb 194 11m γ.204	Pb 195 17m ε (γ.099, .39, .38)	Pb 196 37m ε γ.191, .240, .253, ··	Pb 197 42m ε IT.234 γ.387, .385, .222 | ?	Pb 198 2.4h ε γ.290, .365, .031-.865	Pb 199 12m IT.424 | 1.5h ε, β+2.8 γ.367, .27-1.13 E4.2	Pb 200 21h ε γ.148, .142, .24, .27, .033-.45	Pb 201 6ls IT.63 | 9.4h ε, β+ γ.330, .13-1.10	9- Pb 202 3.6h IT.79, .13 ε γ.96, .24-.66 | ~3x10^5 y ε (γ.44) E.05	Pb 203 6.1s IT.83 | 52h ε γ.28, .40, .68 E.82	9- Pb 204 67m IT.91 γ.29-.90 | 1.48 1.4x10^17 y α 2.6 σ.7 203.97304
81	Tl 191 10m ε	Tl 192 11m ε γ.11, .42 | ≤11m ε	Tl 193 2.1m IT<.025 γ.365 | 23m ε γ.24, .25, .26, .27-.33	Tl 194 ?33m ε γ.097 α 5.8 | 35m ε γ.43	Tl 195 1/+ 3.5s IT.099 γ.393, .383 | 1.2h ε, β+~1.8 γ.037, .23-.88 E3	Tl 196 1.4h IT.120 ε, γ.034-.425 | 1.8h ε γ.425 E4.6	9/- Tl 197 1/+ 0.54s IT.222 γ.387, .385 | 2.8h ε γ.152, .018-1.0 E2.2	7+ Tl 198 2- 1.9h ε, γ.048, ·· IT.261 γ.283, ·· | 5.3h ε, β+2.4, 2.1, ·· γ.412, .195- E3.5 2.8	Tl 199 1/+ 0.027s IT.38, .03 γ.37, .35, .72 E1.1 .158, .050-.492 | 7.4h ε γ.208, .455	Tl 200 2- 0.034s IT γ.20, .52 γ.37, 1.21, .116-2.1 | 26h ε, β γ1.06, 1.44, ·· E2.45	Tl 201 1/+ 0.002s IT.22, .59 γ.33, ·· | 73h ε γ.167, .135, .031-.166 E.4	Tl 202 2- 500μs IT.46 γ.49 | 12.4d ε γ.44, .96, .52 E1.22	Tl 203 1/+ 29.50 σ 11 202.97235
80	Hg 190 20m ε γ.14, .22, .03	Hg 191 57m ε γ.253, .274	Hg 192 4.9h β+1.2 γ.031, .272, .04-.44	Hg 193 3/- 11h ε, β+1.17, ·· IT.101 γ.039, .26, ·· | 3.5h ε γ.038, .187, .76 E2.3	Hg 194 0.4s IT γ.048, .134 | 700d ε (γ.33, .29, ··)	13/+ Hg 195 1/- 40h IT.123 ε γ.016, .037, ·· | 9.5h ε γ.061, .18, .78, .20-1.17	Hg 196 0.146 σ (25 + 880) 195.96572	13/+ Hg 197 1/- 24h IT.165 ε γ.130, .41, .28 | 65h ε γ.077D, .191 E.7	Hg 198 10.02 σ (.02 + ?) 197.96676	13/+ Hg 199 1/- 44m IT.370 γ.158D | 16.84 σ 2000 198.96828	Hg 200 23.13 σ <50 199.96833	Hg 201 3/- ?100μs IT.53 | 13.22 σ <50 200.97031	Hg 202 29.80 σ 4 201.97164
79	Au 189 38m ε γ.29, .14, ··	Au 190 1 43m ε, β+ γ.295, .30-3.5	Au 191 3/+ 3.2h ε γ.091, .048, .16, .13, .03-2.17	Au 192 1- 5.0h ε, β+2.22 γ.32, .29, .05-3.04 E3.24	11/- Au 193 3/+ 4s IT.032, .290, ε γ.258 | 18h ε γ.19, .11, .01-.49	Au 194 1- 39h ε, β+1.16, 1.49 γ.33, .20, .10-2.3 E2.51	11/- Au 195 3/+ 31s IT.057, .318 γ.261 | 183d ε γ.093, .03, .13 E.22	12- Au 196 2- 9.7h IT.175 γ.15, .19, ·· | 6.2d ε, β-.26 γ.356, .333, .43, E+1.48 E-.68	11/- Au 197 3/+ 7.2s IT.130, .409 γ.279 | 100 σ 98.8 196.96654	Au 198 2- 64.8h β-.96, .29, 1.37 γ.412D, .674, ·· σ 26,000 E1.37	Au 199 3/+ 3.15d β-.30, .25, .46 γ.158D, .208D, .050D σ~30 E.46	Au 200 48m β-2.2, .7 γ1.23, .37, ·· E2.2	Au 201 22m β-1.5, ·· γ.53 E1.5
78	Pt 188 10d ε γ.195, .187, .054, .035-.47 α 3.93 E.51	Pt 189 10.9h ε γ.25, .15, ··	Pt 190 0.0127 7x10^11 y α 3.16 σ~150 189.9600	Pt 191 3.0d ε γ.54, .36, .41, ··	Pt 192 0.78 σ (2+ <14) 191.9612	Pt 193 4.4d IT.136 γ.013 E.148 | <500y ε noγ E.05	Pt 194 32.9 σ (.09 + 1.1) 193.9628	13/+ Pt 195 1/- 4.1d IT.030, γ.099 E.129 | 33.8 σ 27 194.9648	Pt 196 25.3 σ (~0 + .05 + .9) 195.9650	Pt 197 2.8h? IT | 1.5h IT.346 γ.05 β-.73, ·· γ.28, .13, ·· | 20h β-.67, .48, γ.077, .19, .27 E.75	Pt 198 7.21 σ (.03 + 4) 197.9679	Pt 199 14s IT.39 γ.032 | 31m β-1.7, .7-1.6 γ.54, .075-.96 σ~15 E1.7	Pt 200 11.5h β-(2.2, ··) (γ1.23-1.60)
77	Ir 187 0.029s IT.12 γ | 11h ε γ.01-1.1	Ir 188 2- 41h ε, β+1.7, 1.2 γ.155D, .63, .48, .32-2.18 E2.8	Ir 189 0.014s IT.11 γ.19, .30 | 13.3d ε γ.070, .245, .031-.798	Ir 190 3.2h ε, β+2.0 γ.19, .36, ·· | 1.2h IT.026 | 12d ε γ.19, .60, .56, .52, .2-1.3 E2.1	Ir 191 3/+ 4.9s IT.042 γ.129, .047, .082 | 37.3 σ (.3+250 +750) 190.9606	9 Ir 192 4- ~600y IT.161 σ(700 +?) E- | 1.4m IT .058 β- 1.45 E+1.2 | 74d β.67, .53, ·· ε, γ .32, .47, .2-2.0	11/- Ir 193 3/+ 12d IT.08 | 62.7 σ110 192.9630	Ir 194 1- 0.05s IT.10 | 19h β-2.35, 2.02, ··· γ.33, .29, .62, ··· E2.35	Ir 195 4.2h β-1.0, .6, ·· γ.19, .29, ·· E1.0	Ir 196 2.0h β-	Ir 197 7m β-2.0, 1.5 γ.50 E2.0	Ir 198 50s β-3.6 γ.78 E4.4	

					At		At 198? 43s α 6.50			At 201 1.5m α 6.35	At 202 3m ε α 6.13, 6.23 E8	At 203 7.2m ε α 6.09 E6	At 204 9m ε α 5.95 E7
84	Po	Po 192 0.5s α 6.58	Po 193 4s α 7.0	Po 194 0.5s α 6.85	Po 195 1s α 6.72, 6.63	Po 196 5.9s α 6.53	Po 197 17s α 6.39, 6.30	Po 198 1.75m α 6.18	Po 199 4.24m α 6.06	Po 200 11.4m α 5.86	Po 201 3/ 17.5m ε α 5.68, 5.57, 5.77	Po 202 45m ε α 5.58	Po 203 5/ 45m ε α 5.48
	Bi 208.980 σ .034							Bi <198 1.7m α 6.2	Bi 198 8m ε α 5.81 E7.4	Bi 199 9/ 24m α 5.52 | 27m ε	Bi 200 7 35m ε γ1.03, .46	Bi 201 9/– 52m ε α5.27 | 1.8h ε (γ.63) E4	Bi 202 5 1.6h ε γ.42, .96
82	Pb 207.19 σ .17					Pb 194 11m γ.204	Pb 195 17m ε (γ.099, .39, .38)	Pb 196 37m ε γ.191, .240, .253, ··	Pb 197 42m ε IT.234 γ.387, .385, .222 | ?	Pb 198 2.4h ε γ.290, .365, .031–.865	Pb 199 12m IT.424 | 1.5h ε, β+ 2.8 γ.367, .27–1.13 E4.2	Pb 200 21h ε γ.148, .142, .24, .27, .033–.45	Pb 201 61s IT.63 | 9.4h ε, β+ γ.330, .13–1.10
	Tl 204.37 σ 3.3			Tl 191 10m ε	Tl 192 11m ε γ.11, .42 | ≤11m ε	Tl 193 2.1m IT<.025 γ.365 | 23m ε γ.24, .25, .26, .27–.33	Tl 194 ?33m ε γ.097 α 5.8 | 35m ε γ.43	Tl 195 1/+ 3.5s IT.099 γ.393, .383 | 1.2h ε, β+~1.8 γ.037, .23–.88 E3	Tl 196 1.4h IT.120 ε, γ.034–.425 | 1.8h ε γ.425 E4.6	9/– Tl 197 1/+ 0.54s IT.222 γ.387, .385 | 2.8h ε γ.152, .018–1.0 E2.2	7/– Tl 198 2– 1.9h ε, γ.048, ·· IT.261 γ.283, ·· | 5.3h ε, β+ 2.4, 2.1, ·· γ.412, .195– E3.5 2.8	Tl 199 1/+ 0.027s IT.38, .03 γ.37, .35, .72 | 7.4h ε γ.208, .455, .158, .050–.492 E1.1	Tl 200 2– 0.034s IT γ.20, .52 | 26h ε, β 1.06, 1.44, ·· γ.37, 1.21, .116–2.1 E2.45
80	Hg 187 3m ε γ.18, .26, .40	Hg 188 or 187 3m α 5.14	Hg 189 9.3m ε γ.029, .14, .22	Hg 190 20m ε γ.14, .22, .03	Hg 191 57m ε γ.253, .274	Hg 192 4.9h β+ 1.2 γ.031, .272, .04–.44	Hg 193 3/– 11h ε, β+ 1.17, ·· IT.101 γ.039, .26, ·· | 3.5h ε γ.038, .187, .76 E2.3	Hg 194 0.4s IT γ.048, .134 | 700d ε (γ.33, .29, ··)	13/+ Hg 195 1/– 40h IT.123 ε γ.016, .037, ·· | 9.5h ε γ.061, .18, .78, .20–1.17	Hg 196 0.146 σ (25+880) 195.96572	13/+ Hg 197 1/– 24h IT.165 ε γ.130, .41, .28 | 65h ε γ.077D, .191 E.7	Hg 198 10.02 σ (.02+?) 197.96676	13/+ Hg 199 1/– 44m IT.370 γ.158D | 16.84 σ 2000 198.96826
	Au 186 12m ε	Au 187 8m ε no γ	Au 188 8m ε E5.2	Au 189 38m ε γ.29, .14, ··	Au 190 1 43m ε, β+ γ.295, .30–3.5	Au 191 3/+ 3.2h ε γ.091, .048, .16, .13, .03–2.17	Au 192 1– 5.0h ε, β+ 2.22 γ.32, .29, .05–3.04 E3.24	1/– Au 193 3/+ 4s IT.032, .290, ε γ.258 | 18h ε γ.19, .11, .01–.49	Au 194 1– 39h ε, β+ 1.16, 1.49 γ.33, .20, .10–2.3 E2.51	11/– Au 195 3/+ 31s IT.057, .318 γ.261 | 183d ε γ.093, .03, .13 E.22	12– Au 196 2– 9.7h IT.175 γ.15, .19, ·· | 6.2d ε, β– .26 γ.356, .333, .43, ·· E+ 1.48 E– .68	11/– Au 197 3/+ 7.2s IT.130, .409 γ.279 | 100 σ 98.8 196.96654	Au 198 2– 64.8h β– .96, .29, 1.37 γ.412D, .674, ·· σ 26,000 E1.37
78	Pt 185 1h ε γ.035, .63, 1.56	Pt 186 2.0h ε α 4.23 γ.065, .14, .19, .68, 1.40	Pt 187 2.1h ε γ.11, .18, 2.01	Pt 188 10d ε γ.195, .187, .054, .035–.47 α 3.93 E.51	Pt 189 10.9h ε γ.25, .15, ··	Pt 190 0.0127 7×10^{11} y α 3.16 σ~150 189.9600	Pt 191 3.0d ε γ.54, .36, .41, ··	Pt 192 0.78 σ (2+<14) 191.9612	Pt 193 4.4d IT.136 γ.013 E.148 | <500y ε no γ E.05	Pt 194 32.9 σ (.09+1.1) 193.9628	13/+ Pt 195 1/– 4.1d IT.030, γ.099 E.129 | 33.8 σ 27 194.9648	Pt 196 25.3 σ (~0+.05+.9) 195.9650	Pt 197 2.8h? IT | 1.5h IT.346 γ.05 β– .73, ·· γ.28, .13, ·· | 20h β– .67, .48, ·· γ.077, .19, .27 E.75
		108		110		112		114		116		118	

25

Z	N 120	121	122	123	124	125	126	127	128	129	130	131	132
	At 205 26m ε α 5.90 E5	At 206 30m ε α 5.70−5.90 E6	At 207 1.8h ε α 5.76 E3.7	At 208 1.6h ε α 5.65 γ.66, 1.75, ·· E5 207.9870	At 209 5.5h ε, α 5.65 γ.195, .091, .55, .78 E3.5	At 210 8.3h ε, α 5.36−5.52 γ.047, .25, .12, .08−2.6 E3.9	At 211 9/− 7.2h ε α 5.87, (7.45, ··) γ.67, (.89, .57) E.75 210.9875	At 212 0.12s: α 7.82, 7.88; γ.06 0.31s: α 7.66, 7.60; γ.06 211.9907	At 213 < 2s α 9.2 212.9931	At 214 < 5s α 8.78 213.9963	At 215 ~ 100μs α 8.00 214.9987	At 216 ~ 300μs α 7.79 216.0024	At 217 0.032s α 7.07 217.0046
84	Po 204 3.5h ε α 5.37	Po 205 5/− 644μs: IT.16; γ.71 1.8h: ε; α 5.2	Po 206 8.8d ε, α 5.22 γ.060, .083−1.32 E 2.03 205.9803	Po 207 5/− 2.5s: IT.26 47μs: IT.31; γ.82 6.2h: ε, β+ 1.14, ··; α 5.10; γ.10−2.06 206.9816	Po 208 2.9y α 5.11 γ.6, .28 207.9813	Po 209 1/− 103y α 4.88, ··, ε γ.91, .26 E1.90 208.9825	Po 210 RaF 138.40d α 5.30, ·· γ.80 σ(<.0005+<.03) 209.9829	Po 211 25s: α 7.1, 8.7, ··; γ 1.06, .57, .89 AcC′ 0.52s: α 7.45, ··; γ.89, 1.06, ·· 210.9866	Po 212 46s: α 11.7, ··; γ 2.6, .6 ThC′ 0.30μs: α 8.78 211.9889	Po 213 4μs α 8.35 212.9928	Po 214 RaC′ 164μs α 7.69 213.9952	Po 215 AcA 0.0018s α 7.38 β− 214.9995	Po 216 ThA 0.15s α 6.78 216.0019
	Bi 203 9/− 11.8h ε, β+ 1.35, .73 α 4.85 γ.06−1.9 E3.19	Bi 204 6+ 11.2h ε γ.08, .38, .22, .91, .10−1.2 E4	Bi 205 9/− 15.3d ε, β+ .98 γ.70, 1.77, .12−2.6, (.026, .99) E2.70	Bi 206 6+ 6.24d ε γ.80, .88, .18, .11−1.9 E3.6	Bi 207 30y ε γ.57, 1.77, ··, (1.06) E2.36	Bi 208 0.0027s: IT.92; γ.51 $3.7x10^5$y: ε; γ 2.62 E2.87	Bi 209 9/− 100 σ(.019+.015) 208.98040	Bi 210 1− RaE 5.0d: β− 1.16; α 4.7, ··; E1.16 $2.6x10^6$y: α 4.95, 4.92, 4.57, ··; γ.26, .30, ··	Bi 211 AcC 2.15m α 6.62, 6.28, ·· β−, γ.35 E−.60 210.9873	Bi 212 1− ThC 60.6m β− 2.25 α 6.05, 6.09, ··, (9.5−10.5) γ.040−2.20 E2.25	Bi 213 47m β− 1.39, .96 γ.44, .32 α 5.86, 5.55 E1.39	Bi 214 RaC 19.7m β− .4−3.2 α 5.5, ·· (8.3−10.5) γ.61, 1.12, 1.76, .45−2.43 E3.2	Bi 215 8m β− E2.24
82	9 Pb 202 3.6h: IT.79, .13; ε; γ.96, .24−.66 ~$3x10^5$y: ε; (γ.44) E.05	Pb 203 6.1s: IT.83 52h: ε; γ.28, .40, .68 E.82	9- Pb 204 67m: IT.91; γ.29−.90 1.48: $1.4x10^{17}$y; α 2.6; σ.7 203.97304	Pb 205 0.004s: IT.026; γ.99, ·· $3x10^7$y: ε E.03	Pb 206 23.6 σ.03 205.97447	13/+ Pb 207 1/− 0.8s: IT 1.06; γ.57 22.6: σ.72 206.97590	Pb 208 52.3 σ.0005 207.97665	Pb 209 3.3h β− .64 no γ E.64	Pb 210 RaD 22y β− .015, .061 γ.047 α 3.57 E.061	Pb 211 AcB 36.1m β− 1.34, .56, ·· γ.83, .41, .06−1.1 E1.34	Pb 212 ThB 10.64h β− .34, .58, ·· γ.239, .30, .11−.41 E.58	Pb 213 10m β−	Pb 214 RaB 26.8m β− .7, ·· γ.352, .295, .050−.259 E1.0
	Tl 201 1/+ 0.002s: IT.22, .59; γ.33, ·· 73h: ε; γ.167, .135, .031−.166 E.4	Tl 202 2− 500μs: IT.46; γ.49 12.4d: ε; γ.44, .96, .52 E1.22	Tl 203 1/+ 29.50 σ 11 202.97235	Tl 204 2− 3.75y β− .765 ε E− .765 E+ .36	Tl 205 1/+ 70.50 σ.11 204.97444	Tl 206 RaE″ 4.3m β− 1.52 no γ E1.52	Tl 207 AcC″ 4.78m β− 1.47, ·· γ.89 E1.47	Tl 208 5+ ThC″ 3.1m β− 1.80, 1.0−2.38 γ 2.61, .58, .51, .23−1.09 E4.99	Tl 209 2.2m β− 1.8 γ.12D, .45, 1.56 E3.9	Tl 210 RaC″ 1.3m β− 1.9, 1.3, 2.3 (n) γ.80, .30, .01−2.43 E5.4			
80	Hg 200 23.13 σ<50 199.96833	Hg 201 3/− ?100μs: IT.53 13.22: σ<50 200.97031	Hg 202 29.80 σ 4 201.97164	Hg 203 46.6d β− .21 γ.279 E.49	Hg 204 6.85 σ.4 203.97349	Hg 205 5.2m β− 1.6, 1.4 γ.20 E1.6	Hg 206 8.2m β− 1.3, 1.1 γ.31 E1.3						
	Au 199 3/+ 3.15d β− .30, .25, .46 γ.158D, .208D, .050D σ~30 E.46	Au 200 48m β− 2.2, .7 γ 1.23, .37, ·· E2.2	Au 201 22m β− 1.5, ·· γ.53 E1.5	Au 202 ~ 25s β− or IT	Au 203 55s β− 1.9 γ.69								
78	Pt 198 7.21 σ(.03+4) 197.9679	Pt 199 14s: IT.39; γ.032 31m: β− 1.7, .7−1.6 γ.54, .075−.96 σ~15 E1.7	Pt 200 11.5h β− (2.2, ··) (γ 1.23−1.60)	Pt 201 2.5m β− .90, 2.66, ·· γ 1.76, .23, ·· E2.66									
	120		122		124		126		128		130		132

Z													
88											Ra 226.0254 σ 20		Ra 213 2.7m α 6.90
			Fr		Fr 204 2s α 7.02	Fr 205 3.7s α 6.91	Fr 206 16s α 6.77	Fr 207 19s α 6.77	Fr 208 38s α 6.65	Fr 209 55s α 6.65	Fr 210 2.65m α 6.55	Fr 211 3.1m α 6.55 210.9953	Fr 212 19m ε α 6.39, 6.41, 6.34 E 5 211.996
86	Rn					Rn 204 3m α 6.28		Rn 206 6.5m α 6.25 ε	Rn 207 11m ε α 6.12	Rn 208 23m ε α 6.15	Rn 209 30m ε α 6.04	Rn 210 2.7h α 6.04 ε 209.9895	Rn 211 16h ε α 5.78, 5.85, 5.61 γ .069, .032−1.82 210.9906
	At			At 201 1.5m α 6.35	At 202 3m ε α 6.13, 6.23 E8	At 203 7.2m ε α 6.09 E6	At 204 9m ε α 5.95 E7	At 205 26m ε α 5.90 E5	At 206 30m ε α 5.70−5.90 E6	At 207 1.8h ε α 5.76 E3.7	At 208 1.6h ε α 5.65 γ .66, 1.75, ·· E5 207.9870	At 209 5.5h ε, α 5.65 γ .195, .091, .55, .78 E3.5	At 210 8.3h ε, α 5.36−5.52 γ .047, .25, .12, .08−2.6 E3.9
84	Po 197 17s α 6.39, 6.30	Po 198 1.75m α 6.18	Po 199 4.24m α 6.06	Po 200 11.4m α 5.86	Po 201 3/ 17.5m ε α 5.68, 5.57, 5.77	Po 202 45m ε α 5.58	Po 203 5/ 45m ε α 5.48	Po 204 3.5h ε α 5.37	Po 205 5/− 644μs IT .16 γ .71 | 1.8h ε α 5.2	Po 206 8.8d ε, α 5.22 γ .060, .083−1.32 E 2.03 205.9803	Po 207 5/− 2.5s IT .26 | 6.2h ε, β+ 1.14, ·· 47μs IT .31 γ .82 | α 5.10 γ .10−2.06 206.9816	Po 208 2.9y α 5.11 γ .6, .28 207.9813	Po 209 1/− 103y α 4.88, ··, ε γ .91, .26 E 1.90 208.9825
	Bi 208.980 σ .034	Bi <198 1.7m α 6.2	Bi 198 8m ε α 5.81 E7.4	Bi 199 9/ 24m α 5.52 | 27m ε	Bi 200 7 35m ε γ 1.03, .46	Bi 201 9/− 52m ε α 5.27 | 1.8h ε (γ .63) E4	Bi 202 5 1.6h ε γ .42, .96	Bi 203 9/− 11.8h ε, β+ 1.35, .73 α 4.85 γ .06−1.9 E 3.19	Bi 204 6+ 11.2h ε γ .08, .38, .22, .91, .10−1.2 E 4	Bi 205 9/− 15.3d ε, β+ .98 γ .70, 1.77, .12−2.6, (.026, .99) E 2.70	Bi 206 6+ 6.24d ε γ .80, .88, .18, .11−1.9 E 3.6	Bi 207 30y ε γ .57, 1.77, ··, (1.06) E 2.36	Bi 208 0.0027s IT .92 γ .51 | 3.7x10^5y ε γ 2.62 E 2.87
82	Pb 195 17m ε (γ .099, .39, .38)	Pb 196 37m ε γ .191, .240, .253, ··	Pb 197 42m ε IT .234 γ .387, .385, .222 | ?	Pb 198 2.4h ε γ .290, .365, .031−.865	Pb 199 12m IT .424 | 1.5h ε, β+ 2.8 γ .367, .27−1.13 E 4.2	Pb 200 21h ε γ .148, .142, .24, .27, .033−.45	Pb 201 61s IT .63 | 9.4h ε, β+ γ .330, .13−1.10	9− Pb 202 3.6h IT .79, .13 ε γ .96, .24−.66 | ~3x10^5y ε (γ .44) E .05	Pb 203 6.1s IT .83 | 52h ε γ .28, .40, .68 E .82	9− Pb 204 67m IT .91 γ .29−.90 | 1.48 14x10^{17}y α 2.6 σ .7 203.97304	Pb 205 0.004s IT .026 γ .99, ·· | 3x10^7y ε E .03	Pb 206 23.6 σ .03 205.97447	13/+ Pb 207 1/− 0.8s IT 1.06 γ .57 | 22.6 σ .72 206.97590
	Tl 194 ?33m ε γ .097 α 5.8 | 35m ε γ .43	Tl 195 1/+ 3.5s IT .099 γ .393, .383 | 1.2h ε, β+ ~1.8 γ .037, .23−.88 E3	Tl 196 1.4h IT .120 ε, γ .034−.425 | 1.8h ε γ .425 E4.6	9/− Tl 197 1/+ 0.54s IT .222 γ .387, .385 | 2.8h ε γ .152, .018−1.0 E2.2	7− Tl 198 2− 1.9h ε, γ .048, ·· IT .261 γ .283, ·· | 5.3h ε, β+ 2.4, 2.1, ·· γ .412, .195− E3.5 2.8	Tl 199 1/+ 0.027s IT .38, .03 γ .37, .35, .72 E1.1 | 7.4h ε γ .208, .455, .158, .050−.492	Tl 200 2− 0.034s IT γ .20, .52 γ .37, 1.21, .116−2.1 | 26h ε, β γ 1.06, 1.44, ·· E2.45	Tl 201 1/+ 0.002s IT .22, .59 γ .33, ·· | 73h ε γ .167, .135, .031 −.166 E .4	Tl 202 2− 500μs IT .46 γ .49 | 12.4d ε γ .44, .96, .52 E 1.22	Tl 203 1/+ 29.50 σ 11 202.97235	Tl 204 2− 3.75y β− .765 ε E− .765 E+ .36	Tl 205 1/+ 70.50 σ .11 204.97444	Tl 206 RaE'' 4.3m β− 1.52 no γ E 1.52
		114		116		118		120		122		124	

Z \ N	126		128		130		132		134		136		138
88						Ra 219 <1m α 8.0 219.0100	Ra220 0.02s α 7.45, 6.90 γ.47 220.0110	Ra221 30s α 6.60,6.75,6.66 γ.089,.150,.176,·· 221.0139	Ra222 37s α 6.56, 6.23 γ.33 222.0154	Ra223 $^{1/4}$ AcX 11.4d α 5.71,5.60,5.53,·· γ.031–.45 σ 130 223.0185	Ra224 ThX 3.64d α 5.68,5.44 γ.24 σ 12 224.0202	Ra225 14.8d β^-.32 γ.040 E.35	Ra226 Ra 1620y α 4.78,4.59,·· γ.187,·· σ 20 226.0254
	Fr 213 34s α 6.77 β^+ 212.9962				Fr 217 < 2s α 8.3 217.0048	Fr 218 < 5s α 7.85 218.0075	Fr 219 0.02s α 7.30 219.0092	Fr 220 28s α 6.69 220.0123	Fr 221 4.8m α 6.34, 6.12 γ.22 221.0142	Fr 222 15m β^- α	Fr 223 AcK 22m α 5.34 β^-1.15 γ.050,.080,·· 223.0198	Fr 224 ~ 2m β^-	
86	Rn 212 25m α 6.27 211.9907			Rn 215 < 1m α 8.6 214.9987	Rn 216 45μs α 8.04 no γ 216.0003	Rn 217 500μs α 7.74 217.0039	Rn 218 0.035s α 7.13, 6.52 γ.61 218.0056	Rn 219 An 4.0s α 6.82 6.55,6.42 γ.27,.40 219.0095	Rn 220 Tn 56s α 6.29, 5.74 γ.54 σ<.2 220.0114	Rn 221 25m β^- α 6.0	Rn 222 Rn 3.823d α 5.49,4.98,·· γ.51 σ.7 222.0175	Rn 223 43m β^-	Rn 224 1.9h β^-
	At 211 $^{9/-}$ 7.2h ε α 5.87,(7.45,··) γ.67,(.89,.57) E.75 210.9875	At 212 0.12s α 7.82, 7.88 γ.06 0.31s α 7.66, 7.60 γ.06 211.9907	At 213 < 2s α 9.2 212.9931	At 214 < 5s α 8.78 213.9963	At 215 ~ 100μs α 8.00 214.9987	At 216 ~ 300μs α 7.79 216.0024	At 217 0.032s α 7.07 217.0046	At 218 1.3s α 6.69, 6.65 218.0086	At 219 0.9m α 6.27 β^- 219.0013				
84	Po 210 RaF 138.40d α 5.30,·· γ.80 σ(<.0005+<.03) 209.9829	Po 211 25s α 7.1,8.7,·· γ1.06, .57,.89 AcC' 0.52s α 7.45,·· γ.89,1.06,·· 210.9866	Po 212 46s α 11.7,·· γ2.6,.6 ThC' 0.30μs α 8.78 211.9889	Po 213 4μs α 8.35 212.9928	Po 214 RaC' 164μs α 7.69 213.9952	Po 215 AcA 0.0018s α 7.38 β^- 214.9995	Po 216 ThA 0.15s α 6.78 216.0019	Po 217 <10s α 6.54	Po 218 RaA 3.05m α 6.00 β^- 218.0089				
	Bi 209 $^{9/-}$ 100 σ(.019+.015) 208.98040	Bi 210 $^{1-}$ RaE 5.0d 2.6×10^6y α 4.95, 4.92,4.57,·· γ.26,.30,·· β^-1.16 α 4.7,·· E1.16	Bi 211 AcC 2.15m α 6.62, 6.28,·· β^-,γ.35 E⁻.60 210.9873	Bi 212 $^{1-}$ ThC 60.6m β^- 2.25 α 6.05,6.09,··, (9.5–10.5) γ.040–2.20 E2.25	Bi 213 47m β^- 1.39,.96 γ.44,.32 α 5.86, 5.55 E1.39	Bi 214 RaC 19.7m β^-.4–3.2 α 5.5,·· (8.3–10.5) γ.61,1.12,1.76,.45–2.43 E3.2	Bi 215 8m β^- E2.24						
82	Pb 208 52.3 σ.0005 207.97665	Pb 209 3.3h β^-.64 no γ E.64	Pb 210 RaD 22y β^-.015,.061 γ.047 α 3.57 E.061	Pb 211 AcB 36.1m β^-1.34,.56,·· γ.83,.41,.06–1.1 E1.34	Pb 212 ThB 10.64h β^-.34,.58,·· γ.239,.30,.11–.41 E.58	Pb 213 10m β^-	Pb 214 RaB 26.8m β^-.7,·· γ.352,.295,.050–.259 E1.0						
	Tl 207 AcC'' 4.78m β^-1.47,·· γ.89 E1.47	Tl 208 $^{5+}$ ThC'' 3.1m β^-1.80,1.0–2.38 γ 2.61,.58,.51, .23–1.09 E4.99	Tl 209 2.2m β^-1.8 γ.120,.45,1.56 E3.9	Tl 210 RaC'' 1.3m β^-1.9,1.3,2.3 (n) γ.80,.30,.01–2.43 E5.4									

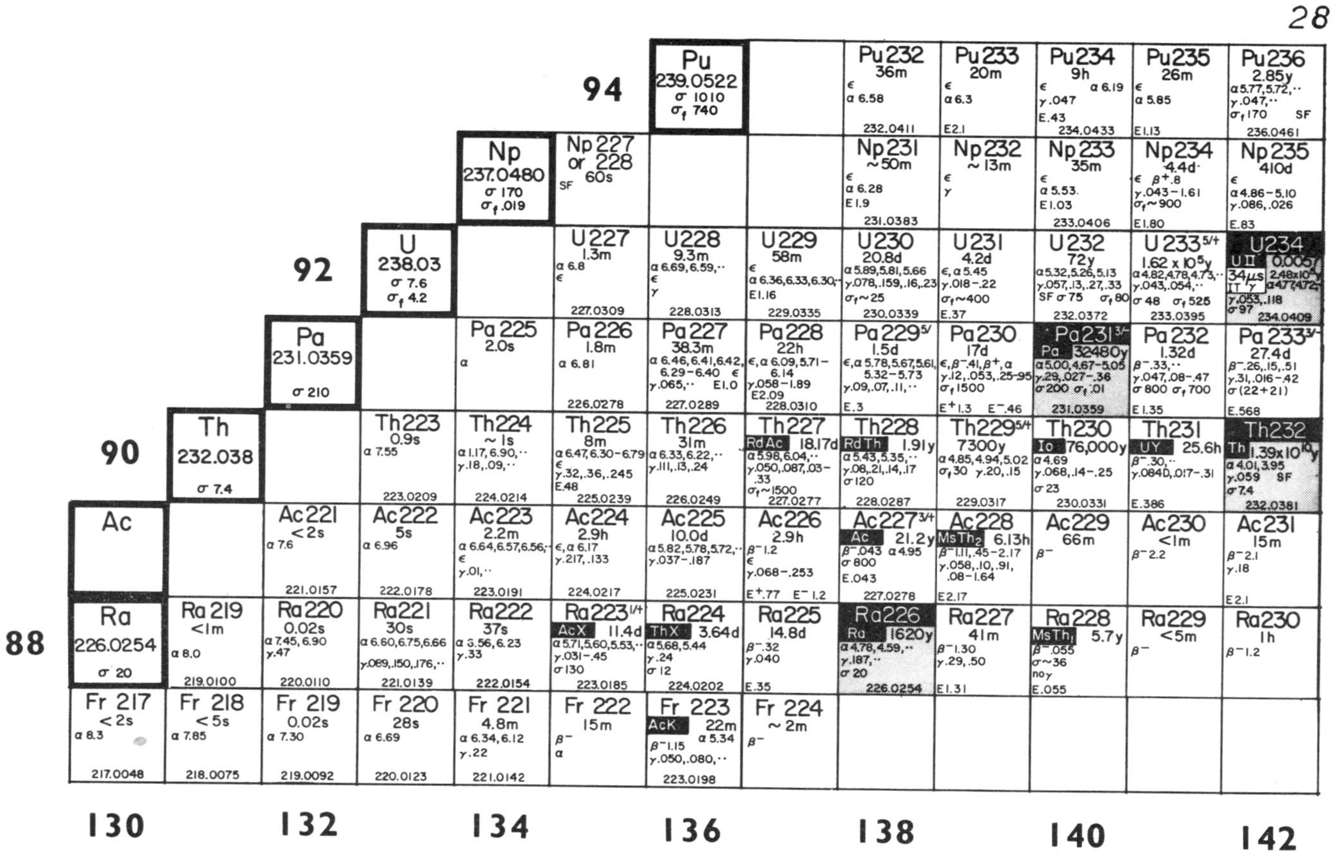

Z	130		132		134		136		138		140		142
94							Pu 239.0522 σ 1010 σ_f 740		Pu232 36m ε α 6.58 232.0411	Pu233 20m ε α 6.3 E2.1	Pu234 9h ε α 6.19 γ.047 E.43 234.0433	Pu235 26m ε α 5.85 E1.13	Pu236 2.85y α 5.77,5.72,·· γ.047,·· σ_f 170 SF 236.0461
					Np 237.0480 σ 170 σ_f .019	Np 227 or 228 60s SF			Np231 ~50m ε α 6.28 E1.9 231.0383	Np232 ~13m ε γ	Np233 35m ε α 5.53 E1.03 233.0406	Np234 4.4d ε β^+.8 γ.043−1.61 σ_f~900 E1.80	Np235 410d ε α 4.86−5.10 γ.086,.026 E.83
92				U 238.03 σ 7.6 σ_f 4.2		U227 1.3m α 6.8 ε 227.0309	U228 9.3m α 6.69,6.59,·· ε γ 228.0313	U229 58m ε α 6.36,6.33,6.30,·· E1.16 229.0335	U230 20.8d α 5.89,5.81,5.66 γ.078,.159,.16,.23 σ_f~25 230.0339	U231 4.2d ε, α 5.45 γ.018−.22 σ_f~400 E.37	U232 72y α 5.32,5.26,5.13 γ.057,.13,.27,.33 SF σ 75 σ_f 80 232.0372	U233 5/+ 1.62×10^5y α 4.82,4.78,4.73,·· γ.043,.054,·· σ 48 σ_f 525 233.0395	U234 U II 0.0057 34μs IT γ 2.48×10^5y α 4.77,4.72,·· γ.053,.118 σ 97 234.0409
			Pa 231.0359 σ 210		Pa225 2.0s α	Pa226 1.8m α 6.81 226.0278	Pa227 38.3m α 6.46,6.41,6.42, 6.29−6.40 ε γ.065,·· E1.0 227.0289	Pa228 22h ε, α 6.09,5.71−6.14 γ.058−1.89 E2.09 228.0310	Pa229 5/ 1.5d ε, α 5.78,5.67,5.61, 5.32−5.73 γ.09,.07,.11,·· E.3	Pa230 17d ε, β^-.41, β^+, α γ.12,.053,.25-.95 σ_f 1500 E^+1.3 E^-.46	Pa231 3/− Pa 32480y α 5.00,4.67−5.05 γ.29,.027−.36 σ 200 σ_f .01 231.0359	Pa232 1.32d β^-.33,·· γ.047,.08−.47 σ 800 σ_f 700 E1.35	Pa233 3/− 27.4d β^-.26,.15,.51 γ.31,.016−.42 σ (22+21) E.568
90		Th 232.038 σ 7.4		Th223 0.9s α 7.55 223.0209	Th224 ~1s α 1.17,6.90,·· γ.18,.09,·· 224.0214	Th225 8m α 6.47,6.30−6.79 ε γ.32,.36,.245 E.48 225.0239	Th226 31m α 6.33,6.22,·· γ.111,.13,.24 226.0249	Th227 RdAc 18.17d α 5.98,6.04,·· γ.050,.087,.03−.33 σ_f~1500 227.0277	Th228 RdTh 1.91y α 5.43,5.35,·· γ.08,.21,.14,.17 σ 120 228.0287	Th229 5/+ 7300y α 4.85,4.94,5.02 σ_f 30 γ.20,.15 229.0317	Th230 Io 76,000y α 4.69 γ.068,.14−.25 σ 23 230.0331	Th231 UY 25.6h β^-.30,·· γ.084D,.017−.31 E.386	Th232 Th 1.39×10^{10}y α 4.01,3.95 γ.059 SF σ 7.4 232.0381
	Ac		Ac221 <2s α 7.6 221.0157	Ac222 5s α 6.96 222.0178	Ac223 2.2m α 6.64,6.57,6.56,·· ε γ.01,·· 223.0191	Ac224 2.9h ε, α 6.17 γ.217,.133 224.0217	Ac225 10.0d α 5.82,5.78,5.72,·· γ.037−.187 225.0231	Ac226 2.9h β^-1.2 ε γ.068−.253 E^+.77 E^-1.2	Ac227 3/+ Ac 21.2y β^-.043 α 4.95 σ 800 E.043 227.0278	Ac228 MsTh2 6.13h β^-1.11,.45−2.17 γ.058,.10,.91, .08−1.64 E2.17	Ac229 66m β^-	Ac230 <1m β^-2.2	Ac231 15m β^-2.1 γ.18 E2.1
88	Ra 226.0254 σ 20	Ra219 <1m α 8.0 219.0100	Ra220 0.02s α 7.45,6.90 γ.47 220.0110	Ra221 30s α 6.60,6.75,6.66 γ.089,.150,.176,·· 221.0139	Ra222 37s α 6.56,6.23 γ.33 222.0154	Ra223 1/+ AcX 11.4d α 5.71,5.60,5.53,·· γ.031−.45 σ 130 223.0185	Ra224 ThX 3.64d α 5.68,5.44 γ.24 σ 12 224.0202	Ra225 14.8d β^-.32 γ.040 E.35	Ra226 Ra 1620y α 4.78,4.59,·· γ.187,·· σ 20 226.0254	Ra227 41m β^-1.30 γ.29,.50 E1.31	Ra228 MsTh1 5.7y β^-.055 σ~36 no γ E.055	Ra229 <5m β^-	Ra230 1h β^-1.2
	Fr 217 <2s α 8.3 217.0048	Fr 218 <5s α 7.85 218.0075	Fr 219 0.02s α 7.30 219.0092	Fr 220 28s α 6.69 220.0123	Fr 221 4.8m α 6.34,6.12 γ.22 221.0142	Fr 222 15m β^- α	Fr 223 AcK 22m α 5.34 β^-1.15 γ.050,.080,·· 223.0198	Fr 224 ~2m β^-					

29

Z	140		142		144		146		148		150		152
94	Pu234 9h ε α 6.19 γ.047 E.43 234.0433	Pu235 26m ε α 5.85 E1.13	Pu236 2.85y α5.77,5.72,·· γ.047,·· σ_f 170 SF 236.0461	Pu237 7/− 0.18s IT.145 SF | 45.6d ε σ_f 2500 α5.2−5.65 γ.03−.076 237.0483	Pu238 89y α5.50,5.46,·· γ.044,.100−.81 σ580 σ_f 17 SF 238.0495	Pu239 1/+ 24,360y α5.15,5.13,5.10,·· γ.013,.038−.42 SF σ 270 σ_f 740 239.0522	Pu240 6760y α5.17,5.12,·· γ.045,·· SF σ290 σ_f <.08 240.0540	Pu241 5/+ 13y β⁻.02 α4.89,4.85 γ.100,.145 σ400 σ_f 950 E.021	Pu242 3.79 x 10^{5}y α4.90,4.86 γ.045 SF σ29 242.0587	Pu243 4.98h β⁻.56,.49 γ.084,.042,.012−.38 σ170 E.56	Pu244 7.6 x 10^{7}y α SF σ1.6	Pu245 10.1h β⁻ σ~260	Pu246 10.85d β⁻.15,.33 γ.047,.027,.08,.175,.215 E.40
	Np233 35m ε α5.53 E1.03 233.0406	Np234 4.4d ε β⁺.8 γ.043−1.61 σ_f~900 E1.80	Np235 410d ε α4.86−5.10 γ.086,.026 E.83	Np236 22h β⁻.52, ε γ.045,.150 E⁻.52 | >5000y β⁻ σ_f 2500 236.0466	Np237 5/+ 2.14 x 10^{6}y α4.52−4.87 γ.087,.029−.20 σ170 σ_f .019 237.0480	Np238 2 2.10d β⁻1.24,.25,·· γ.044,.99,1.03,·· σ_f 1600 E1.29	Np239 5/+ 2.35d β⁻.72,·· γ.045−.504 σ(35+25) E.72	Np240 7.3m β⁻2.18,·· γ.043,.55,.26−1.62 E2.18 | 60m β⁻.84 γ.56,.44,.085−1.16 E2.05	Np241 16m β⁻1.36 E1.36				
92	U232 72y α5.32,5.26,5.13 γ.057,.13,.27,.33 SF σ75 σ_f 80 232.0372	U233 5/+ 1.62 x 10^{5}y α4.82,4.78,4.73,·· γ.043,.054,·· σ48 σ_f 525 233.0395	U234 UII 0.0057 34μs IT γ | 2.48x10^{5}y α4.77,4.72,·· γ.053,.118 σ97 234.0409	1/2+ U235 7/− 26.2m IT<.061 | AcU 0.72 7.13x10^{8}y SF α4.18−4.60 γ.19,·· σ100 σ_f 580 235.0439	U236 2.39 x 10^{7}y α4.50,·· γ.05 SF σ6 236.0456	U237 6.75d β⁻.25,·· γ.060,.21,.026−.37 E.51	U238 UI 99.27 4.51 x 10^{9}y α4.19,·· SF γ.048 σ2.7 238.0508	U239 23.5m β⁻1.21,1.28 γ.074,.043,.031 σ22 σ_f 14 E1.28	U240 14.1h β⁻.36,·· γ.044 E.48				
	Pa231 3/− Pa 32480y α5.00,4.67−5.05 γ.29,.027−.36 σ200 σ_f .01 231.0359	Pa232 1.32d β⁻.33,·· γ.047,.08−.47 σ800 σ_f 700 E1.35	Pa233 3/− 27.4d β⁻.26,.15,.51 γ.31,.016−.42 σ(22+21) E.568	Pa234 UX_2 | UZ 1.18m β⁻2.31,·· IT γ.23−1.8 | 6.66h β⁻.5,·· γ.04−1.7 234.0433	Pa235 24m β⁻1.4 no γ E1.4		Pa237 39m β⁻2.30,1.35,·· γ.090,.91,.145−1.42 E2.30						
90	Th230 Io 76,000y α4.69 γ.068,.14−.25 σ23 230.0331	Th231 UY 25.6h β⁻.30,·· γ.084D,.017−.31 E.386	Th232 Th 1.39x10^{10}y α4.01,3.95 γ.059 SF σ7.4 232.0381	Th233 22.1m β⁻1.24,·· γ.029−.90 σ1500 σ_f 15 E1.24	Th234 UX_1 24.10d β⁻.26,.10 γ.092,.063,.029,·· σ1.8 E.26	Th235 <5m β⁻							
	Ac229 66m β⁻	Ac230 <1m β⁻2.2	Ac231 15m β⁻2.1 γ.18 E2.1										
88	Ra228 $MsTh_1$ 5.7y β⁻.055 σ~36 no γ E.055	Ra229 <5m β⁻	Ra230 1h β⁻1.2										

Z \ N	140		142		144		146		148		150		152
98					**Cf**		Cf244 25m α 7.17 244.0659	Cf245 44m ε α 7.11 E1.52	Cf246 35.7h α 6.76,6.72,·· SF γ.042,.096,.146 246.0688	Cf247 2.4h ε γ.295,.42,.46	Cf248 350d α 6.26 SF 248.0724	Cf249 360y α5.81,5.93,6.20 γ.39,.34,.26 SF σ~270 σ_f 1700 249.0747	Cf250 10y α6.03,5.99, SF γ.043 σ~1500 250.0766
		Bk					Bk243 4.5h α 6.55,6.72,6.20 γ.74,.84,.96,·· 243.0630	Bk244 4.4h ε α 6.67 γ.20,.90,1.06, 1.16−1.72	Bk245 4.95d ε α 5.89−6.37 γ.25,.38,.164− .48 E.84	Bk246 1.8d ε γ.82, 1.09	Bk247 ~10^4y α5.51,5.61,5.30 γ.084,.27 ε? 247.0702	Bk248 23h β^-.65 ε E^-.65	Bk249 314d β^-.125 γ.32 α 5.42,·· SF σ~500 E.13
96	**Cm**		Cm238 2.5h ε α 6.52 238.0530	Cm239 ~3h ε γ.188	Cm240 26.8d α 6.26,·· SF 240.0555	Cm241$^{1/+}$ 35d ε α 5.95 γ.48,.145 E.77	Cm242 163d α 6.11,6.07,·· γ.044,.102,.157 SF σ~20 242.0588	Cm243 32y α 5.78,5.74,5.63− 6.06 ε γ.28,.23, .21,·· σ 250 σ_f 630 243.0614	Cm244 0.034s IT.75,.9,·· γ.04,.10, .15,·· 18.1y α5.80, 5.76,·· γ.043, .10,.15 σ~15 SF 244.0628	Cm245 9320y α 5.36,5.30,···· γ.173,.13 σ~200 σ_f 1900 245.0653	Cm246 5480y α 5.37,·· SF σ~15 246.0672	Cm247 1.67 x 10^7y α σ 180	Cm248 4.7 x 10^5y α 5.05 SF σ 7.2
	Am		Am237 ~1.3h ε α 6.01 E1.4	Am238 1.86h ε γ.98,.58,.37,.95, 1.35	Am239 12h ε, α 5.77 γ.23,.28,.045− .49 E.76	Am240 51h ε γ.042,.099,.89, .99	Am241$^{5/-}$ 458y α 5.48,5.31−5.54,·· γ.060,.027−.37 ·· SF σ(50+600) σ_f 2.9 241.0567	Am242^{1-} ~152y IT.048 σ~1600 σ_f 6400 16h β^-.67,··, ε α, γ.04 σ_f 3000 E^+.72 E^-.67	Am243$^{5/-}$ 7650y α 5.27,5.22,5.17− 5.35,·· γ.075,·· σ 74 243.0614	$^{1-}$Am244^{6-} 25m β^-1.5,·· ε E^-1.5 10h β^-.39 γ.043, .099,.15, .75,.9,·· σ_f 2300	Am245 2.07h β^-.90 γ.036,.230,.255, .153,.140 E.90	Am246 25m β^-1.31,1.60,2.10 γ.035,.10,.245, .18−1.06 E2.29	
94	Pu234 9h ε α 6.19 γ.047 E.43 234.0433	Pu235 26m ε α 5.85 E1.13	Pu236 2.85y α5.77,5.72,·· γ.047,·· σ_f 170 SF 236.0461	Pu237$^{7/-}$ 0.18s IT.145 SF 45.6d ε σ_f 2500 α 5.2−5.65 γ.03−.076 237.0483	Pu238 89y α 5.50,5.46,·· γ.044,.100−.81 σ580 σ_f 17 SF 238.0495	Pu239$^{1/+}$ 24,360y α5.15,5.13,5.10,·· γ.013,.038−.42 SF σ 270 σ_f 740 239.0522	Pu240 6760y α 5.17,5.12,·· γ.045,·· SF σ 290 σ_f <.08 240.0540	Pu241$^{5/+}$ 13y β^-.02 α 4.89,4.85 γ.100,.145 σ 400 σ_f 950 E.021	Pu242 3.79 x 10^5y α 4.90,4.86 γ.045 SF σ 29 242.0587	Pu243 4.98h β^-.56,.49 γ.084,.042,.012− .38 σ170 E.56	Pu244 7.6 x 10^7y α SF σ 1.6	Pu245 10.1h β^- σ~260	Pu246 10.85d β^-.15,.33 γ.047,.027,.08, .175,.215 E.40
	Np233 35m ε α 5.53 E1.03 233.0406	Np234 4.4d ε β^+.8 γ.043−1.61 σ_f~900 E1.80	Np235 410d ε α 4.86−5.10 γ.086,.026 E.83	Np236 22h β^-.52, ε γ.045, .150 E^-.52 >5000y β^- σ_f 2500 236.0466	Np237$^{5/+}$ 2.14 x 10^6y α 4.52−4.87 γ.087,.029−.20 σ170 σ_f .019 237.0480	Np238^{2} 2.10d β^-1.24,.25,·· γ.044,.99,1.03,·· σ_f 1600 E1.29	Np239$^{5/+}$ 2.35d β^-.72,·· γ.045−.504 σ(35+25) E.72	Np240 7.3m β^-2.18,·· γ.043, .55,.26− 1.62 E2.18 60m β^-.84 γ.56,.44, .085−1.16 E2.05	Np241 16m β^-1.36 E1.36				
92	U232 72y α5.32,5.26,5.13 γ.057,.13,.27,.33 SF σ75 σ_f 80 232.0372	U233$^{5/+}$ 1.62 x 10^5y α4.82,4.78,4.73,·· γ.043,.054,·· σ 48 σ_f 525 233.0395	U234 U II 0.0057 34μs IT γ 2.48x10^5y α4.77,4.72,·· γ.053,.118 σ97 234.0409	$^{1/2+}$U235$^{7/-}$ 26.2m IT <.061 AcU 0.72 7.13x10^8y SF α4.18−4.60 γ.19,·· σ100 σ_f 580 235.0439	U236 2.39 x 10^7y α 4.50,·· γ.05 SF σ6 236.0456	U237 6.75d β^-.25,·· γ.060,.21,.026− .37 E.51	U238 UI 99.27 4.51 x 10^9y α4.19,·· SF γ.048 σ2.7 238.0508	U239 23.5m β^-1.21,1.28 γ.074,.043,.031 σ22 σ_f 14 E1.28	U240 14.1h β^-.36,·· γ.044 E.48				
	Pa231$^{3/-}$ Pa 32480y α5.00,4.67−5.05 γ.29,.027−.36 σ200 σ_f .01 231.0359	Pa232 1.32d β^-.33,·· γ.047,.08−.47 σ800 σ_f 700 E1.35	Pa233$^{3/-}$ 27.4d β^-.26,.15,.51 γ.31,.016−.42 σ(22+21) E.568	Pa234 UX_2 1.18m β^-2.31,·· IT γ.23−1.8 UZ 6.66h β^-.5,·· γ.04−1.7 234.0433	Pa235 24m β^-1.4 no γ E1.4		Pa237 39m β^-2.30,1.35,·· γ.090,.91,.145− 1.42 E2.30						

Z \ N	146	147	148	149	150	151	152	153	154	155	156	157	158
104											260 0.3s SF		
				Lw				Lw256 45s	Lw257 8s α 8.6				
102			No				No 254 ~ 50s α 8.8	No 255 ~ 15s α 8.2	No 256 8s α SF				
		Md							Md 255 0.5h ε α 7.34 255.0906	Md 256 1.5h ε α 7.17	Md 257 3h ε α 7.07,·· SF		
100	Fm		Fm 248 0.6m α 248.0772	Fm 249 2.5m α 7.9	Fm 250 30m α 7.43 250.0795	Fm 251 7h ε α 6.89	Fm 252 23h α 7.04 252.0827	Fm 253 ~ 4.5d ε α 6.94 E.19	Fm 254 3.24h α 7.20,7.16,7.06 γ.041,.098 SF 254.0868	Fm 255 20h α 7.02,6.96,·· γ.058,.081 SF σ <100	Fm 256 2.7h α 6.85 SF	Fm 257 94d α 6.56 SF	Fm 258 Short SF
	Es 245 1.2m α 7.7	Es 246 7.3m α 7.35		Es 248 25m ε α 6.87	Es 249 2h ε α 6.76 249.0762	Es 250 8h ε	Es 251 1.5d ε α 6.48 251.0799	Es 252 ~ 140d α 6.64 252.0829	Es 253 20.0d α 6.64,·· SF γ.042,.051,.39,.437 σ ~ 300 253.0847	Es 254 39h: β^-.48,1.13,·· ε (α7.2) SF γ.65,.69,·· σ<40 480d: α 6.44;· γ.062 σ_f ~2700 254.0879	Es 255 40d β^-.185 α 6.30 σ ~ 40	Es 256 Short β^-	
98	Cf 244 25m α 7.17 244.0659	Cf 245 44m ε α 7.11 E1.52	Cf 246 35.7h α 6.76,6.72,·· SF γ.042,.096,.146 246.0688	Cf 247 2.4h ε γ.295,.42,.46	Cf 248 350d α 6.26 SF 248.0724	Cf 249 360y α 5.81,5.93,6.20 γ.39,.34,.26 SF σ ~ 270 σ_f 1700 249.0747	Cf 250 10y α 6.03,5.99, SF γ.043 σ ~ 1500 250.0766	Cf 251 ~ 800y α γ.18 σ_f 3000 σ ~ 3000	Cf 252 2.55y α 6.12,6.08 γ.043,.100 SF σ 30	Cf 253 18d β^-.27 α 5.98 E.27	Cf 254 60d SF σ < 2		
	Bk 243 4.5h α 6.55,6.72,6.20 γ.74,.84,.96,·· 243.0630	Bk 244 4.4h ε α 6.67 γ.20,.90,1.06,1.16–1.72	Bk 245 4.95d ε α 5.89–6.37 γ.25,.38,.164–.48 E.84	Bk 246 1.8d ε γ.82, 1.09	Bk 247 ~ 10^4y α 5.51,5.61,5.30 γ.084,.27 ε? 247.0702	Bk 248 23h β^-.65 ε E^-.65	Bk 249 314d β^-.125 γ.32 α 5.42,·· SF σ ~ 500 E.13	Bk 250 3.2h β^-.72,1.76,·· γ.99,.89,.098,.042 E1.8	Bk 251 57m β ~.5, ~1.0 γ.037,.94,1.40,1.84				

SOLUTIONS

OF PROBLEMS

CHAPTER I

1. (p, n) – $^{56}_{27}$Co; (p, d) – $^{55}_{26}$Fe; (p, 2n) – $^{55}_{27}$Co; (p, α) – $^{53}_{25}$Mn; (n, γ) – $^{57}_{26}$Fe; (n, p) – $^{56}_{25}$Mn; (n, d) – $^{55}_{25}$Mn; (n, 2p) – $^{55}_{24}$Cr; (n, α) – $^{53}_{24}$Cr; (α, n) – $^{59}_{28}$Ni; (α, p) – $^{59}_{27}$Co; (γ, n) – $^{55}_{26}$Fe; (γ, p) – $^{55}_{25}$Mn; (γ, pn) – $^{54}_{25}$Mn; (γ, 2p 3n) – $^{51}_{24}$Cr.

2. β^+: ^{11}C→^{11}B, ^{22}Na→^{22}Ne, ^{48}V→^{48}Ti, ^{52}Mn→^{52}Cr, ^{56}Co→^{56}Fe, ^{65}Zn→^{65}Cu.
β^-: ^{14}C→^{14}N, ^{24}Na→^{24}Mg, ^{32}P→^{32}S, ^{60}Co→^{60}Ni, ^{115}In→^{115}Sn, ^{137}Cs→^{137}Ba, ^{131}I→^{131}Xe.
K-capture: ^{7}Be→^{7}Li, ^{51}Cr→^{51}V, ^{54}Mn→^{54}Cr.
α: ^{12}B→^{8}Li, ^{32}Cl→^{28}P, ^{142}Ce→^{138}Ba, ^{210}Po→^{206}Pb, ^{226}Ra→^{222}Rn.
n: ^{17}N→^{16}N, ^{87}Br→^{86}Br, ^{89}Br→^{88}Br.

3. Disintegration rate: 1.36×10^{3} disintegrations/sec. Activity: 3.68×10^{-8} Ci.

4. ^{11}C: 1.198×10^{-9} g; ^{22}Na: 1.605×10^{-4} g; ^{24}Na: 1.156×10^{-7} g; ^{60}Co: 8.524×10^{-4} g; ^{226}Ra: 1 g.

5. Flux: 7.4×10^{10} particles/cm^{2}sec. Activity: 0.5 mCi and 0.067 mCi. Time: 21.62 h.

6. Percentage: 8.5×10^{-12}.

7. Fraction of target atoms: 1.66×10^{-9}.

8. $A<60$, 0.8 days; $A<101$, 1.0 days; $A<209$, 0.5 days; $t_{\frac{1}{2}}=8.24$ days.

9. a) 1 h: 1; 1 day: 0.728; 1 week: 0.562; 1 month: 0.438; 1 year: 0.230.

b) 1 h: 1; 1 day: 0.662; 1 week: 0.456; 1 month: 0.306; 1 year: 0.092.

10. 12.57 rad/h.

11. U: 2.22 Ci, 1.42×10^6 photons/sec. Cu: 1.51 Ci, 1.41×10^6 photons/sec.

12. ^{132}Te: 80.23 mCi, 7.09×10^{14} atoms. ^{132}I: 41.44 mCi, 1.27×10^{15} atoms.

13. 0.0378 Ci, 1.334 rads, 5.79×10^9 atoms.

14. 1.84×10^4 rem/h.

15.

	C	Al	Fe	W	Pb
0.5 MeV	5.96	3.78	1.31	0.41	0.57
1.25 MeV	10	6.67	2.44	0.88	1.5
2.75 MeV	1.59	10	3.40	1.22	2.1

16.

	Air	Al	C	Pb
1 MeV	3.72×10^3	1.78	2.82	0.425
3 MeV	8.99×10^3	4.3	6.83	1.04

17. Air: 3.2 cm, Carbon, 2.36×10^{-3} cm, Gold, 9.48×10^{-3} cm.

18. 3.94×10^3 particles/sec.

19. Ge: 8.68 mb; Se: 9.18 mb; Ir: 16.68 mb; As: 8.84 mb.

20. 9.26×10^5 rad/h from beta particles,
1.45×10^8 rad/h from gamma rays,
2.53×10^8 rad/h from gamma rays with build-up,
3.31×10^8 rad/h from gamma rays with build-up and Compton electrons.

21. 1.99×10^{-2} rad/h, 50.5 rad/h.

22. $1-b^{-1}(b^2-r^2)^{\frac{1}{2}}$ (b is the distance from the centre and r the moon radius).

23. The radiation field is $\varrho R_0 R$.

24. $D\Phi(1-e^{-\mu T})$.

CHAPTER II

1.

Element	σ_g(mb)	σ_{inel}(mb)
Be	229	216
Al	477	449
Fe	773	728
In	1250	1160
Pb	1860	1750

2. ^{27}Al: 12 mb; ^{56}Fe: 3.2 mb; ^{115}In: 0.43 mb; ^{208}Pb: 0.31 mb.

3. 50 MeV: 2.7×10^{-11} mb; 100 MeV: 4.2×10^{-6} mb; 600 MeV: 0.45 mb; 3 GeV: 3.2 mb.

4. 3.6 rad/h at 1 m.

5. 4×10^5 photons/sec.

6. 8×10^4/sec·cm^2; 9.6 rad/h.

7. 1.1×10^5 particles/cm^{2}sec.

CHAPTER III

1. a) 1 kg, b) 50 kg.

2. $\approx$400 kg ^{235}U or 8×1 MT bombs.

3. 44.5 Ci; 282 Ci; 660 Ci; 1.3 μg; 1.8 mg; 4.5 μg.

4. 5.1×10^6 Ci/MW.

5. a) 7.8×10^{11} Ci, b) 3.9×10^{13} Ci.

6. 220 kg ^{235}U, 1100 kg U fuel; about half a year.

7. a) 5.4×10^4 W/kg ^{235}U; b) 6×10^3 W γ/kg; c) 2.2×10^4 rad/h.

8. 4.8×10^4 MWD; 245 rad/sec.

9. 7.1 rad/h for Cu; 0.04 rad/h for Fe.

10. 4.8 rad/h; 4 Ci.

11. Radiation fields in rad/h per kg at 1 m for 10^7 neutrons/sec cm^2, $t_i = \infty$ and $t_c = 0$.

	Thermal	Epithermal	Fast	High energy
Al	0	0	6.5×10^{-6}	3×10^{-4}
Fe	5.5×10^{-6}	5.8×10^{-7}	5.2×10^{-6}	2.7×10^{-4}
Cu	8×10^{-4}	5.2×10^{-5}	1.9×10^{-6}	2.5×10^{-4}
Pb	0	0	1.2×10^{-7}	4×10^{-4}

CHAPTER IV

1. 3.5, 4, 4.5, 5, 7.3, 9, 10 MeV.

2.

A	σ(p, pn) (mb)
20	150
50	600
100	200
150	80
200	40

3. Theoretical total proton inelastic cross-section: 1700 mb

$\sigma(\text{p,n}) = 250$ mb
$\sigma(\text{p,2n}) = 500$ mb
$\sigma(\text{p,pn}) = 200$ mb
$\sigma(\text{p},\alpha) = 50$ mb

4. 9, 10, 11, 12, 15 MeV.

5.

σ_{max} (mb)	A	Energy range (MeV)
$\sigma(\text{p,pn}) = 600$	45– 65	21–26
$\sigma(\text{p,n}) = 700$	60– 90	9–11
$\sigma(\text{n,2n}) = 1700$	130–200	13–17
$\sigma(\text{d,2n}) = 600$	60–160	12–20
$\sigma(\text{d,p}) = 500$	30– 70	6–10
$\sigma(\alpha,\text{2n}) = 1000$	100–210	25–30

CHAPTER V

1. (γ,n): 20–15 MeV, ^{18}F (20 mb), ^{22}Na (25), ^{54}Mn (70), ^{58}Co (75), ^{114}In (300), ^{196}Au (500).
$(\gamma,\text{2n})$: 24–20 MeV, ^{61}Cu (35 mb), ^{195}Au (200).
(γ,p): 22 MeV, ^{39}Cl (50 mb).
(γ,np): 30 MeV, ^{22}Na (5 mb), ^{38}Cl (12).
(γ,α): 20 MeV, ^{61} Co (1 mb).

2. Use the tables in the appendix and find the aluminium curve drawn on fig. v.28.

3. $Q = 10^9$.

4. $Q = k_{max}^{-1} \int_0^{k_{max}} Ck^{-1} k \, \mathrm{d}k = C$;
$n(k, k_{max}) \, \mathrm{d}k = Ck^{-1} \mathrm{d}k = C \, \mathrm{d}(\ln k) = C = Q$ when $\mathrm{d}(\ln k) = 1$.

CHAPTER VI

1. 2, 3, 4, 5 barn and 1300, 1400, 900, 100 mb.

2. 13, 0.5, 0.15, 1.5, 0.3 mb.

3. 200 and 400 out of 800 mb make 75%.

4. 60 MeV, 20 μb.

5. 43 mb.

6. 27 mb.

7. 30 mb.

8. The independent yields are 1, 7 and 24 mb. After 6 months one has 2×10^8 atoms of ^{89}Y.

CHAPTER VII

1.

	Rad surf. per n/cm^2	Rem surf. per n/cm^2	Rad max. per n/cm^2	Rem max. per n/cm^2
a. Thermal	2.4×10^{-10}	6.2×10^{-10}	3.5×10^{-10}	1.05×10^{-9}
b. E^{-1} spectrum	1.03×10^{-9}	6.88×10^{-9}	1.6×10^{-9}	8.86×10^{-9}
c. Fission	3.02×10^{-9}	2.35×10^{-8}	3.73×10^{-9}	2.85×10^{-8}

2. 27 rad, 2.1×10^{-4} μCi ^{24}Na/mg ^{23}Na.

3. 103 rad, 6×10^{-4} μCi ^{24}Na/mg ^{23}Na.

4. 300 rad, 7.7×10^{-4} μCi ^{24}Na/mg ^{23}Na.

BIBLIOGRAPHY

Introduction

Blewett J. P., Livingston M. S., Particle accelerators (McGraw-Hill, New York 1962).

Eisenbrid M., Environmental activity (McGraw-Hill, New York 1963).

Friedlander G., Kennedy J. W., Miller J. M., Nuclear and radiochemistry, (Wiley, New York 1964).

Kolomensky A. A., Lebedev A. N., Theory of cyclic accelerators (North-Holland Publishing Co., Amsterdam 1966).

Kruger G. P., The sun (University of Chicago Press, Chicago 1958).

Livingwood J. J., Principles of cyclic particle accelerators (Van Nostrand, New York 1961).

Morse P. M., Feshbach H., Methods of theoretical physics (McGraw-Hill, New York 1953).

Pecker, J., Schatzman E., Astrophysique générale (Masson, Paris 1959).

Chapter I

Bleuler E., Zünti W., Zur Absorptionsmethode der Bestimmung von β- und γ-Energien, Helvetica Physica Acta, **19** (1946) 375.

Brucer M., 118 medical radioisotope cows, Isotopes and Radiation Technology **3** (1965) 1.

Dzhelepov B. S., Peker L. K., Decay schemes of radioactive nuclei (Pergamon Press, London 1961).

Evans R. D., The atomic nucleus (McGraw-Hill, New York 1955).

Gladstone S., ed., The effects of nuclear weapons (USAEC, Washington 1957).

Greene M. W., Doering R. F., Hillman M., Milking systems, Isotopes and Radiation Technology, 1, 2, 152.

Goldstein H., The attenuation of gamma rays and neutrons in reactor shields (Nuclear Development Corporation of America, White-Plains, New York, May 1st, 1957).

Gorschkow G. W., Gammastrahlung radioaktiver Körper (Teubner, Leipzig 1960).

Hawkings R. C., Edwards W. J., Mc Leod E. M., Tables of gamma rays from the decay of radionuclides (CRDC-1007, first book, Atomic Energy of Canada Ltd., Chalk-River, Ontario, March 1961).

Hine G. H., Brownell G. L., Radiation dosimetry (Academic Press, New York 1956).

Howerton R. J., Tabulated neutron cross-sections (UCRL 5226 rev., 1959).

Imperial College of Science and Technology, Code of practice against radiation hazards (University of London, Oct. 1962).

Jahnke, Emde, Lösch, Tables of higher functions (Teubner, Stuttgart 1960).

Kamke E., Differential equations (Akad. Verlag, Leipzig 1959).

Katz L., Penfold A., Range–energy relations for electrons and determination of beta ray end point energies by absorption, Rev. Mod. Phys. **24** (1952) 28.

King L. V., Note on the cosine law of radiation, Phil. Mag. ser. 6, **23** (1912) 237.

Lagally M., Franz W., Vorlesungen über Vektorrechnung (Akad. Verlag, Leipzig 1959).

Marion J. B., Fowler J. L., Fast neutron physics, Part I (Interscience Publishers Inc., New York 1960) p. 876.

Masket A. V. H., Rodgers W. C., Tables of solid angles (USAEC TID 14975, July 1962).

National Bureau of Standards, Protection against neutron radiation up to 30 MeV, Handbook 63 (US Department of Commerce, Washington DC, USA).

Nelms A. T., Graphs of the Compton energy–angle relationship and the Klein–Nishina formula from 10 keV to 500 MeV (Nat. Bur. Stds. Circular 542, Washington 1953).

Northcliffe L.C., Passage of heavy ions through matter, Annual Review of Nuclear Science **13** (1963) 67.

Oak Ridge National Laboratory, Radiation Safety and Control Training Manual (Oak Ridge, Tenn., USA).

Pannetier R., Table des Isotopes, Vademecum de radioprotection, tome II, (Maisonneuve SA, Moulin-lès-Metz, Moselle, France, 1965).

Pleinevaux C., Table des fonctions intégrales exponentielles (Euratom 108 F, Ispra 1962).

Pensko J., Bysiek M., Duczynski S., The gamma radioactivity of building materials for the construction of low background laboratories, Report No. CLOR-20 (Central Laboratory for radiological protection, Ed. Office of scientific, technical and economical information, Palace of Culture and Science, Warshaw, Poland 1963).

Price B. T., Horton C. C., Spinney K. T., Radiation Shielding (Pergamon Press, London 1957).

Seelmann-Eggebert W., Pfennig G., Radionuklid Tabellen (Bundesministerium für wissenschaftliche Forschung, Bonn 1964).

Stehn J. F., Table of radioactive nuclides, Nucleonics **18** (1960) 11, 186.

Strominger D., Hollander J. M., Seaborg G. T., Tables of isotopes, Rev. Mod. Phys. **30**, 2, (1958) 2.

Sullivan A. H., Overton T. R., Time variation of dose-rate from radioactivity induced in high energy particle accelerators, Health Physics **11** (1965) 1101.

Sullivan H., Trilinear chart of nuclides (Oak Ridge National Laboratory, Oak Ridge, Tenn., USA, 1957).

Chapter II

Ashmore A., et al., Total cross-sections of protons with momentum between 10 and 28 GeV/c, Phys. Rev. Letters **5** (1960) 576.

Barbier M., Radioactivity induced in materials by high energy particles (CERN 64–9, January 1964, Geneva, Switzerland).

Barbier M., Cooper A., Estimate of induced activity in accelerators (CERN 65–34, 25th October 1965, Geneva, Switzerland).

Barbier M., Induced radioactivity data computed for various materials and compared with that of irradiated samples, Proc. of the special session on accelerator shielding presented at the 1965 Winter Meeting of the American Nuclear Society, Washington, Nov. 15th, 1965, and Shielding Division Report ANS–SD–3.

Barbier M., Hutton A., Pasinetti A., Radioactivity induced in tissues by 600 MeV protons (CERN 66–34, Geneva, Switzerland).

Barbier M., Radioactivity induced in building materials (CERN 67–25, Geneva, Switzerland).

Bellettini G., et al., Proton Nuclei Cross-sections at 20 GeV, Nucl. Phys. **79** (1966) 609.

Dostrovsky I., Fraenkel Z., Friedlander G., Monte-Carlo calculations of nuclear evaporation processes III, Phys. Rev. **116** (1959) 683.

Dostrovsky I., Fraenkel Z., Monte-Carlo calculations of nuclear evaporation processes IV, Phys. Rev. **118** (1960) 781.

Fulmer C. B., Toth K. S., Barbier M., Residual radiation studies for meason factories, Nucl. Instr. and Meth. **31** (1964) 45.

Fulmer C. B., Toth K. S., Barbier M., Residual radiation studies, IEEE Trans. Nucl. Sc. **12** (1965) 673.

Glazov A. A. et al., Relativistic proton cyclotron for an energy of 700 MeV, International Conference on high energy accelerators, Dubna 1963, p. 547.

Goldberger M. L., The interaction of high-energy neutrons and heavy nuclei, Phys. Rev. **74** (1948) 1269.

Horikawa N., Kanada H., On the phase shifts of proton He^4 scattering, J. Phys. Soc. Japan **20** (1965) 1758.

Huizenga J. R., Igo G., Theoretical reaction cross-sections for alpha particles with an optical model, Nucl. Phys. **29** (1962) 462.

Igo G., Optical model analysis of excitation function data and theoretical reaction cross-sections for alpha particles, Phys. Rev. **115** (1959) 1665.

Longo M. J., Moyer B. J., Nucleon and nuclear cross-sections for positive pions and protons above 1.4 BeV/*c* (UCRL-9497 rev. 1st Aug. 1961).

Longo M. J., Moyer B. J., Nucleon and nuclear cross-sections for positive pions and protons above 1.4 BeV/*c*, Phys. Rev. **125** (1962) 701.

Millburn G. P. et al., Nuclear radii from inelastic cross-sections, Phys. Rev. **95** (1954) 1269.

Moyer B. J., Data related to star production by high energy neutrons, LRL, Berkeley, June 20th, 1961, unpublished.

Perret Ch., Faisceau de neutrons rapides du canal 70 MeV, CERN internal report MSC 21/1 5.1.1965, Geneva, unpublished.

Rudstam G., Spallation of elements in the mass range 51–75, Phil. Mag. **46** (1955) 344; Inaugural Dissertation, University of Uppsala, 1956.

Rudstam G., Systematics of spallation yields, Z. Naturforsch. **21**a (1966) 7.

Serber R., Nuclear reactions at high energies, Phys. Rev. **72** (1947) 1114.

Shapiro M. M., Cross-sections for the formation of the compound nucleus by charged particles, Phys. Rev. **90** (1953) 171.

Toth K. S., Fulmer G. B., Barbier M., Estimates of residual radiation levels induced by high energy nucleons, Nucl. Instr. and Meth. **42** (1966) 128.

Wallace R., Shielding and activation considerations for a meson factory, Nucl. Instr. and Meth. **18–19** (1962) 405.

Weisskopf V., Statistics and nuclear reactions, Phys. Rev. **52** (1937) 295.

Wikner F., Nuclear cross-sections for 4.2 BeV/*c* negative pions, (UCRL 3639, Jan. 1957).

Chapter III

Blomeke J. O., Todd M. F., Uranium 235 fission product production as a function of thermal neutron flux, irradiation time and decay time, ORNL 2127, Physics and Mathematics TID 4500 (Oak Ridge Nat. Lab., Oak Ridge, Tenn., USA, June 10, 1958).

Bowman H. R., Thompson S. G., Milton J. C. D., Swiatecki W. J., Prompt neutrons from spontaneous fission of ^{252}Cf, Phys. Rev. **126** (1962) 6.

Burnazyan A. I., Lebedinski A. V., Radiation medicine (Pergamon Press, London 1964).

Burris L., Dillon I. G., Estimation of fission product spectra in discharged fuel from fast reactors (Argonne National Laboratory ANL 5742, July, 1957).

Conference Proceedings Nuclear Data for reactors Vol. I and II (IAEA, Vienna 1967).

Cransberg L., Frye G., Nereson N., Rosen L., Fission neutron spectrum of ^{235}U, Phys. Rev. **103** (1956) 662.

Croall I. F., Collected independent fission yields for thermal neutron fission of ^{233}U, ^{239}Pu and ^{235}U, 14 MeV neutron fission of ^{235}U, Pile neutron fission of ^{238}U, ^{232}Th and ^{241}Am, and spontaneous fission of ^{242}Cm (AERE-R 3209, Harwell 1960).

Culp A. W., Page E. M., Calculational method for predicting the residual dose rates from activated aircraft turbo rotating-machinery, USAF report Astra 301-P-13.5, Astra Inc. Raleigh, N.C., July, 1958 (see also Revised Activation Handbook for Aircraft Designers, USAF report NARF-57-50 T (FZK 9-124), Convair Fort Worth, Tex., Sept. 30, 1957).

Glasstone S., ed., The effects of nuclear weapons (USAEC, June 1957).

Gross E., The absolute yield of low energy neutrons from 190 MeV proton bombardment of gold, silver, nickel, aluminium and carbon (UCRL 3330, Berkeley, Febr. 29, 1956).

Grundl J. A., Study of fission neutron spectra with high energy activation detectors, LAMS 2883, UC-34, Physics, TID 4500 (19th ed.), Los Alamos Scientific Laboratory New-Mexico, May 20, 1963.

Katcoff S., Fission-product yields from neutron-induced fission, Nucleonics **18** (1960) 201.

Keepin G. R., Physics of nuclear kinetics (Addison-Wesley, Reading, Mass. 1964).

Layman D. C., Thornton G., Remote handling of mobile nuclear systems (US Atomic Energy Commission, TID-21719, Division of technical information, 1966).

Leachmann R. B., Determination of fission quantities of importance to reactors, International Conference on the peaceful uses of atomic energy, Geneva 1955, 2, 193.

Lehman R. L., Fekula O. M., Energy spectra of stray neutrons from the betatron, Nucleonics II, Nov. 1964, 35.

Mitler H. E., Particle evaporation from excited nuclei (Smithsonian Institution Astrophysical Observatory, special report No. 204, Cambridge, Mass.).

Moteff J., Fission product decay gamma energy spectrum (APEX 134, USAEC, Techn. Inf. Service, Oak Ridge, Tenn., June 1953).

Niece H. L., Independent yields of ^{75}Zr from thermal neutron fission of ^{235}U and ^{233}U (ORNL-TM-1333, Oak Ridge, Tenn., 1965).

Norris A. E., Wahl A. C., Nuclear charge distribution in fission: ^{92}Y, ^{93}Y, ^{94}Y and ^{95}Y independent yields, Phys. Rev. **146** (1966) 926.

Perkins J. F., Krug R. W., Energy release from decay of fission products, Nucl. Sc. Eng. **3** (1958) 726.

Prawitz J., Löw K., Björnerstedt R., Gamma spectra of gross fission product from thermal reactors, Second UN Int. Conf. on Peaceful Uses of Atomic Energy, Geneva 1958, Paper No A/Conf/15/P/149, Vo. 13, p. 42.

Reactor Physics Constants (ANL-5800, 2nd ed., Argonne Nat. Lab., July 1963).

Report of the United Nations Scientific Committee on effects of atomic radiation, New York 1958, official record of 13th Session, Supplement No. 17 (A/3838).

Romanko J., Dungan W. E., Specification and measurement of reactor neutron spectra, Symposium on neutron detection and standardization, Harwell, England, December 1962, International Atomic Agency preprint No. SM 36/84.

Skyrme D., The evaporation of neutrons from nuclei bombarded by high energy protons, Nucl. Phys. **35** (1962) 177.

Stehn J. R., Clancy E. F., Fission product radioactivity and heat generation, Second UN International Conf. on Peaceful Uses of Atomic Energy, Geneva 1958, paper No A/Conf/15/P/1071.

Strom P. O. et al., Nuclear-charge distribution of fission-product chains of mass numbers 131–133, Phys. Rev. **144** (1966) 984.

Taube M., Plutonium (Pergamon London, and PWN, Warsaw, 1964).

Zobel W., Love T. A., Time and energy spectra of fission product gamma rays measured at short times after uranium sample irradiations (Applied Nuclear Physics Division, ORNL Annual Report 2081, Nov. 20, 1956).

Chapter IV

Audouze J., Epherre M., Reeves H., Spallation of light nuclei, Nucl. Phys. A **97,** (1967) 144 and Survey of available experimental cross-sections for proton induced spallation reactions in ^{4}He, ^{12}C, ^{14}N, ^{16}O (Internal report IPN 91, Institute de Physique Nucléaire, Orsay, 1967).

Barral R. C., MacElroy W. N., Neutron flux spectra determination by foil activation, Vol. II, Experimental and evaluated cross-sections, Air Force Weapons Laboratory, AFWL-TR-65-34, Aug. 1965, USA.

Barral R. C., MacElroy W. N., Nuclear reactions for determining neutron spectrum and dose, a World Literature Search, Personnel Dosimetry for Radiation Accidents (IAEA, Vienna 1965).

Batzel R. E., Crane W. W. T., O'Kelly G. D., The excitation function for the ^{27}Al(d, αp)^{24}Na reaction, Phys. Rev. **91** (1953) 939.

Becker K., Nuclear track registration in dosimeter glasses for neutron dosimetry in mixed radiation fields, Health Phys. **12** (1966) 769.

Becker K., Neutron personnel dosimetry by non-photographic nuclear track registration, ENEA symposium on radiation dose measurement, Stockholm, June 1967.

Bertini H. W., A literature survey of non-elastic reactions for nucleons and pions incident on complex nuclei at energies between 20 MeV and 33 GeV, ORNL-3455, UC-34, Physics, TID 4500 (20th ed. rev.) Oak Ridge, Aug. 23, 1963.

Bertini H. W., Guthrie M. P., Pickell E. H., Bishop B. L., Literature survey of radiochemical cross-section data below 425 MeV, ORNL 3884, UC-34-Physics, Oak Ridge, October 1966.

Bohr N., Neutron capture and nuclear constitution, Nature **137** (1936) 344.

Bramblett R. L., Ewing R. I., Bonner T. W., A new type of neutron spectrometer, Nucl. Inst. Methods **9** (1960) 1.

Brookhaven National Laboratory, Neutron cross-sections, BNL 325, 2nd edition, supplement 2, TID-4500.

Brandt R. et al., Preliminary results on high-energy nuclear fission by means of mica detectors, II. Intern. Conf. on Corpuscular Photography, Firenze 1966.

Bruninx E., High energy nuclear reactions cross-sections III, CERN 64-17.

Burrus W. R., Bonner spheres and treshold detectors for neutron spectroscopy, ORNL 3360, 296, 1963.

Caretto A. A. jr., Literature survey of nucleon-2 nucleon reaction cross-sections at energies above 100 MeV, NYO 10 693, Carnegie Institute of Technology Pittsburgh, Penn., July 1964.

de Carvalho H. G. et al., Fission of uranium, thorium and bismuth by 20 GeV protons, Nuovo Cimento **27** (1963) 468.

Chatterjee A., Shell effects in 14 MeV(n, p) reactions, Nucl. Phys. **60** (1964) 273.

Choppin G. R. and Meyer E. F., jr., Proton-induced fission of ^{238}U, J. Inorg. Nucl. Chem. **28** (1966) 1509.

Christaller G., Aluminium foil activation as beam monitor for deuterons, Intern. report Kernforschungszentrum Karlsruhe.

Crandall W. E., Millburn G. P., Pyle R. V., Birnbaum W., $^{12}C(x, xn)$ and $^{27}Al(x, x2pn)^{24}Na$ cross-sections at high energies, Phys. Rev. **101** (1956) 329.

Cross W. G., The use of magnesium, titanium, iron, nickel and zinc in fast neutron activation dosimetry, Neutron Dosimetry, IAEA, Vienna 1962.

Davis F. J., Neutron dose determination with threshold detectors, selected topics in Radiation Dosimetry, IAEA, Vienna 1961, 399.

Friedländer G. et al., Excitation functions and nuclear charge dispersion in the fission of uranium by 0.1 to 6.2 GeV protons, Phys. Rev. **129** (1963) 1809.

Friedländer G., Fission of heavy elements by high-energy protons, Physics and Chemistry of fission, Vol. II, p. 265, IAEA.

Fulmer C. B., Williams I. R., Ball J. B., Residual radiation studies for medium energy proton accelerators, IEEE Transactions on Nuclear Science, June 1967.

Gardner D. G., Trends in nuclear reaction cross-sections I, the (n, p) reaction induced by 14 MeV neutrons, Nucl. Phys. **29** (1962) 373.

Goshal S. N., An experimental verification of the theory of compound nucleus, Phys. Rev. **80** (1950) 939.

Grover J. R., Caretto A. A., Nucleon, two-nucleon reactions above 100 MeV, Ann. Rev. Nucl. Science **14** (1964) 51.

Grundl J. A., Study of fission neutron spectra with high-energy activation detectors, Los Alamos report LAMS 2883, 1963, also Trans. An. Nucl. Soc. 5, No. 2, 1962.

Hagebø E. et al., Radiochemical studies of charge and mass distribution in the light fragment region in fission of uranium induced by 170 MeV protons, J. Inorg. Nucl. Chem. **26** (1964) 1639.

Hagebø E., Yields and isomeric yield ratios of antimony isotopes from the interaction of 159 MeV to 18.2 GeV protons with uranium (submitted to J. Inorg. Nucl. Chem.).

Hankim D. E., Method of determining the intermediate energy neutron dose, USAEC rep. IDO-16.655, 1961.

Hudis J. and Katcoff S., High-energy fission cross-sections of U, Bi and Au, IV. Internal European Conference on the Interactions of High-Energy Particles and Complex Nuclei, Røros, 1967.

Huizenga J. R., Igo G., Theoretical reaction cross-section for alpha particles with an optical model, Nucl. Phys. **29** (1962) 462.

Jessen P., Bormann M., Dreyer F., Neuert H., Compilation of experimental excitation functions of some fast neutron reactions up to 20 MeV, Report from I. Inst. f. Experimentalphysik, Hamburg, Jan. 1965.

Jodra L. G., Sugarman N., High-energy fission of bismuth, proton energy dependence, Phys. Rev. **99** (1955) 1470.

Khodai-Iwopari A., Fission properties of some elements below radium, UCRL-16489, 1966.

Kohler A. D., An improved method of neutron spectroscopy using threshold detectors, UCRL 11760, Berkeley (California), 1964.

Konshin V. A. et al., Cross-sections for fission of ^{181}Ta, Re, Pt, ^{197}Au, Pb, ^{209}Bi, ^{232}Th, ^{235}U and ^{238}U by 150–660 MeV protons, Sov. J. Nucl. Phys. **2** (1966) 489.

Lange J., Bestimmung einiger Anregungsfunktionen für Deuteronenreaktionen mit ^{141}Pr. KFK 519, Kernforschungszentrum, Karlsruhe 1967.

Lindner M., Osborne R. N., The cross-section for the ^{27}Al(α, αp2n)^{24}Na reaction from threshold to 380 MeV, Phys. Rev. **91** (1953) 342.

Liskien H., Paulsen A., Compilation of cross-sections for some neutron induced threshold reactions, European Atomic Energy Community-Euratom, EUR 119e, Geel, Belgium.

McGowan F. K., Milner W. T., Kim H. J., Nuclear cross-sections for charged particles induced reactions, ORNL-CPX-1, 2, charged cross-section data centre, 1964, Oak Ridge National Laboratory, Oak-Ridge.

Moteff J., Beever E. R., The status of threshold and resonance neutron detectors, Selected topics in radiation dosimetry, IAEA, 383, Vienna 1961.

Neutron Detectors, Bibliographical series no. 18, IAEA, Vienna 1966.

Pappas A. C. and Hagebø E., The charge and mass distribution in fission of uranium induced by 170 MeV protons, J. Inorg. Nucl. Chem. **28** (1966) 1769.

Pate B. D. and Pozkanzer A. M., Spallation of uranium and thorium nuclei with BeV-energy protons, Phys. Rev. **123** (1961) 647.

Ravn H., Brandt R. and Alstad J., private communication.

Ringle J. C., A technique for measuring neutron spectra in the range 2.5–3 MeV using threshold detectors, UCRL 10732, Berkeley, California, 1963.

Rudstam G. and Sørensen G., Fission and spallation yields in the iodine region, J. Inorg. Nucl. Chem. **28** (1966) 771.

Stevenson P. C. et al., Further radiochemical studies of the high-energy fission products, Phys. Rev. **111** (1958) 286.

Tilbury R. S., Activation analysis with charged particles, National Academy of Sciences – National Research Council, Nuclear Science Series NAS-NS-3110, U.S. Dep. of Commerce, Springfield, Virginia.

Widell C. O., Neutron dosimetry by the fission fragment method, Neutron monitoring, IAEA, Vienna 1967, 417.

Williams I. R., Fulmer C. B., Cross-section for formation of ^{7}Be by 20–155 MeV protons in carbon, Phys. Rev. **154** (1967) 1005.

Williams I. R., Fulmer C. B., Excitation functions for radioactive isotopes produced by protons below 60 MeV on Al, Fe, Cu, Phys. Rev. **162** (1967) 1055.

Chapter V

Antufev Yu. P., Miroshnitchenko I. I., Noga V. I., Sorokin P. V., Yields of the photonuclear reactions $^{12}C \to {}^{11}C$, $^{27}Al \to {}^{24}Na$, $^{28}Si \to {}^{24}Na$ at high photon energies, Yadernaya Fisica **6** [2] (1967) 431.

de Carvalho H. G., di Napoli V., Margadonna D., Salvetti F., Tesch K., Studio dei processi (γ, n) su nuclei complessi nell'intervallo tra 1 e 5.5 GeV, to be published.

Dawson W. K., Energy and angular distribution of photo-protons produced by 70 keV X-rays, Can. J. Phys. **34** (1956) 1480.

Dyal P., Hummel J. P., Cross-sections for the $^{11}B(\gamma, \pi^{-})^{11}C$ and $^{11}B(\gamma, \pi^{+})^{11}Be$ reactions, Phys. Rev. **127** (1962) 2217.

Engelman C., Analyse par activation aux photons, CEA – R 3307, Saclay 1967.

Fultz S. C., Bramblett R. L., Caldwell J. T., Harvey R. R., Photoneutron cross-sections for natural Cu, ^{63}Cu and ^{65}Cu, Phys. Rev. **133** (1964) 1149.

Halpern I., Debs R. J., Eisinger J. T., Fairhall A. W., Richter H. G., Yields of photonuclear reactions with 320 MeV X-rays, Phys. Rev. **97** (1955) 1325, 1327.

Jones L., Terwilliger K. M., Photoneutron production excitation functions to 320 MeV, Phys. Rev. **91** (1953) 699.

Jonsson G. G., Forkman B., Lindgren K., Redetermination of some $^{127}I(\gamma, xn)$ yields, Physics Letters **26B** (1968) 508.

Jungerman J. A., Steiner H. M., Photofission cross-sections of ^{235}U, ^{238}U, ^{232}Th, ^{209}Bi and ^{197}Au at energies of 150 to 500 MeV, Phys. Rev. **106** (1957) 585.

Katz L., Daerg A. P., Brown F., Photofission in heavy elements, Proc. Sec. UN Int. Conf. Peaceful Uses of Atomic Energy, Geneva 1958, Vol. 15, 188.

Levinger J., Nuclear Photodisintegration (Oxford University Press 1960).

Masaike A., Investigation of nuclear reactions induced by high energy bremsstrahlung, J. Phys. Soc. Japan **19** (1964) 427.

Meyer R. A., Walters W. B., Hummel J. P., Cross-sections for the $^{16}O(\gamma, \pi^{+})^{16}N$ reaction, Phys. Rev. **138** (1965) B 1421.

Minarik E. V., Novikov V. A., Fission of U, Th, Bi and Tl induced by high energy gamma quanta, Sov. Phys. JETP **5** (1957) 250.

Di Napoli V., Dobici F., Salvetti F., de Carvalho H. G., Radiochemical studies on photonuclear reactions above the mesonic threshold, Nuovo Cimento **37** (1963) 1728; **42** (1966) 358; **48B** (1967) 1.

Photonuclear Reactions, Bibliog. Series 10, IAEA, Vienna 1964.

Photonuclear Data Index, Nat. Bur. Stds, Misc. Publ. 277, 1966.

Roos C. E., Peterson V. Z., Photodissociation of complex nuclei at energies between the mesonic threshold and 1150 MeV, Phys. Rev. **124** (1961) 1610.

Schmitt R. A., Sugarman N., Uranium photofission yields, Phys. Rev. **95** (1954) 1260.

De Staebler H., Transverse radiation shielding for the Stanford two mile linear accelerator, SLAC-9, Nov. 1962.

Stein P. C., Odian A. C., Wattenberg A., Weinstein R., Dependence on the atomic number of the nuclear photoeffect at high energies, Phys. Rev. **119** (1960) 348.

Storm E., Harvey I., Photon cross-sections from 0.01 to 100 MeV for elements 1 through 100, Los Alamos Nov. 15, 1967.

Toms M. E., Bibliography of photo- and electronuclear disintegrations, Naval Research Laboratory 31, 1967, Washington D.C.

Walters W. B., Hummel J. P., Cross-sections for the $^{27}Al(\gamma, \pi^+)^{27}Mg$ reaction, Phys. Rev. **143** (1966) 833.

Williams I. R., Fulmer C. B., Dell G. F., Engebretson M. J., Fission product yields from 1.5 and 3 GeV electrons on uranium (to be published).

Wilson R. R., Precision quantameter for high energy X-rays, Nucl. Instr. **1** (1957) 101.

Wilson R. R., A revision of shielding calculations, CEA-73, 1959.

Wyckoff J. M., Photon energy emitted from 3 to 13 GeV spallation products, Preprint National Bureau of Standards, Washington D.C. 1969, submitted to Health Physics.

Chapter VI

Anderson C. E., Sachs M., Ladenbauer I. M., Preiss I. M., Excitation functions for ^{6}Li induced reactions on ^{197}Au, Nucl. Phys. **45** (1963) 41.

Baraboshkin S. A., Karamain A. S., Flerov G. N., Interaction of nitrogen and gold nuclei, JETP **5** (1957) 1055.

Berezhnoi Y. A., Klyncharev A. P., Ranyuk Y. N., Rutkevich N. Y., On reactions of total decay of nuclei, JETP **43** (1962) 1248.

Blair J. M., Hobbie R. K., Differential cross-sections for the reactions $^{12}C(^{6}Li, p)^{17}O$ and $^{12}C(^{6}Li, d)^{16}O$ from 3.4 to 4 MeV, Phys. Rev. **128** (1962) 2282.

Blann M., Lanzafame F. M., Piscitelli R. A., Reactions induced in ^{54}Fe with 21–63 MeV ^{6}Li ions, Phys. Rev. **133** 3B (1964) 700.

Britt H. C., Quinton A. R., Fission of ^{197}Au and ^{209}Bi by ^{12}C and ^{16}O projectiles, Phys. Rev. **120** 5, (1960) 1768.

Broek H. W., Neutron emission from compound nuclear systems of high angular momentum, Phys. Rev. **124** 1, (1961) 233.

Couch J. G., Mc Intyre J. A., Hiebert J. C., Nucleon transfer in $^{10}B(^{14}N, ^{13}N)^{11}B$, $^{10}B(^{14}N, ^{11}C)^{13}C$, $^{40}Ca(^{14}N, ^{13}N)^{41}Ca$ reactions, Phys. Rev. **152** 3 (1966) 883.

Dar A., Cluster transfer reactions between complex nuclei, Phys. Rev. **139** 5B (1965) 1193.

Fisher D. E., Zucker A., Nuclear reactions induced by the nitrogen bombardment of sulphur, Phys. Rev. **113**, 2 (1959) 542.

Flerov G. N., The fission of nuclei of heavy elements by interaction with carbon, oxygen and nitrogen nuclei, Proc. Conf. on reactions between complex nuclei, Gatlinburg, Tennessee, May 5–7, 1958, 384.

Fleury A., Simonoff G. N., Sur la production de ^{232}Pa par bombardement de ^{232}Th par des ions de ^{6}Li et ^{7}Li accélérés, Physics Letters **14**, 1, (1965) 44.

Fleury A., Mivielle F., Simonoff G. N., Interaction des ions ^{10}B et ^{11}B avec des noyaux d'uranium, Physics Letters **24B**, 11 (1967) 576.

Gaedke R. M., Toth K. S., Williams I. R., Excitation functions for nucleon and alpha particle transfer reactions induced by ^{14}N ions, Phys. Rev. **140**, 2B, (1965) 296.

Gilmore J., Thompson S. G., Perlman I., An investigation of the influence of angular momentum on fission probability, UCRL 9304 Rev., Jan. 8, 1962.

Gordon G. E., Fission and spallation in nuclear reactions induced by heavy ions, UCRL 9083, May 1960.

Hubbard E. L., Main R. M., Pyle R. V., Neutron production by heavy ion bombardment, Phys. Rev. **118**, 2, (1960) 507.

Kibler K. G., Differential and total cross-sections of reaction products from ^{7}Li+^{6}Li between 3.78 and 5.95 MeV, Phys. Rev. **155**, 4, (1967) 1110.

Ladenbauer I. M., Preiss I. L., Anderson C. E., Excitation functions for ^{6}Li induced reactions on ^{27}Al, Phys. Rev. **123**, 4, (1961) 1368.

Ladenbauer I. M., Preiss I. L., Anderson C. E., Excitation functions for heavy ion induced reactions on ^{27}Al, Phys. Rev. **125**, 2, (1962) 606.

Mc Intyre J. A., Neutron reduced widths determined by the ^{14}N(^{14}N, ^{13}N)^{15}N transfer reaction, Congrès International de Physique Nucléaire, Paris, 2–8 Juillet 1964, 1108.

Morse R. I., Preiss I. L., Intermediates in the nucleon transfer process in reactions of ^{14}N with ^{133}Cs, Phys. Rev. **156**, 4, (1967) 1332.

Newman E., Toth K. S., Nuclear reactions induced by the nitrogen bombardment of ^{11}B, F, Al, Si, P, Cl, Phys. Rev. **129**, 2, (1963) 802.

Pinajian J. J., Halbert M. L., Nitrogen induced nuclear reactions in potassium, Phys. Rev. **113**, 2, (1959) 589.

Reynolds H. L., Zucker A., Excitation function for the reaction ^{9}Be(^{14}N, αn)^{18}F, Phys. Rev. **100**, 1, (1955) 226.

Reynolds H. L., Zucker A., Nuclear reactions produced by nitrogen on nitrogen, Phys. Rev. **101**, 1, (1956) 166.

Reynolds H. L., Scott D. W., Zucker A., Nuclear reactions produced by nitrogen on boron and oxygen, Phys. Rev. **102**, 1, (1956) 237.

Sikkeland T., Larsh A. E., Gordon G. E., Fission of ^{238}U with carbon ions, Phys. Rev. **123**, 6, (1961) 2112.

Strudler P. M., Preiss I. L., Wolfgang R., Systematics of some nucleon transfer reactions of complex nuclei, Phys. Rev. **154**, 4, (1967) 1126.

Viola V. E., Sikkeland T., Total cross-sections for fission of ^{238}U induced by ^{4}He and heavy ions, UCRL 10088, Feb. 20, 1962.

Zavara I., Determination of some fragment yields from fission of heavy nuclei by multicharged ions, Proc. Third Conference on reactions between complex nuclei, Asilomar, California, April 14–18, 1963, 389.

Chapter VII

Alsmiller R. G., Moran H. S., Dose rate from high energy electrons and photons, ORNL-TM 2026, 1967, Oak-Ridge.

Anderson J., Osborne S. B., Tomlinson R. W. S., Newton D., Rundo J., Salmon L., Smith J. W., Neutron activation analysis in man in vivo, The Lancet (Dec. 5, 1964) 1201.

Andrews G. A., Radiation accidents and their management, Radiation Research supplement **7** (1967) 390.

Auxier J. A., Sanders F. W., Snyder W. S., ^{24}Na activation in the dosimetry of nuclear accidents, in 'Radioactivity in man', G. R. Meneely Editor, Charles C. Thomas Publ., Springfield (Ill.), USA, 1961.

Auxier J. A., Dosimetric considerations in criticality exposures, Diagnosis and treatment of acute radiation injury, 141, 1961, World Health Organization, Geneva.

Barbier M., Hutton A., Pasinetti A., Radioactivity induced in tissues by 600 MeV protons, CERN 66–34, Geneva.

Barrall R. C., Mac Elroy W. N., Nuclear reactions for determining neutron spectrum and dose, Personnel Dosimetry for Radiation Accidents, 251, IAEA Vienna 1965.

Benson R. E., Thomas R. G., Odland L. T., Proton induced radioactivity in primates, Health Physics **13** (1967) 1123.

Chanteur J., Arnaud Y., Pellerin P., Dosimétrie biologique après exposition à un flux élevé de neutrons thermiques et rapides, Personnel Dosimetry for Radiation Accidents, IAEA Vienna 1965, 235.

Charalambus S., Rindi A., Evaluation de la dose de surexposition auprès des accélérateurs de haute énergie par la réaction $^{12}C(x, nx)^{11}C$ dans le système pileux, Minerva fisica nucleare, Giornale di Fisica Sanitaria, 9, 2, Torino 1966.

Davy D. R., Peshori L. H., Porton J. W., Sodium 24 production in saline-filled phantoms under neutron irradiations, Health Physics **12** (1966) 1353.

Dierckx R., Horn W., Neutron dose determination in criticality accidents by gamma counting of the patients dental fillings, Neutron Monitoring, IAEA Vienna 1967, 563.

Geigy J. R., S. S., Tables scientifiques, Documents Geigy, Département Pharmaceutique, Bale 1963, Suisse.

Harris P. S., Radiation dose estimate in the 1958 Los Alamos criticality accident, Health Physics **5** (1961) 37.

Haxhe J. J., La composition corporelle normale ,Thèse Louvain 1963, Ed. Arsica S.A., Bruxelles.

Hine G. J., Brownell G. L., Radiation dosimetry (Academic Press 1956).

Hurst G. S., Ritchie R. H., Emerson L. C., Accidental radiation excursion at the Oak-Ridge Y-12 plant, III, Determination of radiation doses, Health Physics **2** (1959) 121.

Hurst G. S., Ritchie R. H., Sanders F. W., Reinhardt P. W., Auxier J. A., Wagner E. B., Callihan A. D., Morgan K. Z., Dosimetric investigation of the Yugoslav radiation accident, Health Physics **5** (1961) 179.

ICRP Committee II Report, Elements distribution in total body of the standard man, Health Physics **3** (1960) 146.

Irving D. C., Alsmiller R. G., Moran H. S., Tissue current to dose conversion factors for neutrons with energies from 0.5 to 6 MeV, ORNL 4032 UC-41, Oak Ridge 1967.

Jones T. D., Body sodium activation as a dosimetric tool in nuclear accidents, ORNL 3849, Oak-Ridge.

Kogan A. M., Petrov G. G., Tshudov L. A., Yampolsky P. A., Tissue dose of neutrons, Atomnaya Energiya, **7** (1959) 351.

Komochkov M. M., Estimation of the dose from accidental irradiation by a large amount of high-energy particles, Dubna Report P 2008, 1965.

Legeay G., Court L., Prat L., Jeanmaire L., Daburon M. L., de Kerviler H., Tardy-Joubert P., Dosimetrie des protons de haute énergie par mesure du beryllium 7 formé dans les tissus, Personnel Dosimetry for Radiation Accidents, IAEA Vienna 1965.

Lisco, Standard man, Report ANL 4253, 96, Nov. 1948–Feb. 1949, Argonne National Laboratory, USA.

Mastrocola A., An emergency dosimetry system for neutrons based on body sodium activation, Neutron Monitoring, IAEA, 575, Vienna 1967.

Morawek T., Détermination de l'activité du sodium 24 et de l'autodose gamma à partir des données du spectromètre SNAC, Neutron Monitoring, IAEA, 601, Vienna 1967.

Neufeld J., Snyder W. S., Turner J. E., Wright H., Calculation of radiation doses from protons and neutrons to 400 MeV, Health Physics **12**, 2 (1966) 227.

Palmer H. E., Simplified whole-body counting, Health Physics **12**, 1, (1966) 95.

Parker H. M., Newton C. E., The Hanford criticality accident: dosimetry techniques, interpretations and problems, Personnel Dosimetry for Radiation Accidents, 567, IAEA Vienna 1965.

Petersen D. F., Mitchell V. E., Langham W. H., Estimation of fast neutron doses in man by $^{32}S(n, p)^{32}P$ reaction in body hair, Health Physics **6** (1961) 1.

Price B. T., Horton C. C., Spinney K. T., Radiation shielding, (Pergamon, London 1957).

Sanders F. W., Auxier J. A., Neutron activation of sodium in anthropomorphous phantoms, Health Physics **8** (1962) 371.

Serre C., Evaluation de la perte d'énergie unitaire et du parcours de particules chargées traversant un absorbant quelconque, CERN 67-5, Geneva 1967.

Sklavenitis L., Sur la mesure et l'analyse des rayonnements de haute énergie par détecteurs à activation, Thèse, Toulouse 18 octobre 1967.

Smith J. W., Sodium activation by fast neutrons in man phantoms, Phys. Med. Biol. **7** (1962–63) 341.

Taketa S. T., Castle B. L., Howard W. H., Conley C. C., Haymaker W., Effects of acute exposure to high energy protons on primates, Radiation Research Suppl. **7** (1967) 336.

Tesch K., Dosisleistung und Toleranzflussdichte hochenergetischer Elektronen und Gammastrahlen, Nukleonik **8**, 5 (1966) 264.

Tipton I. H., Cook M. J., Steiner R. L., Boye C. A., Perry H. M., Schroeder H. A., Trace elements in human tissue, Health Physics **9** (1963) 89.

Wunderly C., Elettroforesi su carta, Il Pensiero Scientifico, Roma 1954.

SUBJECT INDEX